THE
EUKARYOTE GENOME
IN DEVELOPMENT
AND EVOLUTION

Mark Carrington
Nov 89

TITLES OF RELATED INTEREST

A biologist's advanced mathematics
D. R. Causton

Bryozoan evolution
F. K. McKinney & J. B. Jackson

Chromosomes today, volume 9
A. Stahl *et al.* (eds)

The essential Darwin
M. Ridley (ed.)

The handling of chromosomes
C. D. Darlington & L. F. La Cour

Invertebrate palaeontology and evolution
E. N. K. Clarkson

The natural history of Nautilus
P. Ward

Rates of evolution
K. S. W. Campbell & M. F. Day (eds)

THE
EUKARYOTE GENOME
IN DEVELOPMENT
AND EVOLUTION

Bernard John and George L. Gabor Miklos

Research School of Biological Sciences,
The Australian National University, Canberra

London
ALLEN & UNWIN
Boston   Sydney   Wellington

Allen & Unwin, the academic imprint of

Unwin Hyman Ltd

PO Box 18, Park Lane, Hemel Hempstead, Herts HP2 4TE, UK
40 Museum Street, London WC1A 1LU, UK
37/39 Queen Elizabeth Street, London SE1 2QB, UK

Allen & Unwin Inc.,
8 Winchester Place, Winchester, Mass. 01890, USA

Allen & Unwin (Australia) Ltd,
8 Napier Street, North Sydney, NSW 2060, Australia

Allen & Unwin (New Zealand) Ltd in association with the Port
Nicholson Press Ltd,
60 Cambridge Terrace, Wellington, New Zealand

First published in 1988

British Library Cataloguing in Publication Data

John B.
    The eukaryote genome in development and evolution
1. Genetics.   2. Eukaryotic cells
I. Title   II. Miklos, George L. Gabor
574.87'322   QH430

ISBN 0-04-575032-7
ISBN 0-04-575033-5 Pbk

Library of Congress Cataloging-in-Publication Data

(applied for)

Typeset in 10 on 11 point Bembo by Oxford Print Associates Ltd
and printed in Great Britain by
Biddles, Guildford, Surrey

Preface

'The mind unlearns with difficulty what has long been
impressed upon it.'

Seneca

Reductionism, is, without question, the most successful analytical approach available to the experimental scientist. With the advent of techniques for cloning and sequencing DNA, and the development of a variety of molecular probes for localizing macromolecules in cells and tissues, the biologist now has available the most powerful reductionist tools ever invented. The application of these new technologies has led to a veritable explosion of facts regarding the types and organization of nucleotide sequences present in the genomes of eukaryotes. These data offer a level of precision and predictability which is unparalleled in biology.

Recombinant DNA techniques were initially developed to gather information about the structure and organization of the DNA sequences within a genome. The power and potential of these techniques, however, extend far beyond simple data collection of this kind. In an attempt to use the new technology as a basis for analyzing development and evolution, attention was first focused on the topic of gene regulation, an approach that had proven so successful in prokaryotes. It is now clear that this has not been an adequate approach. Lewin (1984) has quoted Brenner as stating 'at the beginning it was said that the answer to the understanding of development was going to come from a knowledge of the molecular mechanisms of gene control. I doubt whether anyone believes this any more. The molecular mechanisms look boringly simple, and they don't tell us what we want to know.' Nor is this surprising. The role of genes in specifying the primary structure of protein molecules is usually far removed from the developmental end products to which those molecules contribute. Consequently, the solutions to developmental problems are much more likely to be found at the cellular and intercellular levels than at the level of transcription.

Although the approach was too limited, the philosophy was right. While we appreciate that there are aspects of development that transcend molecular biology, we believe it will not be possible to elucidate these aspects without an understanding of the molecular mechanisms upon which they depend. Thus while cell–cell interactions are a fundamental feature of all developmental programmes, the most direct approach to resolving this form of interaction must come from an understanding of gene products which, like the neural cell adhesion molecules, a series of membrane associated glycoproteins which control cell surface properties (Edelman 1985, reviewed in Rutishauser & Goridis 1986).

As far as evolutionary biology is concerned, Lewontin (1982) has argued that the deepest questions of evolution 'will never be resolved by molecular biology alone.' There is certainly no doubt that some occurrences of the deepest significance to evolution belong to this category. For example, virtually all the species of plants and animals that have existed on Earth are now extinct, so that extinction must rank as the most common evolutionary event. Major extinction events over the past 250 million years show a statistically significant periodicity (Raup & Sepkoski 1984). Although the precise causes of this periodicity are not known, there are grounds for arguing that the forcing agent was environmental rather than biological. If the impact of the physical environment did indeed result in non-random and short-lived mass extinctions, then the direction and type of evolutionary change may well have been uniquely and irrevocably altered, but molecular biology can have nothing to say about such events.

However, in a strict sense, Lewontin's viewpoint is overly pessimistic since the deepest evolutionary questions which are resolvable will certainly require the aid of molecular biology. Indeed, there is no doubt that the findings of molecular biology impinge far more directly on the major problems of both development and evolution than any other approach to these problems. In the chapters that follow we shall try to show how, and why, they do so.

Specifically, we attempt to assess and interpret the facts concerning genome structure in terms of fundamental problems of genome function in relation to development and evolution. Our main aim is to evaluate the impact of the modern molecular work on long-standing biological problems.

B. John & G. L. G. Miklos
Canberra

Acknowledgements

Joan Madden, Marilyn Miklos, Val Rawlings, Jan Richie and Sylvia Stephens have typed and retyped the numerous drafts of this book and we thank them all for their contribution. Gary Brown prepared all the illustrations and our debt to him is obvious. Arnold Bendich, John Campbell, Mel Green, Ian Young and Brian Gunning read earlier drafts of this book while MasaToshi Yamamoto, John Langridge, Gabriel Dover and Ken Campbell drew our attention to relevant genetic, evolutionary and palaeontological works. Michael Rosbash and Steven Henikoff also provided helpful comments. We are grateful to all of them for their advice and comments. We also acknowledge the support of our respective wives, Ena and Marilyn, without whose perseverance this book would never have been written. Finally, we thank Linda Antoniw for her meticulous editorial work.

We are grateful to the following individuals and organizations who have kindly given permission for the reproduction of copyright material (figure numbers in parentheses):

IRL Press (1.5); Figure 1.6 reproduced by permission of Yelton & Scharff and *American Scientist*; Figure 1.9 reproduced by permission from *Nature*, Vol. 284, p. 215, Copyright © 1980 Macmillan Journals Limited; Figure 2.1 reproduced from *From egg to embryo* by permission of J. M. W. Slack and Cambridge University Press; Figure 2.2 reproduced from *Cell Interactions in Differentiation* by permission of Academic Press; Figure 2.4 reproduced from *Results and Problems in Cell Differentiation 5*, by permission of Springer Verlag; Figure 2.5 reproduced by permission from *Nature*, Vol. 265, pp. 211–16, Copyright © 1977 Macmillan Journals Limited; Figure 2.7 reproduced from *37th Symp. Soc. Dev. Biol.* by permission of Academic Press; Figure 2.8 reproduced by permission from *Nature*, Vol. 287, pp. 795–801, Copyright © 1980 Macmillan Journals Limited; Figure 2.10 reproduced by permission from *American Zoologist*; Figure 2.12 reproduced by permission from Bender *et al.*, *Science*, Vol. 221, pp. 23–9, July 1983, Copyright 1983 the American Association for the Advancement of Science; Figure 2.13 reproduced by permission from *Nature*, Vol. 310, pp. 25–31, Copyright © 1984 Macmillan Journals Limited; Figure 2.13 reproduced by permission from *Nature*, Copyright © 1984 Macmillan Journals Limited; Figure 2.15 reproduced from *Dev. Biol. 50* by permission of Academic Press; Figure 2.16 reproduced from *Genetics & Biology of Drosophila*, Vol. 2c by permission of Academic Press; Figure 2.18 reproduced from *Dev. Biol. 89* by permission of Academic Press; Figure 2.20 reproduced from *31st Symp. Soc. Dev. Biol.* by permission of Academic Press; John Wiley & Sons (2.19); *Genetics* (2.21, 3.24, 3.35, 3.43); Figures

2.22 & 5.2 reproduced from *Dahlem Konferenzen* 1982 by permission of Springer Verlag; Figure 2.26 reproduced from *37th Symp. Soc. Dev. Biol.* by permission of Academic Press; Figure 2.27 reproduced from *Mobile Genetic Elements* by permission of Academic Press; Figure 2.28 reproduced from *C.S.H.S.Q.B.* **45**, Copyright 1981 by permission of Cold Spring Harbor Laboratory; Figure 2.29 reproduced by permission from *Nature*, Vol. 290, pp. 625–7, Copyright © 1981 Macmillan Journals Limited; Figures 2.30, 2.31, 2.35, 6.5 reproduced from *Scientific American* by permission of W. H. Freeman; Figure 2.32 reproduced from *Cell* **37**, Copyright M.I.T.; Figure 2.34 reproduced from *Chromosoma* **45** by permission of Springer Verlag; Figure 2.37 reproduced from Varmus, *Mobile genetic elements* by permission of Academic Press; P.N.A.S (3.2, 3.8, 3.21, 3.27, 3.39, 4.13); Figure 3.3 reproduced from *Nature*, Vol. 311, pp. 163–5, Copyright © 1984 Macmillan Journals Limited; Figure 3.4 reproduced from *Cell* **38**, Copyright M.I.T.; Figure 3.5 reproduced from *Cell* **32**, Copyright M.I.T.; Figure 3.6 reproduced from *Trends in Genetics* by permission of Elsevier; Figure 3.7 reproduced from *Nature*, Vol. 297, pp. 365–71, Copyright © 1982 Macmillan Journals Limited; Figure 3.9 reproduced from Rosenfield *et al.*, *Science*, Vol. 225, pp. 1315–20, Copyright 1984 the American Association for the Advancement of Science; Figure 3.15 reproduced from Hall *et al.*, *J. Mol. Biol.*, Vol. 169 by permission of Academic Press; Figure 3.22 reproduced from *Cell* **7**, Copyright M.I.T.; Harvard University Press (3.23); Figure 3.25 reproduced from *Chromosoma* **62** by permission of Springer Verlag; Figure 3.27 reproduced from *Cell* **26**, Copyright M.I.T.; Figure 3.28 reproduced from *C.S.H.S.Q.B.* **42**, Copyright 1978, by permission of Cold Spring Harbor Laboratory; Figure 3.30 reproduced from *Cell* **19**, Copyright M.I.T.; Figure 3.36 reproduced from *Experimental Cell Research* **118**, by permission of Academic Press; Figure 3.37 reproduced from *Cell* **27**, Copyright M.I.T.; Figure 3.38 reproduced from *Cell* **29**, Copyright M.I.T.; Figure 3.40 reproduced from *Cell* **34**, Copyright M.I.T.; Figure 4.1 reproduced by permission from *Heredity* **43**; Figures 4.2 & 4.3 reproduced from R. V. Short, *Reproduction in Mammals* **6**, by permission of Cambridge University Press; Figure 4.8 reproduced from *Genetics & Biology of Drosophila*, Vol. 3B, by permission of Academic Press; Figure 4.9 reproduced from *Chromosoma* **85** by permission of Springer Verlag; Figure 4.10 reproduced by permission from *Genetica* **52/53**; Figure 4.11 reproduced from *Cell* **32**, Copyright M.I.T.; Figure 4.12 reproduced from the *Ann. Rev. Genet.*, Vol. 14, © 1980 by permission of Annual Reviews Inc.; Alan R. Liss Inc. (4.14); Figure 4.16 reproduced from *Chromosomes Today*, Vol. 6, by permission of Elsevier; Figure 5.1 reproduced from *Metamorphosis* by permission of Plenum Press; Figure 5.3 reproduced by permission from *Evolution* **36**; Figure 5.4 reproduced from the Royal Society's Philosophical Transactions by permission of the Royal Society; Figure 6.1 reproduced from *Cell* **45**, Copyright Cell Press; Figure 6.2 reproduced from *Mol. Gen. Genet.* **198** by permission of Springer Verlag; Figure 6.4 reproduced from *Molecular Evolutionary Genetics* by permission of Plenum Press; Figure 6.5 reproduced from *Scientific American* by permission of W. H. Freeman.

Contents

Preface   *page* vii

Acknowledgements   ix

List of tables   xv

1 GENERAL MOLECULAR ORGANIZATION OF GENOMES   1

1.1   Dissecting genomes   1

   1.1.1   *Microdissection and microcloning*   1
   1.1.2   *Chromosome walking and jumping*   4
   1.1.3   *Transposon mutagenesis and transformation*   5
   1.1.4   *Synthetic DNA probes*   7

      1.1.4.1   From protein to gene using monoclonal antibodies   7
      1.1.4.2   From gene to protein   9

1.2   DNA components of genomes   10

   1.2.1   *Non-functional DNA*   10
   1.2.2   *Conserved DNA elements*   11
   1.2.3   *Genomic flux*   15

      1.2.3.1   Transposition   15
      1.2.3.2   DNA replication quirks   17
      1.2.3.3   Unequal exchange   17
      1.2.3.4   Conversion   20
      1.2.3.5   Sequence amplification   22

2 DEVELOPMENTAL ACTIVITIES OF GENOMES   26

2.1   From egg to adult   26
2.2   The genetic control of development in *Drosophila melanogaster* 33

   2.2.1   *Embryonic development*   33
   2.2.2   *Pupal development*   35
   2.2.3   *Genetic control of spatial organization in the* Drosophila melanogaster *embryo*   39

      2.2.3.1   The antero-postero gradient   39
      2.2.3.2   The dorso-ventral gradient   40
      2.2.3.3   Body segmentation   43
      2.2.3.4   Homeotic genes   46

   2.2.4   *Sexual development*   63

2.2.5 *Neurogenesis*   *page* 68
2.2.6 *General conclusions*   88

2.3 General principles of development   90

2.3.1 *Polarity*   90
2.3.2 *Cell lineages*   90

2.4 Genome alterations during development   95

2.4.1 *Nucleotide sequence alterations*   95
       2.4.1.1 The transposable mating type loci of yeast   95
       2.4.1.2 Immunoglobulin class switching   98

2.4.2 *Presomatic diminution*   103
2.4.3 *Macronuclear development in unicellular ciliates*   104
2.4.4 *Antigenic switching in trypanosomes*   109
2.4.5 *The molecular bases of genomic alterations*   111

3 CODING CAPACITIES OF GENOMES   113

3.1 Gene regulation in eukaryotes   113

3.1.1 *Transcriptional controls*   113
       3.1.1.1 Polymerase II systems   114
       3.1.1.2 Polymerase I systems   120
       3.1.1.3 Polymerase III systems   121
       3.1.1.4 DNA methylation and gene activity   122

3.1.2 *Post-transcriptional controls*   124

3.2 *Drosophila* genomes   128

3.2.1 *The general molecular organization of* Drosophila *genomes*   128
3.2.2 *Heterochromatic DNA*   135
       3.2.2.1 Functional studies   137
       3.2.2.2 Distribution of highly repetitive sequences
                within the heterochromatin   137
       3.2.2.3 Satellite DNA binding proteins   140

3.2.3 *Euchromatic DNA*   141

3.3 Comparative genome organization   149

3.3.1 *Size variation between genomes*   149
       3.3.1.1 Chordates   150
       3.3.1.2 Insects   152
       3.3.1.3 Plethodontid salamanders   153

3.3.2 *Interspersion pattern differences*   154
3.3.3 *'Short' dispersed repetitive sequences*   160
3.3.4 *Message complexities in eukaryotes*   163
       3.3.4.1 Ciliate protozoans   164
       3.3.4.2 Fungi   165

3.3.4.3   The sea urchin, *Strongylocentrotus purpuratus*   *page* 166
3.3.4.4   Vertebrates   168
3.3.4.5   Developmental capacities of eukaryote genomes   171

3.3.5   *Differences in heterochromatin content*   173
3.3.6   *Differences in coding DNA*   177

3.3.6.1   Multigene families   177
3.3.6.2   Structural gene systems   189
3.3.6.3   Genome structure and function   193

3.4   Gene dosage relationships   200

3.4.1   *Compensation and magnification*   201
3.4.2   *Selective gene amplification*   208

3.4.2.1   The chorion genes of *Drosophila melanogaster*   208
3.4.2.2   The DNA puffs of *Rhynchosciara americana*   211
3.4.2.3   rDNA amplification   212
3.4.2.4   Somatic endoploidy   214

3.4.3   *Dosage compensation*   216

3.4.3.1   Female X-inactivation   216
3.4.3.2   Male X-compensation   219

3.4.4   *Default transcription*   224
3.4.5   *Genetic balance and development*   226

3.5   The developmental dilemma   230

4   GENOME CHANGE AND EVOLUTIONARY
CHANGE   234

4.1   The basis of evolutionary change   235
4.2   Stability and change in the genome   236

4.2.1   *The spread and fixation of genome change*   237

4.2.1.1   Natural selection   237
4.2.1.2   Neutral drift   238
4.2.1.3   Molecular drive   241

4.2.2   *Genome turnover*   244

4.3   Nucleotype and genotype   245

4.3.1   *Genome size and cell size*   246
4.3.2   *Genome size and metabolic rate*   248
4.3.3   *Genome size and division cycle time*   249
4.3.4   *Genome size and developmental time*   249

4.4   Genome change and speciation   257

4.4.1   *Species differences in mammals*   258

4.4.1.1   Muntjacs   258
4.4.1.2   Equids   260

	4.4.1.3	Mice	*page* 263
	4.4.1.4	Rats	268
4.4.2		*Structural changes in the* Drosophila *genome*	269
4.4.3		*Speciation and morphological change*	272
4.4.4		*Hybrid dysgenesis in* Drosophila melanogaster	274

4.5   Changes in genome size   276

4.5.1	*Modes of change*	277
	4.5.1.1 Polyploidy	277
	4.5.1.2 Duplication	277
	4.5.1.3 Amplification	281
4.5.2	*Genome size, specialization and speciation*	286
4.5.3	*Supernumerary chromatin*	288

4.6   Summary statement   290

5   THE UNSOLVED PROBLEM –
THE ORIGIN OF MORPHOLOGICAL NOVELTY   292

5.1   Timing adjustments   298
5.2   Binary switch mechanisms   302
5.3   Cell interactions   303
5.4   Cell position   305
5.5   The evolutionary dilemma   305

6   CODA   310

6.1   Facts and conclusions   310
6.2   Future prospects   330
6.3   Final statement   334

References   339
Index   386

List of tables

1.1   Haploid DNA contents and percent single copy DNA in animal genomes   *page* 11

1.2   Some examples of the conservation of genes and gene products in living systems   12

2.1   The developmental schedule of *Drosophila melanogaster* at 23–25 °C   34

2.2   Maternal-effect dorsalizing loci of *Drosophila melanogaster*   41

2.3   Gene loci affecting segmentation in *Drosophila melanogaster*   43

2.4   Homeotic genes required for the correct expression of the Bithorax Complex of *Drosophila melanogaster*   48

2.5   Representative homeotic mutations in *Drosophila melanogaster*   56

2.6   Neurogenic mutations leading to hypertrophy of the CNS in *Drosophila melanogaster*   69

2.7   Mutations affecting the giant fibre pathway controlling the indirect flight muscles of *Drosophila melanogaster*   73

2.8   Representative neurological and behavioural mutants of *Drosophila melanogaster*   75

2.9   Cell number variation in the development of *Caenorhabditis elegans*   91

2.10   Chromatin diminution in the copepod genus *Cyclops*   104

2.11   Genome characteristics of ciliates   108

3.1   Characteristics of some mobile dispersed genes in the genome of *Drosophila melanogaster*   129

3.2   A comparison of genome composition in two sibling species of *Drosophila*   133

3.3   Influence of added heterochromatin on male and female mean hatching time in *Drosophila melanogaster*   138

3.4   RNA sequence complexity at different developmental stages in *Drosophila melanogaster*   143

3.5   Transcript lengths of cloned genes in *Drosophila melanogaster*   144–5

3.6   Band number variation in dipteran polytene chromosomes   147

3.7   Genome size variation in eukaryotes   150

3.8   Genome size variation in insects   *page* 152
3.9   Genome sizes in salamanders of the genus
Plethodon   153
3.10   Characteristics of some fungal genomes   156
3.11   Poly (A)$^+$ RNA complexity in *Aspergillus nidulans*   165
3.12   Complexity of poly (A)$^+$ RNA from spore
specific and developmental specific stages of
Aspergillus nidulans   166
3.13   RNA sequence complexity in vertebrates   169
3.14   Heterochromatin and highly repeated content of
Drosophila genomes   174
3.15   Genome variation in the kangeroo rat genus
Dipodomys   176
3.16   Variation in heterochromatin content in placental
mammals   176
3.17   Repetition of 18S + 28S, 5S and tRNA genes in
selected eukaryotes   178
3.18   Numerical variation in the 18S + 28S RNA genes
of eukaryotes   181
3.19   Organization of histone genes in animal genomes   183
3.20   Six types of histone gene containing clones
isolated from the human genome   184
3.21   Histone gene copy numbers in the genomes of
amphibians   186
3.22   The location of known tRNA species in
Drosophila melanogaster   187
3.23   Tubulin gene numbers in eukaryotes   189
3.24   Actin gene numbers in eukaryotes   190
3.25   Characteristics of the major classes of vertebrate
collagen   192
3.26   The location of genes controlling enzymes
involved in five human metabolic pathways   197
3.27   Percent hybridization of rRNA to the DNA of
larval diploid (brain and imaginal discs) and
polytene (salivary gland) tissues in two strains of
Drosophila melanogaster   202
3.28   Percent DNA hybridized with 28S rRNA in adult
flies from different stocks of *Drosophila hydei*   204
3.29   the rDNA content of diploid larval (brain),
polytene larval (salivary gland) and endotetraploid
adult (thoracic muscle) tissues of individuals of
Drosophila hydei with different genotypes   205
3.30   The chorion gene system of *Drosophila melanogaster*   209
3.31   rDNA amplification in oocytes of urodele
amphibians   213
3.32   RNA content of *Xenopus laevis* oocytes   213
3.33   DNA amplification in invertebrate oocytes   214

3.34 Activity of X-linked genes in individuals of
Drosophila melanogaster with differing X-
chromosome:autosome ratios *page* 219
3.35 Phenotypic consequences of polysomy in Man 228
3.36 Types and frequencies of spontaneous
chromosome abnormalities in a sample of (a) 4182
four-day-old chicken embryos and (b) 1926
human abortuses 229
4.1 Nucleotide substitutions in the mRNAs coding
for sequences 24–33 of the histone IV protein in
two species of sea urchin 240
4.2 Frequency changes in twelve families of cloned
repeats between the sea urchins *Strongylocentrotus
purpuratus* and *S. franciscanus* 244
4.3 Cell and nuclear volumes relative to genome size
in nine amphibians 247
4.4 Mitotic cycle time in relation to ploidy in two
wheat genera (*Aegilops* and *Triticum*) 250
4.5 Relationship between genome size and hatch time
in ten amphibians 251
4.6 DNA values of semilarval (paedogenetic) and
permanently larval (neotenic) urodeles 252
4.7 Gestation time in mammals 253
4.8 Relationship between life forms, genome and cell
characteristics in twelve species of Anthemidae 254
4.9 Genome characteristics in the genus *Lolium* 255
4.10 Genome variation in the genus *Xenopus* 258
4.11 Genome variation in barking deer (genus
Muntiacus) 259
4.12 The consequences of inter-specific hybridization
in the genus *Equus* 262
4.13 Distribution of diploid chromosome numbers
in various orders of mammals 264
4.14 Malsegregation rates, estimated from metaphase
II counts, in male fusion heterozygotes of the long
tailed house mouse originating either in
Switzerland or in central Italy 266
4.15 Karyotype differentiation in Australian species of
Rattus 269
4.16 Karyotype variation in endemic Hawaiian
members of the genus *Drosophila* 271
4.17 Location of the main clusters of ribosomal genes
in seven species of *Plethodon* 284
4.18 Discontinuous nature of DNA distribution in 21
species of the plant *Lathyrus* 286
4.19 DNA content of some plant B-chromosomes 289
5.1 Categories of heterochronic timing shifts during
development 298

5.2   Morphological and biochemical changes induced
      by thyroid hormones during amphibian
      metamorphosis
6.1   Buffered genes in *Drosophila melanogaster*

1

General molecular organization of genomes

'Far more critical than what we want to know or do not know
is what we do not want to know.'

Eric Hoffer

1.1 DISSECTING GENOMES

DNA is the most important component of the eukaryote genome in the
sense that it ultimately provides the essential coding information necessary
for specifying the production of all other molecules within the eukaryote
cell. Added to this, it is now the most tractable component in a
technological sense. Much of this technology is now so well known as to
need no introduction (Weinberg 1985). In this book we have chosen to
highlight the fly, *Drosophila melanogaster*, which is, without question, the
most thoroughly studied eukaryote in genetical, developmental and molecular
terms. For this reason we begin by outlining the four principal techniques
currently in use for dissecting the genome of *Drosophila melanogaster*, for
isolating specific genes within that genome and to which we refer in the
other sections of this book.

1.1.1 Microdissection and microcloning

One technique that has enormously speeded up gene isolation in *Drosophila
melanogaster*, and at the same time made the screening of large DNA
libraries effectively obsolete, is termed microcloning (Pirrotta 1984). This
involves the molecular cloning of picogram quantities of DNA. Although
this technique has now been applied to mammals, it was pioneered in
Drosophila melanogaster. Here giant polytene chromosomes are present in a
variety of larval tissues. The remarkable size of these is a consequence of up
to ten rounds of chromosome replication without either chromatid
separation or cell division. Consequently, the DNA helices are duplicated
more than 1000 times. In this polytenization process the different DNA
sequences do not behave uniformly. The highly repeated sequences that
make up the heterochromatic regions of the mitotic chromosomes replicate

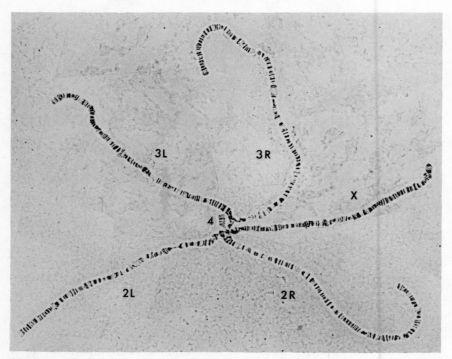

Plate 1 The polytene chromosomes of the larval salivary gland of *Drosophila melanogaster* (photograph kindly supplied by Professor George Lefevre).

little, if at all. These regions coalesce in the polytene nucleus to form a chromocentre, a region consisting of numerous attenuated strands which join the base of each chromosome arm. The result is that the polytene chromosomes are made up almost entirely of the euchromatic portions of the mitotic chromosomes which constitute some 70% of the total DNA. These are laterally duplicated by a factor of 1000 times or more, and the chromosome arms are partitioned into a series of transverse chromatic bands (Plate 1).

These polytene chromosomes provide a cytological map of the genome, and individual polytene bands furnish specific landmarks which allow for the localization of both genes and structural rearrangements. With over 5000 definable bands, the genome is well partitioned even at the level of the light microscope. Consequently, *in situ* hybridization with DNA or RNA probes has a large effective target size even when unique sequences are employed. Figure 1.1 illustrates the chromosomal zip code employed in referring to individual bands. The euchromatin of each major chromosome arm is divided into exactly 20 numbered divisions in the following way: the X-chromosome (1) = 1–20; the left arm chromosome 2 (2L) = 21–40; the right arm of chromosome 2 (2R) = 41–60; the left arm of chromosome 3 (3L) = 61–80; the right arm of chromosome 3 (3R) = 81–100; with two additional divisions for chromosome 4 = 101–102. Each division is allotted six lettered subdivisions, A–F, and individual bands within each subdivision are numbered consecutively from left to right. Thus, no two bands have the same designation and, by following the appropriate terminology, the cytological location of any mutant allele, any structural rearrangement or any particular DNA sequence can be identified within the genome.

Using a glass needle operated by a micromanipulator, a single piece of banded chromosome can be cut out of a polytene arm and transferred to a micro-drop of buffer. Here the DNA is cleaved with restriction enzymes, ligated and then packaged into bacteriophage λ. The resulting plaques from such a single cut together yield a genomic segment from 100 kilobases (kb) upwards in length, and the problem now moves to identifying transcripts within such a landscape (Pirrotta 1984).

Microdissection and microcloning have also been applied to mammalian chromosomes. Here, since chromosomes are not multistranded, it is necessary to pool chromosome fragments obtained from a particular region in order to achieve successful microcloning.

As far as *Drosophila melanogaster* itself is concerned, the entire euchromatic genome of 115 million base pairs (bp) could be microcloned into about 200 or 300 mini-libraries. The only limiting variables are manpower and resources. Since the genetic and cytogenetic localization of genes of developmental significance can usually be narrowed down to ten bands or so, their isolation from particular mini-libraries ought in future to be reasonably straightforward.

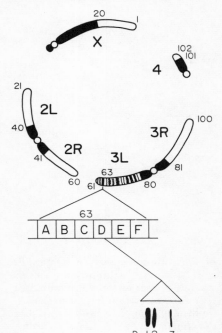

Figure 1.1 The chromosomal zip code for identifying individual polytene bands in the euchromatic regions of *Drosophila melanogaster* chromosomes. Heterochromatic regions are shown solid while the centromeres are represented by open circles. The banded nature of one of the euchromatic arms (3L) is also shown.

1.1.2 Chromosome walking and jumping

The chromosomal origin of any non-repeated DNA segment of *Drosophila melanogaster* can be determined by the *in situ* hybridization of radioactively labelled copies of that DNA to the polytene chromosomes. If a DNA sequence is found within a few bands of a gene of interest then that sequence can be used as a starting point for a chromosome walk to that gene (Bender *et al.* 1983b).

The first step in such a walk involves screening a recombinant DNA library of random large segments of the *Drosophila melanogaster* genome to identify those that overlap the starting point. The sequences that extend farthest from the starting point in either direction can then be used as new starting points. A succession of such steps based on a series of overlapping DNA clones will identify the DNA from a long contiguous region of the chromosome (Fig 1.2).

Jumping, on the other hand, involves a similar approach used in conjunction with structural chromosome rearrangements, such as inversions, translocations or duplications, to shift a walk to a new chromosomal location.

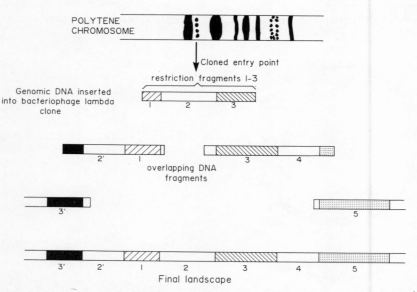

Figure 1.2 Diagrammatic representation of a chromosome walk in *Drosophila melanogaster*. The initial entry point is shown on the polytene chromosome. In this example, restriction fragments 1 and 3 are used as probes to isolate recombinant bacteriophage from genomic libraries. The most distal restriction fragments in each of the newly isolated sets of bacteriophage are then again used as probes to isolate more bacteriophage from the recombinant DNA library. A bi-directional walk thus results in the production of a restriction map for any given chromosomal landscape.

1.1.3 Transposon mutagenesis and transformation

Most eukaryote genomes contain mobile elements or transposons, similar to elements found in bacteria, which move around the genome. In *Drosophila melanogaster* one class of these, termed P elements (Rubin 1985b), has been harnessed in two major ways for gene isolation and characterization.

Under appropriate conditions, resident P elements within the genome can be induced to excise and transpose to a new location. In so doing, they sometimes insert into genes and cause a lethal mutation which can be detected by standard crossing procedures. When a DNA library is constructed from a strain in which a P element has inserted itself into a gene of interest, the use of labelled P-DNA as a probe will yield a number of phages, each of which carries not only the P element itself, but also the flanking DNA belonging to the gene which has been inactivated. Thus, provided it is a lethal, or else has a scorable phenotype, the gene in question can be isolated simply and directly. This, however, is not a particularly useful procedure for identifying genes controlling behavioural phenotypes which are sometimes too complicated for easy screening.

P elements can also be used as vectors for inserting other DNA sequences into the genome (Rubin & Spradling 1982, Spradling & Rubin 1982). Thus if a P element, loaded with virtually any gene sequence, is injected into a preblastoderm egg, it will transform some of the germ cells (Fig. 1.3). The insertion and activity of the novel passenger can then be assayed phenotypically or biochemically.

Using this approach, it has been possible to show that the Minute locus at 99 D in *Drosophila melanogaster* encodes the ribosomal protein 49. Thus, the introduction of an additional wild type copy of the *rp 49* gene, using P element-mediated germ line transformation, both suppresses the mutant Minute phenotype and restores to normal the slow development rate of the 99 D Minute-bearing progeny. The phenotypic similarity of the various *Drosophila melanogaster* Minute loci, when mutated, suggests that many of them are likely to encode ribosomal proteins, so that the Minute mutant phenotype is a consequence of a lowering of ribosomal assembly, and hence the rate of protein synthesis, during development (Kongsuwan *et al.* 1985).

Some vectors have reached very sophisticated levels. In vector pUChsneo (Fig. 1.4), for example, the gene itself is preceded by a heat shock promoter (HSP), so that simply raising the temperature of the flies causes massive induction of the neomycin (neo) gene (Steller & Pirrotta 1985). Thus, a transformed embryo, larva or adult carrying such an insert is able to break down the antibiotic and hence survive. However, non-insert progeny are lethal, so that the selection scheme automatically ensures that only transformants are available for study.

Vectors of this sort, which have been engineered to carry particular genomic sequences, can also be used to 'rescue' mutant phenotypes. For example, mutations at the *K10* locus (2E2–2F1), which is involved in the establishment of the dorso-ventral axis of the *Drosophila melanogaster* embryo, cause female sterility. Using P element-mediated germ line transformation with cosmid clones, involving a 43 kb insert obtained from

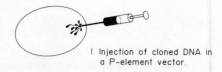

1. Injection of cloned DNA in a P-element vector.

2. Incorporation of DNA into chromosomes of germ cells.

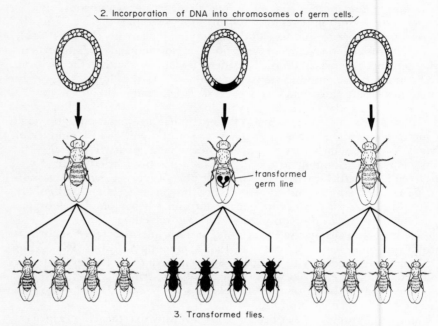

transformed germ line

3. Transformed flies.

Figure 1.3 P element mediated germ line transformation obtained by the injection of DNA into the egg of *Drosophila melanogaster* prior to cellular blastoderm formation.

UNLOADED VECTOR   CONSTRUCT

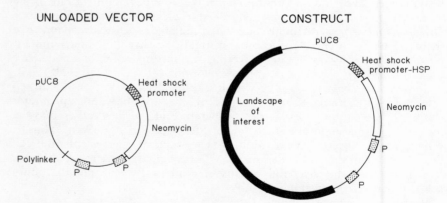

pUC8

Heat shock promoter

Neomycin

Polylinker

P   P

pUC8

Heat shock promoter-HSP

Neomycin

Landscape of interest

P

P

Figure 1.4 The structure of the unloaded and the loaded vector pUChsneo, which incorporates a heat shock promoter (HSP) and a gene for neomycin resistance (neo) as well as two fragments of a P element. Raising the temperature of the transformed flies containing such a construct causes the induction of the neomycin gene, so that flies are able to survive G418 antibiotic treatment and can be automatically selected (data of Steller & Pirrotta 1985).

microdissecting interval 2E2–2F1, it has been possible to rescue *K10* sterility, and so produce transformed females which give rise to normal embryos when homozygous for *K10* (Haenlin *et al.* 1985). Three genes which map genetically close to *K10* (crooked neck, pecanex and kurz) were also rescued. By using smaller and smaller insert fragments, wild type *K10* function was shown to be localized to an 11 kb interval within the 43 kb insert.

1.1.4 Synthetic DNA probes

Sometimes, however, it is necessary to begin analysis even further from the genome itself and at the level of a protein product. Thus, if the amino acid sequence of a protein, or a part of a protein, is known, it is possible to indirectly isolate the gene responsible for its production. In this way, a synthetic DNA probe of the amino acid sequence of rat myelin protein has been used to select two cDNA clones encoding that protein (Roach *et al.* 1983).

The initial problem with this approach is to obtain enough purified protein to carry out a sequencing analysis. Gas phase sequenators are now available which carry out microsequencing on less than 100 pmol of a small protein or peptide. Moreover, for most proteins, they can work on quantities as low as 5 pmol (Hunkapillar *et al.* 1984). Automated, state of the art, machines can yield the sequence of up to 70 amino acids in a single analysis. The success of this technique, however, depends critically on purifying very small amounts of peptide free of any contamination.

Having the amino acid sequence of a peptide allows for the construction of DNA sequences capable of coding for such a molecule. Automated solid phase methods for synthesizing oligonucleotides are also available now. Synthetic DNA probes obtained by oligonucleotide synthesis can therefore be used to screen recombinant libraries even when only a limited amount of protein sequence data is available.

The reverse of this procedure is also practicable. Given a DNA sequence obtained from a genomic DNA clone, one can infer the appropriate protein sequence and so synthesize the peptide. Using automated solid phase systems, single amino acid residues are added successively to a growing chain. In this way long pure peptides, in excess of 70 amino acids, can be synthesized in good yields within one or two days.

Biotechnology has advanced to such a stage that one can infiltrate the genome via a number of macromolecular approaches and produce DNA sequences from proteins, or proteins from DNA sequences.

1.1.4.1 *From protein to gene using monoclonal antibodies*

In order to identify, classify, purify and 'tag' complex proteins, the immune system can be harnessed to produce large quantities of homogeneous antibodies. In the past, the major problem associated with the use of antibodies concerned the heterogeneous response of the immune system. Thus, even when relatively 'pure' immunogens are employed, animals often make antibodies against even minor contaminants in the preparation.

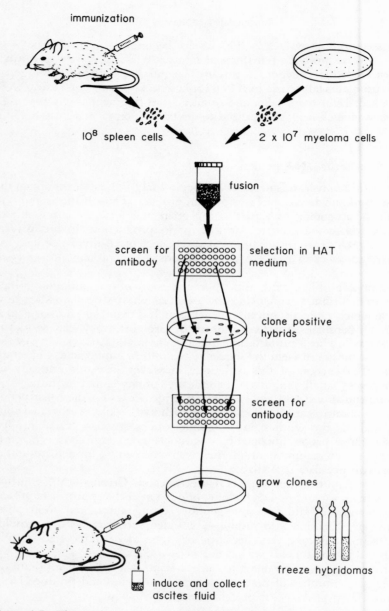

immunization

10^8 spleen cells

2×10^7 myeloma cells

fusion

screen for antibody

selection in HAT medium

clone positive hybrids

screen for antibody

grow clones

induce and collect ascites fluid

freeze hybridomas

Figure 1.5 The production of monoclonal antibodies. The formation of hybridomas involves fusing myeloma cells with spleen cells immunized with a potent antigen. Fused cells are then placed in a selective medium where hybrid clones develop. A radioimmune assay is used to screen for hybrids which produce antibody that will react with the immunogen. Positive hybrids are then recloned, and the subclones again screened for antibody prior to injection into the peritoneal cavity of mice. Here, they induce the accumulation of ascites fluid containing large amounts of monoclonal antibody (data of Yelton & Sharff 1980).

Furthermore, antibody quality can vary from animal to animal so that specific antisera can often contain many different affinities and cross-reactivities to other antigens.

These problems have been overcome by stimulating a single precursor B-cell (see § 2.4.1.2) to divide, so forming a clone of antibody-producing plasma cells which can then be immortalized by fusing it with myeloma cells. The antibody-forming hybrids, or hybridomas, so formed resemble the myeloma cells in that they grow continuously in culture. But they also secrete their own homogeneous immunoglobulin in the form of a monoclonal antibody (Fig. 1.5). These hybrid cells can be amplified clonally, either in tissue culture or by reinjection into an animal where they produce tumors which then secrete vast quantities of antibody into the serum of the animal.

When used in conjunction with the techniques described earlier, monoclonal antibodies make it possible to go from the protein product of a gene back to the gene itself. This route is well illustrated by the isolation of a monoclonal antibody for a neuron-specific glycoprotein in *Drosophila melanogaster* (Zipursky *et al.* 1985). When mice were immunized with retinal homogenates of *Drosophila melanogaster*, a monoclonal antibody was raised which exclusively stained the photoreceptor cells of the retina. The antigen recognized by this antibody was purified by immunoaffinity chromatography and was found to be a glycoprotein of molecular weight 160 000. The first 19 amino acids at the NH_2 terminus of this molecule were identified by microsequencing.

A number of DNA probes corresponding to this 19 amino acid sequence were synthesized and used as hybridization probes to screen a *Drosophila melanogaster* recombinant DNA library in phage λ. A single positive plaque was isolated and shown to encode a 4.3 kb poly $(A)^+$ RNA. The genomic clone was then partially sequenced and was found to be in excellent agreement with the coding sequence at the NH_2 terminus of the original antigen. Radioactively labelled probes of the clone, when used in *in situ* polytene hybridization experiments, revealed that the structural gene for the antigen was localized at the tip of the right arm of chromosome 3.

•1.1.4.2 From gene to protein

The reverse pilgrimage, from gene to protein, is exemplified by the work of Sutcliffe *et al.* (1983). They chose initially to isolate brain-specific mRNAs from the rat, reasoning that many brain-specific proteins were very rare since they probably occurred in only a small subset of neurons. These workers purified poly $(A)^+$ brain RNA from adult rats, converted it into double stranded cDNAs and cloned it. They then characterized a number of such clones by direct DNA sequence analysis. The sense strand of one such a clone was identified and the long open reading frame translated into an amino acid sequence. Peptides of about 30 amino acids were chemically synthesized and used to raise antisera in rabbits. The antisera, in turn, were employed in immunocytological experiments using rat brain sections. In this way the particular protein corresponding to the cDNA clone in question was shown to be localized in certain cell bodies within the brain and to be transported through axons to specific fields in the cerebellum and hippocampus. Further study revealed that the protein in question was a glycoprotein and probably the precursor of a neurotransmitter. Thus from cDNA sequencing analysis it is possible to pass through synthetic peptides and antisera to anatomical sites of protein expression.

1.2   DNA COMPONENTS OF GENOMES

The dissection of genomes has led to three quite unexpected findings:

1.2.1   Non-functional DNA

Although the sizes of eukaryote genomes, and the amount of single copy DNA sequences they contain, vary enormously (Table 1.1), all of them are littered with DNA sequences that appear to be irrelevant to both developmental and evolutionary programmes. Thus, all eukaryotes carry repeated DNA sequences which usually make up a substantial fraction of the genome, and are often non-coding. These are of four main types:

(a) Highly repeated sequences in long tandem arrays (reviewed in Singer 1982a, Miklos 1985). The simplest of these consist of poly (dA-T) with a measurable content of other sequences (Swartz et al. 1962), but most are based on more complex units ranging up to thousands of base pairs.

(b) Dispersed and moderately repeated DNA sequences which are bounded not by copies of themselves but by unrelated sequences (Hardman 1986). In many cases these are dominated by a few major families and a very much larger number of minor ones (Deininger & Daniels 1986). For example, the human genome has three major families, *Alu*, *Kpn* and poly (C-A), which account for 20% of the total DNA (Sun et al. 1984). Included within this category of moderately repeated DNA are mobile or nomadic sequences which are dispersed through, and in most cases move about within, the genome.

(c) Numerous scrambled, clustered arrays of moderately repeated sequences. Here the individual sequence elements are 300–1000 bp in length, while a cluster, which is several kb long, contains many different individual sequences, some of which may be repeated either directly or in inverted orientation. Different clusters within the same genome may share some, though not all, sequences, but the arrangement of shared sequences is not conserved between clusters (Rubin 1983).

(d) Simple sequence DNA which forms tandem arrays of short motifs lying within more complex DNA. The degree of repetition ranges from 2 to 10 nucleotides, and these simple sequences are usually found within introns or else in the non-coding DNA around cloned genes (Tautz et al. 1986).

The genome also includes pseudogenes, that is inactive regions of DNA that display significant homology to functional genes but which include

Table 1.1 Haploid DNA contents (1C) and percent single copy (sc) DNA in animal genomes.

Invertebrates			Vertebrates		
Species	1C(pg)	% sc DNA	Species	1C(pg)	% sc DNA
Protozoa			Protochordata		
Tetrahymena pyriformis	0.2	90	*Ciona intestinalis*	0.2	70
Coelenterata			Pisces		
Aurelia aurita	0.7	70	*Scyliorhinus stellatus*	6.1	39
Nemertini (Rhyncocoela)			*Leuascus cephalus*	5.5	44
Cerebratulus	1.4	60	*Raja montagui*	3.4	47
			Rutilus rutilus	4.8	54
Mollusca			Amphibia		
Aplysia californica	1.8	55	*Necturus masculosus*	83.0	12
Crassostrea virginica	0.7	60	*Bufo bufo*	7.0	20
Spisula solidissima	1.2	75	*Triturus cristatus*	21.0	47
Loligo loligo	2.8	75	*Xenopus laevis*	3.1	75
Arthropoda			Reptilia		
Prosimulium multidentatum	0.18	56	*Natrix natrix*	2.5	47
Drosophila melanogaster	0.18	60	*Terrapene carolina*	4.1	54
Limulus polyphemus	2.8	70	*Caiman crocodylus*	2.6	66
Musca domestica	0.9	90	*Python reticulatus*	1.7	71
Chironomus tentans	0.21	95			
Echinodermata			Aves		
Strongylocentrotus			*Gallus domesticus*	1.2	80
purpuratus	0.9	75	Mammalia		
			Homo sapiens	3.5	64
			Mus musculus	3.5	70

After Straus 1971, Sachs & Clever 1972, Davidson *et al.* 1975, Sohn *et al.* 1975, Crain *et al.* 1976a, Galau *et al.* 1976a, Borschenius *et al.* 1978, Schmidtke *et al.* 1979, Olmo *et al.* 1981, Morescalchi & Olmo 1982.

mutations that prevent, or preclude, their expression (Proudfoot 1980). Thus, while only one gene for glyceraldehyde 3-phosphate dehydrogenase (*GAPDH*) is functional in man, mouse, rat and chicken, there are 10–30 pseudogenes in the human genome and over 200 in the genomes of mice and rats (Piechaczyk *et al.* 1984).

1.2.2 Conserved DNA elements

There are quite striking cases of conservation, both of DNA sequences and gene functions, across the phylogenetic spectrum. This is seen not only in the extent to which identical genes and homologous gene products have been conserved in widely disparate kinds of organisms (Table 1.2), but also in the interchangeability of sequences and mechanisms. This applies not only within eukaryotes, but also between prokaryotes and eukaryotes, emphasizing that there are important implications for eukaryotes within prokaryotic systems. For example, common signal sequences are responsible

Table 1.2 Some examples of the conservation of genes and gene products in living systems.

Element	Level of conservation	Reference
Genes		
dihydrofolate reductase (DHFR)	prokaryotes, Drosophila	Bourouis & Jarry 1983
heat shock protein 70 (hsp 70)	dnaK (E. coli), yeast, Drosophila, chicken	Bardwell & Craig 1984
adenine (ade-8)	yeast, Drosophila	Henikoff et al. 1981
immumoglobulin switch (S) sequences	yeast, Drosophila, sea urchin, mammals	Sakoyama et al. 1982
actin (act)	Dictyostelium, Drosophila, chicken	Fyrberg et al. 1980
oncogenes (onc)	Drosophila, vertebrates	Shilo & Weinberg 1981
homeo box	Drosophila, Xenopus, chicken, man	McGinnis et al. 1984a, b
rudimentary (r)	Drosophila, hamster	Segraves et al. 1983
RNA polymerase II	Drosophila, hamster	Ingles et al. 1983
calmodulin	Drosophila, bovine	Tobin 1984
Gene products		
insulin	prokaryotes, silkworms, mammals	Nagasawa et al. 1984
GAL-repressor	E. coli, yeast	Brent & Ptashne 1984
rhodopsin	Chlamydomonas, bovine	Foster et al. 1984
♀ sex hormone / glucocorticoids }	yeast, mammals	Feldman et al. 1984
muscarinic cholinergic receptors	Drosophila, mammals	Venter et al. 1984
casein kinase II	Drosophila, birds, mammals	Glover et al. 1983
U1 snRNA	Drosophila, mammals	Wieben et al. 1983
neural CAM	all vertebrates	Hoffman et al. 1984

for the transfer of secretory polypeptides across the cell membrane in prokaryotes and eukaryotes (Wiedmann *et al.* 1984). Likewise, a bacterial repressor protein, manufactured in the cytoplasm of a yeast cell can enter the nucleus of that cell, recognize its operator and repress transcription from a yeast promoter (Brent & Ptashne 1984). Similarly, a partial cDNA sequence of a hamster gene is able to correct a defect in a *pyr* B mutant of the bacterium *E. coli* (Davidson & Niswander 1983), while bacteriophage λ DNAs when injected into oocyte nuclei of the toad *Xenopus laevis* undergo recombination of parental markers (Carroll 1983). Finally, the *RpII215* locus in *Drosophila melanogaster* encodes the largest subunit of RNA polymerase II. It makes a 7 kb transcript, and the amino acid sequence predicted by the only long open reading frame of this transcript shows striking homology to the corresponding segments of the β subunit of the RNA polymerase of *E. coli* (Biggs *et al.* 1985).

Comparable homologous mechanisms also exist within eukaryotes. Autonomous replicating sequences, a specific subset of restriction fragments which can replicate independently of the chromosome, are functionally interchangeable between yeast and *Drosophila melanogaster* (Montiel *et al.* 1984), while the yeast *ADH1* promoter also promotes the transcription of the *Drosophila melanogaster ad8* gene and the human leucocyte interferon gene (Henikoff & Furlong 1983). *ADR1* is a transacting regulatory gene of *Saccharomyces cerevisiae* which encodes a protein required for the transcriptional activation of the glucose–repressible alcohol dehydrogenase *ADH2* gene. It encodes a polypeptide chain of 1323 amino acids of which the amino-terminal 302 are sufficient to stimulate *ADH2* transcription. This active region shows amino acid sequence homology with the repetitive DNA-binding domain of TFIIIA, an RNA polymerase III transcription factor of *Xenopus laevis*. Similar domains are also found in the proteins encoded by the Krüppel and Serendipity loci of *Drosophila melanogaster* (Hartshorne *et al.* 1986).

Even more striking examples of .conservation are known between *Drosophila melanogaster* and vertebrates. The *Drosophila melanogaster nina E* gene, encodes the major visual pigment protein, opsin, of the R1–R6 photoreceptors of the compound eye. The amino acid sequence of *Drosophila melanogaster* opsin, as deduced from the nucleotide sequence of *nina E*, shows significant homology to vertebrate opsins, with especially strong conservation of the retinal binding site (O'Tousa *et al.* 1985). Three regions of *Drosophila melanogaster* rhodopsin are also highly conserved compared with the corresponding domains of bovine rhodopsin (Zucker *et al.* 1985). Likewise, the structural properties of a novel 20S cytoplasmic ribonucleoprotein particle present in *Drosophila melanogaster* are similar to those of the prosome of the duck and the mouse (Arrigo *et al.* 1985). Of the several new species of small cytoplasmic RNAs associated with the *Drosophila melanogaster* prosome, one has a strong sequence homology with the U6 mammalian small nuclear RNA. The presence of this RNA, coupled with the fact that, in vertebrates, the prosome is known to be associated with repressed mRNPs (Schmid *et al.* 1984), suggests that this apparently ubiquitous eukaryotic particle is involved in post–transcriptional control.

In *Drosophila melanogaster* the developmentally important Bithorax Gene Complex (BX-C) has been used as a probe at low stringency to isolate equivalent sequences from *Xenopus laevis* (de Robertis *et al.* 1985) and man (Shepherd *et al.* 1984, Bonicelli *et al.* 1985, Joyner *et al.* 1985). Likewise, the calmodulin (Tobin *et al.* 1986) and pyrimidine biosynthesis (Segraves *et al.* 1983) genes of rat have both been successfully employed to isolate their respective counterparts in *Drosophila melanogaster*, while DNA sequences that hybridize with the *onc* genes of vertebrate retroviruses at low stringency have also been found in the fly. These sequences show 50–75% homology with the amino acid sequence of the corresponding vertebrate oncogenes (Neuman-Silberberg *et al.* 1984, Bishop *et al.* 1985). This ability to carry out a cross-genomic raiding of libraries provides a glimpse of what portions of the genome have been highly conserved in an evolutionary context. For example, heat shock genes, which are activated by a wide variety of stress conditions other than temperature elevation, have been characterized in organisms as diverse as bacteria, slime molds, ciliates, insects, sea urchins, vertebrates and angiosperm plants. In each case the major heat shock protein (hsp) produced has a molecular weight close to 70 000 and shows considerable homology between species. Thus, the amino acid sequence of the *Drosophila melanogaster* hsp 70 has 80% homology with that of chicken and 72% homology with that of yeast (Velazquez & Lindquist 1984).

Perhaps the most striking case of conservation is that reported by Kafatos *et al.* (1985) and Mitsialis & Kafatos (1985). The major chorion proteins of moths are encoded by two families of genes which have no obvious homologues in flies. Because choriogenesis is so different in moths and flies, it is not possible to equate specific choriogenic stages in the two groups. Nevertheless, if cloned DNA fragments bearing moth chorion genes are introduced into the *Drosophila melanogaster* germ line by P element-mediated germ line transformation, the genes are expressed with correct sex, tissue and temporal specificity, resulting in the accumulation of abundant moth chorion transcripts in late fly follicles. This implies that both *trans* and *cis* acting regulatory elements must also have been conserved since *trans*-regulators of fly origin must have interacted with the moth *cis*-regulators contained within the transduced DNA fragments.

Finally, Bond *et al.* (1986) have constructed a chimeric β-tubulin gene by combining part of a chicken β-tubulin gene with part of a structurally divergent yeast β-tubulin gene. When introduced into a mouse cell line, this gene codes for a chimeric β-tubulin protein which is incorporated into mouse microtubules. Thus, the mouse cell can tolerate the presence of an alien and 'highly diverged' synthetic tubulin protein with no overt effects on either microtubule structure or function.

As impressive as these cases of conservation are, it is important to recognize that sequence homology comparisons may not be sufficient to reveal all cases of conserved function. White *et al.* (1985) have recently described a particularly telling example which illustrates the point. The nicotinic acetylcholine (AcCho) receptor (AcChoR) is a multisubunit protein complex of four different subunits ($\alpha_2\beta\gamma\delta$) that is involved in vertebrate neuromuscular transmission as well as in the generation of electrical impulses by the electropax of electric fish. Nuclear injections of DNA, or cytoplasmic injection of mRNA, into *Xenopus laevis* oocytes is known to result in the biosynthesis of functional products. Thus, when *Torpedo californica* electric organ mRNA is injected into *Xenopus laevis* oocytes, functional *Torpedo californica* acetylcholine receptors appear in the oocyte membrane. The injection of an equimolar mixture of RNA transcripts made from *Torpedo californica* cDNAs has the same effect. Of particular interest, however, is the fact that a hybrid receptor, containing *Torpedo californica* α, β and γ subunits together with a δ subunit from the mouse, shows a 3–4 fold greater response than does the full *Torpedo californica* complex. No response is seen, however, when a mouse δRNA is injected in place of a *Torpedo californica* γRNA. In terms of amino acid profile comparisons, the mouse δ subunit shows 59% overall homology to the *Torpedo californica* δ subunit but also shows 48% homology with the *Torpedo californica* γ subunit. Thus, the features of primary sequence that determine the functional character of the mouse polypeptide are certainly not revealed by simple homology comparisons.

1.2.3 Genomic flux

Despite the evident long-term constancy of the genic sequences themselves, the eukaryote genome is nevertheless in a constant state of flux.

From a purely practical point of view, DNA as a polymer ought to be subject to a number of chemical processes which it simply cannot avoid. Thus it can be replicated, degraded, nicked, modified, circularized, moved around and generally tinkered with. All the cellular enzymology for carrying out these processes is available even in the simplest eukaryote. The really remarkable thing is how the DNA in a chromosome manages to survive so well in an environment characterized by such open chemical hostility.

There are four principal events that guarantee that sections of the eukaryote genome are subject to continual turnover. Between them these events are responsible for much of the plasticity of the genome.

1.2.3.1 *Transposition*

Transposition is any process which results in the insertion of a DNA sequence found at one site in a genome into a new site. Such a movement may be conservative or duplicative. In the former, the original copy is lost during transposition. In the latter, the original copy is retained at its initial location and a new copy is inserted elsewhere in the genome. When transposition is coupled with such a sequence duplication it leads to an increase in both copy number and location. Conservative transposition depends on the excision of the sequence and its reinsertion elsewhere. If, however, the excised sequence is in the form of a circular intermediate, it is also potentially duplicative, since the excised circle may replicate autonomously (Fig. 1.6). Those DNA sequences with a capacity for replicative transposition evidently possess an inherent mechanism of self-multiplication.

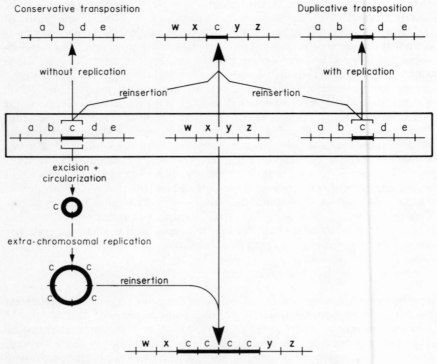

Figure 1.6 Mechanisms of transposition in eukaryotes.

Nomadic elements which have no effect on the phenotype could thus spread through a genome simply on account of their preferential replicative properties. As such, they constitute a class commonly referred to as selfish DNA (Doolittle & Sapienza 1980, Orgel & Crick 1980).

There are good grounds for arguing that most of the middle repeated DNA sequences within a genome are either transposable elements, or else degenerate descendants of such elements, so that no function is necessary to explain either their origin or their maintenance. The *Alu*-like transposable elements of man, for example, are assumed to move around within the genome by cycles of transcription, reverse transcription and reinsertion within the germ line genome (Calabretta *et al.* 1982), and many of the dispersed repetitive sequences contained within eukaryote genomes may have arisen in this fashion (Deininger & Daniels 1986).

The occurrence of processed pseudogenes, which are structured like cDNA copies of cellular mRNAs but which contain mutations that make them unable to encode a functional protein, suggests that any transcribed sequence, and not simply transposable sequences, may undergo reverse transcription and reinsertion (Baltimore 1985). Most cellular transcripts are products of RNA polymerase II, which recognizes promoters 5′ to the site of transcription initiation. Since reintegrated reverse transcripts of polymerase

II genes will, at the very least, lack their appropriate promoter, they are expected to give rise to pseudogenes. In most cases, such pseudogenes will not be able to generate new transcripts and cannot, therefore, produce additional dispersed pseudogenes by the same mechanism as they themselves originated. For example, with few exceptions, eukaryote genomes produce only a single cytochrome c, which is highly conserved in structure and function. Mammalian genomes, however, may contain one or two dozen dispersed sequences homologous to coding regions of the expressed rat cytochrome c gene. The rat itself has at least 25 such sequences. Seven of these show strong homology to the expressed gene, the rest show less. Most of these sequences are pseudogenes.

1.2.3.2 DNA replication quirks

DNA sequencing studies have revealed the existence of 'minisatellites' scattered throughout some eukaryote genomes. One of these, a 33 bp tandem repeat, is found in the first intron of the human myoglobin gene. Here, because it was flanked by a 9 bp direct repeat, it was initially considered to be a novel form of transposable element (Weller *et al.* 1984).

At least some of these minisatellite sequences are related to the generalized recombination sequence of *Escherichia coli* (Jeffreys *et al.* 1985) and, though they appear to have the characteristics of mobile elements, may arise by a replication slippage mechanism. This implies that sequences that are scattered throughout a genome need not necessarily owe their dispersion to conventional mobility mechanisms. Rather, they may arise spontaneously at different sites and then subvert the replicative and recombinative machinery of the cell, to produce localized tandem arrays. The existence of a class of specialized recombination sites, like the crossover hotspot instigator (Chi) of *Escherichia coli*, together with the enzymic machinery required to unwind, rewind, cut and thread DNA (Smith 1983), reminds us that the properties of the exchange junctions involved in recombination may be such as to generate novel sequences within eukaryotes, which may be either unique or repetitive. Thus, not every sequence flanked by a run of A's need necessarily be the result of processed gene insertion. Nor need every sequence flanked by an inverted or direct repeat be the result of mobile element meddlings.

1.2.3.3 Unequal exchange

One generally accepted mechanism of genome expansion is through the localized duplication of small segments of DNA (see § 4.5.1.2). The sequences so generated may either remain in tandem or else become dispersed within the genome. Once generated by duplication, tandem arrays can become not only self-perpetuating but also self-multiplying, since they can be amplified by unequal sister chromatid exchange either within the germ line mitoses or else at meiosis. Thus, exchanges between two sister clusters of tandemly repeated sequences may occur in many registers, only one of which leads to equality. Unequal sister chromatid exchange can then result in either an expansion or a contraction of copy number, and so can generate new length variants of a given repeat sequence

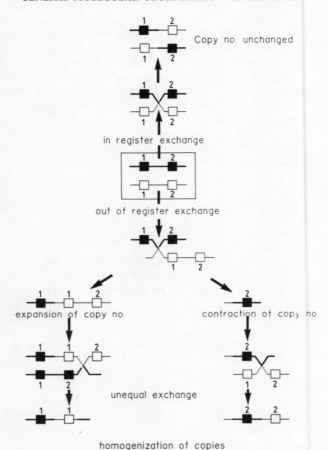

Figure 1.7   Consequences of unequal sister strand exchange between two tandem genes.

(Fig. 1.7). It also provides a basis for homogenization since, in extreme cases, it can result in either the elimination or the fixation of particular variants within a cluster. Finally, by using computer simulation, it has been shown that tandem repeats, when subject to unequal exchange, tend to develop higher order periodicities of the kind commonly found in highly repeated sequences (Smith 1976).

Because of the impracticability of carrying out conventional genetic analysis with repeated sequence DNA, it has, until recently, been difficult to obtain direct evidence for the occurrence of unequal sister chromatid exchange. This has been overcome by using yeast strains carrying an inserted structural gene within the rDNA locus. This involves introducing a plasmid, containing a rDNA fragment and the *Leu-2* gene, back into a yeast cell by transformation. Plasmid integration then takes place by homologous recombination with chromosomal rDNA, and so leads to the insertion of a structural gene into a series of tandemly arranged rDNA repeats. This functional copy of the *Leu-2* gene can then be used to detect deletions or duplications generated within the rDNA cluster by unequal exchange (Fig. 1.8).

With this approach, it has been possible to demonstrate that unequal exchange does indeed take place between sister chromatids during mitotic growth of haploid yeast strains, causing both deletions and duplications of some 6–8 repeat units at an estimated frequency of 5×10^{-4} deletions per generation. If expressed in terms of the entire rDNA cluster, this becomes 10^{-2} per generation, a rate of exchange which would be sufficient to maintain the sequence homogeneity of the rDNA locus of yeast. Insertions of *Leu-2* into the tandem array of rRNA genes in yeast are also unstable at meiosis, and this instability results from a high level of unequal exchange between rDNA clusters located on sister chromatids (Petes 1980).

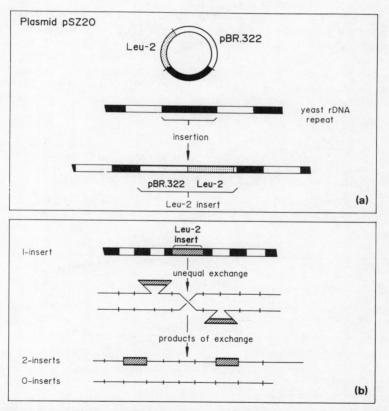

Figure 1.8 (a) Insertion of a *Leu-2* locus from chromosome 3 of yeast into the yeast rDNA locus on chromosome 12 via plasmid pSZ20, and (b) the consequences of unequal exchange following such an insertion (data of Szostak & Wu 1980).

1.2.3.4   Conversion

The concept of conversion was initially devised to explain the exceptional segregation patterns obtained from diploid heterozygotes (A:a) in fungi and mosses. Here, in place of the expected ratio of 2A:2a, observed ratios of 4A:0A, 3A:1a, 1A:3a and 0A:4a were found to occur. According to the original concept, a recessive allele in such a heterozygote could convert to its dominant form, and vice versa. Such conversion might occur with or without bias. Consider, for example, a two gene family, designated as A and B, whose members interact so that the nucleotide sequence of part or all of A is converted into B. In this interaction, while both sequences retain their physical integrity and their location, there is a non-reciprocal and biased alteration in the structure of one of them (Kourilsky 1986). When it includes a bias in the direction of change, conversion leads to the fixation of the favoured allele without any involvement of selection operating on the phenotype.

New emphasis is being given to this phenomenon in terms of change in multiple gene families, because conversion is now known to include both inter- as well as intra-chromosomal effects (Fig. 1.9). Moreover, while unequal exchange is only possible between tandemly arranged sequences, conversion may occur between gene families that are not necessarily arranged in tandem, or else may involve either single or tandem sequences located on non-homologous chromosomes (Nagylaki & Petes 1982). For example, in many cases, the multiple genes coding for one specific tRNA

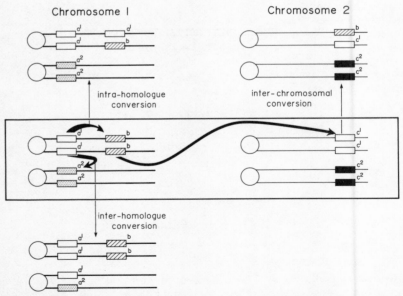

Figure 1.9   Possible modes of gene conversion within and between chromosomes.

are dispersed through the genome. Despite this, the members of such a gene family maintain a common nucleotide sequence. A major mechanism contributing to such sequence conservation involves the transfer of whole segments of genetic information by intergenic conversion. This occurs with a high frequency in the case of three serine tRNA genes located on different chromosomes in the genome of *Schizosaccharomyces pombe* and which share only 200 bp of sequence homology. Information transfer has been demonstrated between all possible pairs of these genes and in both directions (Amstutz *et al.* 1985). This contrasts with the evident directionality that appears to operate in other forms of intergenic conversion.

Most of the well-characterized conversion events come from yeasts. Here it has been possible to show that intergenic conversion, involving a sequence transfer between non-allelic genes, occurs both at mitosis and meiosis. This applies to the two cytochrome c genes of *Saccharomyces cerevisiae*, and to three unlinked tRNA genes of *Schizosaccharomyces pombe* that code in their wild type form for two different isoacceptor tRNAs (Munz *et al.* 1982). It also occurs between *Ty* elements located at different sites in the yeast genome (Roeder & Fink 1982). Furthermore, bacteriophage lambda (λ) sequences inserted into yeast chromosomes can lead to events which are scored as conversions, despite the fact that there is no known homology between the λ sequence and the components of the yeast genome (Stahl, personal communication). Since there is no reason to doubt that equivalent genetic and molecular mechanisms exist in other eukaryotes, it has now become a standard practice to explain the maintenance of sequence homogeneity in families of both clustered and dispersed sequences through conversion mechanisms.

Evidence for gene conversion in organisms other than yeast has been less direct, and comes predominantly from DNA sequence analysis of small families of repeated genes whose sequences have diverged. This includes genes coding for haemoglobins and immunoglobulins. Thus, a stretch of simple sequence DNA which occurs in the large intervening sequence (IVS2) between the $^{G}\gamma$ and $^{A}\gamma$ globin genes of man appears to be a hot spot for exchange leading to conversion. On the 5' side of this simple sequence the allelic $^{A}\gamma$ genes of homologous chromosomes differ considerably in the structure of IVS2, whereas the non-allelic $^{G}\gamma$ and $^{A}\gamma$ genes from the same chromosome differ only slightly in respect of IVS2. On the other hand, on the 3' side of the simple sequence allelic genes differ only slightly, whereas non-allelic genes differ considerably. This has been explained in terms of intergenic conversion (Slightom *et al.* 1980), which results in the 5' two-thirds of the $^{A}\gamma$ gene being more like the $^{G}\gamma$ gene on its own chromosome than the allelic $^{A}\gamma$ gene on its homologue. Gene conversion may also have occurred in the case of the genes responsible for the production of bovine vasopressin and oxytocin, which appear to have arisen by gene duplication (Ruppert *et al.* 1984).

Gene conversion events have evolutionary implications of two kinds. First, in cases where repeated sequences contain sub-sequences that appear to be highly conserved, it has been commonplace to assume that this conservation necessarily implies function. An alternative explanation is that conversion events affect only particular subsets of a sequence, leading to an observed homogeneity which has no functional relevance whatsoever. Second, since conversion, like transposition, can transfer a variant sequence from its chromosome of origin to other members of the chromosome complement it will, if coupled with the conventional processes of random chromosome assortment (meiosis) and sexual fusion (fertilization), allow a new variant to spread in a cohesive fashion, even if that variant confers no advantage. Indeed, if the directional component of such 'molecular drive' (Dover 1982, 1986a, b) is strong enough, the sequence in question may spread deterministically to fixation in a population of diploid sexually reproducing organisms even when it produces deleterious effects on the organism itself (Hickey 1982).

1.2.3.5 *Sequence amplification*

Both duplication and amplification of genes have been observed in bacteria. For example, strains of *Escherichia coli* have been characterized in which the *lac* region, together with varying amounts of surrounding DNA, has been amplified 40–200 fold. Restriction digest patterns of the DNA from over 100 independent strains reveal that the amplification events are different in each case and involve regions of from 7 to 37 kb, resulting in a tandem array of repeated units (Tisty *et al.* 1984). From comparisons of the size and composition of the repeated families shared by closely related taxa, it is evident that there is a continuous process of sequence amplification operative within eukaryote genomes too. This has the effect of boosting individual members of a given family into new sub-families. Highly repeated sequences, for example, are commonly organized into long tracts of tandemly repeated arrays in which the individual repeat units may vary from a few base pairs to several thousand. Moreover, while the genomes of related species may share overlapping families of highly repeated sequences, the reiteration frequencies of these families may be very different.

Interspecies comparisons also reveal differences in the long–order periodicities of repeats which, in some cases, can be shown to result from differential amplification events. Repetitive families with long periodicities can then be constructed from simpler primordial repeats. This is especially clear in the calf (*Bos taurus*). Here many of the eight known satellites have been cloned and sequenced (reviewed by Miklos 1985), and all hybridize to the centric regions of most of the autosomes. In the 1.720 satellite the basic repeat unit is a 23 bp dimer which has been amplified into a 46 bp sequence. The 1.711a satellite also consists of 23 bp repeats into which a 611 bp sequence of foreign DNA has been inserted. This unit, consisting of the 611 bp insertion and its flanking 23 bp primordial repeats, has then been amplified as a 1413 bp repeat. In the 1.715 satellite the basic repeating unit is a 31 bp bite from an array of 23 bp sequences. However, a subsequent 1402 bp bite has been taken from a tandem array of these 31 bp units and then amplified to yield the mature satellite. Finally, the 1.706 satellite has a basic repeat unit of 2350 bp composed, however, of four unequal segments, each of which has a substructure based on either a 23 bp or a 46 bp unit.

Thus, all the calf tandem arrays with long basic repeat units appear to be ultimately based on a 23 bp primordial sequence, and have arisen from smaller tandem arrays by amplification events involving progressively larger repeating units. Moreover, in one case, the unit of amplification has included a sequence which has been inserted into the primordial sequence from elsewhere in the genome.

Likewise, human satellite I includes a Y-chromosome specific, tandemly repeated unit of 2.47 kb, which appears to be of recent origin. This is built up out of three fragments, respectively 775, 875 and 820 bp in length, and the 775 bp fragment includes one *Alu* family member per repeat (Frommer *et al.* 1984).

In more general terms, the satellite sequence patterns present within genomes can be viewed as products of alternating cycles of base pair changes, insertions, deletions and amplification (Fig. 1.10 and see Singer 1982a, Miklos 1985).

Precisely what constraints operate on the sizes and sequences of DNA that can be amplified is not known. Some indirect evidence comes from the changes to genomic DNA which follow the treatment of murine cells with the anti-tumour drug, methotrexate (MTX), an analogue of folic acid (Bostock & Tyler-Smith 1982). The principal mode of action of this drug is to specifically inhibit the activity of the enzyme, dihydrofolate reductase (DHFR). As a result of this inhibition, the cell is unable to synthesise thymidylate, purines or glycine. Continued administration of MTX, however, often results in the emergence of drug-resistant cell lines which contain up to a hundred-fold as much DHFR activity as that normally found. Accompanying this is a corresponding increase in the number of copies of the gene coding for the enzyme. In some cells, the amplified genes are stable after selection pressure is removed; in others they are unstable.

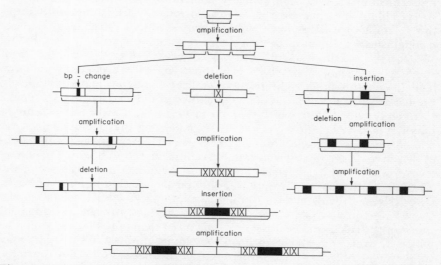

Figure 1.10 The generation of modified sequence patterns by alternating cycles of mutation and amplification.

Unstable lines are characterized by the presence of variable numbers of small chromosome fragments, termed double minutes, of varying size; or else by ring chromosomes or centric chromosome fragments. Stable lines have 4–5 enlarged marker chromosomes, which appear to result from the secondary conversion of ring chromosomes. These marker chromosomes sometimes contain a high proportion of sequences that cross hybridize to highly repeated DNA which has amplified up with the DHFR gene.

Since MTX does not interact with, or become incorporated into, DNA, it appears to act as a selective agent for amplification events which occur within the genome. The DHFR gene itself is some 32 kb in size, whereas the amplified unit is much larger, ranging from 500–3000 kb. Moreover, the structure of the amplified unit is different in cell lines that have been separately selected and, at least in some cases, gives evidence that the sequences at one or both of the ends of the DHFR gene have been rearranged. If this situation provides a sensible model for the production of amplified gene families, then it demonstrates that the amplification process must be quite frequent in occurrence. At least two cases are known where amplification events are developmentally regulated. Both occur in oocytes and involve, respectively, the amplification of rRNA genes in a variety of organisms and the amplification of the chorion genes in *Drosophila melanogaster* (see § 3.4.2).

Models to account for the generation of long tandem repeat arrays have invoked either replication or recombination events as a basis for the amplification of a particular sequence element. The replication models depend upon the multiplication of a single repeat length. If this were to be excised from the chromosome and amplified, then the subsequent reinsertion of the amplified sequence at new chromosomal sites would explain the multichromosomal location of many tandem arrays. For example, in the seven sibling members of the *Drosophila melanogaster* species subgroup, three families of complex, highly repeated DNA sequences can be distinguished on the basis of their major repeat unit lengths. These are the 180, the 360 and the 500 satellite families. The first of these is confined to a single species, *Drosophila orena*, where about 120 000 copies exist within the haploid genome. *In situ* hybridization, however, indicates that this family is present at the centric regions of all the chromosomes (Strachan *et al.* 1982).

In the systems of flux which we have just examined, most of the changes have not involved genic sequences *per se*. In general, they involve repetitive components which are localized, intergenic, or else concern the production of defective gene copies which simply accumulate throughout the genomic landscape without overt effects. In the short term the functional genes play 'fair', with most of the disruptions going on around, but not within, their own restricted regulatory and coding landscapes.

As long ago as 1970, Ohno argued, on the basis of genetic load, that much of the eukaryote genome was little more than junk. This viewpoint, which is still unpalatable to many biologists, now has a substantial supporting DNA data base. More recently, this has led Ohno to conclude that genes in the mammalian genome are like 'oases in a barren desert' (Ohno 1982) and that for every copy of a new gene that has arisen during evolution, hundreds of other copies have 'degenerated' to swell the ranks of junk DNA (Ohno 1985).

This section has emphasized that the genome is extremely labile and can be subject to a variety of changes enforced upon it by the enzymology of the cell. This is a novel way of initially looking at the genome, since it de-emphasizes the conventional and conservative view that one should seek adaptive reasons for all of the changes that go on within a genome.

It has, in the past, been commonplace to assume that most, if not all, aspects of the morphology, physiology and behaviour of an organism represent adaptive responses to the environment in which that organism lives. This assumption, however, is difficult to test objectively and represents more an article of faith than of fact. Indeed, biologists have become addicted to the adaptationist viewpoint not so much because of the compelling evidence in favour of it, but rather because it seems so eminently logical and reasonable. This view, of course, assumes that functional explanations must necessarily exist for all facets of the bewildering diversity we see within and between genomes. An alternative extreme viewpoint is that eukaryote genomes are, in effect, simply larger, more sophisticated and embellished prokaryote genomes, loaded with non-coding DNA sequences which are in a constant state of flux but without any significant short-term impact on phenotype.

To decide which, if either, of these interpretations is the more realistic, we need to determine the number of functional genes within a genome and the proportion of these that are developmentally significant. We also require precise information on the changes that go on within a genome at the molecular level and the extent to which these lead to meaningful evolutionary change. Compared to the differences in structural gene composition between related species, we now know that there are much more striking molecular differences in their repeated DNA components. This raises the question of whether this is because such sequences are important or unimportant. There is also a clear need to distinguish between historical chance and biological necessity as causative factors in determining genome structure.

The principal purpose of this book is to re-evaluate the assumptions that underlie the way biologists think about eukaryote genomes in relation to development and evolution, and to assess the relevance of current molecular approaches to the analyses of these key biological issues. We turn first to the question of development.

2

Developmental activities of genomes

'Some like to understand what they believe in. Others like to
believe in what they understand.'

Lec.

Development is a central theme in biology since it is during
development that genotypic information is translated into phenotype.
The process of development not only defines the operationally
significant portion of a genome, but, additionally, delineates the
functional basis for future evolutionary change. It thus provides an
obvious starting point for both a consideration of genome organization
and an analysis of how that organization has changed during the
course of evolution. In discussing development we need to direct our
attention to two major questions:

Q.1 How does the genome operate during development?
Q.2 How do genes control the formation of a complex organism
from a single fertilized egg cell?

2.1 FROM EGG TO ADULT

Development, in multicellular eukaryotes, involves the production of many
different clonal cell populations from a single fertilized egg cell. Many of
these cell populations are specialized for a particular function. Additionally,
they are organized into distinctive patterns as a result of cell movements
which, coupled with differential growth changes, ultimately control the
shape and the detailed architecture of the body. Following fertilization, eggs
first cleave into a blastula, essentially a monolayer of epithelial cells
(blastoderm) surrounding an acellular blastocoel. This is followed by a
phase of extensive cell movement, termed gastrulation, during which parts

of the blastoderm are carried into the blastocoel by invagination, involution, ingression or else a combination of these (Deuchar 1975). This results in the formation of three distinct cell layers – ectoderm, endoderm and mesoderm – which provide the basic body plan of the developing embryo and from which the embryonic and adult organs are subsequently formed (Fig. 2.1) The disposition of these germ layers within the gastrula is thus crucial for all subsequent development, since the most decisive phases of organogenesis depend on interactions between cell populations originating from these layers.

During its developmental history, each clone of cells passes from the totipotency of the zygote, via a state of determination, to its fully differentiated state. Determined cells are not functionally specialized in a morphological sense but are, nevertheless, committed to limited developmental programmes. By most operational criteria, the genome becomes progressivly restricted in its capacity for expression as cells approach the differentiated state which represents the terminal phase of a determined cell clone. In the differentiated state usually only a part of the total genome is expressed. With the exception of situations in which chromosomes, or parts of chromosomes, are either eliminated from the soma or else rearranged (see § 2.4), it is generally agreed that all the cells of an organism contain basically the same genome and hence the same set of nuclear genes. This implies that there are cell type-specific combinations of active and inactive genes in differentiated cell lineages.

The various cell types within an organism are also highly stabilized in their differentiated states, so that not only do programmes of gene activity differ substantially between different kinds of specialized cells, but switches from one programme to another rarely take place under normal conditions. Minor modulations are certainly possible. Thus, fluctuations in the amount

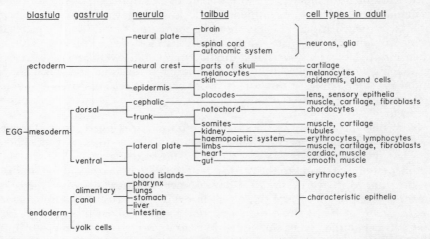

Figure 2.1 Derivation of adult cell types from the three primary germ layers of the vertebrate embryo. Note that some cell types (cartilage, epithelia) originate from more than one cell lineage (data of Slack 1983).

of gene product synthesized by a given cell type are well known, as is the occurrence of hormone-stimulated transcription. These cases aside, however, the differentiated state is a stable state. Nuclei from well-defined cell types of adult frogs have, in some cases, been found to be capable of promoting the development of enucleated eggs into advanced embryos characterized by a variety of well-differentiated cell types. This indicates that the genome of such transplanted nuclei retains many sets of genes in addition to those that function in the donor cell. In all cases, however, the test embryos die prior to the feeding larva stage and, contrary to popular belief, no case is known where an adult frog has been obtained from nuclear transplantation (DiBerardino 1979). Additionally, most nuclear transplants from advanced cell types, including transplanted spermatogonial nuclei, develop pronounced chromosome abnormalities, involving both aneuploidy and extensive structural change, during cleavage, blastula and gastrula stages. These abnormalities, which in many cases originate during the first cell cycle of the transplant nucleus, arise because of an incompatibility between the cell cycle programmes of the transplanted nucleus and the host egg cytoplasm. Thus, most nuclei from advanced cell types are characterized by long generation cycles, and are unable to replicate normally after transplantation into eggs, which have short cell generations.

The importance of the interactions generated between cells as a result of gastrulation cannot be overstated, since they exert a major control over future development. During gastrulation the embryonic genome becomes fully activated, the three germ layers are established and major tissue rearragements occur. Using hybridomas it is possible to show that each nascent germ layer acquires new cell surface antigens, and many of these appear precisely at the time of germ layer differentiation (McClay & Wessel 1985). In many vertebrate groups, cells have already become determined by gastrulation. Additionally, the antero-postero axis of the vertebrate embryo is defined either during, or immediately following, gastrulation – heralding a phase of neurulation. This involves the formation of a thickened plate of ectoderm, the neutral plate, on the dorsal surface of the embryo. This is then induced by the underlying mesoderm to fold over, and sink in, so producing a hollow neural tube, the forerunner of the brain and spinal cord. Inductive interactions of this sort between sheets of cells of different origin are involved in the differentiation of most organ systems in the vertebrate embryo.

A second major event in the axial organization of the vertebrate embryo is the development of metameric segmentation. This involves the production of paired blocks of mesoderm, which subsequently give rise to the segmental musculature, to the cartilages of the vertebral axis and the dermis of the skin. Moreover, during the development of the neural tube, the cells at the margin of the neural folds sink in below the extoderm and become segmentally arranged along the embryonic axis as a series of neural crests, one pair per somite. Cells subsequently migrate laterally and ventrally from these neural crests to produce the spinal ganglia, the autonomic ganglia, the Schwann cells which invest the neurons, the adrenal medulla, the melanophores and some of the cartilages of the head region. Additionally, neural crest cells which migrate into the head region interact with selected areas of the surface ectoderm, the so-called placodes, to give rise to the cranial ganglia. Thus, early contact between the ectoderm and mesoderm sets into motion a whole train of events which culminate in the development of the entire nervous system.

Prior to leaving the neural primordium, the neural crest cells lose their epithelial arrangement as a result of the loss of neural cell adhesion molecules (N-CAM) present on their surfaces, coupled with the rupture of the basement membrane that underlies them. They then send out filopodia and proceed to migrate along the extracellular matrix located in the available cell-free space. This matrix contains at least two kinds of molecules which facilitate the migratory movements (Le Douarin 1984). The most abundant of these, hyaluronic acid, both enlarges the cell-free space and provides an appropriate fibrous material for migration. The second molecular type present in the matrix is fibronectin (Hynes 1985, 1986), along with small amounts of collagen types I and II. Fibronectin fibres are endowed with specific binding properties and provide a direct anchorage for the filopodia produced by the neural crest cells. Once migration is complete, and crest cells aggregate into ganglia, N-CAM reappears on their cell surfaces while hyaluronic acid and fibronectin both disappear from the extracellular matrix.

Cell adhesion molecules of different specificity have now been isolated and partly characterized. They are all glycoproteins which contain sialic acid, and they act as regulators of the overall patterns of morphogenetic movements as well as in organogenesis (Edelman 1985). As far as N-CAM is concerned, both the specificity of its binding region and its basic chemical structure have been highly conserved during the vertebrate evolution (Hoffman et al. 1984). Moreover, it exists in both embryonic and adult forms, the latter arising as conversion products of the former. Mice homozygous for the autosomal recessive staggerer mutation display cerebellar connectional defects, and the conversion of embryonic to adult N-CAM is greatly delayed in the cerebellum at the very time the disordered connections develop (Edelman 1984).

Two other features of development need emphasizing. First, determination, on its own, would simply result in groups of cells which differ in some fundamental respect. Unaided it would not lead to orderly morphogenesis, where shape and pattern are defined in such a way that the formless assumes a definitive form. The problem of how embryonic cells interact in the production of tissues and organs emphasizes the crucial role of the cell surface, both in cell–cell interaction and in selective cell adhesion. These events enable cells in the embryo to identify one another and so integrate into multicellular systems. Second, normal development produces clear alternative states and not an intermediate condition, which implies that the process is canalized.

In effect, the sequences of cell division and cell movement which operate during development are translated into morphological patterns by determining spatial location within the developing embryo. This, in turn, ultimately defines the phenotypic expression of a cell clone. The embryonic axes of symmetry – the antero-postero and the dorso-ventral axes – provide the major pre-patterns to which cells respond in terms of mitotic activity and directed cell movement. Such morphogenetic movements depend either on direct cell–cell interaction or else on the interaction between cells and the extracellular matrix with which they are in contact. Both of these interactions are probably governed by cell surface proteins. Additionally, a special type of cellular interaction that may occur over long distances is that mediated by hormones. The most spectacular developmental event under hormone control is unquestionably metamorphosis, since it involves so many different tissues (see § 5.1).

The events that convert a fertilized egg into a functioning adult thus depend on a precise sequence of gene activity, involving the switching on, or off, of the production of particular proteins in particular cells at particular times. Key decisions are made early in development, since it is at this time that the ground plan for the entire developmental programme is established. The genes whose activities control these early events thus, necessarily, underlie the activity of those that function later. An understanding of the role of the genome in development depends both on identifying the early acting genes and on characterizing the gene circuits they subsequently bring into action.

Earlier attempts to analyze the precise biochemical pathways involved in development concentrated on the sea urchin (Davidson *et al*. 1982) and were largely concerned with characterizing entire RNA populations at successive stages of development. These studies have served to distinguish a core of RNA molecules, present at all stages, which stem from the activity of so-called housekeeping genes. These, however, form only a small proportion of the total pool of active genes, so that there is a large residual RNA population whose developmental significance is difficult to assess. Moreover, while recombinant DNA techniques allow individual RNAs to be isolated in pure form, they do not, on their own, tell us whether development is a modulated product of all the genes within a genome acting *en masse*, or whether there is a particular subset of executive genes whose activity determines critical switches into defined developmental circuits. Thus, despite the considerable investment in the characterization of sea urchin development in terms of changing populations of RNA molecules and in the expression of lineage-specific genes (Davidson *et al*. 1985), the sea urchin system is not ideal for supplying answers to some developmental questions, because of the lack of suitable mutants which provide a ready means of identifying those genes that make key developmental decisions.

Mutants of this type certainly exist in the mouse, where members of the Tailess (*T* locus) complex, for example, have well characterized effects upon early development (Frischauf 1985). Mutations at the *T* locus, when homozygous, produce defects during blastocyst formation, gastrulation and initial axial organization (Fig. 2.2). All of these arise from specific developmental blocks in the differentiation of particular components of the embryonic ectoderm and are lethal, leading to death within 2–3 days of the first visible defect.

Although this example illustrates one class of mouse mutations that lead

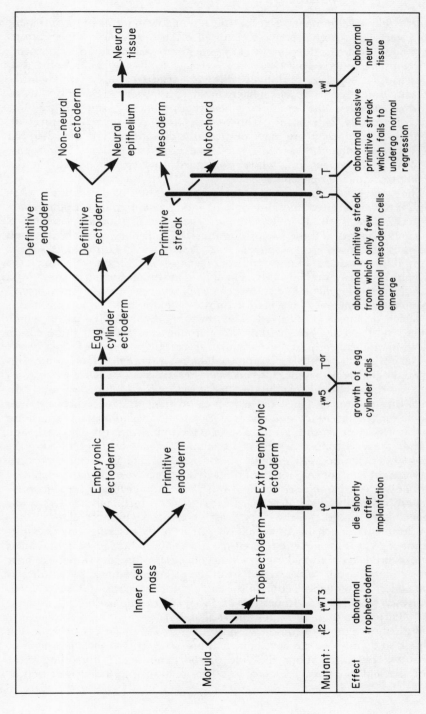

Figure 2.2 T locus mutations of the mouse, showing their time and place of action (data of Bennett *et al.* 1977).

to definable developmental defects, and is tractable in terms of recombinant DNA techniques, what is really required of an organism is the ability to systematically define the subset of key genes involved in early development, as has been pioneered in *Drosophila melanogaster* by Nüsslein-Volhard & Wieschaus (1980). Conventionally, this has involved the use of mutagenesis techniques, the results of which have, in the past, produced the only clues to deciding whether the modified phenotypes obtained are, or are not, developmentally significant. Those that are have several features in common. First, they are developmentally specific. They are also cell specific. Secondly, they are likely to involve genes whose activity begins early in development and whose continued activity is required for the maintenance of the determined state.

The mutagenic approach encounters a number of difficulties when applied to vertebrates. These are only now being overcome through the use of techniques such as microcloning, the use of mobile elements as mutagenic agents and the use of oligonucleotide probes, manufactured on the basis of known protein sequences, to isolate particular genes. A second complication is that walking significant distances in a vertebrate genome is a very time-consuming business. Finally, identifying and characterizing a particular developmental gene is not in itself enough. It is also necessary to obtain complementary information on the developmental circuit or circuits in which that gene is involved. This, in turn, involves identifying which genes, located elsewhere within the genome, interact with the locus in question, and this requires a manipulative genetic system.

The only multicellular eukaryotes which offer these necessary conditions for study are the nematode, *Caenorhabditis elegans*, and *Drosophila melanogaster* and, of these two, *Drosophila melanogaster* is, for obvious reasons, the organism of choice. Thus, the existence of giant polytene chromosomes allows for the precise cytogenetic localization of developmentally significant mutations. More recently, it has also provided a basis for the direct excision of individual chromosome regions, containing genes of interest, using the techniques of microdissection and microcloning. There are many similarities between the early development of quite diverse metazoan phyla, which tend to follow a comparable sequence of morphological change. Consequently, one can reasonably anticipate that the principles which apply to *Drosophila melanogaster* should, at the very least, provide significant clues to the developmental mechanisms which operate in other systems. For example, bithorax, which for genetic reasons has always been held to be a vital locus for executive decision making in *Drosophila melanogaster* development, now has a counterpart in vertebrates. Part of the protein produced by bithorax is also coded for by genes active in the early embryogenesis of *Xenopus laevis* (Carrasco *et al.* 1984, de Robertis *et al.* 1985) and of *Mus musculus* (Jackson *et al.* 1985, Joyner *et al.* 1985, Ruddle *et al.* 1985).

With these points in mind, we now turn to a detailed consideration of what is known about the developmental processes in *Drosophila melanogaster*, what we can infer about the role of the genome in controlling these processes and its general relevance to other organisms, including vertebrates.

2.2 THE GENETIC CONTROL OF DEVELOPMENT IN *DROSOPHILA MELANOGASTER*

Drosophila melanogaster has two distinct phases of development combined in one life cycle – embryonic development, which gives rise to larval organization, and pupal development, which produces the definitive adult organization following a breakdown of larval tissues during metamorphosis. The proportion of lethal effects which occur during these different developmental stages is instructive. Of 450 X-linked zygotic lethal loci studied by Perrimon & Mahowald (1986), 22% died as embryos, 52% as larvae and 26% as pupae. Moreover, two-thirds of these genes are maternally expressed when tested by the dominant female sterile technique.

This major dichotomy in the developmental programme occurs, however, during early embryogenesis. Not only are the larval tissues constructed at this time but, additionally, the primordia of the adult tissues are set aside in the form of imaginal discs.

2.2.1 Embryonic development

The developmental events following fertilization depend on interactions between stored maternal gene products, manufactured during oogenesis, and the products made by the zygote. The contributions of the maternal effect genes to these interacting embryogenic hierarchies cannot be understated, and extensive genetic screens have been employed to isolate those genes which function only during oogenesis (Perrimon *et al.* 1984, Mahowald & Hardy 1985). While it is to be expected that genes active during oogenesis would give rise to female sterility when mutated, an unknown number of important contributors to oogenesis will go undetected in such screens if one of their pleiotropic effects were to lead to zygotic lethality. It is possible, however, to circumvent this problem by using a method known as the 'dominant female sterile technique', in which homozygous clones are produced in an ovary. In this way, it can be determined whether the locus under examination has any specific effects on embryogenesis.

This analysis reveals that at least two-thirds of all zygotic lethal loci are expressed during oogenesis (Perrimon *et al.* 1984). The task ahead is to sort the housekeeping genes from those executive genes whose products, for example, provide the positional information that is used during embryogenesis by later-acting genes.

The embryogenesis of *Drosophila melanogaster* is well known in morphological terms (Table 2.1). It begins with a phase of rapid nuclear division. The first nine of these mitoses are synchronous and occur in a syncytial cluster without cell membranes being laid down. It is this stage that has been successfully exploited in P-element mediated transformation into the germ line, since nuclear uptake of injected DNA is facilitated by the lack of cell membranes at this time. Nuclei now migrate to the surface of the egg where they divide three or four more times. Only then do cell membranes

Table 2.1 The developmental schedule of *Drosophila melanogaster* at 23–25°C (from Wright 1974 after Doane 1967).

Stage No.	Time (h) after fertilization	Major developmental event
1	0–1.5	8 synchronous cleavage mitoses
2	1.5	migration of cleavage nuclei; pole cell formation
3	2.0	syncytial blastoderm
4	2.5	cleavage furrows form
5	3.0	cellular blastoderm present, pregastrula cell movements begin
6	3.5	early gastrulation, ventral furrow forms
7	3.75	cephalic furrow forms, extension of germ band
8	4.0	invagination of anterior and posterior midgut rudiment, embryonic membrane forms
10	5.0	germband extension complete, large neuroblasts present, yolk enclosed by primitive gut
11	5.5	invagination of proctodeum and stomodeum begins, neuroblast mitoses begin
12	6.0	segmentation of mesoderm
13	7.0	tracheal invaginations
14	8.0	segmentation of head and trunk; attachment of muscles
17	11.0	compact gonad primordia present
19	13.0	chitinization begins
20	14.0	condensation of ventral nervous system
22	22–24	hatching of egg

appear, giving rise to a two-dimensional blastoderm, one cell thick (Fig. 2.3) During the early syncytial stage of the embryo, the cleavage nuclei are totipotent. Once the nuclei migrate to the egg surface and become invested by cell membranes they become restricted in their potential for forming both larval and adult structures – according to their position in the embryo. The pattern of cell proliferation and cell movements during embryogenesis has been analyzed in impressive detail by Hartenstein & Campos-Ortega (1985), Technau & Campos-Ortega (1985) and Hartenstein *et al.* (1985), and a fate map of the blastoderm has been derived from their observations.

The single blastoderm layer is subsequently converted into three germ layers (Fullilove *et al.* 1978). The mesoderm derives from a longitudinal infolding at the ventral side of the egg, the ventral furrow. Additional invaginations occur at the anterior ventral side and at the posterior tip of the egg, to give rise to the anlagen of the endodermal gut. Gastrulation is then completed by a stretching of the ventral cell layers along the dorsal side of the egg, giving rise to the so-called germ band extension. This is followed by primary organogenesis.

Embryonic development in *Drosophila melanogaster* also leads to a segmentation of the hatching larva, involving a simple metameric pattern of 11 repeating units, three thoracic and eight abdominal, plus an inconspicuous head consisting of fused segments. The anlagen for these individual segments arise as equally sized subdivisions of the blastoderm, each

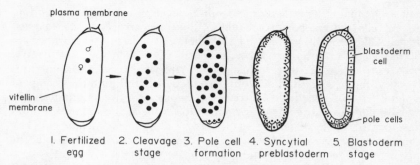

Figure 2.3 Early embryogenesis of *Drosophila melanogaster* from fertilization to the blastoderm stage.

segment represented by a transverse strip of about 3–4 cell diameters. A cell lineage restriction between neighbouring segments is established at, or soon after, the blastoderm stage and segmentation is first visible after gastrulation as a pattern of repeated bulges in the ventral ectoderm.

During the three larval instars, which follow hatching from the egg, little overt morphological change, other than growth, takes place in the somatic tissues, although, as we shall see shortly, the imaginal discs have already embarked on developmental programmes which are necessary for the transition to the adult state. Moreover, although the larva grows in size, there is little increase in cell number. Only the imaginal discs, the histoblasts, the neuroblasts and the primordial germ cells of larvae retain a mitotic capacity. In the bulk of larval tissues, somatic growth is achieved simply by the existing cells enlarging in size, following the production of multiple copies of the genome within individual nuclei. This occurs either by polytenization (salivary glands, midgut epithelium and malpighian tubules) or by somatic endoploidy (see § 3.4.2.4), where the products of chromosome replication do not remain associated (muscles, hind gut epithelium, fat body, tracheal wall and certain cell types in the nervous system). In this sense, *Drosophila melanogaster* departs radically from the more general somatic diploidy of vertebrates.

Estimates of the number of lethal mutations where a lack of function leads to effects detectable in the larval cuticle of *Drosophila melanogaster* indicate that 61 loci on chromosome 2 (Nüsslein-Volhard *et al.* 1984), 45 loci on chromosome 3 (Jürgens *et al.* 1984), 33 loci on the X and one on chromosome 4 (Wieschaus *et al.* 1984) are required zygotically for normal gross larval morphology. This represents about 3% of the presumed vital gene functions.

2.2.2 Pupal development

At pupal metamorphosis most of the larval systems constructed during embryogenesis are broken down and replaced by new structures. Consequently, the adult fly has a more complicated structure and behaviour, as well as a large number of diploid cells. However, polytene nuclei are present in the malpighian tubules and some specialized cell types, including trichogen cells and the epidermal cells of the foot pads of the metatarsus, while testis sheath cells and nurse and follicle cells of the egg chamber are endoploid. The undifferentiated primordial germ line and the central nervous system persist largely unchanged. Moreover, the transition from the larval to the adult central nervous system, associated with the change in form and behaviour which takes place at metamorphosis, occurs progressively throughout the postembryonic period and involves the activity of proliferation centres located posterolaterally in each of the brain hemispheres. The neuroblasts within these centres are active throughout the larval period, but disappear by the second day of the pupal phase (White & Kankel 1978).

The larval segmental pattern is also replaced by an adult segmentation, consisting of a well-defined head, a thorax of three segments, each of which bears a pair of legs, and an abdomen of eight segments. This second phase of development depends upon the activity of specialized nests of cells which are set aside during embryonic development but which do not function in morphological differentiation until the pupal phase. These specialized nests include the imaginal discs and rings of the head and thoracic region, and the histoblast clusters of the abdomen (Gehring 1978). Each imaginal disc forms a specific and precisely defined piece of adult cuticle, a component of the total integument. There are ten pairs of imaginal nests together with a single genital disc (Fig. 2.4). The integument of the adult head arises from three of these. The external parts of the adult thorax are formed by three pairs of leg discs together with the single pairs of prothoracic, wing and haltere discs.

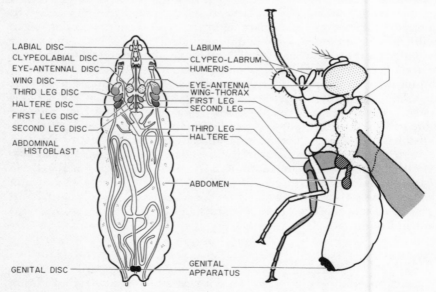

Figure 2.4 Location of imaginal discs and abdominal histoblast nests in the larva of *Drosophila melanogaster* and their corresponding adult derivatives (data of Nöthiger 1972).

Finally, the genital dics gives rise to the integument of the genitalia and the analia, while the integument of each abdominal segment arises from the eight small nests of abdominal histoblasts. Discless mutants are known which give apparently normal larval development and puparium formation (Szabad & Bryant 1982). Evidently, therefore, these developmental events do not require the presence of imaginal discs. On the other hand, extirpation of any of the discs of the larva leads to a loss of structure in the emerging adult. Other discs cannot compensate for such a loss.

Discs are not histologically recognizable as separate structures until the onset of larval life, at which time they are seen to be invaginations of the larval ectoderm, each consisting of a sac-like single layer of epithelial cells. When first evident histologically, each contains between 15 and 60 cells, depending on the precise disc involved (Schneidermann 1976). During the four days of larval life, the epithelium within a disc becomes folded as it grows by cell division. At metamorphosis the tightly folded epithelium is everted to give rise to an extended adult structure. Differentiation and cuticle synthesis then follow. The imaginal discs thus show a particularly clear distinction between determined and differentiated states. Some discs also contain presumptive muscle and nerve cells, and three carry anlagen for internal organs. While the imaginal discs proliferate throughout larval life, the histoblasts, which give rise to the abdominal tergites and sternites, only begin to multiply at puparium formation, by which time most of the disc cells have stopped dividing.

Each disc begins its existence as a group of founder cells, and each external region of the adult is eventually constructed from the descendants of these cells. Little is known concerning the processes that define the formation of such cell clusters. The choice of particular groups of cells to form a disc is probably geographic, depending on polarity and the positional information initially contained within the egg and subsequently within the disc itself. As Garcia-Bellido et al. (1973) first appreciated, growing discs are thus divided consecutively into compartments. First antero-postero, then dorso-ventral, central-peripheral and so on (Fig. 2.5). Cells within a compartment constitute a polyclone (Crick & Lawrence 1975). After each compartmentalization step the cell clones respect the boundary between the compartments involved, and each compartment henceforth follows its own developmental programme. When added to dissected embryos, dissociated and fluorescent-labelled imaginal disc cells obtained from third instar larvae bind preferentially to the epidermis of the embryonic segments from which they were derived (Gauger et al. 1985). This implies that cell recognition and selective adhesion may be important in the establishment of pattern formation and the maintenance of segment boundaries. On the other hand, O'Brochta & Bryant (1985) have identified a narrow zone of non-proliferating cells in the wing discs of third instar larvae of Drosophila melanogaster. This zone, which is 6–10 cells wide, coincides with the presumptive wing margin, and so may account for the observed lineage restriction between the dorsal and ventral surfaces of the wing. The antero-postero compartment boundary of this disc is not coincident with an equivalent non-proliferating cell zone, so that different

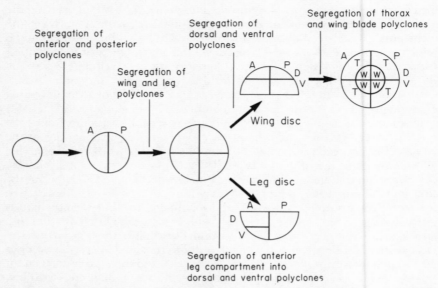

Figure 2.5 Mode of compartmentalization of mesothoracic cuticular structures of *Drosophila melanogaster* (data of Morata & Lawrence 1977). A, anterior; P, posterior; D, dorsal; V, ventral; T, thorax; W, wing blade.

lineages may be maintained by different mechanisms even within the same disc.

Using the technique of fluorescent dye tracers, Weir & Lo (1985) have demonstrated that the engrailed (*en*) mutant of *Drosophila melanogaster* has an anterior-posterior compartment boundary identical to that of wild type. By contrast, the anterior-posterior lineage border of *en* either differs in position from the compartment border, or else is absent altogther. Thus, here the wild type anterior-posterior compartment boundary may not coincide with anterior-posterior lineage boundary. This implies that lineage compartments are not required for the formation of anterior-posterior communication compartments, though it leaves open the possibility that communication compartments provide information for specifying the formation of lineage compartments.

Using intracellular injection of small fluorescent molecules it is possible to show that communication boundaries exist between bands of cells in the wing disc epithelium. Cells within these bands either have few gap junction channels or else the existing channels have reduced permeability (Weir & Lo 1984). Gap junctions are specialized regions of intercellular contact in which the membranes of adjacent cells are closely apposed with a uniform 2 nm gap separating them. Projections from each membrane may extend into this narrow gap. These, by making . contact with one another, define a transcellular channel (Gros *et al.* 1983). Some of the communication boundaries present in the wing disc of *Drosophila melanogaster* coincide with the compartment borders identified from cell lineage studies. This suggests that gap junction communication may play a role in compartment formation.

2.2.3 Genetic control of spatial organization in the *Drosophila melanogaster* embryo

Three major classes of genes are known to be involved in programming the spatial organization of the embryo:

(a)   maternal effect genes which control the polarity and the spatial co-ordinates of the egg;
(b)   segmentation genes which determine the number and polarity of the body segments by interpreting the positional information specified by the maternal effect genes and translating it into a segmental pattern; and
(c)   homeotic genes which define segment identity in terms of the positional information specified by both of the earlier acting classes of genes.

Moreover, the patterning of the entire somatic tissue of the embryo is interpretable in terms of two series of gradients, one antero-postero the other dorso-ventral, which are assumed to be inherent in the organization of the egg and to depend on molecular components preformed during oogenesis. Although the nature, number and type of molecules responsible for these gradients is unknown, some of the relevant genes are now in the process of being characterized.

2.2.3.1   The antero-postero gradient

The recessive mutation bicaudal (Nüsslein-Volhard 1979, Mohler & Wieschaus 1985), results in some of the embryos produced by mutant females developing two abdominal ends arranged in mirror-image symmetry (Fig.2 6). That is, the antero-postero arrangement of segments, their total

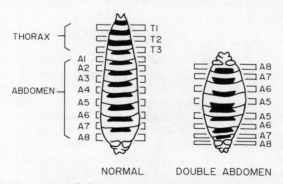

Figure 2.6 Location of thoracic (T) and abdominal (A) segments in a normal embryo and a double abdomen (bicaudal) mutant of *Drosophila melanogaster*.

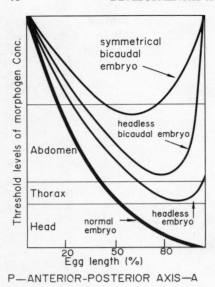

P—ANTERIOR-POSTERIOR AXIS—A

Figure 2.7 Hypothetical gradients of morphogen concentration defining the antero-postero co-ordinate in normal and bicaudal embryos of *Drosophila melanogaster* (data of Nüsslein-Volhard 1979).

number and often their polarity is affected. At gastrulation in mutant bicaudal embryos the posterior midgut invagination occurs not only at the true posterior, but also at the anterior end of the embryo. Coupled with this, malpighian tubules are regularly duplicated at the anterior end, and genital discs are also frequently formed. As a result of the modified development of the anterior end, the posterior region with normal polarity never contains more than five abdominal segments in mutant bicaudal embryos, and the average number is three. The only exception to the perfect symmetry of the gastrulating bicaudal embryos is provided by the pole cells which, as in normal embryos, are found only at the true posterior end. The bicaudal phenotype has been interpreted in terms of a single monotonic antero-postero gradient set up during organogenesis under the control of the maternal genome which controls the basic organization of the egg (Fig. 2.7). The precise number of genes involved in setting up such a gradient remains to be determined.

2.2.3.2 *The dorso-ventral gradient*

The establishment of the dorso-ventral pattern of the embryo requires the activity of at least ten maternal effect genes, collectively known as the dorsal group (Table 2.2). These are widely distributed throughout the genome, yet, in the absence of any one of them, embryos lack a dorso-ventral pattern and all cells differentiate according to a dorsal ground state. Dorsal (*dl*), for example, is a maternal mutation which results in the failure of 'dorsal' embryos to form tissue normally derived from the ventral and lateral regions of the egg. Although an apparently normal cellular blastoderm develops, gastrulation is irregular and neither the ventral furrow nor the anterior midgut invagination form. This mutation has a different effect depending on whether it is homozygous (*dl/dl*) or heterozygous (*dl/+*). In

Table 2.2 Maternal-effect dorsalizing loci of *Drosophila melanogaster* (from Anderson & Nüsslein-Volhard 1984).

Locus	Location	Phenotypic rescue	
		by cytoplasm	by RNA
gastrulation defective	11A1–7	–	nd
dorsal	36C	+	–
nudel	3–17	nd	nd
tube	3–47	+	+
pipe	3–47	–	nd
snake	87D10–12	+	+
easter	88EF	+	+
Toll	97D	+	+
spätzle	97E	+	+
pelle	97F	+	+

nd = not determined.

the former case there is a complete 'dorsalization' of the entire embryo, while in the latter case, ventral pattern elements are missing to various extents. Segmentation is, however, quite normal (Nüsslein-Volhard 1979). The dorsal locus maps to 36C, and a 75 kb DNA sequence from this region has been found to contain *dl*. Of three transcription units produced by this segment, one, which is 8 kb long and encodes a 2.8 kb poly $(A)^+$ RNA, appears to correspond to the dorsal gene (Steward *et al.* 1984, 1985).

Of the ten known mutations responsible for dorso-ventral pattern, one stands out because a number of dominant alleles at this locus ($Toll^D$) show a ventralized phenotype, whereas recessive alleles produce dorsalized embryos. The dominant phenotype is partially epistatic over most of the other dorsalizing loci that have been tested, and only dorsal itself is epistatic over $Toll^D$. This has been interpreted to mean that *Toll* and dorsal are key loci in determining the dorso-ventral gradient (Anderson 1984, 1986, Hudson *et al.* 1986). Double mutant phenotypes indicate that the activity of *Toll* is normally regulated by the other dorsal group genes (Anderson *et al.* 1985), and that the combined effect is to produce a morphogen gradient in the dorso-ventral axis of the wild type embryo. Consequently, in the absence of the $Toll^+$ gene product, the embryo has no inherent polarity in the dorso-ventral axis.

Injection of wild type cytoplasm into young mutant embryos rescues the mutant phenotype for seven of the ten dorsal group loci. Poly $(A)^+$ RNA obtained from wild type embryos generally has a comparable effect (see Table 2.2). In six of these cases the phenotypic rescue results only in a partial restoration of the dorso-ventral pattern. The gene, snake, however, shows a total rescue. Moreover, in snake, over 20% of the rescued embryos hatch into larvae, whereas hatching larvae have not been obtained in the case of the other six loci. The response with *Toll* is also distinctive in the sense that the site of injection is critical. Ventral injection restores normal polarity whereas dorsal injection reverses it.

Mutations at the pelle locus (*pll*), which also belongs to the dorsal group, can be rescued partially by wild type cytoplasm, and this is effective until the late blastoderm stage. In contrast, rescue activity by poly $(A)^+$ RNA ceases by the time of pole cell formation (Müller-Holtkamp *et al.* 1985), indicating that *pll*$^+$ mRNA is lost at an early stage from the pool of maternal mRNA involved in establishing the dorso-ventral pattern.

These cases alert us to the fact that it can be difficult to decide what genes are specifically involved in a given developmental sequence, as opposed to those which may non-specifically interfere with a particular programme. Fortunately, for the maternal mutations affecting the dorso-ventral pattern, a knowledge of the interactions which occur genetically between the loci make it possible to distinguish those genes which are more likely to be of critical importance. This would be a much more difficult operation in a mouse. Here it would be unwise to conclude that a mutant with a particular developmental defect is necessarily representative of a gene with a specific, as opposed to a non-specific, effect. Thus a mutant phenotype could result simply from the inactivity of housekeeping genes, which are so fundamental that their lack of activity automatically leads to embryonic death.

The snake gene has now been cloned and the nucleotide sequence of a cDNA determined (DeLotto & Spierer 1986). An open reading frame yields a predicted protein of 430 amino acids having homology to serine proteases, which are proteolytic enzymes normally produced as zymogens and then activated by proteolytic cleavage. The best homologies are to rat prepro-elastase, rat trypsinogen and bovine chymotrypsin, although there is weaker homology to troponin C from the skeletal muscle of the rabbit. These authors have speculated that some of the products of these maternal effect genes may work in an analogous way to the proteolytic cascades of blood clotting. Such a cascade, if localized and coupled with diffusion processes, might well serve to generate positional information during embryogenesis.

2.2.3.3　Body segmentation

Early in development, the main mass of ectoderm in the *Drosophila melanogaster* embryo is divided up into alternating anterior and posterior polyclones, which generate the anterior and posterior compartments of the body (Garcia-Bellido *et al.* 1979). A combination of one anterior and one posterior clone, an A-P pair, has, traditionally, been assumed to constitute the primordium of a body segment. By contrast, the mesodermal masses, when first formed, are not in register with such primordia. Rather, they appear as parasegments which include a P-A pair of compartments (Martinez-Arias & Lawrence 1985). The Bithorax Complex, as we shall see in the next section, is involved in the segmental determination of both the epidermis and the central nervous system. Tengels & Ghysen (1985) have used a monoclonal antibody that labels a set of uniquely recognizable transverse fasicles to study the pattern of transversal commisures in the central nervous system of adult flies. They find that, while the spatial domains of action of genes of the Bithorax Complex involve segment length units, these units are out of register with the epidermal segments. Rather, defects produced by *abx, bxd* and *iab-2* mutations define regions whose boundaries include the posterior portion of one segment and the anterior portion of the following segment. Finally, while some genes are expressed in a true segmental fashion (e.g. $Antp^+$ and scr^+), others are not (e.g. Ubx^+ and ftz^+). These latter, like the metameric units of the mesoderm and the transversal commisures of the central nervous system, respect parasegmental boundaries.

The first indication of segmentation is the appearance in the 5-7-hour-old embryo of shallow transverse grooves which divide it into 14 uniform metameres, assumed to represent the body segments. The product of the engrailed (en) gene is known to be required only in the posterior compartment of segments (Ali et al. 1985, O'Farrell et al. 1985). Consequently, its transcription can be used as a marker for distinguishing between grooves and true segments. Ingham et al. (1985b) have shown that en^+ transcription is confined to the anterior-most quarter of the metamere defined by each groove. This confirms that metameres are indeed parasegments. It would appear, therefore, that parasegmental, and not segmental, primordia are initially responsible for the design of the larval and adult body. If parasegments are indeed the fundamental units of pattern formation in the young embryo, then they must, in some sense, be replaced by segments as development proceeds.

In addition to the maternal genes which specify dorso–ventral organization, at least 15 loci are known which, in a mutant state, alter the segmental pattern of the embryo but which are effective only in the homozygous state, indicating that the loci in question are active after fertilization. These segmentation mutants fall into three major classes (Table 2.3, Fig. 2.8), none of which alter the overall polarity of the embryo.

Table 2.3 Gene loci affecting segmentation in *Drosophila melanogaster* (from Nüsslein-Volhard & Wieschaus 1980).

Class	Locus	Map position
Class 1, segment polarity mutants	fused (*fu*)	1–59.5
	wingless (*wg*)	2–30
	patch (*pat*)	2–55
	gooseberry (*gsb*)	2–104
	hedgehog (*hh*)	3–90
	cubitus interruptus (*ci^D*)	4–0
Class 2, pair rule mutants	runt (*run*)	1–65
	odd-skipped (*odd*)	2–8
	paired (*prd*)	2–45
	even-skipped (*eve*)	2–55
	engrailed (*en*)	2–62
	barrel (*brr*)	3–27
Class 3, gap mutants	Krüppel (*Kr*)	2–107.6
	knirps (*kni*)	3–47
	hunchback (*hb*)	3–48

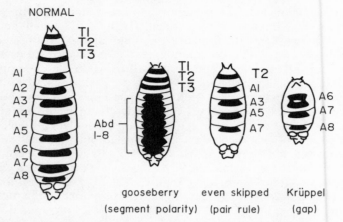

Figure 2.8 Mutations affecting segment number and polarity in *Drosophila melanogaster* (data of Nüsslein-Volhard & Wieschaus 1980).

Class 1, segment polarity mutants. These have a normal number of segments, but in each a defined fraction of the normal pattern is deleted and the remainder is present as a mirror-image duplication showing reversed polarity. These loci are thus involved in specifying the basic pattern of the individual segmental units.

Class 2, pair rule mutants. Here homologous parts of the pattern are deleted in every other segment, which implies that at some stage during normal development the embryo is organized in repeating units corresponding to two segmental anlagen. Hairy (*h*) and fushi tarazu (*ftz*) are two pair-rule loci whose transcripts are detectable in the embryo two cell cycles before cellularization of the blastoderm (Ingham *et al.* 1985a, Ish-Horowicz *et al.* 1985). While *ftz*$^+$ expression is initially restricted to a specific region of the egg, *h*$^+$ transcripts are more or less uniformly distributed throughout the egg at this stage. However, prior to complete cellularization of the blastoderm, *h*$^+$ transcripts have become localized into eight distinct regions along the antero-postero axis. By this stage, *ftz*+ transcripts are similarly localized in seven stripes whose domains are out of register with those of hairy by one metameric unit (Ingham *et al.* 1985a). The eight regions of *h*$^+$ transcript accumulation in the cellularizing blastoderm correspond broadly to the locations of the primordia of missing structures in hairy mutants. Similarly, *ftz*$^+$ transcripts are localized in seven regions, which correspond to structures deleted in *ftz* mutant larvae. Thus, hairy and fushi tarazu products are autonomously required for specifying alternate sets of blastoderm cells. Some cells in the metameric region of the blastoderm, however, express neither gene at significant levels, and these require the expression of one or more of the other pair-rule genes for their specification.

If fushi tarazu is prevented from functioning, larvae lack denticle bands in the mesothorax (T2) and the odd-numbered abdominal (Ab) segments (Ab 1, 3, 5 and 7). By using the *Drosophila melanogaster* heat shock protein 70 gene promoter to drive widespread expression of *ftz*$^+$ transcripts by heat shock treatment, Struhl (1985) has found that unrestricted *ftz*$^+$ activity results in a reciprocal pair-rule phenotype. Here, denticle bands normally derived from T1, T3, Ab2, 4, 6 and 8 are lacking. Thus both 'on' and 'off' states of fushi tarazu play a role in the development of alternating body segments.

Class 3, gap mutants. In these, a single group of up to eight adjacent segments is deleted from the final pattern. For example, embryos homozygous for Krüppel (*Kr*) lack the entire thorax and the first five abdominal segments, while in hunchback (*hb*) mutant embryos the meso- and metathoracic segments are both deleted. These mutations thus appear to be involved in processes in which the position along the antero-postero axis of the embryo is uniquely defined. In the case of Krüppel, the locus has been localized cytogenetically (60F3), microdissected from polytene chromosomes and the DNA microcloned. Krüppel is contained within 39 kb of single copy DNA surrounded by blocks of repetitive DNA, and a 4 kb interval of this region is required for *Kr*$^+$ gene function. This codes for a 2.5 kb poly (A)$^+$ RNA which is transcribed between the syncytial blastoderm stage and the beginning of germ band extension (Jäckle *et al.* 1985, Preiss *et al.* 1985). It is possible to evoke phenotypic rescue of mutant *Kr* embryos after injecting cloned DNA containing the critical 4 kb region. Additionally, by injecting a large dose of *Kr*$^+$ antisense RNA into the wild type of embryo at a very early stage it is possible to phenocopy Krüppel mutants (Rosenberg *et al.* 1985), whereas the injection of sense RNA has no such effect. The mechanism by which antisense RNA inhibits *Kr*$^+$ gene function is not known. It may act in the nucleus by binding to the *Kr*$^+$ transcript and so either prevent its export to the cytoplasm or else lead to its degradation. Alternatively, it may act in the cytoplasm by blocking the translation of *Kr*$^+$ mRNA. There are certainly precedents for this mode of inhibition in prokaryotes.

The protein product of the *Kr*$^+$ gene, as predicted from its DNA sequence, is structurally homologous with the TFIIIA transcription factor which regulates 5S gene expression in *Xenopus laevis* (Rosenberg *et al.* 1986). This suggests that DNA binding may be an important aspect of *Kr*$^+$ function, as has also been proposed both for other segmentation genes and for the homeotic genes of *Drosophila melanogaster*. The structure of the *Kr*$^+$ protein compared to the products of these other genes, however, suggests that its mode of interaction with DNA is likely to be quite different.

The pattern of expression of *Kr*$^+$, as judged from *in situ* hybridization studies, is complex and shows no direct correspondence with the segmentation pattern of Krüppel phenotypes (Knipple *et al.* 1985). In this it contrasts sharply with the behaviour of fushi tarazu and engrailed, two other segmentation genes where the patterns of gene expression are known to correspond to the affected regions of the recessive mutant phenotypes.

Five maternal mutations also partly or completely abolish development of the abdomen (Lehmann & Nüsslein-Volhard 1986). These include oskar (85B), pumilio (85CE), tudor (57B–D), vasa (35B) and staufen (55A). The first four of these also lack pole cells. When wild type cytoplasm is injected into oskar mutant embryos it is possible to rescue the abdominal defect, provided the injection is into the prospective abdominal region of the mutant embryo and that the incoming cytoplasm derives from the posterior pole of the donor (Lehmann & Nüsslein-Volhard 1984, 1986). Furthermore, rescue can also be obtained using cytoplasm from the anterior pole of bicaudal mutant embryos. Since this lacks pole cells, it is clear that the effect is somatically produced.

2.2.3.4   Homeotic genes

The determination of body segmentation in early embryogenesis also depends upon the expression of segment-specific selector genes, principally those comprising the Bithorax Complex and the Antennapedia Complex, both located on chromosome 3 (Fig. 2.9). The differential expression of the genes within these complexes and their associated developmental circuits is responsible for the developmental pathways followed by the derivatives of each segment in both larvae and adults (Akam *et al.* 1985, Bender *et al.* 1985, Gehring 1985a, Hogness *et al.* 1985, Laughon *et al.* 1985, Levine *et al.* 1985, Lewis 1985, Sanchez-Herrero *et al.* 1985). Critical portions of these gene complexes, termed 'homeo boxes' (Gehring 1985b), which code for a

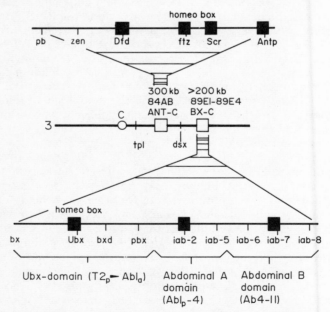

Figure 2.9 The distribution of homeo boxes (shown solid) in and near the Antennapedia and Bithorax Complexes (open boxes) present in chromosome 3 of *Drosophila melanogaster*.

conserved polypeptide sequence, are found in a number of vertebrates which, as we have already seen, also have segmented embryos. This includes *Xenopus laevis*, chicken, mouse and man (Shepherd *et al.* 1984, Boncinelli *et al.* 1985, Colberg-Poley *et al.* 1985a, de Robertis *et al.* 1985, Joyner *et al.* 1985, McGinnis 1985, Ruddle *et al.* 1985). Moreover, in *Xenopus laevis* the homeo box is expressed during early development, starting at gastrulation (Carrasco *et al.* 1984). This case provides direct evidence that much of relevance to other eukaryotes can be gleaned from a consideration of the specific developmental system of *Drosophila melanogaster*. It is important, therefore, to analyze the *Drosophila melanogaster* system in some detail, so that the principles which underlie its development can then be evaluated in vertebrates, which are not as tractable in an experimental sense.

The clearest evidence comes from the Bithorax Complex (BX-C), located at 89E1-89E4 and spanning a region in excess of 200 kb of DNA. When this entire complex is deleted, a condition that leads to death at late embryogenesis, the third thoracic segment (metathorax) and all eight abdominal segments resemble a normal second thoracic segment (meso-thorax). The head and the first thoracic segment, however, are both unaffected, so that the normal activity of the Bithorax Complex is assumed to move posterior segments away from a mesothoracic ground state. A series of individual mutations within the complex have also been identified which give rise to less extreme segmental transformations, and which give an initial impression of there being several distinct genes within the complex. The first of these to be described was that from which the complex takes its name. This recessive mutation, *bx* (bithorax), transforms the anterior portion of the third thoracic segment into an anterior second thoracic segment, so that the anterior portion of the haltere becomes anterior wing tissue, while the anterior portion of the third leg resembles a second leg (Fig. 2.10). A second recessive mutation, *pbx* (postbithorax), affects the posterior half of the third thoracic segment in a similar way, transforming the posterior portion of the haltere to wing, and the posterior part of the third leg to second leg. Double mutations of both *bx* and *pbx* thus carry out both anterior and posterior transformations, and so give rise to a fly with four wings (Fig. 2.11). A series of recessive abdominal mutations, *bxd* (bithoraxoid), *iab-2* (infra abdominal 2), *iab-5* and *iab-8* affect the first, second, fifth and eighth abdominal segments respectively, transforming them into a more anterior state. Dominant mutations are also known within the complex. These, in the main, have a reverse effect to the recessive mutations, transforming a segment or a part of a segment into a more posterior state. Thus *Cbx* (Contra bithorax) transforms posterior wing to posterior haltere and is, therefore, complementary to *pbx*. The most distinctive mutant of the complex is *Ubx* (Ultra bithorax). Animals homozygous for *Ubx* die as larvae or as early pupae, but larval cuticular structure indicates that the third thoracic and first abdominal segments are both transformed into copies of the second thoracic segment.

The bithorax system is itself under the control of the wild type alleles of more than 20 *trans*-regulatory loci which influence the spatial expression of

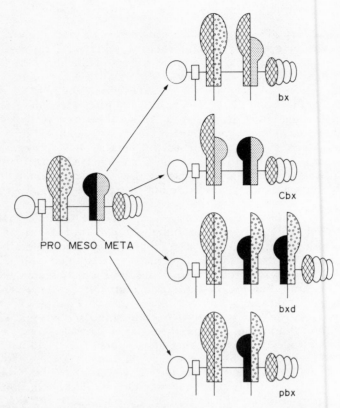

PRO MESO META

bx

Cbx

bxd

pbx

Figure 2.10 Diagrammatic representation of the body pattern associated with four of the mutants of the Bithorax Complex in *Drosophila melanogaster* (data of Lewis 1964).

Table 2.4 Homeotic genes required for the correct expression of the Bithorax Complex of *Drosophila melanogaster*. Note that, since the bithorax genes are normally active only in segments posterior to the mesothorax, there must be other genes which confine the expression of the Bithorax Complex to that body region (from Ingham 1984, Jurgens 1985).

Gene	Location
Haplo-insufficient	
Asx – additional sex combs	51AB
Psc – posterior sex combs	49EF
Pcl – Polycomb-like	55AF
Pc – Polycomb	3–47
Scm – sex comb on mid leg	85EF
ph – polyhomeotic	2D2–4
Haplo-sufficient and maternally rescuable	
esc – extra sex combs	33B1,2
sxc – super sex combs	2–55

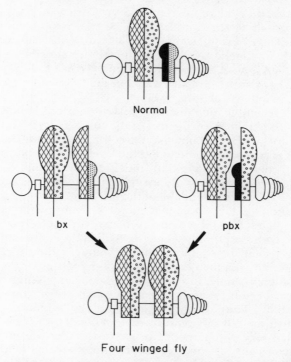

Normal

bx

pbx

Four winged fly

Figure 2.11 Production of a four winged individual of *Drosophila melanogaster* obtained by crossing the *bx* and *pbx* mutants of the Bithorax Complex.

BX-C genes (Table 2.4). For example, in Polycomb, (*Pc*), hemi- and homozygotes for *Pc* mutations have their thoracic and first seven abdominal segments transformed towards abdominal 8. Such animals, which die as late embryos, also have reduced head structures. Pc^+ thus appears to code for a general repression of the bithorax system, and so affects the appropriate bithorax genes in all segments of the body. Mutations of extra sex combs (*esc*) also cause an indiscriminate expression of BX-C, which leads to all body segments being abdominal (Frei *et al.* 1985). By contrast, the *Asx, Pcl, Psc* and *Scm* mutations cause only partial transformation into posterior abdominal development in mutant embryos (Jürgens 1985). However, embryos simultaneously mutant for two or more of these genes show homeotic transformation of all body segments similar to, or even stronger than, that seen in *Pc* or *esc* mutants, indicating that these genes act synergistically in normal development to control BX-C gene expression. Thus, the *Pc* group of genes, like the dorsal group, includes a multi *trans*-regulatory system controlling the spatial expression of this homeotic complex.

Using a monoclonal antibody (1bF12), which recognizes segment specific structures in the central nervous system (CNS) of *Drosphila*, Ghysen *et al.* (1985) have shown that BX-C mutations have independent effects on the

CNS and the larval epidermis. The CNS, in particular, is very sensitive to mild perturbations of the complex, and especially to haplo-insufficiency, i.e. dosage changes brought about by heterozygous deficiencies for the complex. BX-C functions are thus directly responsible for segmental diversity in the CNS. A spatially restricted expression of a mouse homeo box, Hox-3, has also been reported within the CNS of both newborn and adult mice (Awgulewitsch *et al.* 1986).

Classically, recombination distances between various bithorax mutations were used to construct a genetic map of the complex (Fig. 2.12). From this it has, in the past, been assumed that the complex contains a series of genes which code for substances controlling the levels of thoracic and abdominal development, the arrangement of the individual genes corresponding directly to the arrangement of the segments in which these genes are active (Lewis 1978). The wild type alleles of the members of the complex are thus assumed to produce substances that promote the development of simpler posterior segments from the pathway that gives rise to the mesothorax. This is in line with the presumed evolution of dipterans from ancestors with many legs, rather than six, and with four wings, rather than two. In this genetic model, it is assumed that the state of expression of the genes in the Bithorax Complex is able, in some way, to define the segment to which a given epidermal cell belongs, and that cells then differentiate according to this information. Thus, if all the genes in the Bithorax Complex are expressed, then the cell differentiates with a phenotype appropriate to the last abdominal segment. If none are expressed, the cell differentiates as mesothoracic. Finally, if some genes are expressed while others are not, then the cell differentiates in a manner characteristic of the appropriate intervening segment.

This interpretation has been both clarified and complicated by recent studies. Struhl (1984) has shown that it is possible to split the Bithorax Complex into two separate pieces without affecting the development of either the larva or the adult. The complex thus includes at least two autonomous domains that do not, however, control the development of separate sets of segments. Rather, their realms of action appear to intersect at the antero-postero compartment boundary located in the middle of the first abdominal segment.

The whole complex spans 410 kb of DNA, and the left and right halves are similar in genetic organization. The homeotic mutations of the right half, which extends over 215 kb, cause segmental transformations in the second through eighth abdominal segments. These mutants can be grouped into a series of phenotypic classes, one for each abdominal segment, whose order on the chromosome reflects the order of the body segment they affect, namely *iab-2* to *iab-8* (Karch *et al.* 1985).

A molecular examination of the left hand of the complex (Bender *et al.* 1983a), involving a chromosome walk from *abx* to *pbx*, defines a segment of 73 kb which appears to consist predominantly of single copy DNA. The whole of this region is transcribed *in toto* to yield a single, giant, 73 kb primary transcript, which is subsequently processed into RNA molecules of 3.0, 4.3 and 4.7 kb size, and the direction of transcription proceeds from

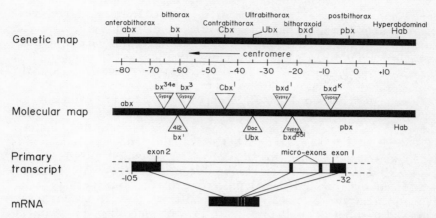

Figure 2.12 A comparison of the genetic and molecular maps of the left half of the Bithorax Complex of *Drosophila melanogaster* and the primary transcript and mRNA produced by the *Ubx⁺* region. The triangles denote DNA insertions (data of Bender *et al.* 1983b). Updated versions of these maps are available in Hogness *et al.* (1985), while the right half of the complex has been mapped by Bender *et al.* (1985) and Karch *et al.* (1985).

pbx through to *abx* (Hogness *et al.* 1985). Moreover, all the spontaneous mutations analyzed, namely bx^1, bx^3, bx^{34e}, Ubx^1, bxd^1, bxd^{sij}, bxd^{5si} and bxd^K, involve mobile element insertions, while the cytologically normal X-ray-induced mutations have either large DNA deletions (Ubx^{9-22}, abx^1, abx^2, Ubx^{6-28}, pbx^1 and pbx^2) or else, in the case of *Cbx*, a large DNA insertion (Bender *et al.* 1983a). The pbx^1 mutation, for example, is associated with the deletion of a 17 kb fragment, while the Cbx^1 mutation results from the reinsertion of this same fragment back into the complex some 40 kb from its original location. This suggests that the 17 kb region in question may encode the information to specify the development of the posterior third thoracic segment. Consequently, the loss of this information causes a switch to the ground state of the second thoracic segment.

The insertion of this region some 40 kb downstream, on the other hand, leads to structures characteristic of the third thoracic segment (T3) being produced in place of those normally formed in the second thoracic segment (T2). What remains to be established is whether these transformations are a result of incorrect transcription, incorrect RNA splicing or other changes at the molecular level. It will also be necessary to characterize in detail the transcripts produced by the right half of the complex, and determine how their products interact with those produced by the left half of the complex.

Cbx^1, and two other dominant mutations in the *Ubx* domain, CBx^3 and *Hm* (Haltere mimic), are now known to cause inappropriate expression of *Ubx⁺* products in T2. Consequently, they transform T2 structures into homologous T3 structures. By examining the distribution of *Ubx⁺* transcripts and proteins in the imaginal discs of larvae, it has been shown that there is little expression of them in wild type wing discs, which give rise to most dorsal T2 structures in the adult. Both, however, were relatively abundant in the homologous discs of T3, which give rise to the halteres, as well as in T3 leg discs. In Cbx^1 homozygotes, however, Ubx^+ transcripts and proteins are expressed in wing discs at levels comparable to those found in T3 discs (White & Akam 1985). Cbx^3 and Hm, similarly, result in the expression of Ubx^+ proteins in the T2 wing discs.

Three major classes of Ubx^+ RNA appear during the first eight hours of embryogenesis. A transient 4.7 kb poly (A)$^-$ RNA forms first, but then disappears near the end of the eight hour period. At about three hours, a 3.2 kb poly (A)$^+$ RNA is produced. This is slightly after the production of the 4.7 kb transcript, but before that of a 4.3 kb poly (A)$^+$ RNA. The 3.2 and 4.3 kb classes are then found at all subsequent stages of development. It is assumed that the differential processing of the long primary Ubx^+ transcript generates a set of mRNAs, and that these are produced and translated in different compartments to provide individual compartment-specific functions (Beachy et al. 1985). Indirect immunofluorescent staining of anti-Ubx^+ antibodies demonstrates a weak anti-Ubx^+ reaction in the posterior part of the second thoracic (T2) segment, followed by a strong staining of the posterior portion of the third thoracic segment (T3p) and the first anterior abdominal segment (Ab1a), with the anterior portions of abdominal segments 2–7 exhibiting progressively weaker staining. Thus, apart from the weak staining of T3a, the pattern produced is that predicted, since the primary determinants of segment identities in the Ab2–7 region derive from other domains of the Bithorax Complex. The distribution of Ubx^+ proteins in homozygous mutant Pc embryos also conforms with the fact that embryos lacking Pc^+ function suffer a homeotic transformation, in which all thoracic and abdominal segments exhibit cuticular structures characteristic of normal Ab8.

The monoclonal antibody, FB 3.38, recognizes an antigenic determinant encoded in the Ubx 5′ exon. It has been used to demonstrate that the subfunction mutations abx, bx, bxd and pbx, which map outside the known Ubx^+ coding region, all affect the distribution of Ubx^+ protein in the imaginal discs (White & Wilcox 1985). This suggests that the subfunction mutations may affect regulatory regions required for the normal pattern of Ubx^+ expression. The regions of BX-C which contain the abdA and AbdB complementation groups also regulate the distribution of Ubx^+ products.

RNA transcripts homologous to the major 5′ exon of the Bithorax Complex have been localized by in situ hybridization to specific tissues. The polypeptide fragment corresponding to this exon is localized predominantly within the nuclei of imaginal disc cells responsible for the third thoracic segment, the haltere and the third leg, but is absent from the cells of the eye, antennal, first leg and genital discs. This protein has properties of known prokaryotic DNA binding proteins, which has important implications for the mode of operation of the locus. Furthermore, the central nervous system is the most prominently labelled tissue in embryos, while in third instar larvae the imaginal discs of the third thoracic segment are the most obviously labelled (Akam 1983). Thus, at least one of the predictions of the genetic model of the functioning of this locus is fulfilled, namely that a concentration of transcripts is present in a defined and predicted segment. This is consistent with the proposal that the wild type product of the complex actually promotes a metathoracic pathway. However, the location of transcripts in the central nervous system of the embryo was not predicted by the genetic model, though it is important to remember that this entire region derives from the ectoderm.

In the mutant Antennapedia (*Antp*), where in extreme cases the antenna is converted into a mesothoracic leg, *in situ* hybridization of cDNA homologous to regions of the Antennapedia locus again reveals that it is localized in neural tissue in embryos, with especially high concentrations in the mesothoracic segment of the ventral nerve cord (Hafen *et al.* 1983). This observation is consistent with the suggestion that $Antp^+$ products also actively promote a mesothoracic developmental pathway and, significantly, Antennapedia, like Bithorax, is also a complex locus which produces an initial giant transcript of 103 kb that gives rise to processed RNAs of 3.5 and 5.0 kb (Scott *et al.* 1983, Scott & Weiner 1984).

The Antennapedia Complex, which has been extensively studied in the laboratories of Kaufman and Gehring, is located in polytene section 84AB and extends over a region of some 300 kb of DNA. It includes at least three homeotic genes, namely proboscipedia (*pb*), Sex combs reduced (*Scr*) and Antennapedia itself (see Fig. 2.9). The members of this complex control segmental development in the head and thorax in a manner analogous to that by which the Bithorax Complex operates in the more posterior body segments. Although these two complexes are spatially separated on chromosome 3, the fact that both specify segmental identity suggests their common origin. There is, in fact, known DNA sequence homology between some of the genes within the two complexes. Thus, the 3' exons of the coding regions of $Antp^+$, ftz^+ and Ubx^+ share a conserved 180 bp region of DNA, which has been termed the homeo box (McGinnis *et al.* 1984a,b). The 60 amino acids coded for by the three homeo boxes are impressively similar. There is 83% homology in amino acid sequence between $Antp^+$ and ftz^+, 75% homology between ftz^+ and Ubx^+, and 87% homology between $Antp^+$ and Ubx^+. This conserved amino acid sequence, moreover, includes a potential DNA binding region of the type present in some bacterial DNA binding proteins (Fig. 2.13). It also shows homology with the a_1 and α_2 proteins of the mating locus of yeast, both of which have been implicated in regulating the transcription of other genes (Laughon & Scott 1984).

In the case of the Antennapedia locus, deletion of the entire complex can lead to recessive, loss of function, mutations which give the familiar homeotic transformation of one segment into a more anterior one. Dominant, gain of function, Antennapedia mutations, on the other hand, lead to transformations in the opposite, posterior, direction, and so produce antennal legs. This latter class of mutation has formerly been difficult to account for, but Schneuwly & Gehring (1986) have solved this problem in an elegant series of experiments. They made a special P element construct which had a wild type Antennapedia cDNA fused to a heat shock promoter. The $hsp70$-$Antp^+$ fusion gene was then introduced into the genome by P

Table 1 — Homeo domain, positions 1–20

Locus		1	2	3	4	5	6	7	8	9	10	11	12	13	14	15	16	17	18	19	20
Drosophila	Antp	Arg	Lys	Arg	Gly	Arg	Gln	Thr	Tyr	Thr	Arg	Tyr	Gln	Thr	Leu	Glu	Leu	Glu	Lys	Glu	Phe
	Ubx	Arg	Arg	Arg	Gly	Arg	Gln	Thr	Tyr	Thr	Arg	Tyr	Gln	Thr	Leu	Glu	Leu	Glu	Lys	Glu	Phe
	ftz	Ser	Lys	Arg	Thr	Arg	Gln	Thr	Tyr	Thr	Arg	Tyr	Gln	Thr	Leu	Glu	Leu	Glu	Lys	Glu	Phe
Human	Hu 1	Gly	Lys	Arg	Ala	Arg	Thr	Ala	Lys	Thr	Arg	Tyr	Gln	Thr	Leu	Glu	Leu	Glu	Lys	Glu	Phe
	Hu 2	Thr	Ala	Arg	Gly	Arg	Thr	Tyr	Thr	Arg	Tyr	Gln	Thr	Leu	Glu	Leu	Glu	Lys	Glu	Phe	
Xenopus	AC1	Arg	Arg	Arg	Gly	Arg	Gln	Ile	Tyr	Ser	Arg	Tyr	Gln	Thr	Leu	Glu	Leu	Glu	Lys	Glu	Phe
Yeast	MATα2	Arg	Gly	His	Arg	Phe	Thr	Lys	Glu	Asn	Val	Arg	Ile	Leu	Glu	Ser	Trp	Phe	Ala	Lys	Asn
	MATa1	Ser	Pro	Lys	Gly	Lys	Ser	Ser	Ile	Ser	Pro	Gln	Ala	Arg	Ala	Phe	Leu	Glu	Gln	Val	Phe
E. coli	Lac Rep																				
	Trp Rep																				
λ phage	Cro																				

Table 2 — positions 21–40 (HELIX 2)

Locus		21	22	23	24	25	26	27	28	29	30	31	32	33	34	35	36	37	38	39	40
Drosophila	Antp	His	Phe	Asn	Arg	Tyr	Leu	Thr	Arg	Arg	Arg	Arg	Ile	Glu	Ile	Ala	His	Ala	Leu	Cys	Leu
	Ubx	His	Thr	Asn	His	Tyr	Leu	Thr	Arg	Arg	Arg	Arg	Ile	Glu	Met	Ala	Tyr	Ala	Leu	Cys	Leu
	ftz	His	Phe	Asn	Arg	Tyr	Ile	Thr	Arg	Arg	Arg	Arg	Ile	Asp	Ile	Ala	Asn	Ala	Leu	Ser	Leu
Human	Hu 1	His	Phe	Asn	Arg	Tyr	Leu	Thr	Arg	Arg	Arg	Arg	Ile	Glu	Ile	Ala	His	Ala	Leu	Cys	Leu
	Hu 2	His	Tyr	Asn	Arg	Tyr	Leu	Thr	Arg	Arg	Arg	Arg	Ile	Glu	Ile	Ala	His	Ala	Leu	Cys	Leu
Xenopus	AC1	His	Phe	Asn	Arg	Tyr	Leu	Thr	Arg	Arg	Arg	Arg	Ile	Glu	Ile	Ala	Asn	Ala	Leu	Cys	Leu
Yeast	MATα2	Ile	Glu	Asn	Pro	Tyr	Leu	Asp	Thr	Lys	Gly	Leu	Glu	Asn	Leu	Met	Lys	Asn	Thr	Ser	Leu
	MATa1	Arg	Arg	Lys	Gln	Ser	Leu	Asn	Ser	Lys	Glu	Leu	Glu	Glu	Val	Ala	Lys	Lys	Cys	Gly	Ile
E. coli	Lac Rep											Leu	Tyr	Asp	Val	Ala	Glu	Tyr	Ala	Gly	Val
	Trp Rep											Gln	Arg	Glu	Leu	Lys	Asn	Glu	Leu	Gly	Ala
λ phage	Cro											Gln	Thr	Lys	Thr	Ala	Lys	Asp	Leu	Gly	Val

H——————— HELIX 2 ———————H

Table 3 — positions 41–60 (HELIX 3)

Locus		41	42	43	44	45	46	47	48	49	50	51	52	53	54	55	56	57	58	59	60
Drosophila	Antp	Thr	Glu	Arg	Gln	Ile	Lys	Ile	Trp	Phe	Gln	Asn	Arg	Arg	Met	Lys	Trp	Lys	Lys	Glu	Asn
	Ubx	Thr	Glu	Arg	Gln	Ile	Glu	Ile	Trp	Phe	Gln	Asn	Arg	Arg	Met	Lys	Leu	Lys	Lys	Glu	Ile
	ftz	Ser	Glu	Arg	Gln	Ile	Lys	Ile	Trp	Phe	Gln	Asn	Arg	Arg	Met	Lys	Ser	Lys	Lys	Asp	Arg
Human	Hu 1	Ser	Glu	Arg	Gln	Ile	Lys	Ile	Trp	Phe	Gln	Asn	Arg	Arg	Met	Lys	Trp	Lys	Lys	Asp	Asn
	Hu 2	Thr	Glu	Arg	Gln	Ile	Lys	Ile	Trp	Phe	Gln	Asn	Arg	Arg	Met	Lys	Trp	Lys	Lys	Glu	Ser
Xenopus	AC1	Thr	Glu	Arg	Gln	Ile	Lys	Ile	Trp	Phe	Gln	Asn	Arg	Arg	Met	Lys	Trp	Lys	Lys	Glu	Asn
Yeast	MATα2	Ser	Arg	Ile	Gln	Ile	Lys	Asn	Trp	Val	Ser	Asn	Arg	Arg	Arg	Lys	Glu	Lys	Thr	Ile	Thr
	MATa1	Thr	Pro	Leu	Gln	Val	Arg	Val	Trp	Val	Cys	Asn	Met	Arg	Ile	Lys	Leu	Lys	Tyr	Ile	Leu
E. coli	Lac Rep	Ser	Tyr	Gln	Thr	Val	Ser	Arg	Val	Val	Asn										
	Trp Rep	Gly	Ile	Ala	Thr	Ile	Thr	Arg	Gly	Ser	Asn										
λ phage	Cro	Tyr	Gln	Ser	Ala	Ile	Asn	Lys	Ala	Ile	His										

H——————— HELIX 3 ———————H

Figure 2.13 Homeo domains of *Drosophila melanogaster*, *Homo sapiens* and *Xenopus laevis*, together with comparisons involving the yeast mating type proteins a_1 and α_2, the *lac* and *trp* repressors of *Escherichia coli* and the *Cro* repressor of λ phage. The positions of the two α-helices of the DNA-binding domain, numbered helix 2 and helix 3 according to the convention for the λ *Cro* protein, are shown at the bottom of the figure (data of Laughon & Scott 1984, Levine *et al.* 1984, Shepherd *et al.* 1984).

element mediated germ line transformation, so that transformed flies now contained three copies of the $Antp^+$ gene, two normal and a third under heat shock control. When such flies were heat shocked during the larval stage there was an overproduction of $Antp^+$ protein and, under these circumstances, the adults developed antennal legs. Thus, if the $Antp^+$ protein is over-expressed in the antennal disc, where it is not normally expressed, homeotic changes can ensue. This, as these authors have pointed out, means that the body plan of *Drosophila melanogaster* can be radically altered by simply changing the expression of a homeotic gene. This emphasizes that for some genes it is the time, the place and the level of gene expression that can be critical for developmental potentiality.

It is worth pointing out that fushi tarazu (*ftz*) (meaning 'lack of enough segments'), though located in the Antennapedia Complex (ANT-C) and containing a homeo box, is not a homeotic gene, although it is necessary for the correct development of segmentation. Rather, it belongs to the class of pair-rule mutants. Located 30 kb to the left of Antennapedia, it encodes a single 1.8 kb poly $(A)^+$ RNA (Weiner *et al.* 1984) which is expressed exclusively from early blastoderm to gastrula (Hafen *et al.* 1984, Kuroiwa *et al.* 1984). Thus, it is one of the first zygotic genes to be transcribed during development and, whereas homeotic genes determine the nature of the segment, ftz^+ seems to be involved in counting the segments. Thus, embryos homozygous for deficiencies of ftz^+ die prior to hatching and lack alternate body segments. The reduction in the number of segments in fact results from the fusion of the posterior portion of one segment with the anterior portion of the next segment.

The bithorax mutants provide a particularly striking example of a more general class of homeotic mutants. The term homeosis, from the Greek *homoios* meaning 'similar', was used classically to describe the replacement of one body part by another, considered homologous and hence of similar design, as a consequence of mutation. That is, a given structure is transformed into another, characteristic of an organ or a segment normally found elsewhere in the body. In this restricted usage, homeotic transformations are known only in insects. However, since the essence of transformation depends upon mutations which switch developmental pathways, the term has been extended to cover the mutational switch controls in cell populations of *Caenorhabditis elegans* (Ambrose & Horvitz 1984). In *Drosophila melanogaster*, homeotic mutations alter the state of determination of individual imaginal discs or parts of discs (Table 2.5). At least 31 different homeotic mutations have been defined in *Drosophila melanogaster*. Collectively, these affect all the known discs, and at least 20 of them affect more than one pair of discs. Most of these homeotics, like the bithorax series, produce intersegmental transformations, the structures formed being normally produced by another disc. Some, however, also lead to intrasegmental changes, where one part of a disc gives rise to a transformation of a different part of the same disc. Since homeotic transformations are never intermediate in character, it is clear that homeotic genes must control switches between alternative developmental circuits.

Table 2.5 Representative homeotic mutations in *Drosophila melanogaster* (from Ouweneel 1976).

Disc	Mutation	Symbol	Locus	Effect
labial	proboscipedia	*pb*	3–47	proboscis $\nearrow$ antenna III + arista $\searrow$ tarsus
antennal	Polycomb	*Pc*	3–47 }	antenna → 2nd leg
	Antennapedia	*Antp*	3–48 }	
eye	Ophthalmoptera	*Opt*^{G}	2–68	eye area → wing
leg	extra sex combs	*esc*	2–54 }	2nd, 3rd leg → 1st leg
wing	Polycomb	*Pc*	3–47 }	
	Metaplasia	*Met*	?	wing → haltere
	podoptera	*pod*	several	wing → leg
	Contrabithorax	*Cbx*	3–58	wing + meso. → haltere + meta.
haltere	bithorax	*bx*	3–58	ant. haltere + ant. meta. → ant. wing + ant. meso.
	Ultrabithorax	*Ubx*	3–58	haltere + meta. → wing + meso.
	postbithorax	*pbx*	3–58	post. haltere + post. meta. → post. wing + post. meso.
	Contrabithoraxoid	*Cbxd*	3–58	haltere + meta. → abl. 1
genital	lethal (3) 703	*(3)703*	3–78	genitalia → leg, antenna

meso. = mesothorax; meta. = metathorax; abl. 1 = 1st abdominal segment.

Not only do homeotic genes switch on early in development, but once activated a majority seem to be required throughout development in order to maintain particular determined states. There is, however, at least one class of maternal effect type homeotics, illustrated by extra sex combs (*esc*) and super sex combs (*sxc*), see Table 2.4, whose activity is essential early in development for initiating the correct determined state of most, if not all, of the segmental primordia, Mutations of the *esc* and *sxc* loci, unlike those of all previously described homeotics, result in transformation only when both mother and zygote are mutant (Bryant 1978, Ingham 1984). Thus, in addition to supplying the egg with spatial information, the mother also provides gene products essential for the correct interpretation of that information during development. For example, when both mother and zygote lack the *esc*$^+$ allele, most, or all, of the body segments develop like that of the eighth abdominal region. Evidently, the product of the *esc*$^+$ allele is necessary for initiating the unique developmental pathway followed by each of the different segments.

Homeotic genes can be subdivided into two major classes – homeotic selector and homeotic regulatory. Genes of the selector class, which include members of the Bithorax and Antennapedia Complexes, are activated in discrete compartments where they determine which cells enter a particular developmental pathway. Genes of the regulator class, which include *Pc* and *esc* (see Table 2.4), are required in most segments to ensure correct expression of the selector genes.

While the effects of homeotic mutations establish that the corresponding wild type alleles influence the determined state of the cells within a given imaginal disc, they do not reveal how that state is controlled. There is now substantial evidence that the individual body segments of *Drosophila melanogaster* can be divided into anterior and posterior compartments, and that cells in these compartments respect a natural boundary between the two. This evidence comes from a clonal analysis of cell behaviour in genetic mosaics produced by X-ray irradiation. This treatment generates a chromatid exchange between homologous mitotic chromosomes. By this method, a cell heterozygous for a recessive mutant marker ($m/+$), and which is phenotypically wild type, will, following mitotic exchange, give rise to two daughter cells, one homozygous for the mutation (m/m), which expresses the mutant phenotype, the other homozygous for the wild type allele ($+/+$). The earlier in development the exchange takes place, the greater will be the number of daughter m/m cells produced. Clonal analysis can be extended by coupling the mutation, m, with another mutation, such as a Minute, or even a cell lethal. For example, when individuals of constitution $m+/+l$ are irradiated then only the $m+/m+$ products survive (Fig. 2.14).

Using this approach it has been established that a previously unrecognized border subdivides both the embryonic and the larval segments of *Drosophila melanogaster* into anterior and posterior compartments. When posterior cells are deficient for the engrailed locus then neither the compartment nor the segment border is maintained. This applies both to the abdominal tergites of the adult, which are produced from the histoblasts, and to the wings, which are produced from discs. The engrailed locus of *Drosophila melanogaster* thus has the characteristics of both a homeotic gene and a segmentation gene (Fjose *et al.* 1985).

As defined by complementation tests, engrailed is a large gene, $c.70$ kb in size. However, a 2.7 kb en^+ transcript is derived from less than 5 kb of genomic DNA (Kuner *et al.* 1985). This has led to the suggestion that the large size of the en^+ complementation unit is due to the fact that regions far distant from the transcription unit may be involved in its regulation. Certainly, the mapping of engrailed mutations suggests that extensive flanking sequences are involved in the spatial and temporal regulation of en^+ expression. Two other loci having complex spatial patterns of activity, BX-C and scute, also interact with genes dispersed over a large region of the genome, giving rise to an extended regulatory region.

In gastrulating engrailed mutant embryos, clear segmental grooves are missing at the border of the pro- and mesothorax, the metathorax and the first abdominal segment, as well as between the second and third, the fourth and fifth, and the sixth and seventh abdominal segments, respectively. This indicates that, even in the embryo, a segment is a composite structure consisting of an anterior compartment, which does not express wild type engrailed function, and a posterior compartment that does (Kornberg 1981). The distinction between cells in these two compartments thus depends on the state of expression of the wild type engrailed gene, which is 'on' in one compartment and 'off' in the other.

On this basis, it is clear that all discs and histoblasts are already

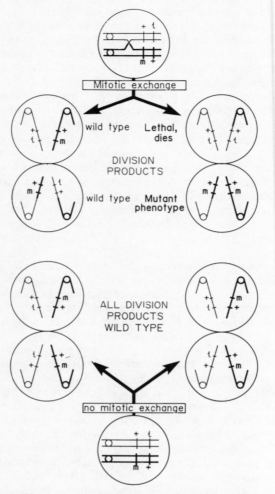

Figure 2.14 Production of a mutant (*mm*) phenotype following mitotic exchange (upper figure) in a heterozygous individual with the constitution (+*l*/*m*+) where *l* is a recessive lethal and *m* a recessive mutant locus. Where no mitotic exchange occurs (lower figure) all division products are wild type.

subdivided into anterior and posterior compartments at the time of their formation. As the founder cells are partitioned into daughter polyclones, it can be hypothesized that there is a permanent activation of a unique switch, or selector gene, in one polyclone but not in its sister. Thus, three such genes, each performing a homologous function within each daughter polyclone, would be capable of generating eight different end products (Fig. 2.15), so that the developmental pathway followed by a series of polyclones would depend on the local state of a small number of controlling selector genes, and each compartment of the adult would be uniquely specified by the precise combination of switch genes active within it. This, in turn, implies that each determinative event would be accompanied by the activation of a specific switch gene.

A cDNA clone obtained from the wild type engrailed locus has been found to include a homeo box (Poole *et al.* 1985). Here, however, it diverges substantially from the other homeo boxes which exist in the *Drosophila melanogaster* genome. Indeed, these show more homology with the human homeo boxes than they do with that of engrailed which, unlike them, is interrupted by an intervening sequence. A neighbouring gene, engrailed-related (now renamed invected (*inv*) by Ali *et al.* 1985), which has a large region showing strong homology with engrailed, contains a similarly divergent and split homeo box.

DNA sequencing studies have indicated that homeo box sequences are both well represented and highly conserved in animals which have a metameric body plan, namely arthropods and vertebrates. However, homologies to such sequences are also present in unsegmented organisms, such as echinoderms and molluscs, but are absent from nematodes, a slime mold and a yeast (Holland & Hogan 1986). While many have assumed that these boxes play a fundamental role in determining body segmentation, it may well be that they have different functions in different organisms.

In the case of the Hawaiian sea urchin, *Tripneustes gratilla*, there are no fewer than five homeo box-containing genes, at least one of which is

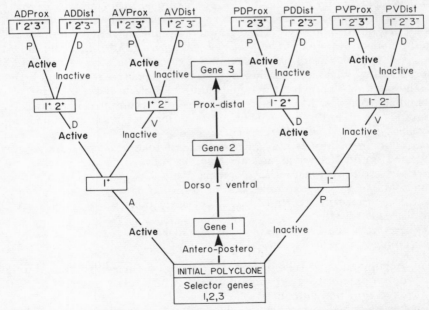

Figure 2.15 Hypothetical scheme for the successive compartmentalization of an imaginal disc involving a series of three switch genes (1, 2 and 3). For each gene the active state is denoted as + and the inactive as − (modified from Lawrence & Morata 1976).

transcribed during the blastula and gastrula stages (Dolecki *et al.* 1986). Since the embryo in this case is non-segmented, it is likely that this gene has a different developmental role to that conventionally assigned to other homeo box-containing genes. This also appears to be the case for some of the homeo box-containing genes of *Drosophila melanogaster*. Thus, in the case of the S67 (= *caudal*) gene, transcription is predominantly in anterior and posterior tissues of the embryo, so that this gene may be implicated in antero-postero positional identity (Hoey *et al.* 1986). In the S60 (= *zen*) gene, on the other hand, transcriptional activity is largely confined to dorsal tissues of the embryo, and this gene may play a role in the dorso-ventral differentiation of the embryo (Doyle *et al.* 1986).

Colberg-Poley *et al.* (1985b) have shown that a novel murine homeo box-containing gene is expressed in embryonal carcinoma stem cells of the mouse. Experimentally, such cells can be induced to differentiate following treatment with retinoic acid, when they give rise to markers characteristic of extra-embryonic tissues. Transcripts of the sequences flanking the novel homeo box can be detected both before and after induced differentiation, whereas a specific homeo box transcript is present only after differentiation, when it peaks transiently. The authors imply that RNA which hybridizes to the same probe is also present in normal mouse embryos, in both embryonic and extra-embryonic tissues. In the mouse embryo, cellular differentiation ocurs at the pre-implantation stage, during the first four days of development. By days 12 and 13, essentially all major organ anlage are present. Jackson *et al.* (1985) report on a murine homeo box, H 24.1, which is located on chromosome 11 and contains a sequence closely homologous to the *Antp* homeo box of *Drosophila melanogaster*. H 24.1 is first transcribed at 7.5 days post coitum, with maximum expression at days 11.5 and 12.5 post coitum – when it is enriched in both embryonal spinal cord and brain. Transcripts are also present in adult kidney. Similarly, poly $(A)^+$ mRNA containing M5 and M6 homeo box sequences accumulate maximally in embryos at day 12. Subsequent to this, they decrease in abundance (Colberg-Poley *et al.* 1985c). These results suggest that homeo box transcripts in mammals may not be confined to times when the serially repeated somite-derived tissues are being established, though it does not preclude a role in regulating development. Thus *Xenopus laevis* has a homeo box-containing gene, homologous to those of *Drosophila melanogaster*, which is maternally expressed and abundantly transcribed in the oocyte (Müller *et al.* 1984).

There are seven important conclusions to be drawn from the data on development:

(a) A population of cells undergoes a series of binary decisions during development, to yield subpopulations which differ in terms of their developmental potential.

(b) It is possible to uncover loci, in the form of homeotic genes, which appear to perform executive developmental decisions. These, however, are relatively few in number, and the bulk of the transcriptively active parts of the genome simply provide the metabolic back-up necessary for implementing the decisions of these executive genes whose activities govern the ultimate expression of large arrays of other loci.

(c) Two of the homeotics, bithorax and Antennapedia, act very early in development, probably soon after blastoderm formation, and at least one of them, bithorax, continues to act during the larval stages. Such prolonged activity implies that the determined state must be maintained until the last irrevocable decision leading to differentiation needs to be made. On the other hand, the antero-postero decision within an imaginal disc can be experimentally altered by making a cell homozygous mutant *en/en*, that is by inactivating engrailed. Thus, despite all the prior decisions that have been taken, cells within a posterior compartment can be made to express an anterior phenotype.

(d) The homeotic loci of *Drosophila* can be grouped into two functional classes. Some, like Ubx^+, $Antp^+$ and en^+, are active only in specific developmental compartments. Such selector genes determine the developmental pathways to be followed by particular polyclones. Others, like Pc^+ and esc^+, regulate selector gene expression, ensuring that specific selectors are activated in appropriate compartments. In contrast to the selector genes, whose products must be absent from particular polyclones, the regulatory class of homeotics are active in all compartments. Because their products have to be confined to particular compartments of the embryos, selector genes are not expected to be expressed during oogenesis, whereas regulatory homeotics are active in the female germ line and contribute functional products to the embryo (Lawrence & Struhl 1983).

(e) The existence of conserved homeo boxes between different homeotic genes indicates not only that gene duplication, followed by divergence, has occurred within the genome, but also that distinct homeotic genes must use a similar regulatory protein in performing their normal duties. Indeed, the existence of homologous homeo boxes in vertebrates implies that this principle can be extended. This is supported by the fact that the conserved protein also shows sequence homology with prokaryotic DNA binding proteins, which play key roles in regulating the activity of prokaryotic loci.

(f) Some of the homeotic genes are, in fact, gene complexes arising as a result of localized tandem duplication. Furthermore, some of the primary transcripts they produce are enormous, and even the final processed transcript can be above average length. Finally, processed transcripts of these complexes are found in the embryonic nervous system. It is of interest, therefore, to note that brain-specific processed transcripts of the rat are also above average in their size. This suggests that special attention should be paid to the origin of such large processed transcripts in terms of their developmental and neurological significance.

(g) When a phase of metamorphosis occurs in the life cycle, the developmental programme does not involve the switching of differen-tiated cell types, but rather the destruction of existing cells and their replacement from reservoirs of existing undifferentiated, but already

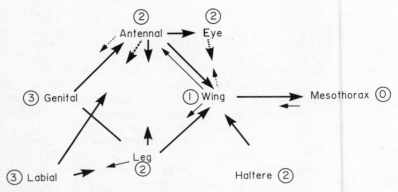

Figure 2.16 Transdetermination pathways in *Drosophila melanogaster*. The length of the arrows is proportional to the frequency of occurrence. Dotted arrows are either very rare or not definitely proven. The encircled numbers indicate the minimum number of steps required to reach the ground, mesothoracic, state (data of Hadorn 1978).

determined, nests of cells. This relates to the fact that programme of gene activity differs substantially among types of differentiated cells, and major switches from one programme to another never take place under normal circumstances. Even under experimental conditions, only limited switches occur and these result from changes in the state of determination. In *Drosophila melanogaster*, for example, imaginal discs from growing third instar larvae not only survive but also increase in size when implanted into the abdomen of adult flies. Moreover, by transferring such discs from ageing adults to younger adults it is possible to establish permanent disc cultures. Such cultures may undergo several different categories of change, included among which is a transformation from one initial disc-specific determined state to another. This type of change is known as transdetermination (Hadorn 1978). Even here, however, the directions and frequencies of the various patterns of transdetermination are not random (Fig. 2.16). Different discs transdetermine at different frequencies and in different directions, with a global tendency towards mesothorax. The important point to note is that, with the exception of the wing disc, it is not possible to go directly from any other state to that of the mesothorax. Developmental pathways in discs thus exist in hierarchies which cannot be short circuited.

Since determination is known to be elective, involving the activation of organ-specific batteries of genes from the total gene inventory, it follows that transdetermination requires a change in the activity of one such a battery to that of another. Most, though not all, of the observed kinds of transdetermination find direct counterparts among the homeotic mutants. Both phenomena are a consequence of the liability of determined states. In the homeotic mutants this instability is a direct consequence of the mutation process, whereas transdetermination occurs in non-mutant genotypes.

2.2.4 Sexual development

Whereas in *Drosophila melanogaster* the patterning of the entire soma is inferred to occur by gradients, that of the germ line precursor is governed by localized determinants (Mahowald *et al.* 1979). At completion of the eighth cleavage division, 3–12 nuclei migrate into a specialized region of ooplasm, located posteriorly in the egg, which is rich in RNA-containing granules. Here, these nuclei form the first cells of the embryo, the pole cells. Pole cells are the primordial germ cells, and at gastrulation they migrate antero-dorsally to the surface of the embryo. They then follow the invagination of the posterior midgut and so become enclosed within its lumen. Finally, they pass through the gut wall to take up their definitive position within the body cavity, where they give rise to the primordial gonad. Not all pole cells become germ cells, since not all of them reach the primordial gonad. Removal or destruction of the pole plasm prevents formation of the pole cells and leads to sterility.

Since the polar plasm of the unfertilized egg is also able to induce the formation of pole cells, the cytoplasmic factors responsible for their determination are evidently produced during oogenesis and become localized in the posterior pole plasm prior to fertilization. Tudor (*tud*) is a maternal effect grandchildless mutation of *Drosophila melanogaster* whose activity during organogenesis is necessary for the production of the primordial germ cells. Progeny of females homozygous for the strong alleles, tud^1 and tud^2, never form pole cells and lack polar granules in the germ plasm. The weak allele tud^4 gives rise to progeny with some functional pole cells, and polar granules of reduced size are formed (Boswell & Mahowald 1985). The properties of these mutations indicate that the gene product of this locus is required for the proper assembly of the germ plasm, and hence for the formation of germ cells.

Thus, at least three determinative factors can be identified in the cytoplasm, of unfertilized eggs of *Drosophila melanogaster*. Two of these are involved in establishing egg polarity, and control the determination of positional information within the embryo by specifying the antero-postero and the dorso-ventral gradients. The third is involved in germ cell determination. Oogenesis in *Drosophila melanogaster* thus involves a transformation of genetic information into a three dimensional structure which is crucial for the development of the fertilized egg.

Male and females of *Drosophila melanogaster* are distinguishable in terms of their sex chromosome constitution. The female has two X-chromosomes while the male has one X and one Y, which differ in size and morphology. Sex determination depends on the ratio of autosomes to X-chromosomes. The Y functions in male fertility but plays no role in primary sex determination. The best evidence for this comes from manufacturing sexual mosaics, or gynandromorphs, by using a ring X-chromosome which is unstable in the early divisions of the zygote. In such X/ringX female embryos, populations of nuclei are produced which are either X0 or X/ringX.

Body parts derived from the X/ringX cell lines are female, whereas those parts derived from X0 lines are male. Such mosaics demonstrate that in somatic cells the ratio of the number of X-chromosomes to the number of autosome sets, the X/A ratio, acts as a critical determinant for establishing the pattern of sexual differentiation. It is customary to assume that this depends on the existence of X-linked female determinants whose products are additive. Consequently, in an XX individual these products reach the threshold necessary to initiate a female mode of sexual differentiation, whereas in an XY individual they fail to do so and male development ensues.

The X/A ratio, however, is neither the only, nor even the primary signal for somatic sexual differentiation (Fig. 2.17). Rather, the wild type allele of the dominant X-linked and male-specific sex lethal (Sxl) is involved both in monitoring the X/A ratio and, as a result of this assessment, then initiating the correct morphogenetic pathway (Cline 1979, Belote *et al.* 1985, Maine *et al.* 1985b). In the development of sexual dimorphism, Sxl is the only locus known to control both sex determination and dosage compensation, and to function in both the germ line and the soma. Evidence in support of this comes from a study of the effects of Sxl on the development of gynandromorphs which indicates that this lethal can cause genetically male, X0 tissue to develop as if it were female. Sxl acts by regulating the expression of the wild type allele of an autosomal gene, transformer (*tra*), whose product is essential for female development. Flies homozygous for the *tra* mutation (*tra/tra*) develop into somatically phenotypic males, regardless of their X/A ratio; tra-2^+ function is also required in the adult for the maintenance of female-specific yolk polypeptide synthesis (Belote *et al.* 1985). Male development thus ensues if the activity of tra^+ is blocked, either by mutation at the locus itself or by the absence of the Sxl^+ allele or its product. The activity of the Sxl^+ itself comes under the control of a maternally synthesized product of the wild type allele of the daughterless (*da*) gene, as well as by a set of genes exemplified by sisterless (Cline 1984, Cline *et al.* 1986). Genetically mosaic females carrying one normal X and a second with various X duplications demonstrate that distal duplications covering some 35% of the X promote female differentiation, whereas much larger proximal duplications of some 60% of the X result in male differentiation (Steinmann-Zwicky & Nöthiger 1985). The strong feminizing effect of the distal duplications originate from a small segment (3E8-4F11) which, when present in two doses, activates Sxl. Thus, both Sxl itself and the distal element are needed in two doses for female differentiation. When the wild type DNA landscape of Sxl is compared with that of 17 female-specific loss of function mutations, DNA lesions in twelve of these identify an 11 kb DNA region required for proper Sxl function (Maine *et al.* 1985a).

Whether the Sxl^+ gene is active or inactive represents the first step in a sequence of binary circuits which begin with the tra^+ switch gene (Belote & McKeown 1986). The developmental pathway is then extended by the action of the wild type alleles of two other autosomal loci – doublesex (*dsx*) and intersex (*ix*) (Baker & Ridge 1980). Doublesex is a bifunctional locus that can be expressed in either of two alternative ways. In male individuals, dsx^+ functions to repress female sexual differentiation, whereas in female individuals it represses male sexual differentiation. Flies homozygous for

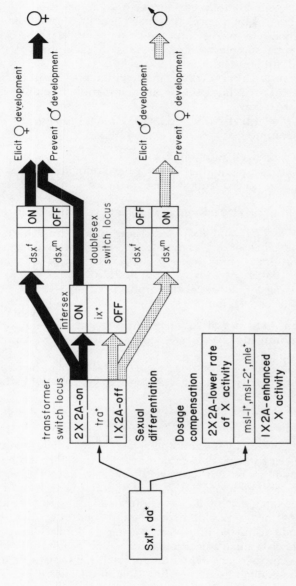

Figure 2.17 Interaction of the regulatory loci concerned with sexual differentiation and dosage compensation in *Drosophila melanogaster*.

the mutant *dsx* mutation transform both potential males and potential females into intersexes. Mutant homozygotes for *ix*, on the other hand, transform potential females into intersexes but have no effect on males. Females simultaneously homozygous for the mutation *dsx* and either of the mutations *tra* and *ix* exhibit a *dsx* phenotype. These three genes thus exhibit a clear epistatic hierarchy of the form *dsx* > *tra* > *ix*, which is consistent with their sequential involvement in a single developmental pathway regulating sexual differentiation. In *Drosophila melanogaster* then, three major regulatory systems are involved in the differentiation of females and males. These control germ line development, somatic sexual differentiation and dosage compensation.

Earlier we have seen that homeotic loci in *Drosophila melanogaster* share three features in common:

(a)   they function in a cell autonomous manner;
(b)   while they begin functioning early in development, their wild type products are also needed late in development for normal differentiation; and
(c)   each controls a binary decision in a given developmental pathway.

The regulatory loci *tra*$^+$ and *dsx*$^+$ share these same three features. Thus, both function in a cell autonomous manner and the transcriptive activity of both is required at pupariation. In addition, at least one of these loci, *dsx*, has a homeo box (Baker, personal communication).

In *Drosophila melanogaster* sexual dimorphism is most striking in three regions of the adult cuticle, namely:

(a)   the fifth and sixth dorsal segments (the tergites) of the abdomen, which are uniformly darkly pigmented in the male but not in the female;
(b)   the most proximal segment of the tarsus of the foreleg (the basitarsus), which in the male possesses a sex comb, a row of 9–14 distinctive bristles; and
(c)   the terminal segments of the abdomen and the genitalia, which differ in structure.

The tergites are produced by the abdominal histoblasts, while the basitarsus arises from the foreleg imaginal disc. Significantly, therefore, both *tra*$^+$ and *dsx*$^+$ can be shown to be transcriptively active in both the abdominal histoblasts and the foreleg imaginal disc, and in both cases the transcription of *dsx*$^+$ continues after that of *tra*$^+$ is completed. Likewise, the effect of temperature shifts on the development of the genital disc show that *tra*$^+$ function is required from the mid-second instar to the early pupal stage. Finally, *tra*$^+$ controls a binary decision as to which programme of sexual differentiation is to be followed. Furthermore, while the somatic dimorphism determined by the genital disc is clearly visible in the external genitila, no less important are the internal derivatives of this disc. In the male, if the internal accessory structures produced by the genital disc do not contact the developing testes, then the testes will not change from an oval to a coiled state, and sterility will ensue. Thus, there is an important inductive

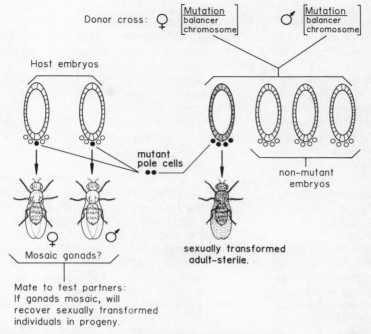

Figure 2.18 Protocol for studying the effect of mutations, known to interfere with normal sexual development of somatic cells in *Drosophila melanogaster*, on germ cell development. Mutant pole cells, are produced from an appropriate cross, and then transplanted into host embryos (data of Schüpbach 1982).

interaction between the products of the disc and the germ line cells themselves.

In monitoring the X/A ratio, the Sxl^+ gene ensures not only that the major gene responsible for male sexual differentiation (dsx^+) remains in the male mode of expression if that ratio is less than 1, but, additionally, that the genes controlling dosage compensation are also activated (Lucchesi & Skripsky 1981). Indeed, the lethal effects of both *da* and *Sxl* mutations can be accounted for through the involvement of both loci in the control of dosage compensation. The genes tra^+, dsx^+ and ix^+, which we have seen to be also necessary for sexual differentiation, exert no influence on dosage compensation. Likewise, the wild type products of the male-specific lethals, $msl-1^+$, $msl-2^+$ and mle^+, which are necessary for dosage compensation in males, have no effect on the development of sexual characteristics (Schüpbach 1982).

Individuals homozygous for all three mutations (*tra, dsx* and *ix*) are sterile and contain only rudimentary gonads. By experimentally transplanting donor mutant pole cells into suitable host embryos, it is possible to construct embryos with mutant germ cells carrying homozygous combinations of the *dsx* locus in an otherwise wild type somatic background (Fig. 2.18). In these mosaics, the germ cells behave according to their chromosomal sex and are in no way affected by the presence of the *dsx* mutant allele. Thus, XX cells give rise only to oocytes, and XY cells only to sperm. Likewise, using irradiation to induce mitotic exchange, it is possible to produce germ cell clones homozygous for the mutations *tra* or *ix*. Here, too, germ cells which are XX and homozygous for either *tra* or *ix* mutations follow the normal female pathway and give rise to normal eggs. Thus, mutations which prevent somatic cells from pursuing normal female development, or from undergoing normal male development, still allow normal differentiation of the germ cells, which implies that the wild type functions of these genes are not needed in the germ line. It also implies that sex determination of the germ cells takes a quite distinct route from that of the somatic cells.

In the nematode *Caenorhabditis elegans*, primary sex determination depends upon the ratio of X-chromosomes to autosomes, though in this case XX individuals are self-fertilizing hermaphrodites, while X0 individuals are male. Recessive mutations of the autosomal gene, *tra-1*, located on chromosome 3, convert XX hermaphrodites into phenotypic males which are often fertile, while dominant mutations at this same locus convert both XX and X0 individuals into fertile females. Stable mutant strains, in which sex is determined not by X-chromosome dosage but by the presence or absence of active *tra-1* genes, can be constructed in the laboratory. In such strains, the chromosome carrying the dominant *tra-1* mutation is, in effect, a W sex chromosome, while a chromosome carrying the recessive *tra-1* mutation is a Z sex chromosome. ZZ individuals are male and ZW individuals female, regardless of the X-chromosome dosage (Hodgkin 1983). Thus, all aspects of the sexual phenotype of *Caenorhabditis elegans*, both somatic and gonadal, can be controlled by the state of the *tra-1* switch locus.

2.2.5 Neurogenesis

The nervous system is, without doubt, the most complex structure to arise during development in *Drosophila melanogaster*. The primary role of a large number of the genes in the *Drosophila melanogaster* genome is to specify the cellular events which lead eventually to the production of what must be regarded as an architecturally, biochemically and electrically precise computer, which combines stereotyped reflex programmes with a variety of more labile learning responses and an information retrieval system which functions as a memory. The central problem in neurogenesis, therefore, is to define how neurons make specific synaptic connections with one another. There are some 10^3 neurons in each of the segmental ganglia of insects and many of these are involved in specific patterns of synapsis. This extraordinary pattern of neuronal specificity is largely achieved during embryonic development by a sequential series of cell recognition events (Goodman & Bastiani 1984, Goodman *et al.* 1984, Bastiani *et al.* 1985).

The central nervous system (CNS) of *Drosophila melanogaster* develops from precursor cells termed neuroblasts. These segregate early in development from a defined neurogenic region of the ectoderm which contains precursors of both neural and epidermal cells. Thus, of the 1800 cells within the neurogenic region of the embryo some 450 actually give rise to neuroblasts (Campos-Ortega 1985). Neuroblasts are first recognizable in the antero dorso-lateral ectoderm, the future brain region, about 4 hours after fertilization. A short time later they appear also on either side of the ventral and dorsal midlines. When these neuroblasts divide, they lose their connection with the ectoderm and produce a small cell, which divides once more to give rise to two ganglion cells, and a large cell, which continues to function as a neuroblast for a further eight divisions. Thus each original neuroblast produces 18 ganglion cells and these form the basis of the CNS. While neuroblast production is normally restricted to particular regions of the embryonic ectoderm, in the presence of a number of neurogenic mutations an extensive area of ectoderm differentiates into nueroblasts (Table 2.6). The net result is a severe hypertrophy of the CNS. This hypertrophy occurs at the expense of epidermal structures and can be interpreted as a switch in ectodermal differentiation.

These neurogenic loci have been shown to be functionally inter-related in an epistatic network of the following kind

$$amx \rightarrow mam \rightarrow neu \overset{\rightarrow Dl \rightarrow}{\underset{\rightarrow N \rightarrow}{}} E(spl) \rightarrow bib$$

with loss of function mutations resulting in the re-routing of ectodermal cells into neurogenesis, so producing an enlargement of the CNS (Brand et al. 1986). A further class of mutations have been described which also lead to hypertrophy of the CNS (Smouse et al. 1986). Here the developmental lesions result from the continued division of either neuroblasts or ganglion mother cells. At least ten such loci have been isolated, including kayak, canoe, topless and punt. All of these subsequently lead to problems in the dorsal closure of the embryo, and it is from this that they derive their names.

Table 2.6 Neurogenic mutations leading to hypertrophy of the CNS in *Drosophila melanogaster*. Note, all loci other than big brain have maternal and zygotic expression. In big brain there is no maternal expression. (After Lehmann *et al.* 1981, Knust *et al.* 1984.)

Locus	Location	Acetylcholinesterase activity (arbitrary units)	Approx. no. of nerve cells per embryo
Notch (N)	(3C7)	310	17 000
almondex (amx)	(8D)	–	–
big brain (bib)	(30A9, 30F)	240	15 000
mastermind (mam)	(50C)	200	11 000
neuralized (neu)	(85C)	380	16 000
Delta (Dl)	(92A2)	380	18 000
Enhancer of split (E(spl))	(96F5–7)	–	–
wild type		100	6000

Additionally, there are antineurogenic loci which map predominantly to the tip of the X-chromosome (1B). These are associated with the Achaete-Scute Complex, lethal EC4, embryonic lethal abnormal vision (*elav*) and ventral nervous system defective (*vnd*). There is also an autosomal antineurogenic locus, denervated, in region 31AB. Loss of function mutations at these antineurogenic loci lead to misrouting of putative neuroblasts into epidermogenesis, and so result in a reduced CNS.

In general, the wild type functions of these neurogenic loci are required not only during embryogenesis when neuroblasts are undergoing active division, but also in larval and pupal stages when the neuroblasts are again active. In the case of Notch (*N*), molecular cloning studies suggest that the entire locus spans some 40 kb of DNA, and produces a 10.5 kb poly (A)$^+$ RNA which is present both during the first half of embryogenesis and subsequently in the larval and pupal phases (Artavanis-Tsakonas *et al*. 1983, Yedvobnik *et al*. 1985). Big brain (*bib*) is exceptional in this respect, since its function appears to be restricted to early neurogenesis (Knust *et al*. 1984).

Part of the Notch locus encodes multiple repeats of a motif, termed *opa*, which is equivalent to the M repeat found in the Antennapedia Complex and is also present in a number of loci containing homeo boxes. Computer searches reveal that another Notch repeated unit shares sequence homology with the epidermal growth factor (EGF) of mammals (Wharton *et al*. 1985a,b). While the precise function of this factor is not known, it does exert pleiotropic effects, which include the induction of mitogenic activity as well as the stimulation of differentiation in cells of ectodermal and mesodermal origin.

Detailed DNA sequence analyses reveal that the major Notch transcript is 10 148 nucleotides long, with a 8109 nucleotide open reading frame. This codes for a 2703 amino acid transmembrane protein (Wharton *et al*. 1985a,b). Part of the extracellular domain of this protein has homology to mammalian epidermal growth factor, while part of the intracellular domain is homologous to nucleotide phosphate binding sites (Artavanis-Tsakonas *et al*. 1986). The sequence data also reveal that the *opa* repetitive element is included in the open reading frame, and would thus be translated into a glutamine-rich run.

The most striking feature of the protein is that nearly half of it consists of 36 tandemly arranged 40 amino acid sequence units in which there are six cysteines at specific intervals. It is this cysteine-rich module that picks up homology to eight mammalian proteins as well as to a viral protein (vaccinia viral growth factor, VVGF). The mammalian homologies are to the EGF precursor, transforming growth factor (TGF), blood coagulation factors IX, X and protein C, plasminogen activator, urokinase and the LDL receptor protein. The functional significance of such homologies is at present unknown, but it may be that these invertebrate and vertebrate sequences shared common ancestors. If Notch is indeed a transmembrane protein, it may well be involved in cell–cell interactions that determine the normal differentiation of ectoderm.

Finally, a number of other neurogenic genes such as mastermind (*mam*) Delta (*Dl*) and Enhancer of split [*E(spl)*] have also now been cloned (Knust *et al.* 1987a,b, Vassin & Campos-Ortega 1987). All show homologies to Notch. Not only are *opa* sequences found in *mam, Dl* and *E(spl)*, but *Dl* and *N* share epidermal growth factor-like sequences. *E(spl)* and *mam*, however, do not appear to possess the EGF-like sequences.

Knust *et al.* (1987b) have also used parts of the Delta sequence to screen genomic and cDNA libraries for cross homologies, and a number of clones have been isolated which carry putative EGF sequences. One clone maps by *in situ* hybridization to 95F and is near to, or maybe even at, the site of a previously reported gene affecting neurogenesis, termed Enhancer of split mimic.

The CNS of the larva is bilaterally symmetrical, consisting of brain hemispheres and a central nerve cord. Thirteen major ganglia can be distinguished, including eight abdominal and three thoracic, together with a sub- and a supra-oesophageal ganglion. While the CNS does not change in shape during the three larval stages, it does increase some 30 times in volume. Gross morphological changes occur in the CNS during the pupal phase. At this time the components of the ventral nerve cord coalesce, to give rise to a pair of compound ventral ganglia. This process, which combines the central neurons of the embryonic thorax and abdomen, involves a marked shortening of the nerve cord.

These events are evidently under genic control, since in flies where the Bithorax Complex is lacking, for example, the embryonic segmental ganglia fail to condense. The mutation, ventral nervous system defective (*vnd*), also results in a non-condensed and poorly organized ventral nervous system (Hall 1982). Two other mutations, fushi tarazu and paired, give rise to a ventral nervous system which has, at most, only half the number of ganglia normally present in early embryos.

The CNS continues to develop throughout the larval phase. This involves the addition, deletion and spatial redistribution of neurons, leading to a remodelling of the nervous system, and especially of the brain, in anticipation of its adult functions, some of which are strikingly different from those of the larva. This is especially clear in the optic centres of the brain, and this despite the fact that the compound eyes, which these centres serve, are as yet undeveloped in the larva. Evidently, the information required for the establishment of adult connections is, in some sense, already programmed in the larval CNS.

In the tobacco moth, *Manduca sexta*, some of the neurons which have developed to serve larval functions do not degenerate during metamorphosis. Instead they redifferentiate, undergo morphological and synaptic reorganization, and assume a role in the innervation of new adult muscles. In one particular identified motor neuron, this switch has been shown to be accompanied by a marked alteration in dendritic morphology (Truman & Reiss 1976, Levine & Truman 1982).

The sensory neurons, as well as the cells that produce their corresponding cuticular sense organs, arise peripherally from epidermal cells. Since the cell bodies of these neurons remain immediately below the epidermal region from which they originate, their segmental nature is at once obvious. Moreover, under normal circumstances, the axons of these peripheral neurons join with others from neighbouring sensory neurons of the same body segment to form a nerve trunk which passes to the CNS. In grasshopper embryos, the first pair of nerve cells to appear in the limb buds project axons along the length of the limb to the CNS. The route navigated by these pioneer neurons (Bate 1976) is subsequently followed by other neurons and eventually becomes a major nerve trunk. The guidance cues which delineate this route are provided by a system of guide-post cells, which the pioneers use to navigate their channel to the CNS (Bentley & Keshishian 1982, Ho & Goodman 1982). Given the conservative nature of the arthropod nervous system, which seems to be constructed using a common embryonic plan, equivalent pioneer neurons must also operate in the fly. Thus, as Thomas *et al.* (1984) have emphasized, the early *Drosophila melanogaster* embryo is a miniature replica of the grasshopper embryo. Moreover, in *Drosophila melanogaster* the larval nerves, which connect each imaginal disc to the CNS, may provide equivalent guides when the adult sensory axons differentiate at metamorphosis (Palka & Ghysen 1982).

At the molecular level, we do not know how major neuronal connections are specified during development. A first obvious step is to search for mutations which differentially disrupt specific synapses in a defined pathway. In a number of neural mutants in *Drosophila melanogaster*, only a small number of cell types are frequently affected. Such a specificity implies that each mutation affects only a subset of the whole. This is clearest in the case of the giant fibre system which is activated by a 'light-off' visual stimulus and mediates the early events of an escape response consisting of a jump and subsequent flight (Wyman *et al.* 1984, Bacon & Strausfeld 1986). This giant fibre is thus a command interneuron which drives the mesothoracic jump and flight muscles via connections with identified neurons. This system involves a simple network of eight neurons. Each has its cell body in the brain, but sends a long axon that passes unbranched into the mesothorax. Here it makes an electrical synapse with another interneuron, the peripherally synapsing interneuron (PSI), which, in turn, makes chemical synapses with the five motor neurons that innervate the contralateral dorsal longitudinal flight muscle (DLM). It also makes another electrical synapse with the motor neuron that enervates the tergotrochanteral (TTM) or jump muscle (Fig. 2.19).

At least three distinct mutations are known which affect this system and lead to the failure of the fly to carry out an escape response (Thomas & Wyman 1984a). In bendless (*ben*) the defect is at the giant fibre-tergotrochanteral motor neuron junction. In *gfA* the tergotrochanteral muscle is driven normally, but the dorsal longitudinal flight muscle is not. Here, therefore, the defect lies either in the giant fibre output to the peripherally synapsing interneuron, or else in the interneuron itself. Finally, in Passover (*Pas*) neither muscle response is normal (Table 2.7).

The information available for this system at the structural and genetic levels is now being refined genetically and molecularly (Baird *et al.* 1986, Miklos *et al.* 1987b), and the loci of interest are being finely mapped at the DNA level. The regions so described are likely to contain a number of RNA transcripts, only some of which will code for the particular proteins

Key: —cell body
     —⊢ electrical synapse
     ◀ chemical synapse

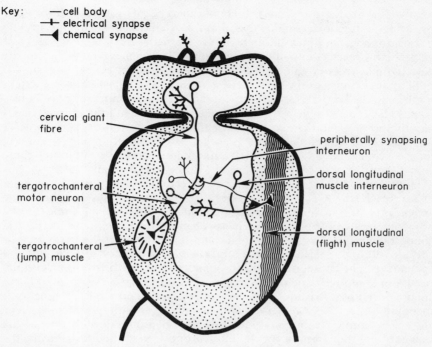

cervical giant
fibre

peripherally synapsing
interneuron

dorsal longitudinal
muscle interneuron

tergotrochanteral
motor neuron

tergotrochanteral
(jump) muscle

dorsal longitudinal
(flight) muscle

Figure 2.19 The giant fibre pathway in *Drosophila melanogaster*, in which the cervical giant fibre makes synaptic contact with both a tergotrochanteral motor neuron and a peripherally synapsing interneuron. Only one side of the bilaterally symmetrical system is shown (data of King & Wyman 1980).

Table 2.7 Mutations affecting the giant fibre pathway controlling the indirect flight muscles of *Drosophila melanogaster* (from Thomas & Wyman 1984a).

		Giant fibre (GF) outpath defect	
Mutant	Locus	tergotrochanteral muscle (TTM) innervation	dorsal longitudinal muscle (DLM) innervation
bendless (*ben*)	12A6–12D3	GF–TTM motor neuron junction defective	–
gfA	18A5–18D1	–	transmission between either GF and the peripherally synapsing interneuron (PSI) or between PSI and DLM defective
Passover (*Pas*)	19D1–20A2	GF–TTM motor neuron junction defective	transmission between either GF and PSI or PSI and DLM defective

involved in the jump response. It should then be possible to identify these by using a rescue technique based on P element mediated germ line transformation, in which different DNA segments from the landscapes of interest are injected into mutant embryos. The clones which rescue the defective phenotypes will then define the RNA transcripts of interest.

The connectivity of the nervous system must also depend to some extent on the cues laid down during segmentation. Thus, in mutant bithorax flies (abx, bx^3/DF(3)P2) where the metathoracic segment has the characteristics of a mesothorax, so that the fly has in effect two mesothoraces in tandem, the giant fibre makes its normal branches in the mesothoracic region but then continues beyond its normal termination point and duplicates these branches in the converted metathorax (Thomas & Wyman 1984b).

Homeotic mutations lead to groups of ectopic sensory neurons whose axons reach the CNS at abnormal sites. If wild type and homeotic appendages are situated close together in the body, then some of the homeotic sensory fibres are usually able to find their correct central connections. If, however, they are located at a distance from their wild type site, the fibres terminate in the central region appropriate to their replaced appendage. The homeotic mutation, spineless aristapedia (ss^a), for example, results in a replacement of the antennal arista by tarsal segments. These antennal tarsi bear bristles, at the base of which are nerve cells. Axons from these join to form a small nerve bundle which joins with the antennal nerve and runs into the brain. Not only do neurons develop from the homeotically transformed cuticle, but a proportion of them behave like the tarsal chemoreceptors of normal flies. Thus, when stimulated by a sucrose solution they may initiate a reflex which leads to extension of the proboscis (Deak 1976). That is, the homeotic tarsus responds as if it were a normal leg tarsus, and ss^a is evidently capable of producing a sensory apparatus able to transmit information which, in turn, can be correctly interpreted and used in the CNS of the fly.

Over 200 mutations have been described which influence most aspects of neural development in *Drosophila melanogaster*, a small sample of which are shown in Table 2.8. These include mutations affecting neurotransmitter metabolism, sensory transduction and the protein channels used to regulate the flow of ions between cells and along axons. Since the development of all these circuit components of the neural system, as well as the circuits themselves, are under genic control, it is to be expected that a mutation which affects the structure or function of any of these components may also lead to altered behaviour.

The precise anatomical site at which such behavioural mutations exert their primary effect can be determined by the use of appropriate genetic mosaics. Such mosaics are composite individuals in which some tissues are mutant but the rest have a normal genotype and phenotype. It is, therefore, possible to decide which part of the body needs to be mutant to produce the altered behaviour. Genetic mosaics also allow tissues carrying a lethal mutation to survive past the stage at which the organism would normally die.

In *Drosophila melanogaster*, commonly exploited mosaics are sex mosaics,

Table 2.8 Representative neurological and behavioural mutants of *Drosophila melanogaster* (from Hall 1982).

Mutant	Location	Effect
Early neural development		
Notch (*N*)	X–3	
big brain (*bib*)	2–34	Neural hypertrophy and reduction of hypodermis
mastermind (*mam*)	2–70	
Visual system structure		
glass (*gl*)	3–63	Ommatidia disarrayed and rhabdomeres of photoreceptor cells absent
optomotor-blind (*omb^{H31}*)	X–7	Giant arborizing fibres in lobula plate optic lobe absent or greatly reduced; adults impaired in optomotor turning responses
small optic lobes (*sol*)	X–60	Medulla, lobula and lobula plate optic lobes reduced in volume and cell number; specific cell types missing in medulla
Physiological and motor		
comatose (*com*)	X–40	Larvae or adults paralyzed when exposed to high temperature
ether-a-go-go (*eag*)	X–50	Leg shaking induced by ether; aberrant potassium conductances
Shaker (*Sh*)	X–58	
Neurotransmitters		
Acetylcholinesterase (*Ace*)	3–52	Reduced or no acetylcholinesterase (AChE) activity
Dopa decarboxylase (*Ddc*)	2–54	Reduced or no dopa decarboxylase activity; do not synthesize serotonin or dopamine in CNS
General behaviour		
Giant fiber-A (*gfA*)	X–62	Unable to jump or fly; physiology of 'giant fiber' pathway is impaired
Learning and memory		
dunce (*dnc*)	X–4	Seemingly blocked or impaired in learning, in reality learn then rapidly forget
amnesiac (*amn*)	X–63	Abnormal memory
Reproductive behaviour		
cacophony (*cac*)	X–34	Males have aberrant courtship song and court abnormally
celibate (*cel*)	X–48	Males court vigorously but rarely even attempt to copulate and are behaviourally sterile

or gynandromorphs. As we have seen, one method used to generate these depends upon the properties of a specially constructed and unstable ring X-chromosome. This chromosome is frequently lost during the early cleavage divisions and this leads to the production of embryonic nuclei of two kinds, X/ringX and X0. Following further division these ultimately migrate to the surface of the egg to form the blastoderm. Since nuclei tend to retain proximity to their sister mitotic products, female, X/ringX, cells tend to populate one part of the blastoderm with male, X0, cells occupying the remainder. Since the site occupied by a cell in the blastoderm largely determines its ultimate fate in the developing embryo, a variety of arrangements of male and female parts can be produced in the adult gynandromorph, both externally and internally. The external divisions often follow intersegmental boundaries and the longitudinal midline of the fly (Fig. 2.20), since each body part is produced independently from a specific imaginal disc which, in turn, is derived from a specific part of the blastoderm.

If the X-chromosome of such gynandromorphs is marked genetically with both a morphological mutation and a behavioural mutation, while the ringX carries wild type alleles at both loci, then X/ringX body parts will appear not only female but also wild type, while X0 body parts will appear male and can be tested for phenotypic normalcy. By selecting from a random pool of gynandromorphs, it is possible to choose individual flies in which, for example, the head is normal and the thorax mutant, or vice versa. Alternatively, the organism may be a complete or a partial bilateral mosaic.

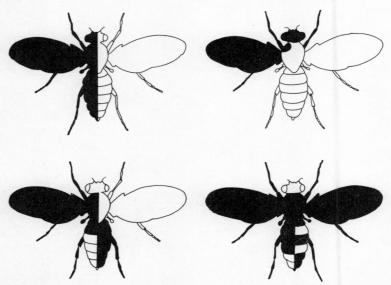

Figure 2.20   Typical *Drosophila melanogaster* gynandromorphs. Body parts derived from XX cell lines (solid) are female. Those derived from X0 lines (open) are male (data of Hotta & Benzer 1973).

The value of such mosaics is that they allow a decision to be made about what part of the body needs to be mutant for the abnormal behaviour to be expressed. Thus, the mutant, stoned (*stn*), exhibits aberrant locomotion at low temperatures and is completely immobilized at high temperatures. Furthermore, it has an abnormal jump response when exposed to a sudden 'light-off' stimulus (Kelly 1983). When only the head is mutant in a mosaic stoned fly, the jump response can be shown to be abnormal, indicating that the primary source of the abnormal behaviour lies in the head and not in the legs. Similarly, in Hyperkinetic (*Hk*) flies, which display shaking of the legs under ether anaesthesia, it is not the brain that is abnormal but the neurons of the ventral ganglia, since mosaic flies with normal head tissue but mutant thoracic tissue display characteristic hyperkinetic activity. Neurophysiological studies of this mutant indeed identify abnormal recordings from the ventral ganglia of hyperkinetic flies (Ikeda & Kaplan 1970). Since the shaking response of each leg is independent of all others, it is clear that each leg has its own separate focus of control. Sex mosaics are especially instructive in cases where the same behavioural response may result from different causes. In flies which are flightless, for example, the first decision to make is whether it is the wing musculature which is defective or the innervation of that musculature. If the primary defect is in innervation, it is then necessary to decide whether this is due to the lack of correct input from the brain or to a defect in the thoracic nerves.

A combination of behavioural mutants and sex mosaics have been used to analyze the neurobiological basis of mating in *Drosophila melanogaster* (Fig. 2.21). Both visual and olfactory cues, provided by the female, are important in initiating male courtship behaviour (Hall 1977, 1985). Thus, when the female is immobilized through the use of the temperature-sensitive shibhire (*shi*) mutation, or when the male carries the optomotor-blind mutation (*omb*), which leads to generally defective turning responses, there is a dramatic depression in male courtship response. Moreover, gynandromorphs are courted by males only if these mosaics have abdomens that are wholly or partly female. Mutant smellblind (*smb*) males, which are unable to respond to pheromones produced by the female abdomen, also show an abnormal courtship response similar to that of visually deprived males. Courtship begins by the male orienting towards and tapping the female. As he follows her, he carries out a ritualized wing display. This involves a unilateral extension of a wing and vibration of it to produce a courtship song. A sex mosaic only exhibits male courtship behaviour if the dorsal posterior brain is male on one side or both. Even so, the courtship sequence is not completed if only this portion of the CNS is male. Using the mutant, cacophony (*cac*), which produces an abnormal courtship song, it is possible to show that X0 neurons are required on one side or other of the thorax for normal song production by either wing. Male proboscis extension, which occurs after wing display and song but before attempted copulation, is also controlled by the brain. However, male tissue is almost always required to be present on both sides of the brain if licking is to occur.

A further pair of mutations, celibate (*cel*) and fruitless (*fru*), both block attempted copulation, while in the presence of the mutation, coitus

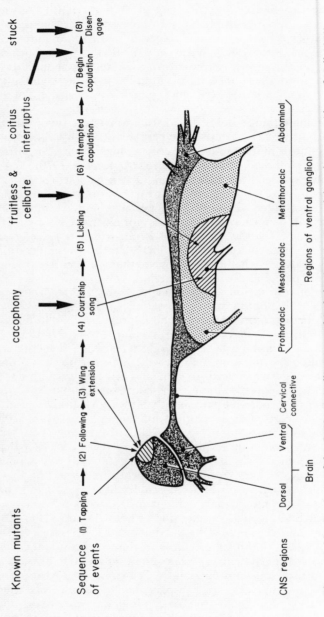

Figure 2.21 Regions of the CNS controlling courtship behaviour in *Drosophila melanogaster* (data of Hall 1979).

interruptus (*coi*), males copulate for only about 60% of the normal time. Finally, mutant stuck (*sk*) males frequently cannot disengage after copulation, or take much longer to do so. Analysis of female receptivity in gynandromorphs suggests that anterior tissues must be XX for a mosaic having female genitalia to accept copulation. However, a completely female brain as well as totally female genitalia are not sufficient for normal receptivity in a gynandromorph (Hall & Greenspan 1979). Whether there are sex-specific differences in the synaptic connections involved in the neural foci responsible for the differential behaviour of the two sexes during courtship is not known. What is known, however, is that sexual differences do occur. For example, the connection of the male lobula giant neuron is different in male and female flies (Strausfeld 1976).

One of the key general questions that needs to be answered in developmental neurobiology is how specific neuronal connections are made within the CNS. In vertebrates, monoclonal antibodies have already shown that neurons which look alike need not be functionally identical (Rougon *et al.* 1982). Moreover, as will be seen later, the high complexity of brain RNA populations (see § 3.3.4.4), does not result simply from the summation of many different cell types but rather from differences between the neuronal nuclei themselves (Ozawa *et al.* 1980). The cloning and sequencing of neurologically important genes will allow not only for a range of neural defects to be put into a molecular perspective but, additionally, will make it possible to genetically dissect the molecular subunits of different neural circuits. Furthermore, the availability of defined transformation systems in *Drosophila melanogaster* means that specific segments of genes which have been mutagenized *in vitro* can be reintroduced into the genome and their effects systematically tested.

One well-defined aspect of the specificity of neurons is their distinctive signalling capabilities. These derive from two families of specialized membrane proteins, called channels and pumps, that control the passage of ions across the cell membrane. Pumps actively transport ions against an electrical gradient. Channels allow ions to move rapidly along an existing electrochemical gradient (Salkoff & Tanouye 1986). Neurons vary in the type of channels they possess; even different regions of a single neuron can have different types of channel (Kandel 1983). The mutation, Shaker (*Sh*), for example, yields flies which produce jerky and erratic movements in which there is a loss of motor control. This is because the Shaker locus is thought to code for one molecular component of the potassium channel that carries a fast transient outward current, I_A (Wu *et al.* 1983). Thus, although there are three separate potassium currents, as well as a calcium current and a glutamate-dependent synaptic current, in the dorsal longitudinal flight muscles of *Drosophila melanogaster* only I_A is perturbed by *Sh*. A second mutation, ether-a-go-go (*eag*), affects the potassium current, I_i. Both *Sh* and *eag* affect transmitter release at the neuromuscular junctions (Salkoff & Wyman 1983).

A large number of genes affecting the construction and function of the *Drosophila melanogaster* nervous system at different developmental time windows have now become amenable to recombinant DNA analyses, either because of the more extensive use of mutagenic screens, the use of microdissection and microcloning techniques, or else because of complementation of mutant phenotypes using the P element mediated germ line transformation methods developed by Rubin and Spradling. This powerful combination of genetic and moleculer genomic dissection, coupled with routine genomic re-entry techniques, has meant that the nervous system of *Drosophila melanogaster* is now, literally, poised for exploitation. The following examples provide a cross-section of the loci and approaches being used, and illustrate why the molecular neurobiology of *Drosophila melanogaster* has become such an explosive and exciting field.

The visual system and photoreceptor function – rhodopsins. The first *Drosophila melanogaster* rhodopsin gene to be cloned was denoted *Rh1* which is expressed in photoreceptor cells R1–R6 (O'Tousa *et al.* 1985, Zuker *et al.* 1985). Subsequent P element mediated transfer of a wild type *Rh1* gene into the germ line of *Drosophila melanogaster* restored the phenotype of the mutant, *ninaE*, to wild type, so that *ninaE* was the locus for *Rh1*. By using *Rh1* sequences as a probe, a second rhodopsin gene (*Rh2*) was isolated and found to be restricted in its activity to photoreceptor cell R8. The production of oligonucleotide probes, based on the sequences of *Rh1* and *Rh2*, then allowed the isolation of two further opsin genes, *Rh3* and *Rh4*, which were found to be expressed predominantly in the ultraviolet sensitive photoreceptor cell, R7 (Cowman *et al.* 1986, Montell *et al.* 1986, Zuker *et al.* 1986). The isolation of all four genes has meant that they can be genetically engineered *in vitro* and reintroduced into the genome. For example, genotypes have now been constructed in which *Rh2*, instead of being exclusively expressed in photoreceptor cell R8, is expressed in cells R1–R6. This combination of *in vitro* mutagenesis and germ line transformation has thus opened the way for detailed analysis and manipulation of the spectral properties of the compound eye.

The *trp* gene (transient receptor potential). The *trp* mutant behaves as if blind in bright light, although it has normal vision in low ambient light. However, the amount of rhodopsin present in the eye, and its absorption characteristics, are normal. The ultrastructure of the photoreceptors also appears normal at eclosion, but the rhabdome progressively develops signs of degeneration. It is thought that the *trp*$^+$ protein may be involved in catalyzing the reassociation of transducin subunits, which dissociate after binding photoactivated rhodopsin. To test such an hypothesis, however, requires detailed molecular characterization of the locus, some of which has now been achieved. The *trp* gene (99C) has been cloned, and the mutant phenotype restored to wild type function by P element mediated germ line transformation of a DNA fragment carrying the *trp*$^+$ region (Montell *et al.* 1985). The amino acid sequence of most of the protein has been determined, showing that it contains a number of regions of repetition. One sequence, Asp-Lys-Asp-Lys-Lys-Pro-Gly/Ala-Asp, is repeated nine times in a tandem array. The protein is localized specifically in photoreceptor cells (Montell *et al.* 1986), and the way is now open to modify its structure and proceed with elucidating its biological roles.

The *ninaC* gene (neither inactivation nor after potential). The *ninaA* to *ninaE* mutants were originally described as having abnormal electro-retinograms and reduced rhodopsin contents, and the *ninaC* mutation (27E–28B), in particular, was found to interfere with rhabdomere structure. Specifically, the cytoskeletal filament system, which usually is found in the axis of each microvillus, is greatly reduced in *ninaC* mutants. The gene may thus be involved in the formation of the cytoskeletal structure of rhabdomeral microvilli (Matsumoto *et al.* 1986). The gene was cloned by rescue of the *ninaC* mutant phenotype via the P element mediated germ line transformation method. Antisera to the *ninaC*$^+$ product have revealed two proteins, both of which are restricted to photoreceptor cells. Each of these *ninaC* proteins has been found to contain two domains, one with homology to protein kinases, and the other to myosin heavy chains (Montell *et al.* 1986).

Biological clocks – the period gene (*per*). Despite intensive research, little is known about the nature and function of biological oscillators which determine rhythmic activities in biological systems. However, genetic and molecular analysis of the *Drosophila melanogaster* clock locus, termed period (*per*), has now provided a solid basis for the analyses of biological rhythms. Mutations at this locus (3B1–2) cause abnormally short (*per*^{s}) or long rhythm periods, or else lead to arrhythmic phenotypes (*per*^{o}). The use of P element mediated germ line transformation has allowed the *per*^{o} phenotype to be restored to wild type using DNA from the *per*$^+$ region (Young *et al.* 1985). The rescuing fragment encodes a 4.5 kb RNA which controls the array of rhythmic functions that come under the influence of this clock gene. The *per* locus has been cloned and sequenced (Reddy *et al.* 1984, 1986), and the 4.5 kb RNA has been found to localize in the embryonic nervous system, as well as appearing later in pupae and adults. Interestingly, part of the *per*$^+$ transcript codes for 20 consecutive pairs of Gly-Thr residues and this part of it has significant homology to a chondroitin sulphate proteoglycan (Jackson *et al.* 1986, Reddy *et al.* 1986).

The generation of mosaics has also provided crucial insights into the system. Mosaic flies in which the brain genotype is *per*$^+$ but in which the thorax is genotypically *per*^{s} have a normal 24-hour activity rhythm but a short courtship song cycle. This demonstrates that the two rhythm phenotypes can be uncoupled, and indicates that the brain and thorax may have independent oscillators (Kyriacou & Hall 1987).

Mutations causing shaking or paralysis. A number of mutants in *Drosophila melanogaster* are paralysed for several minutes when subjected to mechanical shock and this 'bang sensitivity' is thought to be due to hyperexcitability of nerves. Some paralytic phenotypes can be suppressed by other unlinked mutations, such as *nap* (no action potential) in which action potentials can be blocked. Three bang sensitive mutations, easily shocked (*eas*), bang senseless (*bss*), and technical knockout (*tko*), have now been cloned (Jones *et al.* 1986, Royden & Jan 1986). The first two have been localized to 14BC and have landscapes of about 20 and 30 kb respectively, while a lethal technical knockout allele has been rescued by P element mediated germ line transformation. Genomic DNA, as well as cDNAs, have been sequenced from the *tko* region, and the predicted open reading frame has revealed excellent homology to the S12 ribosomal protein of both the *Euglena gracilis* chloroplast gene and the equivalent *Escherichia coli* gene (Royden & Jan 1986).

The Shaker gene (*Sh*).   Mutations at this locus (16F) cause spasmodic tremors of the legs and abdomen when mutant individuals are subjected to ether anaesthesia. The Shaker Complex (*ShC*) is thought to be a prime candidate for a voltage-dependent potassium channel, since it has been implicated in the normal functioning of a potassium current in nervous tissue and muscle (Jan *et al.* 1985). It has been subjected to extensive genetic analysis and found to consist of at least two parts, a maternal effect area and a haplolethal region (Ferrus *et al.* 1987). Although it has been analyzed physiologically and molecularly, the locus has so far resisted total molecular characterization, possibly because of its large size. By using both cosmid and bacteriophage libraries, Tanouye (1986) and Tempel *et al.* (1986) have characterized genomic areas of about 300 kb and 200 kb around the Shaker Complex, and have found that Shaker translocation breakpoints map over at least a 65 kb interval. The analyses of cDNA clones are at present being carried out in a number of laboratories.

Genetic studies have led to the isolation of an unlinked gene, designated tetanic (*tta*), at 14AB. This may well be a positive regulator of the haplo lethal region of the Shaker Complex. While tetanic is a viable allele, it is sensitive to *ShC* dosage, so that *tta ShC/ShC* males, in which the *ShC* is present as a duplication, are lethal (Ferrus *et al.* 1987). Clearly, owing to the manipulative genetic system of the fly, the beginnings of an interactive network are already at hand. Finally, a structural brain mutant, minibrain (*mnb*) has recently been isolated using P element mutagenesis (Fischbach *et al.* 1987). In this mutant the volume of the optic lobes, as well as that of the central brain, is greatly reduced. Interestingly, minibrain maps to the maternal effect region of the Shaker Complex and adds further complexity to this area.

Another locus, shaking-B (Homyk *et al.* 1980), with a phenotype similar to Shaker, namely ether anaesthesia–induced leg shaking, has been mapped to 19E3 (Miklos *et al.* 1987a), and entry points to the gene region have been isolated following microdissection and microcloning (Davies *et al.* 1986).

Sodium channel functions.   By using a portion of a cloned vertebrate sodium channel gene as a probe, a gene, which has close homologies to its vertebrate counterpart, has been isolated from genomic and cDNA libraries of *Drosophila melanogaster*. The gene has an autosomal location at 60DE and is near to, but not at, the site of the seizure locus which causes temperature-induced paralysis. Approximately half of this *Drosophila melanogaster* sodium channel gene has already been sequenced, and the amino acid homology between it and its vertebrate counterpart is near 80% (Salkoff *et al.* 1986). Future reintroduction of *in vitro* mutagenized forms of this gene will undoubtedly permit detailed *in vivo* dissection of sodium channel functions.

The paralytic gene (*para*). Mutations at this X-linked locus can cause temperature-induced paralysis, and the mutant gene is thought to disturb some component of the normal functioning of sodium channels. The gene has been cloned using P element induced mutagenesis, and over 100 kb of the *para* landscape has been isolated (Ganetsky *et al.* 1986). Lesions which give rise to the paralytic phenotype extend over at least 90 kb of DNA, so the gene appears to be large. Interestingly, when double mutant combinations are made between *para* and either of two other genes which are believed to be involved in sodium channel functions, *nap* (no action potential) and *tip-E* (temperature induced paralysis), the double homozygotes are lethal.

Genes involved in synaptic transmission. The process of cholinergic synaptic transmission depends predominantly on the co-ordinated action of the neurotransmitter biosynthetic enzyme choline acetyltransferase, acetylcholinesterase and the acetylcholine receptor.

The choline acetyltransferase gene (*ChAT*) has been cloned as a cDNA, and parts of the sequence exhibit good homology with the rat neuronal α-acetylcholine receptor subunit (Salvaterra *et al.* 1986). Since there is good homology between the acetylcholinesterase gene of *Torpedo californica* and *ChAT*, it may well be that these catabolic and anabolic enzymes are duplication products of the same ancestral gene.

The acetylcholinesterase gene (*Ace*) codes for one of the components of the cholinergic synapse, and is located in the 315 kb zone of the rosy–Ace walk at 87DE (Bender *et al.* 1983a, Spierer *et al.* 1983, Gausz *et al.* 1986). Its genomic landscape was thus already available and this facilitated the isolation of relevant cDNAs. DNA sequence analyses of such cDNAs have revealed a 650 amino acid open reading frame which has extensive sequence homology to the acetylcholinesterase of *Torpedo californica*. The *Drosophila melanogaster* sequence also contains the active site Phe-Gly-Glu-Ser-Ala-Gly, which establishes that *Ace* is indeed the structural gene for acetylcholinesterase (Bossy *et al.* 1986).

A nicotinic acetylcholine receptor gene has been obtained by screening an adult head cDNA library of *Drosophila melanogaster*, using a cDNA from the γ-subunit of *Torpedo californica* as a probe. The cDNAs which have been isolated code for a *Drosophila melanogaster* neuronal acetylcholine receptor protein. DNA sequencing analysis reveals a predicted protein of 497 amino acids which has excellent homology to a rat neural α-acetylcholine receptor (Hermans–Borgmeyer *et al.* 1986). However, since the *Drosophila melanogaster* protein does not have the two adjacent cysteines, which are characteristic of all known ligand binding α-subunits, it is probably a non-α acetylcholine receptor subunit.

A gene for what may be another subunit of the acetycholine receptor has been isolated by using a chicken α-subunit as a probe. The predicted protein from this *Drosophila melanogaster* cDNA does have extensive sequence homology to the α-subunits of vertebrates, including homologies at the presumed acetycholine binding site. The gene has been mapped *in situ* to region 96A (Bossy *et al.* 1986).

The Dopa decarboxylase gene (Ddc). The *Ddc* gene is found in region 37C1,2 of the genome, where there is a cluster of approximately 10 genes in a 25 kb interval. These genes are mainly engaged in cuticle production and catecholamine metabolism. Over 100 kb of DNA surrounding the *Ddc* gene has been cloned (Gilbert *et al.* 1984). The primary *Ddc* transcript has been characterized and is known to be spliced to yield two different RNAs. The *Ddc* RNA used in the central nervous system has an additional exon to the *Ddc* RNA used in other tissues, and consequently the protein it produces may well be functionally different to that made in non-nervous tissue (Morgan & Hirsh 1986).

The molecular map of the genes adjacent to *Ddc* has been determined by Eveleth & Marsh (1986), and the order from the centromere is *Cc, Ddc, Cs* and *amd*. The *amd* gene (alpha-methyldopa) is functionally related to *Ddc* and is presumed to play a role in cuticle formation. The *Cc* gene, whose DNA sequence and pattern of expression has been analyzed by these same authors, is also likely to function in cuticle synthesis.

The dunce gene (dnc). This gene, in its mutant state, interferes with learning and memory processes and perturbs cAMP metabolism and female fertility. It is the structural form of a cAMP phosphodiesterase, has been almost completely cloned, and has at least 13 exons extending over 100 kb (Chen *et al.* 1986, R. L. Davis, personal communication). The predicted open reading frame has homologies to certain proteins of rodents and humans, but particularly to a calcium/calmodulin-dependent cyclic nucleotide phosphodiesterase. It also has weak homology in one of its exons to the precursor of the *Aplysia californica* egg-laying hormone. Other parts of its amino acid sequence exhibit homologies to the RII regulatory subunit of cAMP-dependent protein kinase.

The cloning and characterization of the loci discussed above has been facilitated by a number of factors. The loci involved have either had an extensive genetic and cytological data base, or else it has been possible to tag them using vertebrate sequences to which they exhibit homology. While the molecular characterization of genes which are amenable to cross raiding of vertebrate and invertebrate libraries can be expected to proceed rapidly, many genes involved in neuronal pathfinding, neuroanatomy and behaviour cannot be isolated by such strategies. For these genes, we do not as yet know what phenotypes to expect from the null condition of the locus. For others, we simply lack the paradigms with which to analyze the organism. For still others, we lack a systematic biochemical inroad, even when we know the relevant neuroanatomy. For example, only one part of one neuron in a group of 1000 may exhibit an antigen at a particular time in development, and the system may thus remain biochemically inaccessible for technical reasons. The following examples provide data on the molecular inroads that have been made in these more difficult areas.

Mutations affecting the embryonic peripheral nervous system.   The *Drosophila melanogaster* embryonic peripheral nervous system (PNS) has a highly invariant neuronal pattern. In the thorax and abdomen there are 373 neurons which innervate external sensory structures, and 162 neurons which innervate chordotonal organs (Ghysen *et al.* 1986). The search for mutations which perturb the PNS has been initiated by using either chromosomal deficiencies or mutations, which together cover nearly one-third of the *Drosophila melanogaster* genome. Embryos thus produced are totally deficient for a given part of the genome, and can be examined directly for PNS abnormalities. So far, Jan *et al.* (1986) have identified at least 20 genomic regions which, when mutated or deleted, result in quite specific abnormalities in the PNS. These mutations fall into four groups:

(a)   those which increase or decrease the number of neurons;
(b)   those which affect particular cell types;
(c)   those which transform one sensory neuron into another; and
(d)   those which cause abnormal pathway formation.

As an example of the third class, Jan *et al.* (1986) found that two lethal complementation groups in the *cut* locus caused certain neurons associated with external sensory structures to now transform to chordontonal neurons. Since the *cut* locus has already been cloned (Jack 1986), the molecular basis of these sensory neuron transformations should be close at hand.

Mutations affecting development of embryonic midline neurons. The isolation of mutations which have somewhat subtle effects on the embryonic nervous system has been carried out by direct visual examination, using either antibodies against horseradish peroxidase (which binds to a surface antigen present in all neurons of the central and peripheral nervous systems of *Drosophila melanogaster* [Jan & Jan 1982]), or else a panel of monoclonal antibodies with various degrees of specificity on the developing nervous system.

With this approach, mutant embryos containing chromosomal deficiencies are produced as part of a crossing programme and are then assayed for nervous system defects. In this way, Crews *et al.* (1987) have isolated a gene (termed *S8*), the mutant form of which eliminates certain neuronal precursor cells and their progeny, as well as neighbouring ectodermal cells, in the midline of the embryonic central nervous system. The DNA of the *S8* locus had, in fact, already been isolated (in an unrelated walk) and so was available for detailed analysis. The gene landscape was refined to an area of about 5 kb, and was found to have two transcripts, one of which is expressed between the third and sixth hour of embryonic development and the other mainly between the sixth and ninth hour. *In situ* analyses of tissue sections reveal that gene expression is specific for the midline cells of the embryonic nervous system. Thus, it appears that genes such as *S8* may be required in order for precursor cells to form, or to form in the correct location in the nervous system. The detailed analyses of the protein product should be illuminating.

Cell surface glycoproteins. Direct immunological procedures are also helpful in isolating neuronal surface recognition molecules, particularly when there is evidence that axon fascicles are differentially labelled on their surfaces by such recognition molecules. Patel *et al.* (1986) used a particular monoclonal antibody (7G10) which picked out an embryonic surface antigen expressed in only a small number of neurons (such as the RP1 cell) in the developing embryonic nervous system. This monoclonal antibody immunoprecipitates four membrane glycoproteins of molecular weights 55 000, 60 000, 65 000 and 75 000. By using antisera against these proteins to challenge an embryonic central nervous system library, made in the expression vector, λgt11, a single gene has now been isolated. Molecular analysis should determine if this *Drosophila melanogaster* gene has homology to the fasciclin genes of the grasshopper. Mapping of the gene and genetic characterization of the appropriate landscape should also reveal whether it bears any relation to known lethal or visible loci in that area.

The disconnected gene (*disco*). Genes involved in neuronal pathfinding are a particularly important group to identify, and Steller *et al.* (1987) have begun to characterize one such a candidate, *disco*, which is necessary for the navigational properties of a particular subset of neurons in the peripheral nervous sytem. In *disco* individuals there is a problem with Bolwig's nerve, which is a larval pioneering visual nerve consisting of the axons of the larval photoreceptor cells. While the mutant forms the correct number of photoreceptor cells, the axons of Bolwig's nerve explore the cellular terrain blindly, and so do not establish their correct synaptic connections with the larval brain. Neurons which develop subsequently, and which use Bolwig's nerve as a cue in their navigational procedures, are also misrouted, leading to secondary defects. The locus has been mapped to 14B3–4, and an area of over 170 kb has been defined by chromosomal walking. The gene itself has been restricted to a 110 kb subregion of this walk by deficiency analysis.

Structural brain mutants. A novel way of analyzing the nervous system has been pioneered in the laboratories of Heisenberg and Fischbach, who isolated structural brain mutants by a combination of genetic and histological techniques and then characterized them in detail by Golgi staining and by subjecting the mutants to sophisticated behavioural tests. The direct screening of internal brain morphologies has always been considered too tedious, but once the histological techniques were streamlined, such screens became more routine, and wealth of brain mutants have been uncovered. These have been placed into four categories:

(a) low fidelity mutants in which the brains are imprecisely constructed;
(b) brain shape mutants with reduction in size;

(c)   architectonic mutants that have defects in the repetitive architecture; and

(d)   vacuolar mutants with holes in defined areas of the brain.

Clearly these classes of mutants have defects in various areas of neuronal development such as cell proliferation, wiring and cell death (Heisenberg & Böhl 1979). Mutants in these categories were subsequently tested in behavioural paradigms, and correlations were sought between neuro-anatomical defects and behavioural responses (Fischbach & Heisenberg 1981, 1984).

Small optic lobes (*sol*) is one product of such a screen, and here the mutation leads to degeneration of specific ganglion cells in the cortex of the medulla at a certain stage in development. The volume of the medulla and lobula complex neuropils is thus greatly reduced. The structural defects in this mutant lead to problems in visual orientation behaviour, so that the landing response, figure-ground discrimination and orientation behaviour are all modified. The wild type fly needs to convert, for example, movement-specific sensory signals into optomotor responses of wingbeat amplitudes. There also needs to be a stabilization not only of altitude and course, but fixation of an object in the frontal zone of the visual field. Götz (1987) has analyzed the flexibility of orientation control in wild type and *sol* mutants, using a flight simulator, and has concluded that this flexibility is due to a subset of neurons which is missing in *sol* individuals. However, the optomotor yaw response in *sol* is well preserved even under threshold conditions (Fischbach & Heisenberg 1981). The beauty of such mutations is that they influence only some functional subsystems of the brain, while leaving others unaffected. Molecular entry points have now been isolated in the environs of this locus by microdissection and microcloning (Miklos *et al.* 1987b, Yamamoto *et al.* 1986).

A number of other mutants, such as reduced optic lobes (*rol*) and minibrain (*mnb*), have also been harnessed for neuroanatomical studies. By constructing mutant flies of the genotypes *rol sol* or *rol sol mnb* for example, it is possible to determine which cells are more stable against genetic dismantling of the brain (Buchner *et al.* 1987). The *rol sol* double mutant, which is entirely motion blind, has also been utilized by Wolf & Heisenberg (1986) in their behavioural studies on optomotor balance and operant strategies. These authors have found that operant behaviour is a basic constituent of visual orientation in flies.

Finally, a number of neurologically interesting genes have been unearthed fortuitously when specific, rather large, portions of the genome have been analyzed in great detail by genetic and molecular means. Thus, as already mentioned, the acetylcholinesterase and S8 genes had already been cloned in the 315 kb rosy–Ace walk. Divisions 19 and 20 a the base of the X-chromosome also represent an area of significant potential. This region has been subjected to near saturation mutagenesis for lethals (Schalet & Lefevre 1976), and, in the process, genes with effects on behaviour have been mapped to this area with great accuracy. In the maroonlike to suppressor of forked interval (19E, 19F and 20), which spans 30 lethal and visible complementation groups, the following genes are found: shaking-B, Passover, uncoordinated, flightless O and flightless I, small optic lobes, sluggish-A, uncoordinated-like, stoned and stress sensitive-C (Miklos *et al.* 1987a,b). This interval has been microdissected and microcloned so that a number of these phenotypically diverse loci are now amenable to molecular targeting. Chromosomal walks have already been initiated in a number of these neurologically interesting loci: shaking-B (Davies *et al.* 1987); uncoordinated (Miklos *et al.* 1984); flightless I (De Couet *et al.* 1987); and the small optic lobes/sluggish-A region (Yamamoto *et al.* 1987).

Memory remains a black box in neurobiology (Lynch & Baudry 1984). Nor is it easy to see how one might approach the topic from a molecular point of view in an organism as complex as a mammal. Like mammals, *Drosophila melanogaster* has both a short-term and a long-term memory. For example, flies can be taught to avoid specific odours which they have experienced in conjunction with appropriate shock treatment. Several single-gene mutants are now known which fail to learn such a task. These include dunce, rutabaga and turnip, all three of which affect some aspect of cyclic AMP metabolism (Livingstone *et al.* 1984). An additional mutant, amnesiac (19A), learns normally but forgets four times more quickly than do wild type flies (Quinn *et al.* 1979).

Thus, there exist in *Drosophila melanogaster* a large number of mutations affecting many different aspects of the development and functioning of the nervous system. The principles by which these operate are not likely to be different from the principles which apply to neurological mutants in vertebrates. The way is now open to rapidly determine the molecular bases which underlie a variety of developmental problems, including circuit determination, circuit structure, multimodal convergence and complex behaviour pathways, by using the large spectrum of *Drosophila melanogaster* mutants which impinge on these problems.

2.2.6 General conclusions

The conclusions which emerge from an analysis of development in *Drosophila* are of six kinds:

(a)   In early embryogenesis a reference system of spatial co-ordinates is laid down which provides fundamental positional information for all subsequent developmental events.

(b)   Cell lineages result from the early partitioning of fixed regions of the egg cytoplasm into defined cell lines. This is most clearly seen in the case of the imaginal discs.

(c)   Switch genes, through their involvement in binary developmental decisions, may lead to the wholesale reprogramming of development in both soma and germ line. Simple switch mechanisms of this type appear, for example, to underlie sex determination.

(d)   It is possible to uncover loci, in the form of homeotic genes, which appear to perform executive developmental decisions. These, however, are relatively few in number, and the bulk of the transcriptively active

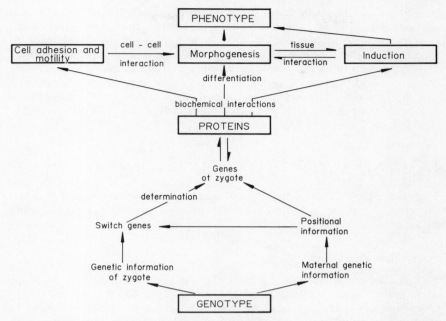

Figure 2.22 The developmental interactions which govern the production of phenotype from genotype (data of Alberch 1982).

parts of the genome simply provide the metabolic back-up necessary for implementing the decisions of these executive genes, whose activities govern the ultimate expression of large arrays of other loci.

(e)   These homeotic genes function as major switch genes which control further developmental pathways. Even the most complex system can now be seen to involve a cascade of switch decisions. The earlier the switch the more fundamental its role. However, no matter how many binary decisions are made, without controlled cell movements all that can be produced is a determined and differentiated cell mass. The production of form and pattern requires cell interaction coupled with cell division. Morphogenesis thus also hinges on the control of cell cycle, cell surface and cell mobility (Fig. 2.22). These are multi-dimensional matters in the sense that they involve both temporal and spatial co-ordination.

(f)   Analyses of the underlying principles which govern nervous system construction and function in *Drosophila melanogaster* must soon outstrip the amount and type of data available in vertebrate systems. The reason for this is not difficult to appreciate. Once a genetic or molecular bridge-head has been secured for a locus, it is possible to isolate other genes involved in a common network, simply because of the genetic flexibility available in the *Drosophila melanogaster* genome. By contrast, the obvious lack of genetic manipulability of vertebrate genomes is likely to remain an intractable problem for the foreseeable future. The routine use of chromosomal deficiencies, covering much of the genome, and of embryo screens, in combination with monoclonal antibodies, is an advantage as yet available only to *Drosophila melanogaster*. This, coupled with a routine genomic re-entry system, in the form of P element mediated germ line transformation, and direct microdissection and microcloning technologies, means that an awesome number of significant gene systems which influence neurogenesis are certain to be molecularly defined in the next few years.

2.3   GENERAL PRINCIPLES OF DEVELOPMENT

To what extent, then, can we anticipate that developmental principles equivalent to those seen in *Drosophila melanogaster* will operate in other eukaryotes?

2.3.1 Polarity

The establishment of polarity is certainly of general significance, and is one of the earliest and most critical events of embryogenesis. In some cases, as in *Drosophila melanogaster*, this polarity is inherent in the unfertilized egg. In other cases, as in mammals, polarity is not fixed until late in cleavage. Whenever it occurs, and whatever the co-ordinate system it establishes, the existence of an overall positional reference system within the embryo plays a fundamental role in morphogenesis and pattern formation.

2.3.2 Cell lineages

The principle of lineage segregation, which underlies the formation of the initial polyclones involved in imaginal disc formation, is also inherent in most developmental pathways, though it is frequently obscured by the complexities of other developmental events. A particularly clear case of this is evident, however, in the roundworm *Caenorhabditis elegans* (Hedgecock 1985). Nematodes have a small and invariant cell number and cell location. The adult of *Caenorhabditis elegans* has only 959 somatic cells and these originate as a series of founder cells within the embryo. Most of the somatic cells of the adult are generated during the first half of embryogenesis and follow well-defined and invariant cell lineages (Fig. 2.23). One in eight of all cells which are produced in these lineages subsequently die. Genetic analysis has already revealed two genes, *ced-3* and *ced-4*, whose wild type products are needed for the initiation of programmed cell death. If these wild type products are not produced or if they are made in lowered amounts, then the cells now survive, differentiate and adopt new cell fates (Ellis & Horvitz 1986). Such programmed cell death is especially prevalent in the nervous system, where it forms an integral part of determination. The first-stage larva has 558 cells. This difference in cell number, compared to that of the adult, is the result of blast cells which divide and differentiate after hatching (Table 2.9).

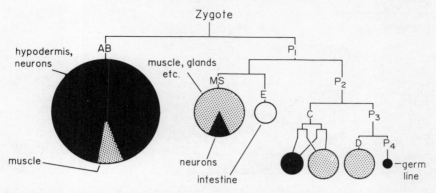

Figure 2.23 The origin of founder cell lineages in the embryo of *Caenorhabditis elegans* (data of Sulston 1983).

At least four loci affecting neural cell lineages are analogous to homeotic genes, in the sense that they cause particular cells to differentiate in an alternative pathway. In normal development, a series of ectoblasts (V1–V6) are produced. Five of these (V1, V2, V3, V4 and V6) give rise to hypodermal cells which form part of the cell surface of the organism. V5, however, gives rise to both hypodermal and neural cells, and in such a way that programmed cell death is apparent in the neural lineage. In *lin-22* mutants, ectoblasts V1–V4 adopt the same cell fate as V5 (Fig. 2.24). This situation is analogous to the alteration of cell fates that operates in homeotic mutants of *Drosophila melanogaster*.

Mutations in a second series of loci, exemplified by *unc-86*, show that not only can cell fates within a lineage be transformed but, additionally, extra cell divisions can occur giving rise to supernumerary dopaminergic neurons (Fig. 2.25). This reinforces the principle that single gene effects can produce extra cell divisions in a manner analogous to the mutations kayak, canoe,

Table 2.9 Cell number variation in the development of *Caenorhabditis elegans* (from Edgar 1980, Sulston *et al*. 1983).

Cell type	Cell number	
	1st stage larva	Adult
hypodermal (→ cuticle)	85*	213
neuronal and glial	248	338
mesodermal	91+	122
alimentary track	130	143
gonad { somatic tissue	2	143
gonad { germ line	2	indefinite
totals	558	959

* includes 35 blast cells which give rise to hypodermis and neurons
† includes one mesoblast

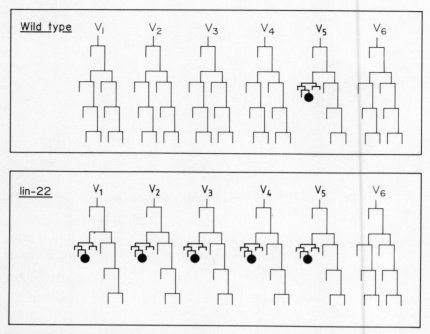

Figure 2.24 The effect of the heterochronic mutation *lin-22* on cell lineage fates in the ectoblasts (V1–V6) of *Caenorhabditis elegans*. In normal development ectoblasts V1–V4 and V6 give rise to hypodermal cells, whereas V5 produces both hypodermal and neural cells. In *lin-22* mutants V1–V4 adopt the same cell fate as V5 (data of Horvitz *et al.* 1983).

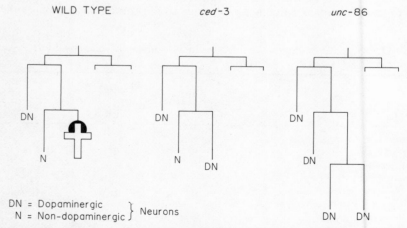

Figure 2.25 Programmed cell death and resurrection in specific cell lineages of the *ced*-3 and *unc*-86 mutants of *Caenorhabditis elegans* (data of Horvitz *et al.* 1983).

punt and topless of *Drosophila melanogaster*, which lead to extra divisions, and hence to the hypertrophy of the embryonic nervous system. This, in turn, can have consequences for the timing of development. Thus, a recessive *lin-14* mutation accelerates development by causing cells to express fates normally shown only by their descendants. Dominant *lin-14* mutations, on the other hand, retard development by causing cells to repeat the lineages of their progenitors. The importance of such heterochronic mutations is that the differences in developmental timing to which they give rise may lead to radical morphological differences in the developmental programme.

The *lin-12* locus of *Caenorhabditis elegans* is a homeotic gene controlling binary developmental decisions in cells of several different tissues and at different times. Many of these cells appear to be multipotential at birth but have their fates specified by subsequent cell–cell interactions. Part of the *lin-12+* gene product is a peptide with 11 tandem repeats of a sequence of some 38 amino acids. Each repeat copy is homologous to part of a 53 amino acid peptide of mouse and human EGF (Greenwald 1985). It is even more homologous to the Notch repeat of *Drosophila melanogaster* and, like it, causes well-defined changes in cell lineages.

Ascidians also have exceptionally well-defined and invariant cell lineages, resulting from the early partitioning of fixed regions of the egg cytoplasm into particular cell lines by cleavage (Fig. 2.26). Consequently, cell–cell interactions play only a limited role in ascidian embryogenesis, and differentiation is programmed by factors intrinsic to the embryonic cells themselves (Jeffrey 1985). The need for such an early specification and differentiation of cells in this instance is associated with the fact that the locomotory 'tadpole' larva of the ascidian is but a transitory phase which undergoes retrogressive metamorphosis to give rise to a sessile adult. Since the ascidian larva is commonly accepted as reflecting an evolutionary progenitor of the vertebrate body plan, this case is clearly one of more general interest. Indeed, most novel vertebrate characters can be considered to have originated by the modification of pre-existing embryonic tissues of the ascidian larva type. Thus, many of the sensory, integrative and motor systems of vertebrates, as well as their supportive skeletal structures, are derived embryologically from the neural crests and from the neurogenic epidermal placodes; the former giving rise to motor neurons and the latter to those sensory structures involved in cephalization. The major developmental changes involved in vertebrate evolution thus involve the ectoderm and its interactions with other tissues, while the basic developmental sequence, including gastrulation and its attendant redistribution of mesodermal tissues, requires far less fundamental reorganization (Gans & Northcutt 1983).

The concept of stem cells, which persist into adulthood, where they serve as clonal sources for major differentiated cell populations which are developed at that time, is also clearly related to that of specific cell lineages and extends the principle into the adult organism, a novel form of heterochrony. In mammals, for example, the stem cells of the hemopoietic, germ line and pigment systems all originate in the embryo at considerable

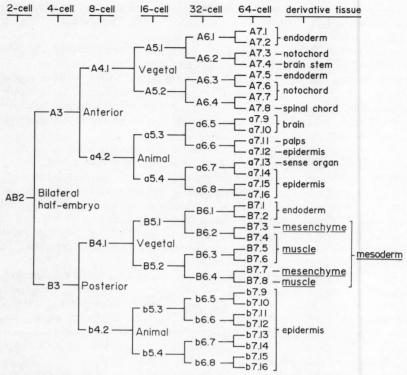

Figure 2.26 Cell lineage fate map of a half embryo of bilaterally symmetrical ascidians (data of Whittaker 1979). Mesodermal structures are underlined.

distances from the sites at which they function in the adult. Migration of these stem cells to the correct microenvironment is, thus, a prerequisite for their normal differentiation. Active hemopoiesis begins extra-embryonically with the formation of blood islands in the yolk sac of the embryo. The cells then migrate to the foetal liver subsequently to the bone marrow. Germ cells also originate in extra-embryonic tissue and are first observed in the yolk sac. Following multiplication, they migrate to the genital ridges which later develop into the adult gonad. Finally, the primordial melanoblasts appear during the formation of the neural crest and subsequently migrate laterally through the skin.

Two mouse mutations exhibit pleiotropic effects which suggest that all three categories of stem cells may be influenced by similar developmental cues (Hall 1983). Thus, homozygous mutations at either the 'dominant spotting' (W) or the steel (S) locus produce a phenotype known as 'black-eyed whites'. In these mutants the coat is completely devoid of pigment, the animals show severe macrocytic anaemia and a marked deficiency of primordial germ cells. Thus, despite the considerable differences in the developmental end products of vertebrates and invertebrates, there seems little doubt that the early embryonic decisions that underlie development are indeed based on common principles. These include polarity, the establishment of cell lineages and the possibility of heterochronic shifts in these lineages. There is, then, every expectation that the fundamental molecular mechanisms which underlie development will not only be common to *Drosophila melanogaster* and vertebrates, but will rapidly be unearthed in the former and then quickly evaluated in the latter.

2.4 GENOME ALTERATIONS DURING DEVELOPMENT

Two novel modes of genome alteration, not represented in the developmental programme of *Drosophila melanogaster*, are known to occur in other eukaryotes. Although uncommon, they serve to illustrate the important principle that the components of the genome can be rearranged, or else the genome itself can be fragmented, without any deleterious effects on its capacity to function effectively in somatic development.

2.4.1 Nucleotide sequence alterations

Traditionally, differentiation has been regarded as a problem in which an unchanged nucleotide sequence is either expressed or not expressed. While this is probably true for a majority of developmental changes, it is certainly not true of all. At least two forms of cell differentiation are now known which involve alterations in nucleotide sequence arrangement.

The principle on which both these eukaryote systems is based is clearly exemplified by the phenomenon of phase variation in the prokaryote *Salmonella typhimurium*. This refers to the ability of the bacterial cell to switch between the expression of the two structural genes, H_1 and H_2, encoding the protein flagellin, which determines the antigenic properties of the flagella produced by the cell. These two genes are located in different regions of the *Salmonella typhimurium* genome. One of them H_2, is adjacent to a 970 bp DNA sequence bounded by two inverted repeats (IRL and IRR), each of 14 bp. Within this region is located the promoter (P) element which regulates the co-ordinate transcription of H_2 and an adjacent rH_1 gene, whose product acts as a repressor of H_1. A second locus, *hin*, within the region bounded by IRL and IRR, mediates the specific inversion of the sequence delimited by the two repeats, which act as switching points. In the inverted state, P becomes disconnected from H_2 and rH_1, so that it is H_1 that now transcribes (Fig. 2.27). Since the site-specific inversion is reversible, it behaves as a flip-flop switch mechanism for activating and inactivating the H_2 gene (Simon *et al.* 1980). Phase variation is thus the result of an interaction between repetitive elements and resident genes within the *Salmonella typhimurium* genome.

2.4.1.1 *The transposable mating type loci of yeast*

Two distinct haploid cell types are recognized in the budding yeast, *Saccharomyces cerevisiae*. These are termed (a) and (α) and, in many respects,

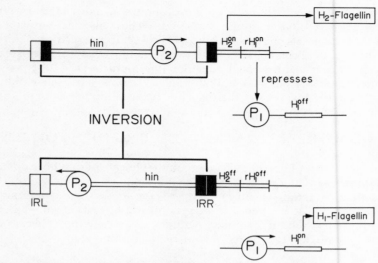

Figure 2.27 The phase variation system of *Salmonella typhimurium*. H_1 and H_2 represent the two structural genes that encode the variant form of the protein flagellin. A promoter site, P_2, adjacent to the H_2 gene is involved in regulating its expression. When this site is in the H_2 'on' state the H_2 gene is expressed, leading to the production of H_2-flagellin. An adjacent rH_1 gene is also co-ordinately expressed. The product of this gene acts as a repressor of the P_1 promoter that controls expression of H_1. When the P_2 controlling site is in the H_2 'off' configuration, neither H_2 nor rH_1 is expressed. This allows H_1 to transcribe and so produce H_1-flagellin. The switch from H_2 'on' to H_2 'off' is governed by the inversion of the P_2 sequence. This depends on a recombinational-type event between two 14 bp inverted repeat sequences, IRL and IRR, which represent switching points. The inversion, thus, behaves like a flip-flop switch, activating and inactivating H_2 gene transcription (data of Silverman & Simon 1983).

they behave as though determined by a pair of Mendelian alleles at the mating type (MAT) locus, which is present on the right arm of chromosome 3. These presumptive alleles are thus referred to as MATa and MATα. As its name suggests, the MAT locus controls the ability to mate. Thus (a) cells mate efficiently with (α) cells to give heterozygous (aα) diploids. Cells of like mating type, on the other hand, rarely mate with each other. When they do, they form homozygous, (aa) or (αα), diploids. Such homozygous diploids behave like (a) and (α) haploids, and will mate with each other or with a haploid of opposite mating type, but not with a diploid of the same constitution. The resulting cells, namely the tetraploid (aaαα), and triploids (aaα) and (aαα), are all heterozygous at MAT and so, like (aα) diploids, are unable to mate.

The MAT locus also controls meiosis and sporulation. These processes occur only in cells containing at least one representative of MATa and MATα. Such heterozygotes turn off mating functions and allow cells to sporulate. Thus (aα), (aaα) and (aαα) cells sporulate, whereas (a), (α), (aa) and (αα) cells do not. It is, then, the genetic composition of the yeast cell at the MAT locus, rather than the overall ploidy, which is critical for determining cell type and developmental capability.

Yeast strains also differ in the stability of their mating types. Those which are stable are termed heterothallic, in contrast to the unstable homothallic strains. These differ by a single determinant which is not linked with MAT. Homothallic strains carry the dominant *Ho* allele, while heterothallic strains are characterized by the recessive *ho* allele. In strains carrying *Ho* both MATa and MATα are interconvertible and, in fact, switch mating type at almost every cell division until conjugation leads to the production of a diploid zygote in which the capacity to switch ceases. This situation depends upon the existence, in standard laboratory strains, of silent or cryptic copies of a and α which are located at some distance from the actual mating type locus itself. These silent copies, designated as HMLα and HMRa, are situated respectively near the ends of the left and right arms of chromosome 3, with the mating type locus itself (MAT) lying between them (Fig. 2.28). Heteroduplex analysis shows that while there is some homology between HMLα and HMRa they are distinct and functionally non-homologous blocks of DNA.

The actual operative mating type thus depends upon which of the two silent copies is inserted into MAT. This pattern of activity is referred to as the casette mechanism, since it likens the transposition of specific instructions from HML or HMR loci into MAT to that of inserting a casette into the playback head of a tape recorder. The casettes are silent unless plugged into the head. Mating type switching then depends on a silent locus of opposite type replacing the existing active locus present at MAT by a

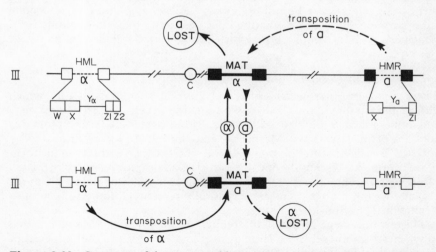

Figure 2.28 Structure of the transposable mating type system of yeast. Here, the reciprocal switching of HMRa and HMLα sequences into, and out of, the mating type (MAT) locus on chromosome III depends on a transposition event in which one sequence replaces the other (data of Rine *et al.* 1981).

form of non-reciprocal gene conversion. The rules governing such switching can be summarized as follows (Nasmyth & Tatchell 1980):

(a) Only cells that have undergone division are able to switch mating type, and the frequency of switching is about once per cell division.
(b) A switching event produces pairs of cells with changed mating type.
(c) The switching process occurs preferentially to the opposite mating type.
(d) When a casette previously at MAT is replaced by one of opposite type the original is lost.
(e) Homologous casette replacement can occur.

The HMLα and HMRa loci of yeast are, thus, essentially transposable elements which integrate specifically into a defined segment of DNA, the MAT locus, and by so doing determine the mode of expression of that locus. Other transposable elements occur in yeast, as well as in *Drosophila melanogaster* and man (§ 3.2.1.). These, however, differ from the switch loci of yeast in lacking the specificity of integration which these yeast loci show.

If, then, genes can be switched on and off by specific transposable elements, is it likely that elements of this sort are regularly involved in the patterns of gene expression seen during development and differentiation? The answer is unknown. However, the lack of specificity of most of the nomadic elements characterized to date argues against such a role. Thus, in *Drosophila melanogaster* it is known that the distribution of the nomadic elements *copia, 412* and *297* is different in the chromosomes of the same tissue from individuals of different laboratory strains, while the distribution of *412* is the same in DNAs from embryos, larval brains and adults of the same strain (Finnegan *et al.* 1982). On the other hand, as has already been mentioned, significant homology exists between the amino acids coded from parts of the a_1 and α_2 mating type genes and the homeo domains of *Drosophila melanogaster*, both of which control the stable determination of cell types.

2.4.1.2 Immunoglobulin class switching

There are two primary classes of lymphoctyes in mammals. T-cells which differentiate in the thymus and are concerned with cell-mediated immunity, and bone marrow derived or B-cells. Both classes carry receptors for antigens (Tonegawa 1985, Marrack & Kappler 1986). When stimulated by antigen, B-cells turn into plasma cells and secrete these receptors in the form of free antibody. T-cells do not. In both classes, developmentally regulated DNA rearrangements occur, with the result that gene segments which are separate within the germ line are somatically rearranged to bring them into a functional relationship.

The terminal stage of B-cell differentiation, the plasma cell, is committed to synthesizing and secreting large quantities of a single molecular species of immunoglobulin protein. Such proteins, conventionally referred to as antibodies, are composed of two types of polypeptide chain, designated respectively as 'light' and 'heavy', which are the products of separate sets of genes. A given antibody consists of two identical light chains and two

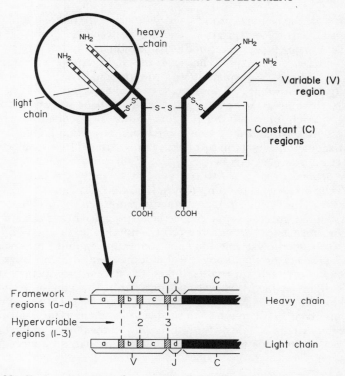

Figure 2.29 Basic structure of an antibody molecule which consists of four polypeptide chains, two heavy (H) and two light (L), held together by disulphide (−s−s−) bonds. Each chain is functionally divided into constant (C) and variable (V) regions. The V regions include four framework and three hypervariable segments which form the actual antigen binding site. Each light chain is encoded by three separate genes: a C gene codes for the constant region; a J gene codes for the fourth framework segment and a part of the third hypervariable segment; while a V gene codes for the first three framework segments, as well as hypervariable segments 1 and 2 and part of 3. In the heavy chain, an additional D gene codes for the third hypervariable segment (modified after Robertson 1981).

indentical heavy chains, folded and interconnected to form a split 'Y'. Both light and heavy chains include an aminoterminal segment forming the variable region of the molecule, and a carboxyterminal segment forming the constant region (Fig. 2.29).

There are two types of light chain in most vertebrates, kappa (κ) and lambda (λ). Every antibody molecule has one or other of these. Additionally, each antibody has one of five heavy chain types: mu (μ), delta (δ), gamma (γ), epsilon (ε) and alpha (α). The precise heavy chain type then defines five major classes of immunoglobulin, namely IgM, IgD, IgG, IgE and IgA. The variable region serves to recognize and bind to a specific antigen, while the constant region determines precisely how the antibody carries out its immunological task, that is whether it does so by remaining associated with the surface of the cell that produces it, whether it circulates in the blood, or whether it binds to the surface of another cell type.

The two major types of antibody chain, light and heavy, are encoded by distinct sequence families. Those of the light chains consist of three categories (Joho & Weissman 1980), known respectively as variable (V), joining (J) and constant (C). In the case of the kappa light chain of man, for example, there are a large number, perhaps 150 or so, or alternative V-sequences, each preceded by a short leader (L) sequence. These L/V sequences are separated, by a long stretch of non-coding DNA, from a run of five J-sequences which, in turn, are again separated by a long non-coding stretch from a single C-sequence. During lymphocyte development one of the L/V sequences is joined to one J-sequence and the single C-sequence, through the deletion of the intervening DNA, to form an active kappa gene (Fig. 2.30). This is then transcribed into a messenger which is, in turn, translated into a specific kappa chain, the L-sequence being cleaved off from the chain as it passes through the cell membrane, leaving a VJC protein – where VJ forms the variable and C the constant region of the chain. In the human lambda chain the arrangement is somewhat different, since there are six C-sequences each apparently linked to its own J-sequence. Immunoglobulin light chain diversity thus arises from the precise array of VJC sequences that are combined into one active gene. The DNA rearrangement events involved in VJ joining are mediated by recognition signals that lie immediately to the 3′ side of the V gene and to the 5′ side of the J gene. As Simon *et al.* (1980) point out, there is a marked similarity between the DNA

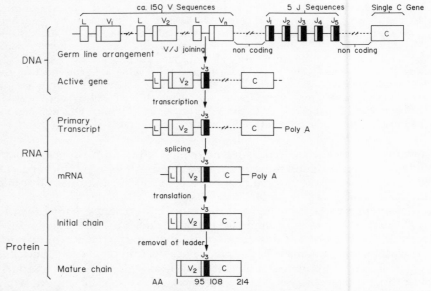

Figure 2.30 Production of human kappa light chain protein by VJ joining (data of Leder 1982).

sequences found at these recognition sites and the sequences present at the switching points in the *Salmonella typhimurium* phase variation system.

Each active heavy chain gene is assembled from four, rather than three, sets of sequences, namely L/V, D (diversity), J and C. Rejoining brings together one of the numerous L/V sequences, together with a single D and a single J to code for the variable region of the chain. There are, however, eight distinct C-sequences (μ, δ, γ^3, γ^1, γ^{2b}, γ^{2a}, ϵ and α) separated from one another by long non-coding sequences. Each C-sequence is, in turn, divided into from 3–6 coding domains separated by short non-coding intervening sequences. Moreover, each constant region gene domain is preceded by a switching (S) site consisting of a tandem repetition of short unit sequences the simplest of which is Sμ, which involves repeat units of two kinds, GAGCT and GGGGT. Sμ-like sequences have also been shown to be present in a wide variety of organisms, including yeast, *Drosophila melanogaster* and sea urchins, which do not produce immunoglobulins (Sakoyama *et al.* 1982).

The final assembly of coding sequences in the active heavy chain gene depends on a second category of switching which involves the heavy chain constant sequences (Fig. 2.31). It is generally agreed that IgM is the earliest immunoglobulin class to be expressed in the differentiation of a B-cell, so that initially the cell transcribes μ-chains. If transcription ends at the fourth coding domain of the mu gene, the mRNA eventually produced is translated to give an IgM that is secreted. If, on the other hand, the transcript includes mu domains 5 and 6, then splicing removes the 'stop' codon at the end of domain 4 and brings all six domains into the one messenger. This, when translated, gives rise to a membrane-bound IgM. If the primary transcript also includes the delta sequence, then splicing sometimes connects it directly to the J-sequence, making a messenger that encodes only a delta chain and produces IgD. So long as a lymphocyte is synthesizing mu or delta chains the gamma, epsilon and alpha sequences play no part in the active gene. However, the immature B-cell has the capacity to differentiate into a variety of alternative pathways by switching from IgM expression to any of the other classes of immunoglobulin. Thus, the switch from mu chain (IgM) to alpha chain (IgA) is achieved by the joining of the VDJ sequence to the switch site (S) preceding the alpha sequence, and deleting all the intervening DNA.

The T-cell antigen receptor is a heterodimer composed of two disulphide-linked polypeptide chains, α and β. Like the immunoglobulin heavy chain V genes, the receptor V_β gene of T-cells appears to be assembled from four gene segments, V_β, D_β, J_β, and C_β which rearrange to generate a continuous V_β (Fig. 2.32).

Thus, somatic joining is a central event in the activation of immunoglobulin molecules and leads to a vast increase in the information encoded within the genome.

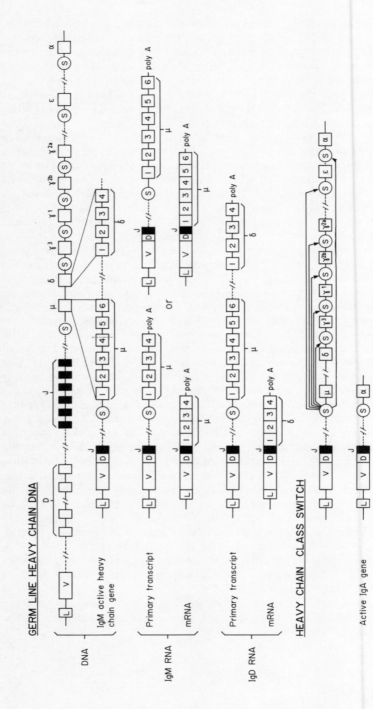

Figure 2.31 Production of active heavy chain immunoglobulin genes by VDJ joining (data of Leder 1982).

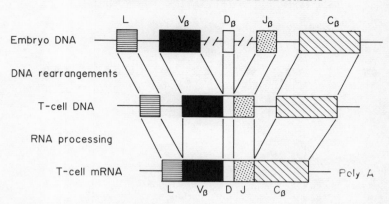

Figure 2.32 Events involved in the expression of the β mRNA of the T-cell antigen receptor of humans (data of Siu *et al.* 1984).

2.4.2 Presomatic diminution

In a number of species of ascarid roundworms, the copepod *Cyclops* and insects belonging to the families Chironomidae, Sciaridae and Cecidomyidae, DNA is eliminated specifically from the genomes of all prospective somatic cells during the early cleavage divisions of the embryo. As a result of this, the somatic nuclei involved in development have a much reduced genome compared to that of the germ line.

In both ascarids and *Cyclops* this process of presomatic diminution involves the controlled molecular excision of all heterochromatic segments from the chromosomes. These segments are terminal in *Parascaris univalens*, where they take up some 70% of the total chromosome length, but terminal and interstitial in *Parascaris equorum* (Goday & Pimpinelli 1984). In the latter species, Roth (1979) showed that the diminution process removes nearly all of the highly repeated DNA from the genome, which implies that this DNA is irrelevant to subsequent somatic development.

Only terminal heterochromatic segments are involved in diminution in *Cyclops divulsus*, but these segments are both terminal and procentric in *Cyclops furcifer*, while in *Cyclops strenuus* the heterochromatin that is lost is distributed along the entire length of each chromosome (Fig. 2.33, Table 2.10). In both *Cyclops divulsus* and *Cyclops furcifer* numerous free DNA rings, ranging from 0.6 to >16 μm in the former and from 0.4 to 40 μm in the latter, are present at the time of diminution (Beermann 1984).

The diminution process in the three insect families involves entire chromosomes, parts of which are molecularly degraded. Consequently, there are pronounced differences in the number of chromosomes, and hence the amount of DNA, present in the somatic and the germ line genomes. In the cecidomyids it has been shown that if the chromosomes lost from the soma are also experimentally removed from the presumptive germ line, then the gonads that develop contain germ cells of reduced size which do not give rise to mature eggs or sperm (Bantock 1970). Evidently, those

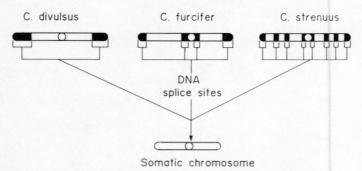

Figure 2.33 Modes of presomatic heterochromatin (solid) elimination in three species of the copepod *Cyclops* (data of Beermann 1977).

Table 2.10 Chromatin diminution in the copepod genus *Cyclops* (from Beermann 1977).

Species	Heterochromatin distribution	DNA content (pg) Germ line	Soma	DNA lost pg	%
C. divulsus	terminal	3.1	1.8	1.3	42
C. furcifer	terminal and procentric	2.9	1.4	1.5	52
C. strenuus	interspersed	2.2	0.9	1.3	59

chromosomes that are somatically dispensable are in some way indispensable for fertility. Whether this applies to the other cases of presomatic diminution is not known.

2.4.3 Macronuclear development in unicellular ciliates

A majority of unicellular ciliates have a unique segregation of nuclear functions in which the transcriptive activity necessary for vegetative development and growth is restricted to DNA sequences contained in one or more large and impermanent macronuclei (*ma*). Genome continuity, on the other hand, is maintained by a small micronucleus (*mi*) which is active primarily in promoting sexual reproduction, since it, and it alone, is capable of undergoing meiosis. During vegetative growth the *mi* divides mitotically while the *ma* divides amitotically. Sexual conjugation results in degeneration of the *ma*, while the *mi* undergoes meiosis. Thus, the macronucleus does not contribute any genetic information to the sexual progeny. This separation of nuclear functions within a single cell parallels the soma–germ line differentiation which characterizes multicellular eukaryotes. The one exception to this situation in ciliates is found in a single group of asexual marine organisms, which have a multinuclear organization in which all the nuclei are equivalent and divide asynchronously only by mitosis (Raikov

1976). Where nuclear dimorphism is present the *ma* and *mi* differ commonly in DNA content, in the mechanism by which they divide and the time at which they do so, despite the fact that they cohabit a common cytoplasm. These two categories of nuclei thus serve as a unique model for analyzing the molecular mechanisms by which genetic information can be maintained in different structural and functional states.

Nuclear dimorphism in ciliates may take two principal forms:

Species with non-dividing *ma*. Here there are two to several hundred *ma* which are segregated into daughter cells at division without themselves ever dividing. A given *ma* then persists through a number of cell generations, but eventually becomes non-functional and is resorbed and replaced by a new *ma*, which develops from the mitotic products of the division of the *mi*.

Although the *ma* in such species are described as 'diploid', they do differ from the *mi* in gross morphology. Moreover, in *Trachelonema sulcata*, the one case that has been examined in any detail, there is an abrupt decrease in DNA content at the onset of *ma* development (Kovaleva & Raikov 1978). This leads to the *ma* anlagen containing only about half the DNA content of the *mi* from which they develop. While the DNA content of the young *ma* remains at the same level as that of the anlagen, the old *ma* has a highly variable genome size, ranging from values equal to that of the *mi* to about six times that level. Thus, part of the genome of the *mi* is evidently lost at the beginning of the transformation into a *ma* and there is a subsequent partial replication of some fractions in the maturing and aging *ma*.

Species with polyploid *ma*. This group includes several major types of *ma*. Some are clearly polygenomic. Thus, eight chromatin aggregates form before division of the *ma* in *Colpoda steini*, which is known to contain eight times more DNA than the *mi*. Moreover, these aggregates behave as diploid segregation units and thus correspond to diploid subnuclei, each of which is presumed to be genetically identical to the *mi*. In *Tetrahymena*, and *Paramecium* too, the *ma* appears to contain a large number of diploid subnuclei. In the former, there are some 45 haploid genomes in a *ma*, each of which carries at least 90% of the sequences found in the *mi* genome (Gorovsky 1980).

More complex *ma*, with fragmented chromosomes, chromatin diminution and differential gene amplification, characterize hypotrich ciliates such as *Stylonychia* (Ammerman *et al.* 1974). Here, only a small percentage of the sequences present in the *mi* are also present in significant amounts in the *ma*, and there is extensive loss of DNA from the *ma*, which develops through two polyploidization cycles (Fig. 2.34). In the first of these cycles, polytene-like chromosomes are formed. Unlike conventional polytene chromosomes, however, the bands are formed from aggregates of 30 nm loops laterally attached to 10 nm interband fibrils. The dimensions of these loops correspond to an average DNA content of 40 kb. Immediately following their production, these 'polytene' chromosomes undergo multiple fragmentation during which the loops are eliminated in the form of 30 nm

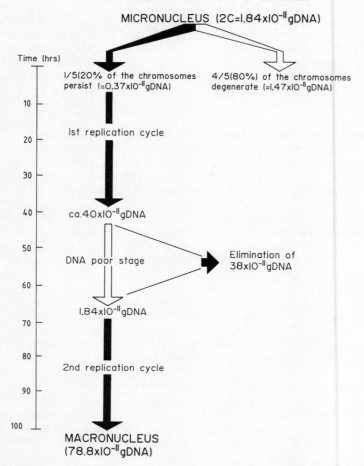

Figure 2.34 Genome changes involved in the development of the macronucleus in *Stylonychia lemnae* (= *S. mytilus*) (data of Ammerman *et al.* 1974 and unpublished).

rings of DNA, which are subsequently degraded (Meyer & Lipps 1984). As a result of this, more than 90% of the DNA is lost from the genome and the nucleus now contains a mass of chromatin-containing vesicles, rather than individual chromosomes. About 3000 such vesicles are present per nucleus. A second replication phase now ensues, leading ultimately to a mature *ma* with a DNA content of about 4000C. In both cycles, however, a differential replication occurs which drastically rearranges the relative proportions of most of the DNA sequences present in the *mi* genome. Thus, *mi* DNA has several highly repeated satellite sequences, none of which are represented in *ma* DNA. There is also a reduction in the sequence complexity of the genome. This takes place in two steps. One of these occurs before the formation of polytene chromosomes, the other after their breakdown. The development of the *ma* thus involves a reduction in the size of the DNA molecules within the genome to produce gene-size pieces (Swanton *et al.* 1980), though there are indications that these can be organized into higher order structures *in vivo* (Boswell *et al.* 1982). Each linear macronuclear 'gene' is terminated at both ends by the addition of tandem repeats of the hexanucleotide C_4A_2 (Cherry & Blackburn 1985), which forms a highly conserved telomere in simpler eukaryotes. Renaturation kinetics of the *ma* DNA reveals that the frequency of most DNA sequences within the macronucleus is about the same, with less than 2% of the DNA present in higher copy numbers. The ribosomal genes, however, are subject to differential amplification, so that in all species they occur in higher copy numbers (Steinbrück 1984).

In sum, the mature *ma* of *Stylonychia* is composed of a highly heterogeneous DNA with a kinetic complexity roughly 10 times that of *Escherichia coli*. Since the haploid *mi* genome contains an amount of DNA equivalent to 1600 times that of *Escherichia coli*, it is evident that *ma* DNA must be derived from an extensive repeated replication of only a small portion of *mi* DNA. Thus, although the *ma* has a ploidy level of 4096C, it is formed from only 1.6% of the sequences represented in the *mi* genome.

A comparison of the genome characteristics of these several ciliates is revealing (Table 2.11) and provides information on three major matters relating to gene number and gene expression:

(a) If all the gene pieces in the macronucleus of hypotrichs are transcriptively active and each produces a single mRNA, there would be about 12 000 genes in *Oxytricha similis*, the species with the lowest macronuclear complexity. In *Paramecium bursaria*, where the complexity is only 18 million bp, the number of genes would be in the order of 8000, if the mRNA producing sequences were distributed in a similar way to *Oxytricha similis*, that is if there is one average coding stretch of DNA in about 2200 bp. The 'somatic' gene number in these relatively complex unicells is thus likely to be about 10 000.

(b) Micronuclear content in ciliates ranges from 200 million bp (*Tetrahymena*) to 7200 million bp (*Stylonychia*), while macronuclear complexity ranges from 18 million bp (*Paramecium*) to 49 million bp (*Stylonychia*). Thus, both 'germline' and 'somatic' genomes vary widely between species with similarly restricted phenotypes. Since *Stylonychia* has some 300 chromosomes compared with the 10 present in *Tetrahymena*, there seems little doubt that its micronucleus is polyploid, and the same appears to be true for *Paramecium*, where different species have haploid numbers ranging from approximately 40–165. Therefore, if we take the *mi* complexity of 200 million bp found in *Tetrahymena* to represent something approaching a minimum genome for this group of unicells, then it is evident that ciliates provide a clear forerunner for the occurrence of excess DNA over that required for both conventional 'somatic' and 'germline' activities.

(c) Hypotrich ciliates, such as *Stylonychia*, eliminate up to 98% of their DNA sequences, both repetitive and unique, from the macronucleus,

Table 2.11 Genome characteristics of ciliates (from Steinbruck *et al.* 1981, Steinbruck 1986).

Species	Micronuclear content (mbp)	Macronucleus content (mbp)	complexity (mbp)	gene sized pieces (bp)	rep. DNA content
Hypotrichous					
Stylonychia lemnae (= *mytilus*)	7200	756 000	49	2500	
Euplotes aediculatus	1900	365 000	45	1836	<2% rep. DNA
Oxytricha similis	600	49 000	27	2200	
Holotrichous					
Paramecium bursaria	7200	266 000	18	N/A	No extensive
Tetrahymena pyriformis	180–200	45 ploid	200–240	N/A	under replication

N/A = not applicable.

leaving the remainder as amplified gene-sized pieces. Having dispensed with nearly all of its middle and highly repeated DNA sequences, the genome is still able to maintain metabolic normalcy. This implies that gene regulation in this organism cannot depend on the integrity of the chromosome, or even on DNA domains larger than an average sized macronuclear piece of DNA. Metabolically, therefore, there must be an enormous amount of junk DNA in the somatic genome of this organism.

One final point of interest concerning the relationship between the micro- and the macronucleus of ciliates relates to the occurrence in some species of distinct mating types whose determination involves a stable macronuclear change. This change selects one specific mating type from a genetically transmitted spectrum of potential types, which is carried by the micronucleus. In *Tetrahymena thermophila*, for example, there are seven distinct mating types, which behave as a self-incompatibility system. Consequently, sexual conjugation is possible only between clones of different mating type. Individual mating types are not inherited through the sexual stage of the life cycle, because the micronucleus remains undifferentiated and is capable of regenerating the entire spectrum of types within the new macronucleus formed at the start of each sexual generation. The mating type of a clone, however, is not expressed until sexual maturity, which occurs some 50 fissions after conjugation, but is, nevertheless, propagated faithfully on each occasion that the cell divides vegetatively. This case is of particular interest because it has been suggested that mating type determination may involve a developmental alteration of somatic DNA during macronuclear development, which may be analogous to the somatic differentiation of immunoglobulin genes (Orias 1981).

2.4.4 Antigenic switching in trypanosomes

Trypanosomes are unicellular flagellates, parasitic in the blood of man and other mammals. Following initial infection the parasites go through a multiplicative phase, as a result of which the host produces antibodies against the surface coat of the unicell. This consists mainly of a single glycoprotein, and is the only antigenic structure exposed to the host's immune system. Antibody production results in the death of the majority of the parasites, but a small fraction evade immunodestruction by producing a variant form of surface glycoprotein with different antigenic properties. A second cycle of parasite multiplication then ensues. This is again followed by the destruction of a majority of the flagellates, but with yet a further switch in the surface glycoprotein of a small number. This cycle is then repeated again and again, giving rise to an oscillating parasitaemia in which, despite a massive antibody response by the host, the parasite is able to circumvent the immune response by continually changing its surface coat antigen.

This repetitive switching is not induced by the antibody of the host but is programmed within the genome of the parasite. When a trypanosome first enters the host the switching rate is high, but progressively it decreases.

Moreover, different variant antigen types appear in a predictable, though somewhat imprecise, order. Trypanosomes are diploid organisms with a 2C genome size of 0.091 pg DNA, 32% of which is repetitive DNA, and a genome complexity about six times that of *Escherichia coli*. There are around 10^3 variable surface glycoprotein (VSG) genes per genome, so that about 10% of the genome is taken up by VSG genes and their flanking sequences. These genes are tightly linked within the genome, and consist of a number of different families which are related in sequence (Borst 1983).

Only one of these many VSG genes is expressed at any one time and this commonly depends on a form of duplicative transposition. In this mechanism an extra copy of a VSG gene is activated by transposition into a specific expression site, located near to a telomere. Only one expression site is functional at a time, and this site is able to accommodate only one VSG at a time, although two or more different VSG genes may be transposed to the same expression sites at different times. VSG gene expression is then controlled at the transcriptional level by promoter addition at the expression-linked site. This allows a large number of different genes to use the same expression site and the same promoter (Fig. 2.35).

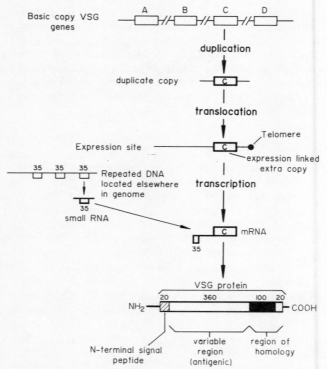

Figure 2.35 Duplication, translocation and transcription of the VSG genes involved in the production of the variable surface glycoprotein (VSG) which determines antigenic switching in trypanosomes (data of Donelson & Turner 1985).

Not all VSG genes show simple transposition activation of this kind. Some, and especially those which are expressed early in infection, are already located at a site near a telomere. For at least some of these, there is evidence to suggest that a functional copy may be generated by the recombination of sequences of at least two genes, each of which codes for a part of the resulting VSG.

Phase variation switching points of Salmonella

956 ● 985 ●
C A C A G G T . . .23 nucleotide. . . .T C A A A A A C C
gap

Switch points for VJ joining

C A C A G T G . . .23 nucleotide. . . .A C A A A A A C C
gap

Figure 2.36 Sequence homology between the switching points involved in the phase variation system of *Salmonella typhimurium* and the VJ joining regions of mammalian immunoglobulins (data of Simon *et al.* 1980).

2.4.5   The molecular bases of genomic alterations

Underlying the diversity of the various forms of genomic alteration that occur during eukaryote development are some elements of molecular commonality. As Simon *et al.* (1980) point out, the switch sites used in VJ joining in the immunoglobulin system of mammals show a remarkable similarity to the DNA sequences involved in the flip-flop transitions which characterize the phase variation mechanism of *Salmonella typhimurium* (Fig. 2.36). Additionally, as will become apparent in Section 3.2.1, there is an evident evolutionary link between certain prokaryotic and eukaryotic transposable elements, which are able to rearrange sequences within a genome, and the vertebrate retroviruses, which can rearrange sequences between genomes. All share an internal coding domain of from one to several kilobases, large terminal direct repeats of several hundred base pairs, usually with short inverted terminal repeats, and short flanking direct

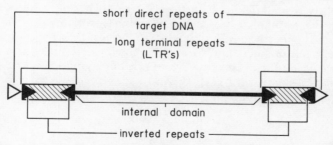

Figure 2.37 Structurally related nomadic elements of the kind found in *Tn9* (*Escherichia coli*), *copia* and *412* (*Drosophila melanogaster*) and two retroviral proviruses of vertebrates, Roux Sarcoma Virus (RSV) and Mouse Mammary Tumor Virus (MMTV) (data of Varmus 1983).

repeats of target DNA (Fig. 2.37). Moreover, all share the functional properties of insertion mutagenesis, the ability to transduce other genes and the capacity for deletion.

Given the presence of site-specific DNA sequences of this type, together with the necessary recombinases, the potential for patchworking the DNA of the genome by various forms of switching automatically exists within the genome. In the phase variation system of *Salmonella typhimurium* and the mating type system of yeast, switching allows for the alternate expression of only one of two genes. In *Salmonella typhimurium* the switch is effected by the inversion of a DNA sequence that makes or breaks the link between a gene and its promoter. In the mating type switch of yeast, a single expression site can be filled by either of two coding sequences. In the case of the VSG genes of trypanosomes the gene family involved is very large, and the duplicative transposition mechanism of control far more complex, since it allows for multiple switching.

Alternatively, the deletion of DNA segments by controlled activation of recombinases at specific times and specific sites makes possible the removal of large parts of the genome, either with (ciliates) or without (*Cyclops*) wholesale fracture of the genome itself. In the cyanobacterium *Anabaena* both deletion and rearrangement are involved in the differentiation of heterocysts. These are specialized structures which provide an anaerobic microenvironment which allows nitrogen fixation to proceed at times of nitrogen deprivation. Both categories of alteration occur adjacent to the nitrogen fixation (*nif*) genes. The first involves the excision of an 11 kb DNA segment, in the form of a circular molecule. The second results in a rearrangement of the DNA adjacent to the *nif* gene (Golden *et al.* 1985).

All these various forms of genomic modification depend on the existence of suitable exchange sites within the genome, whether these are involved as switch points or as excision points. One of the strongest candidates for this role are the simple sequences which appear to be ubiquitous repetitive components of the eukaryote genome (Tautz & Renz 1984). Such sequences consist of short stretches of DNA involving (poly [dA] · poly [dT]), (poly [dG-dT] · poly [dC-dA]) or (poly CAG · poly GTC) tandemly repeated nucleotides. They occur interspersed within the genome, and so are distinguished from satellite sequences. Since such simple sequences are present in the metabolically inactive micronucleus of *Stylonychia*, but are absent from the metabolically active macronucleus, they appear to have no general function in gene expression. We return to consider them in more detail when dealing with comparative aspects of genome organization in the next chapter of the book.

3

Coding capacities of genomes

*'It is the uncompromisingness with which dogma is held and
not in the dogma itself that the danger lies.'*

Samuel Butler

Now that we have an overview of the major features of eukaryote
development we can ask four key questions about the molecular
control of the processes involved. All of these relate to matters of
genome organization and the way in which the components of the
genome interact in a functional sense:

Q.1 Does all the DNA in a genome play a role in development. If
not, how much of the genome has a coding function?

Q.2 How is gene activity regulated in eukaryotes?

Q.3 Does the large-scale arrangement of genes in chromosomes
have any functional significance?

Q.4 How important is the amount of gene product for the
canalization of the phenotype?

3.1 GENE REGULATION IN EUKARYOTES

From the genetic analysis of development it is clear that genes are turned on
and off at specific times during development, which implies that gene
activity is precisely regulated. Such regulation can occur at a variety of
levels.

3.1.1 Transcriptional controls

Gene expression always begins with RNA transcription, and the available
evidence indicates that the most common mode of eukaryote gene control is
at the transcriptive level. To understand this mode of control we need to
know:

(a) what signals turn genes on and off;
(b) how genes are exposed to the on–off signals; and
(c) to what extent the response of a given gene to these signals is influenced by the DNA domain in which it occurs.

Transcription involves using the base sequence of one DNA strand of the double helix as a template for a polymerization reaction which produces a complementary RNA molecule. In eukaryotes three distinct kinds of nuclear RNA polymerases are involved in the transcription of discrete kinds of RNA. Polymerase I, found in the nucleoli, controls the transcription of the 28S and 18S ribosomal (r) RNA genes. Polymerase II transcribes the genes that encode proteins, and results in the initial production of messenger (m) RNA molecules, while polymerase III transcribes those genes responsible for the production of 5S rRNA and the transfer (t) RNAs (Darnell 1985).

The first step in transcription involves the binding of RNA polymerase to the DNA molecule. Thus, the initiation of RNA synthesis is a selective process, in the sense that the specific transcription of a given class of eukaryotic genes depends on recognition by the correct polymerase. Binding occurs at specific sites, termed promoters. Evidence indicates that eukaryote promoters are first recognized by DNA-binding protein factors which are distinct from the polymerase itself. Polymerase then interacts with the DNA-factor complex to initiate transcription. On binding to the promoter site, the polymerase determines not only where transcription is to begin, by exposing a stretch of single stranded DNA, but also which of the two DNA strands is to serve as a template for complementary base pairing with incoming ribonucleotides. The polymerase then moves along the DNA template in a $5'\rightarrow3'$ direction, extending the RNA chain one nucleotide at a time. It continues transcription someway past a second control signal, the termination sequence. At this point the polymerase is released from the template strand, which also dissociates itself from the newly formed RNA chain.

3.1.1.1 Polymerase II systems

Figure 3.1 summarizes the general landscape of a protein-producing gene. Notice that transcription does not terminate at the eventual mRNA 3' sequence but proceeds through this site and terminates some distance beyond it. However, although eukaryote cells produce initial transcripts with heterogeneous 3' ends they also incorporate a surprisingly efficient machinery to produce mRNAs with well-defined 3' ends (Birnstiel *et al.* 1985).

The 5' flanking regions of many protein-producing genes contain one or both of two conserved sequences whose location and structural similarity to prokaryotic promoter elements suggest their involvement in promoting the transcription of polymerase-II systems. The first of these is an AT rich sequence, the TATA box, which is generally located some 20–30 bp upstream from the transcription initiation site and is characterized by the consensus sequence

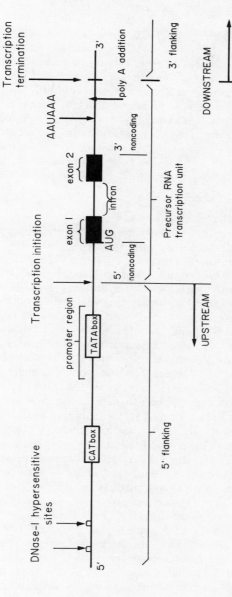

Figure 3.1 Molecular landscape of a generalized mRNA producing gene.

$$5'\text{-}TATA^A_TAT^T_N\text{-}3',$$

which is reminiscent of the Pribonow box of prokaryotic promoters

(5'-TATAAT-3')

There are many variants of this consensus sequence, with the first four nucleotides being the most conserved. Hence the term 'TATA box' is often used to describe the sequence. Some promoters either lack such a box or can function *in vivo* when the sequence is deleted, though following such deletion the start points of transcription are often heterogeneous. Thus, the TATA box appears to accurately position the polymerase so that the initiation event occurs precisely. The second conserved sequence is the CCAAT box, which is located 70–90 bp upstream from the transcription initiation site of some polymerase II genes. This has the canonical sequence $5'\text{-}GG^C_TCAATCT\text{-}3'$, reminiscent of the –35 region of some *Escherichia coli* promoters, and is referred to as the CAT box.

Some housekeeping genes, that is genes which function in all cell types, can have promoters that deviate radically from what, until now, has been considered the norm for polymerase II systems. These housekeeping gene promoters have high GC contents and may lack both the TATA sequence and the CAT sequence. Such GC-rich promoters have so far been found in the upstream landscapes of at least ten housekeeping genes, including the hypoxanthine phosphoribosyl transferase gene of the mouse (Melton *et al.* 1986), the human HPRT gene, the adenine phosphoribosyl transferase gene of the mouse, the dihydrofolate reductase gene of both human and hamster, the human adenosine deaminase gene and the human phosphoglycerate kinase gene (reviewed by Dynan 1986). In a detailed analysis of the human glucose-6-phosphate dehydrogenase gene, Martini *et al.* (1986) have confirmed the upstream presence of multiple copies of the hexanucleotide GGGCGG, and have compared this upstream area to those from ten other housekeeping genes drawn from human, chicken, mouse and hamster sources. These authors have noted that a consensus sequence can be found which is located between the TATA box (when it is present) and the cap site. This particular consensus sequence has not as yet been found in non-housekeeping genes. It will be of interest to determine whether there are fundamental differences, at the cellular level, in the way in which the transcriptional machinery discriminates between the control sequences of familiar genes with the TATA type landscapes, such as histones, globins and immunoglobulins, and those of the housekeeping type.

Positive control of transcription in prokaryotes often involves the binding of activator proteins to specific DNA segments located 5' to the promoter. Similarly, *Drosophila melanogaster* RNA polymerase II requires at least two distinct transcription factors, A and B, to initiate accurate *in vitro* transcription of histone H3, H4 and actin genes. The B factor contains a sequence-specific component which binds to a 65 bp region of DNA that includes the TATA box, the start point of transcription and a portion of the leader sequence (Parker & Topol 1984a, b).

The sigma initiation factor of bacteria is a polypeptide that binds to RNA polymerase, conferring on it the ability to initiate the transcription of genes whose promoters conform to the -35 and -10 consensus sequences. In at least some bacteria, multiple sigma factors are present, each of which directs the core RNA polymerase to initiate at a different class of promoters (Travers 1985). In *Escherichia coli* a sudden increase in temperature stops the production of normal gene transcripts and initiates the transcription of 15 heat shock genes which require the product of the *htpR* (*hin*) gene for their expression. This gene has strong homology with two regions of the sigma factor of vegetative *Escherichia coli*, one of which has the characteristics of a DNA binding site.

In certain cases, DNA sequences which extend 66 bp or more upstream from the initiation site, and which are additional to the TATA and CAT boxes, may also play a role in controlling the transcription of polymerase II eukaryote genes. When individuals of *Drosophila melanogaster* are exposed to thermal stress, a series of novel proteins, the heat shock proteins, are produced and the synthesis of most other proteins is repressed. This response, which provides protection against thermal damage, depends upon the activity of a series of heat shock genes which begin vigorous transcription of the mRNAs responsible for the production of the heat shock proteins within minutes of temperature elevation. The most prominent of these is a molecule of molecular weight 70 000 (hsp70) encoded by five closely related genes located at 87A and 87C of the polytene map. It is also one of the few proteins to be highly conserved between prokaryotes and eukaryotes. Moreover, the -35 consensus sequence CNN<u>CTTGAA</u> of the heat shock responsive promoters of *Escherichia coli* genes contains the first half of the <u>CTNGAA</u>, NN<u>TTCN</u>AG of eukaryotic heat shock responsive genes.

Copies of the *Drosophila melanogaster* hsp 70 genes from both these sites have been cloned and sequenced. They exhibit close homology of both the coding and the 5' flanking sequences extending about 250 bp upstream. Using P element germ line transformation, Dudler & Travers (1984) have introduced *in vitro* mutated hsp 70 genes, carrying deletions in their promoter regions, into the fly genome. To distinguish between the RNA product formed under the control of the introduced hsp 70 promoters and that produced by the endogenous hsp 70 promoters, the 5' part of the hsp 70 gene was fused to a cloned *Drosophila melanogaster Adh* gene. Different variants of this fusion gene contained respectively 186, 130, 97, 68 and 44 bp of hsp 70 flanking DNA. The fusion genes were then inserted into the fly genome, using a P element vector (Carnegie 20) which also contained a functional rosy gene (ry^+) serving as a marker for transformed flies. The result was that 97 bp of the hsp upstream sequence were sufficient for the promoter to function at a level similar to that of the wild type gene. By contrast, the activity of the -68 deletion gene was 50–100 times lower. Thus, the region between -68 and -97 must contain an additional sequence necessary for efficient promoter utilization. A comparison of the hsp 70

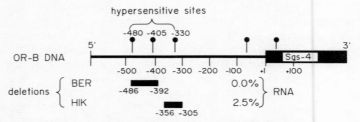

Figure 3.2 Reduction in RNA production caused by two deletions in the DNase I-hypersensitive sites 5′ to the *Sgs-4* gene of *Drosophila melanogaster* (data of McGinnis *et al.* 1983).

sequences between −68 and −97 with the corresponding regions of other *Drosophila melanogaster* heat shock genes reveals no homology common to all of them, though a sequence with an 8 out of 10 match to a consensus sequence is located between −72 and −85.

A second region which appears to be important in the control of polymerase II gene systems is defined by one or more DNase I-hypersensitive sites (Elgin 1981, Eissenberg *et al.* 1985). These are short, nucleosome free, regions, some 50–400 bp long, which mark sites at which regulatory proteins are thought to gain access to specific DNA sequences. Such sites are generally located within 1000 bp of the 5′ ends of genes that are either active or else capable of being activated. The *Sgs-4* locus of *Drosophila melanogaster* (3C11) is one of a number of genes responsible for producing the glue protein used to attach the pupal case to a dry surface during metamorphosis. The activity of this gene varies between different strains of *Drosophila melanogaster*, many of which produce little or no *Sgs-4* protein. McGinnis *et al.* (1983) have shown that this is due to an almost total abolition of transcription, a consequence of the deletion of, or else

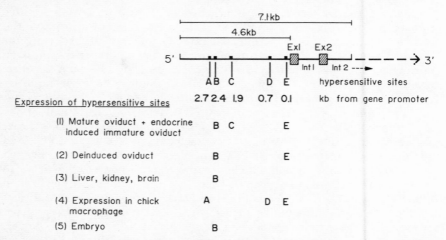

Figure 3.3 Developmental expression of upstream DNase I-hypersensitive sites (A–E) involved in the control of the chicken lysozyme gene (data of Fritton *et al.* 1984).

alterations in, a complex of upstream hypersensitive sites located 300–500 bp upstream from the transcription initiation site (Fig. 3.2).

That these hypersensitive sites may be relevant not only to the control of gene expression but also to its tissue specificity, is indicated by the observations of Fritton *et al*. (1984) who have demonstrated a complex pattern of DNase I-hypersensitivity, involving multiple upstream sites, in the case of the chicken lysozyme gene. The synthesis of lysozyme mRNA in the adult oviduct is under the control of steroid hormones, and can be reversibly induced in immature birds by oestrogen administration and withdrawal. Most of the cell types in adult chickens do not express lysozyme mRNA, though it is constitutively transcribed at a low rate by mature macrophages. In cells of the mature oviduct, the region upstream from the promoter of the lysozyme gene is marked by three hypersensitive sites. Relative to their distance in kilobases from the promoter these are referred to as HS-2.4, HS-1.9 and HS-0.1. These same sites can also be demonstrated in the cells of the oviducts of immature birds when these are treated with a synthetic oestrogen. At deinduction, following interruption of hormone treatment, the HS-1.9 site disappears, but reappears again following secondary induction. None of these sites are present in erythrocytes, and only one, HS-2.4, is present in liver, kidney and brain. Finally, in mature macrophages, which constitutively express the gene, neither of the two oviduct sites, HS-2.4 and HS-1.9, are present, but two new sites are formed at HS-2.7 and HS-0.7 (Fig. 3.3). These observations indicate that chromatin structure varies with the functional state of the gene. Thus, the 0.1 site is present only in cells which express, or have the potential to express, the gene, namely the oviduct (whether mature or immature) and the mature macrophage.

Yet a third region capable of influencing polymerase II gene activity in *cis* consists of DNA sequences that markedly increase the transcription of a gene, in a manner which is relatively independent of their position and orientation relative to the gene, i.e. their activity can be exerted over large distances and from a position either 5′ or 3′ to the gene. Like the DNase I-hypersensitive sites, these enhancers may also exhibit cell-type specificity, which implies they may be of importance in regulating cellular differentiation. They occur, for example, in the intron between the VDJ coding region and the C region exons of rearranged heavy chain immunoglobulin genes (reviewed in Marx 1983) and play a role in turning on immunoglobulin genes in antibody-producing cells. Enhancer sequences of this type also turn on insulin genes in insulin-producing cells, and chymotrypsin genes in cells that produce this enzyme. Using an *in vitro* transcription assay, it has been possible to show that the simian virus 40 enhancer acts on linked human β-globin genes, transiently introduced into HeLa cells, by increasing the number of RNA polymerase II molecules transcribing the linked gene (Treisman & Maniatis 1985, Weber & Schaffner 1985). Enhancers, like promoters, may turn out to be regions of interaction for site-specific DNA binding proteins (Dynan & Tjian 1985).

The systems we have dealt with so far all operate *cis* to the genes in question. *Trans* acting regulators of polymerase II transcription are also known. Section 68C of the polytene system of *Drosophila melanogaster* includes three genes involved in the production of salivary gland glue secretion proteins, namely *Sgs-3*, *Sgs-7* and *Sgs-8*. The *l(1)npr-1* gene is a *trans* acting regulator of the 68C region. In the presence of this X-chromosome mutation none of the 68C glue protein RNAs accumulate (Meyerowitz *et al.* 1984).

3.1.1.2   Polymerase I systems

The genes responsible for the production of the large ribosomal RNAs in eukaryotes occur as tandem arrays. Each transcriptional unit is separated from its neighbour by several kilobases of spacer DNA, composed primarily of varying numbers and kinds of repeated DNA sequences. In *Xenopus laevis*, for example, each spacer is punctuated by 2–7, or more, imperfect duplications of the gene promoter, the so-called *Bam* islands. Separating these spacer promoters from each other are regions of intermingled 60 and 81 bp repeating elements. The 81 bp elements represent 60 bp units with an additional 21 bp added. Spacers of different length may be intermingled in the same tandem unit. This length variation results from a variable number of super repeats, comprising one spacer promoter plus 6–12 copies of the 60/81 bp elements. Less commonly, the expansion or contraction of repetitive regions 0 and 1 (Fig. 3.4) may contribute to spacer length variation.

Both the spacer promoters and the 60/81 bp repeats contain sequences homologous to the gene promoter. Thus, each spacer promoter shows 90% homology to the gene promoter, while each of the 60/81 bp elements contain an imperfect array of a 42 bp sequence present in nucleotides -72 to -114 of the gene promoter (Reeder 1984). The 60/81 bp repeats act as enhancers of the gene promoter, presumably as a consequence of the 42 bp elements held in common. The spacer promoters, on the other hand, do not appear to act as enhancers, since their removal has no influence on transcription.

In interspecies hybrids between *Xenopus laevis* and *Xenopus borealis* the *Xenopus laevis* ribosomal genes are transcriptively dominant to those of *Xenopus borealis*, regardless of which parent introduces them into the hybrid. This can be explained by the fact that the *Xenopus laevis* spacer, on average, has many more enhancer elements than does a *Xenopus borealis* spacer (Reeder & Roan 1984).

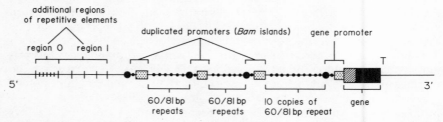

Figure 3.4   Structure of a generalized cloned repeat unit of the upstream rDNA landscape of *Xenopus laevis*, to illustrate the presence of duplicated promoters (data of Reeder 1984).

3.1.1.3 Polymerase III systems

In eukaryote genes transcribed by RNA polymerase III, the promoter is contained within the coding sequence itself. RNA polymerase III, like RNA polymerase II, cannot recognize DNA sequences on their own, but can do so when aided by a sequence-specific binding protein. Thus, the RNA polymerase III transcription factor of *Xenopus laevis* (TF III A), like the B factor promoter region of *Drosophila melanogaster* polymerase II, binds specifically to the control region of 5S genes.

As far as tRNA genes are concerned, the promoter, which is highly conserved, is split into two boxes, A and B, each of about 10 nucleotides and separated by a 30–40 bp interval. Initiation of transcription begins 11–18 bp upstream from the 5′ element of the A box, but continuity of the 34 bp intervening sequence is not necessary for efficient transcription (Ciliberto *et al.* 1983). The gene is delimited at its 3′ boundary by at least one run of four or more T residues on the non-coding strand, which are the signals required for efficient termination of transcription.

In the case of the 5S RNA gene, the internal control region (ICR) is a sequence of 34 nucleotides, and initiation of transcription occurs 50 bp upstream from the 5′ boundary of this region. The ICR of *Xenopus laevis* can be divided into two components, the first of which is homologous and functionally equivalent to the A box of tRNA genes (Fig. 3.5). It is possible, therefore, to construct functional hybrids between 5S and tRNA genes.

Two kinds of 5S RNA multigene families are present in *Xenopus laevis*, a somatic type and an oocyte type. These provide a neat internal control system for studying differential gene activity, since while the somatic type comprises only 2% of the total 5S RNA genes, it encodes more than 95% of the 5S RNA synthesized in somatic cells. In oocytes, on the other hand, mainly oocyte-type 5S RNA is synthesized. This difference is due to control at the level of transcription (Brown 1984). Polymerase III does not recognize the internal promoter sequence of the 5S gene directly, rather it recognizes a transcription complex between the DNA of the promoter and three or more proteins. While somatic-type 5S RNA genes are able to form

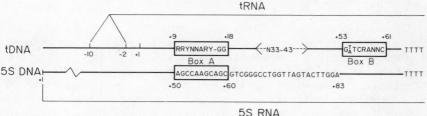

Figure 3.5 Sequence homology between the internal control regions of polymerase III systems. R, purine; Y, pyrimidine; N, any nucleotide; −, variable presence of 1–2 unspecified nucleotides, or else no nucleotide (data of Ciliberto *et al.* 1983).

such complexes in somatic cells, the oocyte-type 5S RNA genes are not, because they exist in a nucleosome configuration which prevents the formation of stable transcription complexes.

3.1.1.4   DNA methylation and gene activity

Over and above the transcriptional controls that operate through individual promoter systems, it has been suggested that methylation may play a general role in the control of gene expression. There are two kinds of argument that can be used in support of this suggestion. First, methyl cytosine (MeCy) is the only major modified base known in eukaryotes, and certainly represents a stable DNA modification which is known to be clonally inherited. Secondly, a number of developmentally regulated genes have been found to be hypomethylated in cells where those genes are expressed, and to be methylated in cells where they are not (Razin & Riggs 1980).

There are two situations, however, which indicate that, at best, any involvement of methylation in the regulation of gene activity is likely to be variable:

(a)   No clear example of a heavily methylated gene is known in any invertebrate. DNA from the brine shrimp *Artemia salina*, for example, contains less than one 5MeCy residue per 59 kb of DNA (Warner & Bagshaw 1984). Likewise, Urieli-Shoval *et al.* (1982) were unable to find any detectable 5MeCy in *Drosophila melanogaster* DNA, despite having used several very sensitive analytical techniques. In vertebrates, the identification of methylation in specific gene regions has been based on the use of restriction enzymes that are blocked in their cleaving activity by the methylation of cytosine residues at their specific recognition sequences. Since only a minority of CpG sequences are within a segment that can be recognized by one of the methyl-sensitive restriction endonucleases (Bird 1984) this assay is clearly biased.

(b)   Instances are known where specific genes can be transcribed even when they are fully methylated. This applies to the vitellogenin A1 and A2 genes (Gerber-Huber *et al.* 1983) and the α2(I) collagen gene (McKeon *et al.* 1982). Changes in methylation pattern cannot, therefore, be a general prerequisite for gene activation.

There is a dramatic and global undermethylation of DNA sequences in the two terminal extra-embryonic cell lineages derived from the trophecto-derm and primitive endoderm of the mouse (Fig. 3.6). This applies, however, to both repetitive and low copy number sequences, whether transcribed or non-transcribed (Sanford *et al.* 1985). Moreover, the mesodermal layer of the visceral yolk sac exhibits high levels of methylation, despite its extra-embryonic nature, and so follows the pattern of the primitive ectoderm, the pluripotent lineage from which it is derived. The extra-embryonic tissues form the interface between the foetus and the maternal environment. As such, they function as nutritive, excretory and

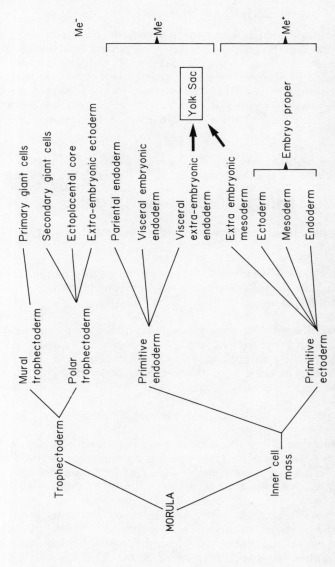

Figure 3.6 Undermethylation (Me⁻) in the terminal extra-embryonic cell lineages of the mouse which form the interface between the foetus and the maternal tissue (data of Sanford *et al.* 1985).

secretory exchange surfaces, as well as providing an immunological barrier, and so carry out functions performed by a variety of different adult organ systems. Undermethylation may therefore be necessary to allow for the flexibility of gene regulation required for these multiple activities.

3.1.2 Post-transcriptional controls

The biogenesis of RNA requires not only the transcription of a particular gene, but also the proper processing of the primary transcript to yield a mature RNA. The coding sequences of many eukaryotic genes are interrupted by stretches of non-coding DNA, and these are transcribed as part of the precursor RNA but are subsequently deleted by a cleavage-ligation process – termed 'RNA splicing'. There are three general classes of splicing reaction, one for each of the major transcription systems (Cech 1983).

Nuclear mRNA.   Here splice sites are designated largely or entirely by short sequences at the intron-exon junctions with the consensus form:

$$5'\text{-}^{A}_{C}AG\underline{GT}GAGT. \ldots .intron. \ldots .(Py)_6XC\underline{AG}G^{G}_{T}\text{-}3'$$

where splice points are designated by arrows and the underlined residues are invariant.

Nuclear pre-tRNAs.   Splicing is directed mainly by the structure of the exons but, unlike mRNA precursors, there is no known consensus sequence for the splice sites.

Nuclear pre-rRNA.   Splicing is directed largely by the structure of the intron. Indeed, the excised intron is itself capable of undergoing a second splicing reaction to produce a covalent circular form. Homology at the splice sites is modest, with only one nucleotide conserved at each border, namely:

$$UX. \ldots . . intron. . . . . GX$$

Added to this, the cleavage-ligation activity responsible for the splicing of the rRNA precursor in *Tetrahymena* is known to be intrinsic to the RNA molecule, so that there is no requirement for additional enzymes *in vitro* (Bass & Cech 1984). That is, the RNA itself functions as a ribozyme.

In the case of mRNA, processing involves two other events (Fig. 3.7). First, the transcript is 'capped' by the enzymic addition of the methylated sequence at the first nucleotide. Secondly, a poly (A) sequence of about 200 adenylated residues is usually added to the 3' terminus of the transcript, though this does not occur in the case of histone gene transcripts. Only after processing is a transcript able to exit from the nucleus. With the interposition of several biochemical events between transcription and

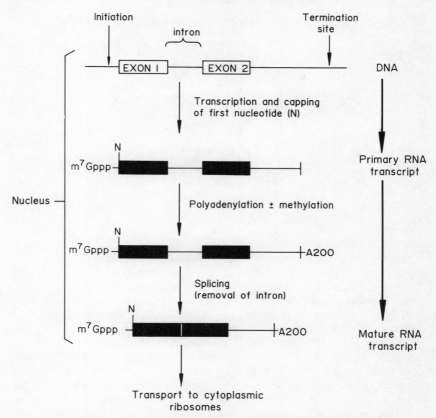

Figure 3.7 Steps in mRNA formation (data of Darnell 1982).

eventual mRNA production, the potential arises for post-transcriptional regulation of mRNA supply, since this can be achieved through the modulation of any of the processing steps. This is most obvious in the case of complex transcription units whose primary transcript can give rise to two or more mRNAs. There are three rather different ways of achieving this, all of which are capable of generating distinct gene products from a common transcription unit in a cell- or tissue-specific fashion:

(a) There may be alternative 5′ initiation sites, as in the case of the α-amylase gene of the mouse (Young *et al.* 1981).
(b) There may be alternative termination sites in the 3′ region. This occurs in the chicken vimentin gene (Zehner & Paterson 1983) and the β2-microglobulin gene of the mouse (Parnes & Robinson 1983).
(c) There may be differential splicing of the coding exons. Fibronectins, for example, are a class of high molecular weight glycoproteins that play a key role in a variety of contact processes, including cell attachment and cell migration (Hynes 1986). They include both

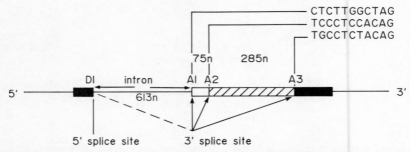

Figure 3.8 Alternative splicing patterns responsible for the production of the three fibronectin mRNAs found in rat liver. These are generated from a single coding sequence by the pairing of one 5' splice site (D1) with any one of the three 3' sites. One of these, A1, is at the beginning of, and two, A2 and A3, are within, a single complex exon (data of Tamkun *et al*. 1984).

cellular and plasma forms, which have distinctive binding characteristics. They are heterodimers of similar, but not identical, polypeptides. The diverse forms are generated by the transcription of a single gene into a common precursor which then undergoes alternative splicing to yield at least ten different polypeptides (Kornblihth *et al*. 1985). The three fibronectin mRNAs found in the rat liver, for example, are generated from a single coding sequence by joining a single 5' splice site with any one of the three 3' splice sites, one at the beginning of, and two within, a single complex exon (Fig. 3.8). Differential selection of alternative exons is also used to generate tissue-specific protein isoforms of the calcitonin gene (Fig. 3.9) and the tropomyosin gene (Medford *et al*. 1984), as well as stage-specific variants of the myosin heavy-chain protein of *Drosophila melanogaster* (Rozek & Davidson 1983).

Splicing is an intramolecular event involving the deletion of intron sequences from the primary transcript of a given gene. *In vitro* experiments indicate, however, that exons from two different mRNA precursors can be spliced together, a process referred to as trans splicing (Konarska *et al*. 1985, Solnick 1985). Thus, in the genesis of the chloroplast ribosomal protein S12 of the liverwort, *Marchantia polymorpha*, two transcripts, *on opposite strands and more than 60 kb apart*, are spliced together to yield the mature RNA that codes for this protein (Fukuzawa *et al*. 1986).

Having been successfully transported to the cytoplasm, mRNA molecules are subject to a number of controls which further affect their stability. We will meet an example of selective translational regulation in the case of the mRNA coding for chorion protein S15-1 of *Drosophila melanogaster*. This appears prematurely with respect to ·the known period of C15 protein synthesis and is then selectively excluded from polysomes and degraded, only to be retranscribed and translated later in oogenesis (see § 3.4.2.1). A second case involves the heat shock response of *Drosophila melanogaster*.

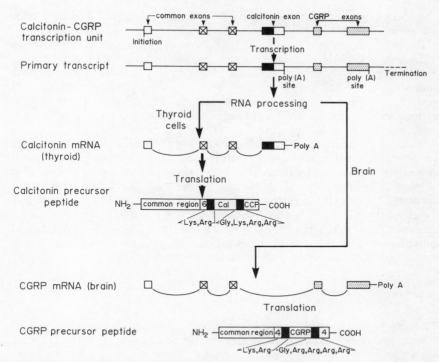

Figure 3.9 Alternative pattern of RNA processing of the primary transcript of the rat calcitonin–CGRP gene. The thyroid actually contains both calcitonin and CGRP in a ratio of 95:1–98:1. In brain, most, or all, of the mature transcript is present as CGRP mRNA (data of Rosenfeld *et al.* 1984).

Here the proteins produced under normal growth conditions are immediately repressed and nine new heat shock polypeptides are produced almost instantaneously to replace them. This requires dramatic changes in the specificity of protein synthesis in which pre-existing messages are sequestered from translation, while heat shock protein mRNAs are translated very efficiently. Thus, controls exerted on both transcription and translation are co-ordinated to ensure that heat shock proteins are produced as rapidly as possible (DiDomenico *et al.* 1982). The third case concerns an abundant mRNA, termed T1, which is produced by a unique site (39CD) very close to the histone genes in *Drosophila melanogaster*, and which codes for a small acidic protein. It is first synthesized during oogenesis, when it becomes associated with polysomes and is presumably translated. It is also present in early embryos, but is selectively excluded from polysomes in 3 hour and 5 hour embryos where it remains sequestered as an untranslated ribonucleoprotein. This mRNA species becomes polysome-associated again in 18 hour embryos and is then translated (Fruscoloni *et al.* 1983). T1 mRNA is, thus, not translated at all developmental stages and is clearly under specific translational regulation. Unlike the S15-1 mRNA which codes for a chorion protein in *Drosophila melanogaster*, however, it is not degraded when not translated.

Finally, following translation, the polypeptide product may be further modified. The various collagens have similar mechanisms of transcription, processing and secretion, but their biosynthesis involves several post-translational modifications including hydroxylation, oxidative deamination and glycosylation of a precursor polypeptide, procollagen. These modifications occur extracellularly through the mediation of highly specific enzymes, the procollagen peptidases (Bernfield 1981).

Most of the neuroactive peptides so far characterized in eukaryotes appear to be synthesized from a large polyprotein in which individual peptide sequences are flanked by internal cleavage sites, providing the potential for generating multiple small peptides. Thus, egg laying in molluscs is mediated by a small multigene family of some nine members which express functionally related, but non-overlapping, sets of neuroactive peptides in different tissues (Scheller *et al.* 1983). This ability to alter the processing of polyproteins provides a mechanism for changing the pattern of specific gene expression without altering the pattern of transcription.

3.2   *DROSOPHILA* GENOMES

The genome of *Drosophila melanogaster* has been better characterized at both genetic and molecular levels than that of any other eukaryote. It, therefore, allows us to ask and answer the question – how much of what we know about the structure of this genome actually relates to function? In the first place, the haploid genome is only 165 million bp in size. In the second place, this small genome is partitioned into only four chromosomes. How then is this genome organized in a molecular sense?

3.2.1   The general molecular organization of *Drosophila* genomes

The *Drosophila melanogaster* genome was classically partitioned by reassociation kinetics into three major components (Crain *et al.* 1976b). The largest of these (c. 60%) consists of single copy sequences. The remainder includes both highly repeated sequences (c. 25%), and a very heterogeneous moderately repeated fraction which contains a significant number of nomadic sequences.

In situ hybridization experiments indicate that the highly repeated sequences are present as long tandem arrays located in the heterochromatic regions of all the chromosomes (Peacock *et al.* 1974). They thus have a very confined distribution within the genome, occurring in a small number of large blocks. The moderately repeated sequences are, in general, interspersed with the unique sequences, and chromosomal walks have revealed that, on average, there are between one and three nomadic elements per 100 kb of euchromatic DNA. However, some parts of the genome contain localized, moderately repeated sequences. This includes, for example, the rRNA genes on the X- and Y-chromosomes, the 5S genes (56F) and the histone genes (39DE).

Table 3.1 Characteristics of some mobile dispersed genes in the genome of *Drosophila melanogaster* (from Rubin 1983).

Element	Number of copies per genome	Length total (kb)	Length direct repeat (bp)	Number of base pairs duplicated at insertion
copia	20–60	5	276	5
412	≈40	7	481/571	4
297	≈30	6.5	412	4
mdg 1	20–30	7.2	442/444	4
mdg 3	15	5.5	269	3 or 5
B104	100	8.5	429	5
gypsy	≈10	7.3	500	nd

nd = not determined.

A large proportion of the moderately repeated DNA is composed of different sequence families. *In situ* hybridization to polytene chromosomes indicates that these are dispersed through, and in most cases move around within, the genome. The first of these mobile, or nomadic, sequences to be discovered were the *copia*-like elements, a group of families whose members average 5 kb in size. There are over 30 of these families, seven of which have been studied in some detail (Table 3.1). Although non-homologous in sequence, these various mobile families share five diagnostic properties (Rubin 1983):

(a) They have nucleotide sequences at their ends which are terminally redundant and which are flanked by direct or inverted repeats of various lengths. Figure 3.10 illustrates the structure of a common mobile element with two long terminal repeats (LTRs), consisting of direct repeats flanked by small inverted repeats. These LTRs harbour between them a stretch of some 6 kb of DNA, which can code for a number of RNAs.

(b) They each occur at a large number of widely scattered and highly variable sites within the genome, though only a single element occurs at each site. They have been found within other middle repeated DNA sequences, including those that are tandemly repeated, in highly repeated DNA and within unique DNA.

(c) They undergo transposition within the genome, as evidenced by the striking differences in both their number and their location in different strains of *Drosophila melanogaster* (Strobel *et al.* 1979), and they can be

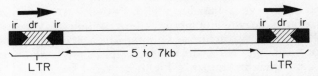

Figure 3.10 Structure of a common nomadic element of *Drosophila melanogaster*. LTR, long terminal repeat; ir, inverted repeat; dr, direct repeat.

induced to move under different experimental protocols. Different families of nomadic elements have variable copy numbers and different terminal repeat lengths. When they insert into the genome they generate different length small duplications of the flanking DNA at the point of insertion. When they excise they may do so precisely or they may generate varying length deficiencies of the surrounding DNA in either or both directions, depending on the element.

(d) At least two members of the *copia* family insert preferentially into the genome. *Gypsy* is incorporated specifically into the sequence TACATA (Freund & Meselson 1984), while *17.6*, a 7 kb element closely related to *297*, shows preferential insertion into the target sequence 5′ATAT which corresponds to the major portion of the consensus TATA box, TATA$_\text{T}^\text{A}$A$_\text{T}^\text{A}$ (Inouye *et al.* 1984).

(e) Some of these families, such as *copia*, are known to be transcribed, though it is not known if all copies within the genome are active in transcription. It is, indeed, from the fact that they often produce copious amounts of RNA that members of this family derive their name. About 90% of the RNA produced by the transcription of *copia* is located within the nucleus, and less than 10% exits into the cytoplasm where it can be found in both poly (A)$^+$ and poly (A)$^-$ forms. The major size class of poly (A)$^+$ RNA is 5.2 kb. This results from transcription which begins and ends in the terminal repeats of the element. Likewise, in the case of mobile element *412*, which is 7.5 kb long, a major size class of poly (A)$^-$ RNA occurs which is also 7.5 kb in length. *B104*, too, produces full-length transcripts. This type of transcription, which begins and ends in LTRs, also characterizes the integrated (proviral) form of vertebrate retroviruses. Indeed, this class of *copia* elements is strikingly similar in structure and behaviour to both the integrated proviruses of RNA tumour viruses and the *Ty1* nomadic elements of yeast. The *copia*-like element *17.6*, for example, has long terminal repeats homologous in nucleotide sequence to those of avian leukaemia sarcoma virus. It also contains three long open reading frames comparable with the *gag*, *pol* and *env* genes of the retrovirus, and the longest of these includes a sequence similar to that of reverse transcriptase (Saigo *et al.* 1984).

A rather distinctive class of mobile elements is provided by the foldback (*FB*) family (Truett *et al.* 1981). This, as its name indicates, consists of DNA sequences which, under experimental conditions, 'snap back' following denaturation at low DNA concentrations. A common component of this family is a pair of inverted repeats. Members of the family are, however, heterogeneous and show variability both in their total length and in their inverted terminal repeats, though the ends of the inverted repeats are closely conserved in sequence structure and are similar in all of them. In at least one member of this family, *FB3*, the inverted repeats are composed of tandem runs of short sequences similar to highly repeated DNA.

An important consequence of having nomadic elements spread through-out the genome is that two such elements can mobilize normally

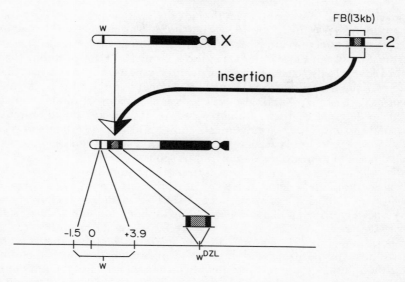

Figure 3.11 Mobilization of a chromosome 2 segment of *Drosophila melanogaster* by two fold back (*FB*) elements, leading to the introduction of this segment into the X-chromosome where it produces a highly unstable mutation (w^{DZL}), white-·Dominant Zeste-like (data of Levis *et al.* 1982, Levis & Rubin 1982, Zachar & Bingham 1982).

untransposable euchromatic segments. *FB* sequences, for example, have the capacity to move unrelated genomic segments *en bloc* to new locations. Moreover, the wanderings of mobile elements within a genome have unavoidable consequences when they are inserted into, or removed from (Collins & Rubin 1983), either a structural gene or its immediate flanking sequences. The highly unstable mutations, white-crimson (w^c) and white-Dominant Zeste-like (w^{DZL}), for example, are caused by insertions of nomadic DNA near to the white locus (Zachar & Bingham 1982). Both these insertions are partially homologous to one another and to the *FB* family. The w^c insertion corresponds to a single *FB* element, whereas the w^{DZL} insertion contains two *FB* elements between which there are unique sequences complementary to those normally found near the tip of chromosome 2 (Levis *et al.* 1982, Levis & Rubin 1982). It appears that these unique sequences have been mobilized by flanking *FB* elements and transposed from their normal position to the domain of the white locus (Fig. 3.11).

At least some of the mutant phenotypes associated with the insertion of the *gypsy* element into the Achaete-Scute Complex are due to transcriptional defects. Transcripts initiate correctly upstream of the T5 gene of this complex, but terminate within the mobile element that has inserted into it. The truncated transcripts thus contain the 5′ part of the gene, followed by *gypsy* sequences and a poly (A) tail. In this particular case, the chimeric RNAs still retain some wild type function and there is a tenfold increase in the level of RNA produced, even though the *gypsy* element is more than 400 bp downstream from the site of initiation. The mechanism by which this particular effect occurs is unknown (Campuzano *et al.* 1986).

A third family of mobile elements, the P family (O'Hare & Rubin 1983, Rubin 1985b), is also characterized by inverted repeats, but in this case they are only 31 bp in length. The complete P element is some 3 kb in size and, as we shall see later (§ 4.4.4), also functions as a transposon, though only in the germ line. Here their mobility leads to increased mutation rates, structural chromosome rearrangements and hybrid sterility (Bingham *et al.* 1982). As we have already mentioned, these P elements can be experimentally tailored so as to reintroduce suitable DNA sequences into new sites within the *Drosophila melanogaster* genome by microinjection into preblastoderm embryos. Although P elements are absent from sibling species of *Drosophila melanogaster* itself, similar, though not identical, elements are present in every species group in the subgenus, Sophophora, to which *Drosophila melanogaster* belongs (Lansman *et al.* 1985). They differ, however, in distribution within the genome. In *Drosophila melanogaster* they can be inserted at many positions within the euchromatic portion of the genome and, although they can insert into heterochromatin, are rarely found there. By contrast, in *Drosophila nebulosa*, *Drosophila saltans* and *Drosophila willistoni* they are found predominantly within the β–heterochromatin of the chromocentre. While *Drosophila hawaiiensis*, a distant relative of *Drosophila melanogaster*, is not known to contain mobile sequences, a P element of *Drosophila melanogaster*, introduced into young embryos of *Drosophila hawaiiensis* by microinjection, undergoes germ line transposition and numerical increase in subsequent generations (Brennan *et al.* 1984).

The *TE* element, a very large mobile DNA sequence, which includes some 100–200 kb of DNA (Ising & Block 1981), has been recovered at over 120 sites scattered throughout the genome, and induces lethal mutations at approximately half of the sites into which it inserts. Many *TE* elements contain sequences homologous with *copia*, and *FB* elements are also found at or near the ends of *TE* elements.

It is worth pointing out that in *Drosophila melanogaster* the interspersion pattern defined crudely by reassociation kinetics is undoubtedly an inescapable consequence of the generation, dispersion and loss of the numerous mobile families present within the genome.

Drosophila melanogaster and its sibling, *Drosophila simulans*, are near identical in both a metabolic and a developmental sense. However, they differ strikingly in their content of mobile sequences, which implies that these sequences are unlikely to play any important functional role in the gross organization of the genome. When plasmid libraries are constructed from DNA of both species, and clones are withdrawn from them at random and tested for their repetitiousness and mobility, the two genomes can be partitioned as in Table 3.2. These two genomes can be more directly compared by using the low copy number sequences as a constant (Fig. 3.12). Since the single copy complexities of the two species are probably not very different, and working on the reasonable assumption that only euchromatic DNA has been cloned, it has been concluded that

Table 3.2 A comparison of genome composition in two sibling species of *Drosophila* (from Dowsett & Young 1982).

Genome component	D. melanogaster		D. simulans	
	%	mbp	%	mbp
dispersed repeats (nomadics)	21	24	3	3.5
localized repeats	7	8	7	8
total repeats	28	32	10	11.5
low copy number sequences (unique)	72	83	90	83
total euchromatin		115		95
total heterochromatin		50		50
total genome		165		145

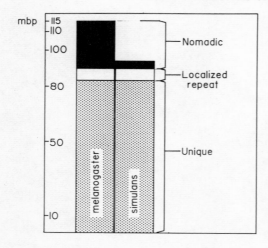

Figure 3.12 A comparison of genome organization in *Drosophila melanogaster* and *Drosophila simulans*, in which the nomadic component is shown solid. The centromeric heterochromatin, which is comparable in quantity in both species, has not been included (data of Dowsett & Young 1982).

Drosophila melanogaster has at least eight times more nomadic sequences, by mass, in its euchromatin than does *Drosophila simulans*. Thus, these two closely related species, with strikingly similar morphologies, can differ by some 20 million bp of nomadic DNA without any overt developmental or metabolic effects.

In a wider evolutionary context, five other members of the *Drosophila melanogaster* species subgroup have been examined for sequences homologous to 28 different middle repetitive sequences of *Drosophila melanogaster* itself, most of which were nomadic. Only two were held in common (Dowsett 1983). Thus, nomadic sequences provide no essential specific functions for either development or metabolism. The conclusion from all these studies is that nomadic elements are unstable genomic components found in different numbers and locations, not only between related species but even between strains of any one species. The implication of this, from the point of view of *Drosophila melanogaster* itself, is that at least 20% of the genome is dispensable in a developmental context. Whether these nomadic elements are also dispensable in an evolutionary context will be evaluated later (see § 4.4.4).

The fact that a significant portion of the *Drosophila melanogaster* genome consists of sequences that move around in the genome extends a principle that has long been recognized in prokaryotes, where transposons both produce mutants and lead to DNA rearrangements. In many respects, these nomadic elements of *Drosophila melanogaster* resemble the proviruses of vertebrate RNA tumour viruses. When a cell is infected with such oncogenic retroviruses the RNA strand of the virus is reverse transcribed, to finally yield a linear double stranded DNA molecule which is subsequently circularized. Such circular DNA templates then integrate into a host genome in the so-called proviral form, and the sarcomas and leukaemias produced by oncogenic viruses result from the inclusion of such a transforming oncogene (v-*onc*) into the genome. The cellular counterparts of these v-*onc* sequences in non-transformed cells are termed c-*onc* sequences (Bishop 1983, 1985).

DNA sequences that hybridize with *onc* genes have been found in *Drosophila melanogaster* (Shilo & Weinberg 1981, Bishop *et al.* 1985). Thus, of the 15 known oncogenes, sequences homologous to v-*abl* and v-*src1* clones map to 73B while a second clone, v-*src2*, maps to 64B (Hoffman-Falk *et al.* 1983). Human epidermal growth factor is a mitogenic polypeptide which binds with high affinity to a specific membrane glycoprotein receptor possessing intrinsic tyrosine-kinase activity, and so initiates a chain of events leading to DNA replication. This receptor molecule shares a close similarity with v-*erbB*, a member of the *src* family of oncogenes which also have tyrosine-kinase activity. By probing the *Drosophila melanogaster* genome with v-*erbB*, Livnen *et al.* (1985) have identified a gene at 57F which has three functional domains similar to the human epidermal growth factor receptor. Additionally, virus-like particles in cells of *Drosophila melanogaster* contain sequences homologous to those of *copia* (Shiba & Saigo 1983, Emori *et al.* 1985), and are associated with reverse transcriptase-like activity of the type that characterizes vertebrate retroviruses. Both proviruses and *copia*-like elements are transcribed into long polyadenylated RNAs which are initiated in one terminal direct repeat and terminated in the other, and the RNAs of both have 3' polyadenylation. There are, however, some differences between the two. The virus-like particles (VLPs) of *Drosophila melanogaster* are concentrated in the nucleus, whereas in vertebrates they are assembled in the cytoplasm. The *Drosophila* VLPs may well be more related to the intracisternal A-type particles found in mouse cells (Leuders & Kuff 1980) – which are non-infectious RNA-containing molecules, the poly-adenylated RNAs of which code for subunits of the particles themselves.

3.2.2 Heterochromatic DNA

The heterochromatic portions of *Drosophila melanogaster* genome, that is those chromosome regions which remain condensed during interphase, have long been recognized as 'inert' and very deficient in conventional genes (Gershenson 1940). About 29% of the female (XX) genome is composed of heterochromatin (Gatti *et al.* 1976), whereas in the male (XY) the value is slightly higher (35%). These regions are now known to be the predominant sites that house the highly repeated DNA sequences which collectively make up some 22% of the genome. The three autosomal pairs carry large pericentric blocks of heterochromatin. Additionally, the Y-chromosome is almost entirely heterochromatic and a substantial part of the X is too (Fig. 3.13). The actual base pair investment in the heterochromatin of the Y amounts to some 40 million bp with a further 20 million bp in the X heterochromatin. There are some 200 copies of the 18S and 28S ribosomal genes in the middle of the satellite-rich heterochromatic region of the X, and about the same number in the short arm of the Y (Y^s). Otherwise, the only known mappable functions present in the sex heterochromatin are the ABO and compensatory responder (cr^+) loci of the X (see § 3.4.1) and the six fertility loci of the Y (Pimpinelli *et al.* 1986). There are also a number of meiotic pairing sites on both the X- and the Y-chromosomes whose nature remains to be determined. A small number of conventional loci may map to the heterochromatic region of chromosome 2. These, however, occur at a

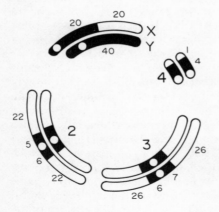

Figure 3.13 Relative proportions, in million base pairs, of heterochromatic (solid) and euchromatic (open) regions of the chromosomes of *Drosophila melanogaster*.

much lower density than in a euchromatic segment of comparable length. The third chromosome has no known gene loci within its heterochromatin (Hilliker *et al.* 1980).

Sequence analysis has revealed that three of the basic repeating units which make up the highly repeated DNA of *Drosophila melanogaster* are quite short, comprising, respectively, tandem repeat runs of the following sequences – 5'AATAT3', 5'AAGAG3' and 5'AATAACATAG3'. The fourth has a long repeating unit of 359 bp. Although small sites of hybridization also occur in particular euchromatic regions of the autosomes, a majority of highly repeated sequences are localized within the hetero-chromatin as long tandem arrays, which are not transcriptively active in somatic cells. A large number of other simple sequences also occur in heterochromatic regions, but these, in total, make up only a minor fraction of the total 'satellite' content. The total amount of heterochromatic DNA in the haploid *Drosophila melanogaster* genome is about 50 million bp, of which the major highly repeated DNA families make up at least 37 million bp (Lohe & Brutlag 1986).

It is clear from the results of *in situ* hybridization, as well as from clones which contain nomadic sequences juxtaposed to satellite sequences, that many nomadic families are likely to have representatives within the chromocentre. Additionally, the type 1 insertion sequences which interrupt the rRNA genes have extensive representation in the distal heterochromatin of the X. Thus, in addition to the highly repeated DNA, nomadic elements and other moderately repeated sequences must make a substantial contri-bution to the composition of heterochromatin in *Drosophila melanogaster*. Pimpinelli *et al.* (1986) have argued that the six fertility factors on the Y-chromosome are of enormous size, with kl–5 occupying some 4 million bp. If this turns out to be correct, then the 50 million bp of heterochromatic DNA in a haploid genome includes at least 37 million bp of satellite sequences, 3 million bp of ribosomal genes, with the remainder being accounted for by the six large fertility loci on the Y, sundry nomadic elements and the few genic sequences sequestered within the autosomal heterochromatin.

3.2.2.1 Functional studies

The highly repeated DNA sequences of *Drosophila melanogaster* have no known direct influence on phenotype, and whatever effects seem to stem from the heterochromatin which houses them are most likely due to the few transcriptionally active sequences embedded within it (Miklos 1982, 1985). Consequently, large deficiencies of heterochromatin are readily tolerated, as are excesses of added heterochromatin. Thus X0 males, with a deficiency of 40 million bp of DNA from the diploid genome, are somatically normal and fully viable, as are XYY males which carry an additional 40 million bp of DNA. Similarly, when the heterochromatin content is systematically altered, using chromosome duplications, it is possible to show that the developmental process is unaltered.

The normal diploid male (XY) genome contains approximately 214 million bp of euchromatin and 116 million bp of heterochromatin. In such a genome the average time to complete embryogenesis and hatch from the egg is 1250 min (21 h). Comparisons have been made of the time to hatching between normal individuals and a series of experimentally constructed individuals carrying a variety of heterochromatin mini-X-fragments in addition to the normal complement (Miklos 1982). In these partial polysomics the heterochromatin content ranged from 118 to 136 million bp, whereas the euchromatin content was standard. Hatching times were also determined for X0 and XYY males with heterochromatin contents of 76 and 156 million bp respectively. All genotypes, both control and experimental, were from the same cross so that only the heterochromatin component varied in the different comparisons. Additionally, the mini-chromosomes in question were introduced both through males and females. The data are summarized in Table 3.3, and indicate that in all cases the differences in developmental times were less than 2.5% of the average time. A similar series of comparisons involving the emergence times of females gave comparable results.

Thus, substantial increases and decreases in the amount of heterochromatin (± 34%), in otherwise rigidly controlled genetic backgrounds, produce a negligible effect on developmental time in *Drosophila melanogaster*. Furthermore, there are no overt effects on viability or morphology that correlate with increasing or decreasing amounts of heterochromatin. The one case where there is a disproportionate effect relative to size is in mini 1337. Significantly, this chromosome also has the largest amount of euchromatin, indicating that genic effects can have a disproportionate influence on developmental time.

3.2.2.2 Distribution of highly repetitive sequences within the heterochromatin

The organization of satellite sequences within the heterochromatin of different chromosomes has been examined in *Drosophila melanogaster*, and each chromosome has been found to have its own particular arrangements of satellites. It is symptomatic of the tendency to first seek a functional explanation, that chromosome-specific satellite patterns were argued to be critical for recognition processes between homologous as well as non-

Table 3.3 Influence of added heterochromatin on male and female mean hatching time in *Drosophila melanogaster*. The normal male genome is 330 mbp in size and includes 116 mbp of heterochromatin. The normal average hatch time is 1250 min. (From Miklos 1982.)

Added heterochromatin	Size (mbp) of addition	Percent difference in median hatch time			
		(XY + addition) − XY		(XX + addition) − XX	
		♀ transmitted	♂ transmitted	♀ transmitted	♂ transmitted
mini 1187	2	−0.08	0	−0.32	+0.56
mini 164	5	—	0	—	0
mini 118	6	—	+0.16	—	+0.32
mini 1337	8	—	+1.44	—	+0.72
mini 1492	10	+0.16	+0.24	+0.08	−0.40
mini 1514	10	+0.08	+0.48	+0.80	−0.08
mini 3	20	+0.88	+0.24	+0.88	+0.40
extra Y	40	+2.40	—	+1.20	—

homologous chromosomes (Peacock *et al.* 1977). The alternative, and equally valid viewpoint, is that the specificity has arisen as a result of non-random enzymic processes which the DNA cannot avoid.

This concept of satellite relevance to chromosome recognition phenomena received quite a boost when a DNA binding protein was isolated for the 359 bp 1.688 sequence (Hsieh & Brutlag 1979). Since this uncloned satellite was shown to predominantly hybridize to the X- and Y-chromosomes, it was seen as mediating in meiotic recognition processes.

On the other hand, the recombinant DNA and genetic data bases have provided a very different perspective. First, cloned 359 bp sequences hybridize *only* to the proximal heterochromatin of the X, indicating that the original Y-location was due to contaminating sequences in the initial uncloned probes (Hilliker & Appels 1982). Irrespective of the protein-DNA binding studies, therefore, it is clear that the X and Y are unlikely to pair on the basis of DNA sequence homologies due to the 359 bp arrays when these sequences are only on the X.

A similar situation holds for the major highly repetitive DNA of *Drosophila hydei*. There, a 1.696 satellite comprises 13% of the diploid genome and is located exclusively in the X-heterochromatin (Renkawitz 1978). Thus, the X and Y in *Drosophila hydei* are unlikely to utilize these X-specific sequences for chromosome recognition phenomena. Furthermore, since this major repeat compromises 13% of the haploid genome, as indeed does the X-heterochromatin, it is unlikely that pairing sites or other sundry elements occupy a significant portion of this heterochromatin.

The genetic data on chromosome recognition phenomena complement the molecular biology. The use of massive heterochromatin deletions, chromosomal rearrangements, and free duplications (reviewed by Miklos 1985), all indicate that satellite sequences *per se* are functionally irrelevant to all hypotheses dealing with chromosome pairing (Yamamoto 1979).

In coming to terms with the postulated roles of highly repetitive DNA sequences in cell size control, in growth and differentiation, in gene expression and in speciation events via chromosomal rearrangements, it is necessary to realize that the galaxy of attributes which has been showered on heterochromatin (reviewed by Cooper 1959) has also been forced onto satellite sequences *per se*. What has become abundantly clear is that where genetic effects are attributable to heterochromatic regions, the causal agents are generally other sequences sequestered within and between satellite sequences *per se*.

From data of this kind there is every reason to believe that the fly would be metabolically and developmentally normal were all the highly repeated sequences to be removed from the genome. Indeed, nature has carried out experiments of this kind, since related species of *Drosophila* can have enormous variations in their heterochromatin content and yet show no overt differences in development. What this means for *Drosophila melanogaster* is that the heterochromatin content of the haploid genome can effectively be pruned by 50 million bp, less the contribution made by the ribosomal genes, the few fertility factors and the fewer genic sequences, to the total heterochromatin content. In other words, some 40 million bp of the haploid genome is dispensable in terms of its contribution to normal development. Coupled with the 25 million bp of nomadic sequences, which are also developmentally dispensable, and not counting the 10 million bp contributed by the fertility factors sequestered on the Y-chromosome, this leaves 90 million bp of euchromatic DNA, which we can examine from the point of view of its coding capacity.

3.2.2.3 Satellite DNA binding proteins

With most of the hypotheses relating satellite DNA to somatic functions, either disproven or languishing due to inadequate data bases (Miklos 1985), one topic continues to attract attention – namely the roles of proteins which bind to such DNAs.

Since the initial demonstration of a particular embryonic protein which bound to the 359 bp sequences of *Drosophila melanogaster* (Hsieh & Brutlag 1979), Alfageme *et al.* (1980) and Levinger & Varshavsky (1982) have characterized a protein, termed D1, which binds to AT-rich sequences, such as the AATAT 1.672 satellite sequences, as well as binding to other sites throughout the genome. Alfageme *et al.* (1980) argued that such a protein could be involved in the control of a large number of genes, possibly as a repressor, since it bound to a number of different sites in polytene chromosomes.

Levinger & Varshavsky (1982) have further demonstrated that the D1 protein also binds not only to the 359 bp 1.688 sequences but even more tightly to the AATAT 1.672 sequences. Since AT-rich stretches occur throughout *Drosophila* and mammalian genomes, it was suggested that such a protein may:

(a) participate in higher order chromatin organization, such as chromatin compaction;
(b) participate in microtubule–centromere processes;
(c) prevent non-specific binding of other nuclear proteins, such as RNA polymerase or regulatory proteins, to AT-rich satellite sequences.

Strauss & Varshavsky (1984) have characterized a protein in the African green monkey (AGM), which binds preferentially to the 172 bp AGM repeat. These authors have argued that such a protein may aid both in nucleosome positioning and in higher order chromatin structure.

Finally, it has been demonstrated in *Xenopus laevis* that satellite I is a competitor to the oocyte-type 5S RNA genes, in that 5S RNA transcription is remarkably reduced when satellite I sequences and 5S sequences are juxtaposed on a plasmid whose transcription is examined in the *Xenopus laevis* oocyte system. This competitive reduction in transcription is due to a polymerase III type promoter present within the satellite I sequence itself (Andrews *et al.* 1984).

It is clear from these examples that satellite DNA binding proteins are revealing important insights into chromatin structure, but the question remains as to how germane such studies will be to phenotype. For example, in a metabolic context, repetitive DNA sequences could act as functional sinks for either specific or non-specific protein-DNA binding. The story really goes back to *Escherichia coli* where regulatory proteins bind very tightly to their target DNAs but also bind much more loosely to the chromosome itself. This equilibrium of tight, loose and unbound molecules underlies the regulatory activities of gene circuits (Kao-Huang *et al.* 1977, Berg *et al.* 1982, Takeda *et al.* 1983). Originally it was found that *lac* repressor binding was very sensitive to total DNA content (Lin & Riggs 1975), and these authors pointed out that the junk DNA in eukaryotes would have acted as a functional sink for all regulatory proteins. Hence, if the DNA amounts were altered, the regulatory circuits would operate suboptimally.

Hypotheses such as these can be readily evaluated in *Drosophila melanogaster* where large differences in heterochromatin content can be manufactured experimentally. Thus XO, XYY and XYYY males can be constructed, and all of them prove to be somatically undisturbed. This means that removal of 40 million bp of repetitive DNA from the genome, or the addition of 90 million bp, the majority of which is simple sequence DNA, leaves somatic development intact. Furthermore, massive deficiencies of the X-heterochromatin, which remove most, or maybe even all, of the heterochromatic 359 bp 1.688 satellite to which the Hsieh & Brutlag (1979) and the D1 proteins bind, also leave the organism perfectly viable and fertile. Thus, it is unlikely that these two proteins play a major role in any of the hypotheses which relate to higher order chromatin structure, microtubule–centromere processes or functional sinks for regulatory proteins.

3.2.3 Euchromatic DNA

We have already referred to the fact that most messenger RNAs carry a polyadenylated 3′ segment which is added to the initial transcript prior to its movement to the cytoplasmic ribosomes. At least 60% of the RNA found in polysomes of *Drosophila melanogaster*, however, is poly $(A)^-$ and there is no reason to doubt that this comprises mRNA molecules with the 3′ poly (A) segment extensively shortened. Most of the poly $(A)^+$ nuclear RNAs exit to the cytoplasm and form poly $(A)^+$ and poly $(A)^-$ RNAs on polysomes (Zimmerman *et al.* 1982). With the exception of histone gene transcripts, the nuclear poly $(A)^-$ RNAs apparently do not exit into the cytoplasm (Fig. 3.14). It is, therefore, possible to use the RNAs released from polysomes (poly $(A)^+$ and poly $(A)^-$ mRNA) to obtain a measure of gene transcription, either by employing reassociation techniques or else by analyzing the transcription products of specific cloned fragments of the genome. Let us begin with the reassociation approach.

Consider an idealized mRNA population consisting of ten species of RNA each 1000 nucleotides long, such that RNA-1 has a copy number of 1000, RNAs-2, 3 and 4 each have a copy number of 10, while the remainder, 5–10, all have a copy number of 1. Thus, there are 1036 mRNAs in the cell giving a total RNA content of 1 036 000 nucleotides. The mRNA complexity of this population is, however, only 10 000 nucleotides, since there are 10 distinct mRNAs each of 1000 nucleotides. Their individual frequencies are thus irrelevant from the point of view of complexity. In a

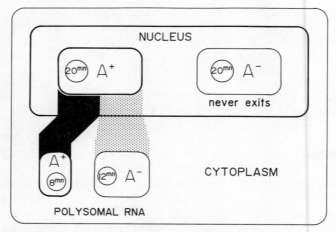

Figure 3.14   The partitioning of primary poly (A)$^+$ and poly (A)$^-$ gene transcripts in the nucleus and cytoplasm of *Drosophila melanogaster* (data of Zimmerman *et al.* 1982).

homogeneous population of cells the mRNA is operationally placed into three very general classes. The first of these, the abundant class, has some 10^3–10^6 copies per cell. An intermediate class includes 10^2–10^3 copies. Most of the sequence complexity is, however, contained in the rare copy class ($< 10^2$), also referred to as the complex class. In a highly heterogeneous cell population, the rare class is expected to be composed of cell-specific species of mRNA as well as those species that are present as few copies in most, or all, cells.

When single-copy DNA is fractionated from *Drosophila melanogaster* pupae, and incubated with a large excess of RNA from mature eggs, it is found that the total egg RNA complexity is about 12×10^6 nucleotides (Hough-Evans *et al.* 1980). This is sufficient for 10 000 messages each of an average length of 1250 nucleotides. However, one question that needs to be answered is – are all of these sequences actually mRNA? Using total polysomal RNA, and poly (A)$^+$ mRNA in vast excess, in reassociation reactions with single copy DNA, the RNA complexities at different life cycle stages have also been examined (Table 3.4). From such a study it has been concluded that about 16 000 different messages are expressed during the entire life cycle, and about 5500 or so of these appear as polyadenylated transcripts. There is a high degree of overlap between the poly (A)$^+$ mRNAs of larvae, pupae and adults, and a similar degree of overlap of the poly (A)$^-$ mRNAs of all three stages. These data tell us that the two stages which we see as very distinct morphological types, namely larval and adult, are running on virtually the same genes. In *Drosophila melanogaster* the major branch point in the developmental programme that partitions different tissue primordia takes place during embryogenesis. Not only are larval tissues constructed at this time but, additionally, the primordia for the adult tissues are also set aside. Thus postembryonic development in *Drosophila*

Table 3.4 RNA sequence complexity at different developmental stages compared to adult heads in *Drosophila melanogaster* (from Levy & Manning 1981).

Stage	RNA	Complexity in million nucleotides
larval	total polysomal	19
	poly (A)$^+$ mRNA	6.7
pupal	total polysomal	19
	poly (A)$^+$ mRNA	8.2
adult	total polysomal	20
	poly (A)$^+$ mRNA	7.6
adult heads	total polysomal	14

melanogaster is not expected to be accompanied by sweeping changes in gene expression, since the RNA populations of larvae, pupae and adults evidently overlap extensively. A better comparison, however, would be between the RNA populations of individual imaginal discs, which in the larval stage represent the developmentally significant groups of cells, with those of the remainder of the larva. The discless mutants seemingly provide an excellent tool for such a comparison.

In terms of the actual nucleotides required to produce mRNAs, the figure is about 16 000 sequences of average size 1250 nucleotides, though there are approximately 2000 sequences which are specific to pupae and adults (Levy & Manning 1981). The *Drosophila melanogaster* genome thus produces some 16 000 polysomal RNAs of average length 1250 nucleotides, giving a complexity of 20 million nucleotides. This implies that only about 20% of the euchromatic portion of the genome is expressed. A major uncertainty in this calculation concerns transcript length. However, we now have reasonable estimates of this from data on clones which indicate that 1500 nucleotides represent a good average value (Table 3.5).

The amount of euchromatin which is transcribed is certainly more than these calculations suggest, since transcription units include sequences, such as leaders, trailers and introns, which are not represented in the mRNAs they eventually give rise to. We know from Table 3.5 that a processed transcript averages 1.5 kb. If we allow a generous 3–5 kb to include introns, leaders, trailers and the necessary flanking landscapes for correct expression, then 10 000 transcripts with an average necessary domain of 5.0 kb means that 50 million bp of the *Drosophila melanogaster* genome would be adequate for the total control of development and metabolism. The only qualification that needs to be made is in relation to transcript number. If a given tissue, such as the brain, were to contain a disproportionately high number of large processed transcripts, then the transcript number could easily fall below 10 000.

These uncertainties can be overcome by using cloned DNA sequences

Table 3.5 Transcript lengths of cloned genes in *Drosophila melanogaster*.

Gene	Location	Length of transcript (kb)	Number of introns
Heat shock proteins			
hsp 80	63BC	3.7	1
hsp 70	87A7, 87C1	2.25	0
hsp 68	95D	2.1	0
hsp 28		1.25	0
hsp 26	67B	1.05	0
hsp 23		0.96	0
hsp 22		1.03	0
Alcohol dehydrogenase			
Adh	35B2–3	1.3	2
Glue proteins			
Sgs-3	68C	1.1	0?
Sgs-4	3C11–12	0.9	0
Chorion proteins			
S38-1	7F1–2	1.5	
S36-1		1.1	
S16-1		0.75	1
S19-1	66D11–15	0.74	1
S15-1		0.67	1
S18-1		0.82	1
Larval serum proteins			
LSP-1α	11A7–B9	2.85	>1
LSP-1β	21D2–22A1	2.85	>1
LSP-1γ	61A1–A6	2.85	>1

Ribosomal protein	99D	0.65	1
Vitellogenin proteins			
YP-1	8F–9A	1.6	1
YP-2	8F–9A	1.6	1
YP-3	12B–C	1.5	2
Tubulin proteins			
α-1	84B3–6	1.8	
α-2	85E6–15	1.65, 1.8	
α-3	84D5–8	1.8, 2	
α-4	67C4–6	1.5, 1.7, 1.9, 1.95	
Rudimentary locus	15A1	7.03	
Actin proteins			
A-1	88F	1.7	1
A-2	5C	3.3	1
A-3	42A	1.6	0
A-4	57A	2.2	1
A-5	87E	1.6	0
A-6	79B	1.6	1
Collagen protein	25A	6.4	
Tropomyosin I ⎱	88F2–5	1.5, 1.8, 2.8	
II ⎰			
Myosin heavy chain	36B	19	>9
H, D, L genes (3 genes)	44D	1.7 per gene	
Bithorax locus	89E	>70	>3
Antennapedia locus	84AB	>100	

After Riddell et al. 1981, Spradling & Rubin 1981, Bautch et al. 1982, Griffin-Shea et al. 1982, Hoveman & Galler 1982, Meyerowitz & Hogness 1982, Segraves et al. 1983, Snyder & Davidson 1983, North 1984, Hogness et al. 1985.

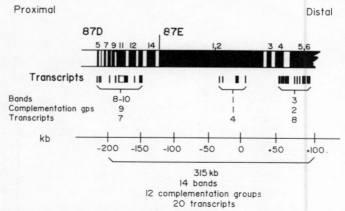

Figure 3.15 Relationship between band number, complementation groups and transcripts in a 315 kb walk through regions 87D–87E of the genome of *Drosophila melanogaster* (data of Hall *et al*. 1984). Transcript number is now known to be at least 43 (Bossy *et al*. 1984).

which have been sufficiently characterized in terms of their message coding capability to provide independent estimates of the number of RNA coding sequences per unit length of DNA. Using a combination of genetic and molecular approaches, the structure of region 87D,E of the *Drosophila melanogaster* genome has been analyzed, both in terms of genes in which mutation produces either a lethal or a visible phenotype and in terms of those sections of DNA that are transcribed into a product with the characteristics of mRNA. This region contains the genes *rosy* (xanthine dehydrogenase) and *Ace* (acetylcholinesterase), and a 315 kb walk (Fig. 3.15) has been made through 14 polytene bands (Hall *et al*. 1984, Gausz *et al*. 1986). This walk encompasses 12 lethal complementation units and is now known to give rise to 43 transcripts (Bossy *et al*. 1984). The sum of the transcript sizes is 105 kb and the transcription units are unevenly spaced throughout the walk. Thus, the average transcript length in this walk is 105/43 = 2.5 kb. This figure, if extrapolated, suggests that approximately one-third of the euchromatic genome, that is 30 million bp, could consist of coding sequences. This figure, though based on only one gene region, is nevertheless not too different from that obtained from RNA/DNA reassociation experiments. A second point to note is that the 43 transcripts per 315 kb of DNA yield an average estimate of 1 transcript per 7 kb DNA. Thus, a unique genome of 90 million bp would produce about 12 000 transcripts.

Notice that the data from this cloned walk indicates that the number of transcripts exceeds the polytene band number by a considerable margin. While the *Drosophila melanogaster* genome may have about 5000 complementation units and about the same number of bands, it certainly has more RNA transcripts than bands. While it cannot be doubted that the polytene banding sequence must, in some sense, reflect the genetic organization of the chromosome, it seems increasingly unlikely (Lefevre & Watkins 1986) that one band consistently reflects one complementation unit as has been formerly believed. Such a belief rested largely on the assumption that the array of detected mutations was able to saturate the functional elements of the genome. Since mutations leading either to overt morphological change or else to lethality are the most easily isolated, the validity of this assumption depends on what proportion of genes are capable of mutating to either a recessive lethal form or else to one which results in easily detectable morphological change. There is certainly considerable variation in the concentration of loci able to mutate to a lethal form in different regions of the X-chromosome of *Drosophila melanogaster*. Some regions (1B, 1F–2A, 10A, 11A and 19EF) are densely populated with such vital loci, while others (6EF, 10B–10E) are either sparsely populated or else (3C8–3D4) appear to have none at all. Moreover, in some of the densely populated regions, such as 1B and 3AB, the number of vital loci may actually exceed the number of bands (Lefevre 1981). Added to this, wide variation is known to occur in the number of polytene bands in different species of dipterans (Table 3.6). There is, however, no rational basis for assuming that the number of genes should show comparable variation.

Both the extrapolated cloning data (11 000 transcripts) and the RNA reassociation data (13 000 transcripts) have given us comparable estimates of transcript numbers within the *Drosophila melanogaster* genome. These estimates imply that only 20 million nucleotides, or about 25% of the euchromatic component of this genome, function in processed transcription. Even if another 20% were to constitute introns, leaders and trailers, what of the remaining 55%? An evaluation of this fraction of the genome is possible from cloning and transcription experiments in which coding sequences have been reintroduced at new sites into the *Drosophila melanogaster* genome via P elements, as a result of microinjection into preblastoderm embryos. With this approach, a segment of DNA containing a specific gene can be introduced into the germ line of individuals carrying an identical genetic background, so that the only variable is the site at which the single intact copy of the gene is integrated. The ability to experimentally transpose a gene to a new genomic location is thus a powerful tool for studying the effects of genomic position on the expression of that gene.

In a series of experiments the *Xdh* (xanthine dehydrogenase, Spradling &

Table 3.6 Band number variation in dipteran polytene chromosomes.

Species	1C (pg)	Approximate band number
Prodiamesa olivacea	0.13	1200
Drosophila lebanonensis		1200
Chironomus tentans	0.21	1900
Drosophila hydei	0.23	2000
Drosophila melanogaster	0.18	5000

After Beermann 1952, Berendes 1963, Daneholt & Edstrom 1967, Berendes & Thijssen 1971, Lefevre 1976, Zacharias 1979.

Rubin 1983), *Ddc* (dopa decarboxylase, Scholnick *et al.* 1983) and *Adh* (alcohol dehydrogenase, Goldberg *et al.* 1983, Savakis & Ashburner 1985) genes, all of which exist in the haploid genome as single copies, have been inserted at a variety of sites into different strains of *Drosophila melanogaster*, and their behaviour then analyzed in terms of levels of enzyme activity and the tissue and time of expression. In the case of *Xdh*, the vector used contained 7.2 kb of *Xdh* DNA landscape of which at least 4.5 kb is required to code for the XDH polypeptide, so that, at most, only 2–3 kb of flanking sequences would have been included in the transposons used. Even so, flies containing an inserted transposon expressed normal, or near normal, amounts of XDH at a variety of autosomal sites, though insertions into the X-chromosome exhibited the same dosage compensation that characterizes normal X-linked genes in *Drosophila melanogaster* (see § 3.4.3.2). Evidently the *Xdh* gene does not require a large surrounding landscape for its normal or near normal expression, and the 2–3 kb of DNA flanking the gene appears sufficient to ensure its normal activity at varying autosomal sites within the genome. Additionally, with the exception of X-chromosome insertions, chromosomal domains capable of overriding the local controls on *Xdh* expression appear uncommon, though the line with the lowest specific activity did, in fact, contain an insert on chromosome 4, which polytene analysis indicated was near to, if not within, the heterochromatic chromocentre.

There are two sets of cytogenetic data which are relevant to the question of the functional restraints imposed by gene domains. In the case of induced euchromatic (eu) rearrangements, it has been found that those with lethal consequences have breakpoints which are located at or near a single gene (Lefevre 1974). Thus, eu–eu rearrangements do not show a sphere of influence which extends beyond a single complementation unit. Similarly, an insertional translocation of a euchromatic segment, carrying the *Sgs-4* glue protein gene, within 12 kb of the Y-heterochromatin leaves its activity sufficiently unaffected to produce a wild type phenotype (McGinnis *et al.* 1980). Evidently, the domain of the *Sgs-4* locus is not influenced by being close to Y-heterochromatin. Both points testify to the fact that, in many cases, gene domains in *Drosophila* are no larger than single genetic units of function.

The *Ddc* gene behaves in an identical manner when reintegrated into the *Drosophila melanogaster* genome by P element mediated germ line transformation. In this case the transforming vector carried a 7.5 kb *Ddc* restriction fragment inserted into the *Xdh* vector described above, so that the transposon included both *Ddc* and *Xdh* structural genes. The *Ddc* fragment used included a 4.5 kb transcription unit, plus 2.5 kb of 5′ and 1 kb of 3′ flanking DNA. Thus, the 14.5 kb vector contained two gene landscapes, of which 9.0 kb was transcriptively active. Not only were reintegrated *Ddc* genes expressed at the proper stage of development, but all strains showed levels of *Ddc* activity consistent with correct temporal regulation. One point of particular interest, however, was that the same inserts gave more activity in Oregon-R than in Canton-S, their strain of origin.

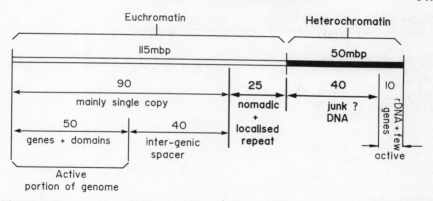

Figure 3.16 Partitioning of the genome of *Drosophila melanogaster* into active and inactive portions.

If we extrapolate from the results of these elegant experiments, then, with an average processed transcript of say 1.5 kb, a generous 3.5 kb of flanking landscape plus introns and an assumed transcript number of 10 000, we can calculate that the total control of development and metabolism may require no more than 50 million bp of the *Drosophila melanogaster* genome. It, thus, seems probable that a substantial portion of the euchromatic portion of 90 million bp is also irrelevant to development and metabolism. Given that 40 million bp of highly repetitive DNA and 25 million bp of nomadic DNA plus localized repeats are also irrelevant to these processes, this leaves us with a figure of 55 million bp from which to construct a fly (Fig. 3.16). This figure is only slightly more than ten times the *Escherichia coli* genome, though we certainly have a bargain compared to *Escherichia coli* in terms of morphological and biochemical sophistication. Added to this, the total complexity of transcripts produced by the fly is, as we shall shortly discover, not very different from that found in a single celled protozoan.

3.3 COMPARATIVE GENOME ORGANIZATION

The message we obtain from *Drosophila melanogaster* is that a significant portion of the genome is irrelevant to development because it has neither a coding nor a regulatory function. How general is this likely to be? To answer this question we need first to briefly examine genome sizes in different eukaryotes.

3.3.1 Size variation between genomes

From a sample of available measurements (Table 3.7), it is evident that genome size varies markedly even between similar kinds of eukaryotes. The chordates are an especially useful group in which to examine comparative genome size, because their evolutionary history is so well documented.

Table 3.7 Genome size variation in eukaryotes.

Group		Genome size 1C (pg)	Approximate size range
Animals			
Protozoa		0.06–350	5800
Coelenterata		0.35–0.73	2
Nematoda		0.08–0.66	8
Annelida		0.7–7.2	10
Arthropoda	Crustacea	0.7–22.6	30
	Insecta	0.05–12.7	250
Mollusca		0.4–5.4	12
Echinodermata		0.5–4.4	8
Protochordata		0.2–0.6	3
Agnatha		1.4–2.8	2
Pisces	Chondrichthyes	2.8–7.4	3
	Osteichthyes	0.4–142	350
Amphibia	Anura	1.0–10.8	10
	Urodela	15.1–83.5	5
Reptilia		2.0–5.4	3
Aves		1.2–2.1	2
Mammalia	Marsupialia	3.0–4.7	2
	Placentalia	2.5–5.9	2
Plants			
Fungi		0.01–0.19	19
Algae		0.04–200	5000
Bryophyta		0.64–4.30	7
Pteridophyta		1.00–310	310
Gymnospermae		4.20–50	12
Angiospermae		0.1–127	>1000

After Atkin *et al.* 1965, Benirschke *et al.* 1970, Conner *et al.* 1972, Sparrow *et al.* 1972, Wurster & Atkin 1972, Hinegardner 1973 & 1974, Ohno 1974, Rheinsmith *et al.* 1974, Sulston & Brenner 1974, Olmo & Morescalchi 1975 & 1978, Vlad 1977, Cavalier-Smith 1978, Eden *et al.* 1978, Rees *et al.* 1978, Davis *et al.* 1979, Olmo 1981, Olmo & Odierna 1982, Leutwiler *et al.* 1984.

Note: genome size is conveniently defined as the total amount of DNA present in a haploid gametic (1C) set of chromosomes. There are a variety of ways of representing genome size in terms of DNA content. It is usually expressed in picograms (1 pg = 10^{-12} g), daltons (1 d = 1.66×10^{-24} g), or else in terms of base pairs (bp). These units are interconvertible in the following manner: 1 bp = 630 d; 1 pg = 0.96×10^9 bp.

3.3.1.1 Chordates

For the purpose of comparison (Fig. 3.17) the human genome, which has been taken as 3300 million bp, provides a convenient standard. In the protochordates the measured values range from 68% of human (urochordates) to 20% human (hemichordates and cephalochordates). The values are higher in the jawless chordates (Agnatha) where they range from 40% to 80% human. In fish there is a staggering 350-fold range, which stretches from a mere two times that of *Drosophila melanogaster* to 40 times that of man. There is a tenfold range of size variation in anuran amphibians (frogs and toads), and a further fivefold range is found in urodele amphibians (newts and salamanders), giving an overall range of some 80-fold within the Amphibia as a whole. Genome size is far less variable in amniote vertebrates (reptiles, birds and mammals). Here, the enormous variation seen in fish and in amphibia disappears. In reptiles, for example, the size of the genome varies from 50% to 150% human, a mere threefold range. An even smaller range is found in birds, where the minimum genome size is less than half that of man. In mammals the largest known genome is found in the aardvark (1.7 human) whereas the smallest mammalian genome is that of the deer, *Muntiacus muntjak vaginalis*, (0.7 human). The overall size range in mammals is thus approximately twofold.

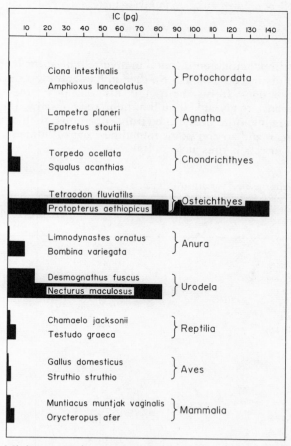

Figure 3.17 Minimum and maximum genome sizes in chordates (compare Table 3.7). Note 1 pg = 0.96×10^9 bp and 1 bp = 630 daltons.

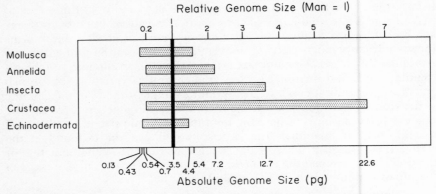

Figure 3.18 Genome size variation in invertebrates (compare Table 3.7).

3.3.1.2 Insects

A not too dissimilar situation is found in invertebrate animals (Fig. 3.18). If we consider only insects (Table 3.8), then in flies and mosquitoes (Diptera) genome size ranges from about 130 million bp in the chironomid, *Prodiamesa olivacea*, to about 900 million bp in the housefly, *Musca domestica*. In grasshoppers, genome sizes can be truly enormous, with a starting point near that of man and an end point four times that of humans. The overall range among insects is thus at least 100-fold.

Table 3.8 Genome size variation in insects.

Species	1C (pg)	Proportion of human genome
Diptera		
Prodiamesa olivacea	0.13	0.04
Drosophila melanogaster	0.18	0.05
Chironomus tentans	0.21	0.06
Drosophila hydei	0.23	0.07
Drosophila virilis	0.34	0.10
Sarcophaga bullata	0.60	0.17
Culex pipiens	0.87	0.24
Musca domestica	0.89	0.25
Lepidoptera		
Bombyx mori	0.52	0.14
Orthoptera		
Morabine P151	3.3	0.9
Locusta migratoria	5.3	1.5
Omocestus viridulus	12.7	3.6

After Rasch *et al.* 1971, Jost & Mameli 1972, Laird 1973, Rasch 1974, Rees *et al.* 1978, Samois & Swift 1979, Zacharias 1979.

Table 3.9 Genome sizes in salamanders of the genus *Plethodon* (all $2n = 28$). (From Mizuno & Macgregor 1974.)

Group	Species	1C (pg)
Eastern small	*P. nettingi shenandoah*	18
	P. cinereus cinereus	20
	P. richmondi	20
	P. hoffmani	21
Eastern large	*P. wehrlei*	20
	P. glutinosus	23
	P. jordani	23
	P. ouachitae	34
	P. yonahlossee	36
New Mexican	*P. neomexicanus*	31
Western	*P. elongatus*	34
	P. vehiculum	37
	P. dunni	39
	P. larselli	48
	P. vandykei	69

3.3.1.3 Plethodontid salamanders

In the plethodont salamanders of North America, genome size varies from five times to nearly 20 times that of man (Table 3.9). Despite this large differential in DNA content, species within this genus are remarkably uniform in morphology, and there is certainly no obvious developmental differences which would require one species to have four times the DNA content of another. This highlights the major apparent dilemma posed by genome size variation in eukaryotes. Namely, why do organisms of comparable morphological and anatomical complexity possess such radically different amounts of DNA? This so-called C-value paradox can be even more forcefully expressed by summarizing the known spectra of genome sizes found in particular kinds of animals relative to their presumed evolutionary relationships (Fig. 3.19). Thus, even before we systematically partition the various categories of DNA present within different genomes, there is clear evidence that in any one group the minimum genome size required to produce a given grade of morphogenetic organization can be quite small when compared to the maximum genome size found in that same group.

There has been a persistent reluctance to accept the obvious conclusion from these data; namely that in both a metabolic and a developmental sense, much of the excess DNA is somatic junk.

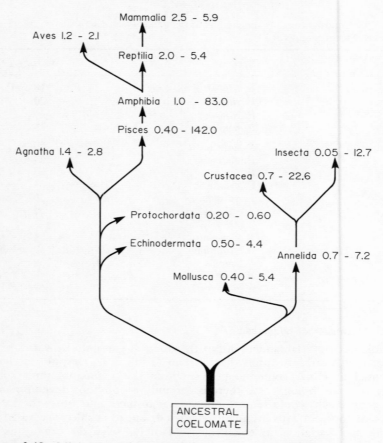

Figure 3.19 Minimum and maximum genomes sizes (in pg) in the major animal groups (compare Table 3.7).

3.3.2 Interspersion pattern differences

Interspersed repetitive DNA has been found to some extent in the genomes of most eukaryotes. Typically, one group consists of short DNA elements – one to a few hundred base pairs long. Such a short period interspersion pattern, first shown in *Xenopus laevis*, is widespread among animal genomes. A second group of diverse repetitive elements consists of components several kilobases in length. This long period pattern, which characterizes *Drosophila melanogaster*, though much less frequently found, is no less widespread throughout the animal kingdom (Fig. 3.20).

As we have already seen, most of the moderately repeated sequences in *Drosophila melanogaster* belong to different forms of mobile elements that are scattered throughout the genome. Thus, the pattern of interspersion seen in reassociation experiments in *Drosophila melanogaster* is unlikely to have any

functional significance, and is simply an outcome of the processes which
move nomadic elements throughout the genome. It has, however, been
postulated that DNA sequence arrangement plays a critical role in the
control of gene expression, and specifically that the short interspersed
middle repetitive sequences serve an important regulatory function (Davidson
& Britten 1973, 1979). The non–universality of this pattern argues against
this, as do the unusual forms of interspersion that have been identified in
some animals. Birds are a particularly informative group to consider, since
their genomes are composed predominantly of single copy DNA. The
chicken and the ostrich, which differ almost twofold in genome size, do so
largely at the expense of their single copy component. Both have a very
long interspersion pattern, since most of the genome of both species is

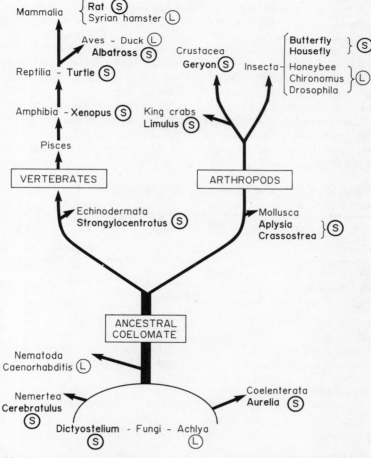

Figure 3.20   Interspersion patterns (L = long, S = short) of unique and repetitive
DNA sequences in eukaryote genomes (data of Schmidtke & Epplen 1980).

single copy DNA. Given the scarcity of repeated sequences in the chicken, it would seem an ideal vertebrate in which to conduct chromosomal walks in order to obtain estimates of the number and length of transcription units.

Finally, fungi are a heterogeneous group of eukaryotes which have very small genomes, usually less than 0.05 pg, and often contain extraordinarily small amounts of repetitive DNA (Table 3.10). Moreover, while in the slime mold, *Dictyostelium discoideum*, the interspersion pattern consists of a 0.3 kb repeat adjacent to unique DNA (Firtel & Kindle 1975), in a manner similar to the *Xenopus laevis* pattern, in the water mold, *Achlya bisexualis*, the small amount of repeated sequences are clustered and the genome effectively consists of long uninterrupted single copy sequences (Hudspeth *et al.* 1977). Thus, even within fungi the interspersion pattern can be very variable. Comparable differences are found in insects and mammals (see Fig. 3.20).

The fact that some fungal genomes evidently do not contain interspersed middle repetitive sequences of the types that have been described, indicates that models of gene regulation which invoke such sequences as regulatory elements are inadmissable in these cases. They are also generally inadmissable in birds, and there is no compelling reason for believing they apply to any eukaryote. In cases where co-ordinate induction of dispersed non-homologous genes does occur, the responsible repetitive elements have now been shown to be short sequences, of from 9 to 24 bp, which are common to the 5′ flanking region (reviewed by Davidson *et al.* 1983). Sequences of this length are so short as to be undetectable by the classical methodology pioneered by Britten and Davidson, which formed the basis for their initial hypothesis (Britten & Davidson 1971, Davidson & Britten 1979). These authors were,

Table 3.10 Characteristics of some fungal genomes (after Dusenberry 1975, Hudspeth *et al.* 1977).

Species	1C (pg)	%DNA repetitive	%DNA single copy
Oomycetes			
Achlya bisexualis	0.044	18	82
Zygomycetes			
Phycomyces blakesleeanus	0.032	35	65
Ascomycetes			
Saccharomyces carlsbergensis	0.016	11	89
Torulopsis holmii	0.023	16	84
Hansenula holstii	0.011	8	92
Candida macedoniensis	0.014	7	93
Candida catenulata	0.021	15	85
Phycomycetes			
Aspergillus nidulans	0.027	3	97
Myxomycetes			
Dictyostelium discoideum	0.050	30	70

however, absolutely correct in their argument that a common variety of signal sequences must be required for the co-ordinate expression of genes. The realm of action of these signal sequences simply moves from the 300 bp–many kb type of middle repeat to the very short, 20 bp type, which is of the same order of magnitude as the sequences recognized by bacterial and phage regulatory proteins.

A wide range of dispersed and moderately repeated DNA sequences has been identified in a variety of eukaryote genomes. These can be classified into families within which the individual members share a related, though not necessarily identical, nucleotide sequence. A number of different families of this type have now been studied in sufficient detail to indicate that many contain members that are, or at one time were, mobile. For example, a major long, highly repetitive, sequence family, the HpaII repeats, dominates the genome of *Physarum polycephalum*. It takes the form of scrambled clusters, and occupies about 20% of the genome, accounting for about a half of all the repetitive DNA (Pearston *et al.* 1986). The HpaII repeat forms part of a larger repetitive element, c. 8.6 kb long, which shares common characteristics with recognized transposable elements. Scrambled clusters of this sort have, therefore, presumably arisen from transposition-like events, in which the element inserts preferentially into sites located in other copies of the same repeated sequence. This is an extremely instructive example because it indicates that clustered-scrambled arrays can arise by nomadics inserting into nomadics.

In some ciliates, macronuclear development involves the programmed fragmentation of micronuclear chromosomes into small pieces – with an average length of 1800–2500 bp and terminated at both ends by 50–70 tandem repeats which form telomeres (see § 2.4.3). Because the macronucleus is so rich in telomeres, it is possible to purify telomeric DNA and so determine its structure. In this way, it has been shown that the telomeric DNA sequences of several simpler eukaryotes are built up of simple repeating units which fit the consensus formula $C_{1-8}(A/T)_{1-4}$ (Blackburn 1984). Identical repeats are, however, also found internally in the micronuclear chromosomes of ciliates. Thus, *Tetrahymena thermophila* contains 100 copies of non-telomeric C_4A_2 repeats, referred to as mic C_4A_2 (Cherry & Blackburn 1985), while a few hundred internal mic C_4A_4 repeats are present in *Oxytricha fallax* (Herrick *et al.* 1985). In both cases these internal telomere-bearing elements arise at the ends of transposon-like elements, and telomere-type repeats are added to the ends of the free forms of the transposable elements prior to reintegration.

Yeast is also relatively easily monitored for the movement of DNA sequences, and two mobile families have been characterized to-date. One, the *Ty1* family, has been shown to be capable of relatively rapid movement throughout the genome (Eibel *et al.* 1981). The 5.9 kb segment of *Ty1* exists in some 35 copies in the genome, and each contains a 0.3 kb delta (δ) sequence at either end. These δ units consist of identical 338 bp sequences. There are some 1200 copies of such δ sequences in the genome, so that not all of them are connected to *Ty1* elements. While *Ty1* elements are capable of direct transposition within the genome, they can also be moved by gene conversion (Roeder & Fink 1982). This occurs, for example, between *Ty1* elements which differ by large insertion and substitution mutations, and without any alteration in the flanking DNA sequences.

The *Ty1* element is transcribed to produce a virtual full length RNA which begins and ends in the δ sequences. In this it resembles retroviral RNAs. Furthermore, *Ty1* has a region homologous to yeast tRNA^{Met} just inside the δ sequence and, significantly, retroviral RNAs use just such a region as a site for DNA synthesis. Thus *Ty1*, like the nomadic elements of *Drosophila melanogaster*, shares properties in common with mammalian retroviruses (Elder *et al.* 1983). Indeed, there is now compelling evidence for the existence of an RNA intermediate during *Ty* transposition, which implies that the event involves reverse transcription. By utilizing a donor *Ty* element whose expression was under the control of a GAL-1 promoter, Boeke *et al.* (1985) have shown that transposition can be induced by galactose. This establishes a link between RNA synthesis and the transposition event. Moreover, by engineering an intron into the marked *Ty* element it was demonstrated that the intron was correctly spliced out during transposition, despite the fact that *Ty* elements do not have their own introns. Since splicing is characteristic of RNA, this proves that *Ty* elements must pass through an RNA stage during transposition. An important implication of these findings is that yeast cells contain reverse transcriptase. Moreover, such transposition-induced yeast cells also contain virus-like particles, Ty-VLP, that are very similar in appearance to the intracisternal A-type particles of the mouse and the *copia* particles of *Drosophila melanogaster* (Garfinkel *et al.* 1985).

Sigma, a second family of nomadic elements in yeast, is an exceptionally short, mobile element of 340 bp in length and has 8 bp inverted repeats at its ends, flanked by 5 bp direct repeats. It has homologies at its ends both with *Drosophila melanogaster copia* nomadic elements and spleen necrosis

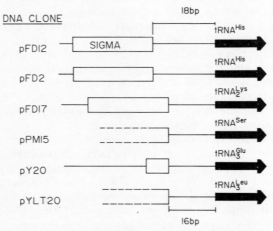

Figure 3.21 Location of *sigma* sequences relative to tRNA genes in six clones isolated from the yeast genome (data of Del Rey *et al.* 1982).

virus. It is found in different locations in closely related yeast strains, but is almost always located within 16–18 bp of the 5′ end of yeast tRNA genes (Fig. 3.21). *Ty1* and delta elements may also occur adjacent to several tRNA genes in yeast (Eigel & Feldman 1982).

While *Drosophila melanogaster* and yeast are easily monitored for movement of mobile sequences, mammals, on account of their larger genome sizes and longer generation times, pose more problems. They are, however, characterized by a variety of interspersed and moderately repeated segments (Singer 1982b). A particularly prevalent family of short repeats is that, first identified in man, which is cut by the restriction enzyme *Alu* I (Ullu 1982). Most λ clones from human genomic libraries carry an *Alu* repeat, and the family has a total size of not less than 300 000 members. The largest of the three introns of the human β-tubulin gene contains no less than ten members of the *Alu* family. Together these account for 60% of the total intron length of 4826 nucleotides. The ten sequences are substantially divergent, both from each other and from the *Alu* consensus sequence (Lee *et al.* 1984). Likewise, the 56 kb human globin gene locus contains seven *Alu* family members with an average spacing of 8 kb.

Alu sequences share a number of characteristics, including short terminal repeats, which indicate that they can move, or be moved, in the genome (Calabretta *et al.* 1982). Most *Alu* elements are flanked by direct repeats of sequences some 7–20 nucleotides long. Such flanking direct repeats tend to characterize sequences that have been integrated into a genome. The sequence, dispersion pattern and copy number of the *Alu* family members are very similar in the genomes of both man and the African green monkey, *Cercopithecus aethiops*, and under stringent conditions more than 75% of the λ phage of a recombinant DNA library from *Cercopithecus aethiops* hybridizes with a human *Alu* sequence (Grimaldi *et al.* 1981). Great apes and man also contain long interspersed repeated segments that are cleaved by *Kpn*-1 and which are also remarkably constant in segment lengths, relative amounts per genome and distribution patterns (Shafit-Zagardo *et al.* 1982). These families are much longer than *Alu*, with lengths ranging from 1.2 to over 6 kb.

Alu-equivalent sequences have been identified in both the Chinese hamster (Haynes *et al.* 1981) and the mouse (Kominami *et al.* 1983), though here they are only about half the size of those found in man. The chicken genome, too, contains some 1500–7000 copies of a 165 bp long sequence, CR1, flanked by imperfect but homologous short repeats which show several regions of extensive homology to both human *Alu* and mouse B1 families (Stumph *et al.* 1981), and which possess several features associated with the LTRs of avian retroviruses (Stumph *et al.* 1984). In addition to the short interspersed repeated segments, the mouse genome contains a long interspersed family, MIF-1 (Gebhard *et al.* 1982), together with 1000 copies of a family of middle repeated DNA sequences, which have 11 bp inverted repeats at their termini but also contain signals implicated in RNA polymerase II transcription and regulation. These show structural characteristics of both insertion elements (IS) and solitary long terminal repeats (LTR), and are, therefore, referred to as LTR-IS elements (Wirth *et al.*

1984). Over 100 copies of VL30 elements are dispersed in the mouse genome, but it also contains solo VL30 long terminal repeats, i.e. single LTRs detached from the rest of the element (Rotman *et al.* 1984).

Alu sequences have been found in *Drosophila melanogaster*, though only as part of the 7SL-DNA, of which there are two copies in the genome (Gundelfinger *et al.* 1983). It has been suggested, therefore, that the *Alu* sequences of vertebrates may represent defective 7SL-RNA molecules that have been reverse transcribed into DNA and reinserted into the genome (Ullu & Tschudi 1984, Deininger & Daniels 1986). 7SL-RNA is involved in the organization of the signal recognition particle concerned with the translocation of proteins across the membranes of the endoplasmic reticulum.

Finally, approximately half of the DNA sequences in the genome of *Xenopus laevis* are repetitive, and many of these are transcribed both during oogenesis and in the embryo. This repetitive component includes a highly repeated *1723* sequence that has the characteristics of a transposable element and is interspersed throughout the genome. The *1723* sequence is over 6 kb in length, with an internal expandable region consisting of a variable number of 185 bp repeats and terminated by inverted repeats flanked by short sequence duplications (Kay *et al.* 1984). In oocytes, approximately 100 pairs of lampbrush loops are active in the transcription of *1723* elements, and RNA homologous to *1723* is present in total cellular RNA from ovaries, embryos and liver cells. Transcripts of both strands are present in similar concentrations, and are evenly distributed into poly $(A)^+$ and poly $(A)^-$ fractions in ovaries and tadpoles. They are, however, heterogeneous in length.

The overall picture we end up with is that dispersed, moderately repeated DNA sequences of varying size, some short (less than 500 bp) others long (up to 6 kb), constitute a not uncommon component of all eukaryote genomes that have been analyzed. These have been referred to as SINEs and LINEs respectively (Singer 1982b). Collectively, these may well account for the varying patterns of interspersion present in different species, and indicate that nomadic elements are undoubtedly ubiquitous in eukaryotes. Several of the SINE families of the prosimian, *Galago crassicaudatus*, have promoter regions similar to tRNA genes. Additionally, the monomer and type II SINE families of *Galago crassicaudatus*, both of which are *Alu*-related, are 68% and 62% homologous, respectively, to a human methionine tRNA gene, while rat identifier and mouse B2 families have sequences closely related to alanine and serine tRNA genes, respectively. It appears likely, therefore, that SINE families represent amplified tRNA pseudogenes (Daniels & Deininger, 1985).

3.3.3 'Short' dispersed repetitive sequences

The repetitive DNA components of the genome to which we have, so far, given most emphasis have been the highly repetitive and localized tandem arrays (satellite DNAs) and the dispersed mobile elements. Once we leave these reasonably well-defined sequences and examine 'simple' dispersed sequences, we enter a grey area where evaluation of function is far less tractable.

The problem can be put into perspective by considering a sequence found in the human globin gene cluster. During an analysis of the $^{G}\gamma$ and $^{A}\gamma$ globin genes of man, Slightom *et al.* (1980) detected a stretch of about 50 bp of simple sequence DNA, predominantly of the form poly (TG), in the second intron of these genes. Owing to the sequence characteristics on either side of this small island of repetitive elements, these authors argued that this simple sequence constituted a hot spot for the initiation of recombinant-type events. They hypothesized further that intergenic conversion events initiated at such islands would probably be between non-homologous as well as homologous chromosomes.

This hypothesis has far-reaching consequences since, if small islands of this type were to occur throughout the genome, they would provide multiple hot spots for conversion events. Moreover, if such islands were to arise either in introns or else in trailer or leader regions of genes, then they would be automatically transcribed by default, and some mature transcripts would undoubtedly contain such sequences at their 5' or 3' ends.

How then do such islands arise? Are they mobile or mobilizable, or are they a consequence of unavoidable quirks of the replication machinery of the cell? Studies of the human myoglobin gene (Weller *et al.* 1984) help to resolve some of these issues. Here, about 1800 bp upstream from the gene, there is a 652 bp tandem repetitive region of core sequence, $(GGAT)_{165}$. There is also a second tandemly repeated sequence, based on a 33 bp run, within the first intron of the gene. Initially, this second sequence was thought to be a mobile element, since it was flanked by a 9 bp direct repeat reminiscent of the target site duplication found in many nomadics. Subsequent analyses, however, have revealed that such 'minisatellite' regions can be partially homologous to the generalized recombination signal, crossover hotspot instigator (*Chi*), of *Escherichia coli*.

These small tandem arrays vary in length, and are hypervariable regions which arise either by unequal exchange or by replication slippage. The core region of this particular myoglobin minisatellite may, in fact, aid in generating its own amplification, precisely because it is related to *Chi*. Consequently, when such sequences arise by chance within a genome, they may have the propensity to non-randomly subvert the replicative and recombinative machinery of the cell in such a way as to locally ensure their own amplification. Additionally, they may facilitate conversion events within and between chromosomes, as they appear to have done between the $^{G}\gamma$ and $^{A}\gamma$ globin genes of man.

The finding that a simple repeated sequence is flanked by a direct repeat, or even by a poly (A) sequence, obviously need not necessarily imply that it is mobile or mobilizable. It may be that the replicative machinery which generates, maintains or alters such mini-tandem arrays is also capable of generating peculiar boundary conditions which simulate the consequences of insertion/excision events performed by *bona fide* nomadic elements.

Many dispersed simple sequences are not of sufficient complexity to assume the properties of elements like the *Chi*-type sequence or its relatives.

Thus, sequences of the form poly (TG), poly (TTTGC) and poly (GCCTCT) which are found near mouse immunoglobin genes and near putatively mobile elements, such as the R, B1 and B2 sequences of the mouse (Gebhard & Zachau 1983), are too simple to be envisaged in this light. In the case of alternating purine–pyrimidine tracts, however, it has been argued that they might conceivably act as modulators of DNA conformation, or else in protein binding. The fact that many of these short dispersed sequences have been maintained throughout evolution may mean either that they are genuinely conserved for functional reasons or, alternatively, that mechanisms exist in cells which generate such sequences quite frequently. Since Gebhard & Zachau (1983) have shown that poly (TTTGC) and poly (GCCTCT) commonly occur in mouse and salmon DNA, but not in human DNA, this cautions us against initially attaching too much functional significance to the sequences *per se*.

A further series of experiments produces a comparable conclusion. The hybridization of simple polynucleotide probes of the type TG, GA, AA, GG and CAG, to a variety of different genomes, viz. human, *Drosophila*, sea urchin, *Stylonychia* and yeast, revealed that these sequences occur in most of the genomes examined (Tautz & Renz 1984). The most interesting and relevant point in a functional context arises from a comparison of the occurrence of such sequences in the macronuclei and micronuclei of the protozoan *Stylonychia*. While these simple sequences are all found in the micronucleus, they are absent from the macronucleus, yet it is the macronucleus which carries out all the metabolic functions of the cell. Clearly, these simple sequences are not required for everyday gene expression, and this despite the fact that they are transcribed. Are simple repeats then transcribed by default or is there a functional facet to their existence?

In *Drosophila virilis*, genomic clones have been isolated by screening with radiolabelled poly $(A)^+$ RNA from larval salivary glands. Five of these clones were then subjected to sequence analyses, and all turned out to harbour simple repetitive sequences (Tautz & Renz 1984). Repetitive sequences of the form poly (GATA) and poly (GACA) have also been described in both vertebrates and invertebrates (Epplen *et al.* 1982, Alonso *et al.* 1983). Here, not only are they transcribed but Singh *et al.* (1984) have claimed that they are directly involved in sex determination, a claim which has been based, in part, on the large numbers of cross-hybridizing sequences found on the W- and Y-chromosomes of snakes and mice, respectively. Levinson *et al.* (1985) have compared sequence data obtained from cross-hybridizing clones of snakes, mice and *Drosophila melanogaster* and find no evidence for the evolutionary conservation of the two tandem repeats in question, which are differently organized in the different species. Added to this, cattle totally lack such sequences (Miklos *et al.* 1987, in manuscript), so that a primary role in sex determination seems highly improbable.

Finally, in *Drosophila melanogaster*, a novel family of sequences, the *opa* repeats, have been found in the Notch locus (Wharton *et al.* 1985a,b), a gene involved in early neurogenesis. The repetitive element concerned is 93 bp in length, is predominantly constructed out of the triplets CAG and CAA and is translated as part of the Notch protein. The *opa* repeat, unlike the poly (TG) myoglobin repeat, is in an exon. Sequences homologous to *opa* are also found in the bithorax, Antennapedia and engrailed genes, and are distinct from the homeo boxes found in these same genes.

Another sequence in *Drosophila melanogaster*, with properties akin to *opa*, is the suffix sequence of Tchurikov *et al.* (1982). This occurs in about 300–400 copies per genome and has been claimed to reside at the 3' terminus of a number of transcription units. It has been suggested that sequences of this dispersed, transcribed simple sequence may function in the co-ordinate control of genes during development. In the rat brain there is an 82 bp identifier (ID) sequence which exists predominantly in brain-specific transcripts. This sequence has been claimed to be important in tissue-specific expression (Sutcliffe *et al.* 1983).

In sum, there is very probably a continuum of dispersed repeat types, both in terms of overall length and in terms of repeat frequencies. The available evidence on their mobility and genesis is still equivocal, although, at least for the simpler sequences, there appears no need to endow them with any developmental significance. In the case of the longer ones we still lack sufficient data to make a meaningful decision relating to function. Beyond this length, however, come *bona fide* dispersed nomadic elements of the *Alu* type, which also lack developmental significance.

3.3.4   Message complexities in eukaryotes

In *Drosophila melanogaster* both reassociation data and the use of cloned probes gave comparable estimates of gene numbers, and indicated that over the entire life cycle there were approximately 16 000 mRNA sequences, of which some 7000 appeared as polyadenylated transcripts. That is, about 25% of the genome gave processed transcripts. How representative then is this figure for eukaryotes in general?

Because of the large size of many eukaryotes genomes, the use of chromosomal walks for estimating gene numbers is still in its infancy, whereas there is a much larger body of data available from reassociation experiments. With large complex populations of RNA, which tend to characterize species with larger genomes, methodological problems inevitably arise in obtaining accurate estimates of message complexity from reassociation data but, in the absence of suitable measurements using recombinant DNA technology, we have no recourse but to consider the reassociation data. Two main reassociation methods have been employed to measure message complexity:

(a)   The hybridization of radioactive single copy DNA to a vast excess of cellular RNA. The amount of DNA annealed at saturation then provides an estimate of transcription from active single copy sequences. Such estimates can be obtained from both nuclear and cytoplasmic RNA populations, though only about half of the nuclear RNA complexity is likely to be represented in cytoplasmic RNA.

(b)   A second approach exploits the ability of reverse transcriptase to synthesize complementary DNA copies of polyadenylated RNA molecules. An analysis of the kinetics of hybridization of cDNA copies of mRNA to a vast excess of template RNA then allows one to determine not only the number of different sequences present as mRNA, but also the relative abundance of these sequences within the message population.

The second approach does not address itself to poly $(A)^-$ RNA which, as we saw in *Drosophila melanogaster*, in fact makes up more than half of the total RNA population. A combination of the two approaches, however, provides the best data for obtaining a gross overview of genomic transcription.

When measuring message complexity it is standard practise to subdivide the total RNA population of a cell or tissue into three general sub-classes, containing what is termed complex, moderately prevalent and super-prevalent mRNA. This division provides a workable basis for examining RNA populations. Complex RNAs are taken as having only one to several copies per cell; moderately prevalent RNAs have up to a few hundred copies, while in the superprevalent category there is in excess of 10 000 copies. The last class characterizes terminally differentiated tissues, for example, the globin mRNAs in reticulocytes of mammals, or the ovalbumin mRNAs of avian oviducts.

3.3.4.1   Ciliate protozoans

A majority of unicellular ciliates have a unique segregation of nuclear functions in which the transcriptive activity necessary for vegetative development and growth is restricted to DNA sequences contained in one or more large and impermanent polyploid macronuclei, whereas meiosis and sexual reproduction are functions of a small diploid micronucleus (see § 2.4.3). This separation of nuclear activities within a single cell parallels the soma-germ line differential found in multicellular eukaryotes. In unicellular ciliates the micronuclear ($\equiv$ germ line) genome varies enormously between species, with a range of from 200 million bp to 7200 million bp. The complexity of the macronuclear ($\equiv$ somatic) genome also varies from 17 million bp to at least 47 million bp in species where the macronuclear genome is in gene-sized pieces (see Table 2.11). In *Stylonychia lemnae* (= *mytilus*) there are between 12 000 and 15 000 different mRNA species in a cell, where each mRNA averages between 1200 and 1500 nucleotides in length. Whether all the DNA pieces within a macronucleus are transcribed is not known. If all of them were transcriptively active, and each produced a single mRNA, there would be about 12 000 genes in *Oxytricha similis*, the hypotrich with the lowest macronuclear complexity. In *Paramecium bursaria*, where the macronuclear complexity is only 18 million bp, the estimated number of genes would be only 8000 if messenger sequences were distributed in a similar way to *Oxytricha similis*, that is given an average coding length of DNA of about 2200 bp. Since it is unlikely that *Stylonychia*, *Oxytricha* and *Paramecium* require very different gene numbers for somatic functions, an estimate of 10 000 different metabolic genes provides a sensible average value.

3.3.4.2 Fungi

The water mold, *Achlya bisexualis*, an aquatic oomycete, has a genome size of 42 million bp, nine known different cell types, and is well-characterized in terms of conventional genome organization and messenger RNA expression. Some 82% of the total genome, or 34 million bp, is in the unique class, while the remainder includes middle repetitive sequences as well as a foldback fraction. The organization of the genome is such that the single copy sequences are not interspersed with moderately repeated sequences. The complexity of single copy DNA in *Achlya bisexualis* is 34 million bp and, if all of this consisted of domains of roughly 2200 bp, the maximum gene number would not exceed 15 000.

The transcriptive activity of this small genome is also revealing (Timberlake *et al.* 1977). Vegetatively growing cells express an mRNA complexity of 2.1 million nucleotides, which is only a small fraction (6%) of the single copy complexity of 34 million bp, and which corresponds to about 1800 genes with an average length of 1150 bases. Only minimal differences exist between the lengths of whole cell RNA, nuclear RNA, poly (A)$^+$ RNA and mRNA, so that in *Achlya bisexualis* there is no detectable difference between the size distribution of nuclear and messenger RNA molecules. This contrasts with the situation often found in more complex eukaryotes where, as we have already seen in the extreme cases of the bithorax and Antennapedia genes of *Drosophila melanogaster*, transcripts may be enormous relative to the finally processed mRNA.

The filamentous fungus *Aspergillus nidulans* has a single copy DNA complexity of 26 million bp, with a genome size even smaller than that of *Achlya bisexualis*. Here, a comparison has been made between the poly (A)$^+$ RNA complexities of somatic hyphae, conidiophores and conidia, and conidiospores (Table 3.11). Vegetative hyphae express about 27% of their single copy DNA, which equates to 5600 transcripts with an average length of 1200 nucleotides. About 400 sequences can be shown to be spore-specific and another 1000 are non-spore but developmentally specific (Table 3.12). In total, then, *Aspergillus nidulans* expresses approximately 7000 genes as poly (A)$^+$ RNA. This is still only one third of the total single copy complexity. However, as we have seen in *Drosophila melanogaster*, a

Table 3.11 Poly (A)$^+$ RNA complexity in *Aspergillus nidulans* (from Timberlake 1980).

Cell type	Poly (A)$^+$ RNA complexity in 1200 nucleotide sequences
somatic hyphae (vegetative cells)	5600
conidiophores and conidia (spore-producing structures)	6700
spores	6500

Table 3.12 Complexity of poly (A)$^+$ RNA from spore specific and development specific stages of *Aspergillus nidulans* (from Timberlake 1980).

cDNA source	Poly (A)$^+$ RNA complexity in 1200 nucleotide sequences
spore specific	298
non-spore, developmental specific	1052

significant proportion of the genome is expressed as a poly (A)$^-$ fraction which is, nevertheless, found in polysomes. Hence the poly (A)$^+$ RNA complexity can significantly underestimate the proportion of the genome which is expressed as mRNA.

Finally, the slime mold, *Dictyostelium discoideum*, has a haploid genome size of 45 million bp, with single copy content of 30 million bp. Although its single copy content is close to that of *Achlya bisexualis*, the organization of its unique DNA sequences is quite different. There is a high degree of interspersion of moderately repeated sequences, of average length 250–400 bp, with unique stretches averaging 1500 bp. In further contrast to *Achlya bisexualis*, it has nuclear RNA molecules some 25% longer than the equivalent mRNA. It has been estimated that in *Dictyostelium discoideum* about half of the single copy DNA is transcribed during development (Firtel *et al.* 1976), which equals 12 500 transcripts of average length 1200 nucleotides. Thus, in the two fungi in which the total polysomal RNA population has been estimated, the total transcribed single copy sequences account respectively for between 1800 and 12 500 genes. In *Aspergillus nidulans* only the poly (A)$^+$ population is known, and this gives a minimum estimate of 7000 genes, a figure that may be higher when the total polysomal population is examined.

3.3.4.3　*The sea urchin,* Strongylocentrotus purpuratus

Sea urchins have long been favoured organisms for developmental analyses. Here, the fertilized egg undergoes cleavage to produce a blastula of about 400 cells. Gastrulation follows and results in a free-swimming pluteus larva of about 1500 cells some 3 days after fertilization. After a further 40 days of growth, this larva undergoes metamorphosis to produce the adult. Most molecular work on sea urchin development has used *Strongylocentrotus purpuratus*, a species which has a genome size of approximately 900 million bp, of which 675 million bp (75%) is classified as unique (Davidson *et al.* 1982). When the RNA populations of a variety of developmental stages and a range of adult tissues are compared, three clear conclusions emerge:

(a)　The highest complexity is found in the oocyte and is equivalent to 37 million nucleotides (Fig. 3.22). This amounts to about 6% of the total single copy DNA. The average transcripts in sea urchin have been found to be 2500 nucleotides in length, so that the complexity of the oocyte accounts for 15 000 diverse structural genes. The diversity of transcripts decreases, however, as development proceeds. By gastru-

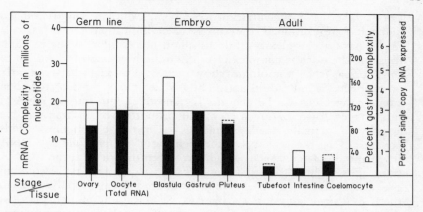

Figure 3.22   Sets of structural genes active in sea urchin embryos and adult tissues. The solid portion of each bar indicates the amount of single copy DNA shared between the gastrula mRNA and the other tissues. The open portions show the amount of single copy sequence present in the various mRNAs studied but absent from gastrula mRNA (data of Galau *et al.* 1976b).

lation, the complexity of polysomal RNA has fallen to 17 million nucleotides, and by the pluteus stage it is 14 million nucleotides. In non-reproductive adult tissues it is 5.8 million nucleotides in the tubefeet, 3.7 million nucleotides in the circulating coelomocytes and 2.9 million nucleotides in the intestine. This implies that these adult cell types require fewer than 5000 transcripts of average length to operate normally. The process of embryonic development, by contrast, is much more demanding in terms of gene activity. These changes in the RNA population contrast markedly with the situation in *Drosophila melanogaster*, where some 70% of the transcripts are present throughout all stages of the life cycle. In sea urchins it appears as if most transcripts are present at fertilization, and that the RNA complexity of the populations then subsequently decreases. On the other hand, metamorphosis in the pluteus is known to involve an imaginal disc which, amongst other things, is responsible for the replacement of the ectodermal epithelium of the pluteus and the production of the larval spines (Czihak 1971). Since the mRNA complexity of this structure has not been taken into account, the existing data on the adult must underestimate the total RNA complexity.

(b)   Also in contrast to the results obtained in *Drosophila melanogaster*, it is evident that different developmental stages in *Strongylocentrotus purpuratus* share only a proportion of their total mRNA sequence diversity. Of the 14 000 mRNA species present in the gastrula, some 11 000 are expressed in the pluteus but only 1000–5000 are found in the three adult tissues examined. Qualitatively distinct, though partially overlapping, sets of single copy transcripts thus appear on the polysomes during sea urchin development (Fig. 3.22).

(c)   On the other hand, when nuclear RNA complexities in *Strongylocentrotus purpuratus* are examined, they prove to be high in all cases. Gastrula nuclei have a complexity of 170 million nucleotides, while in adult intestine the nuclear complexity is 190 million nucleotides. Added to this, nuclear RNA homologies show very large overlaps between early embryogenesis and adulthood. Thus, at least one third of the total single copy DNA of the genome is transcribed into nuclear RNA at most stages. Even so, only a very small proportion of this transcriptive activity is actually made available for message functions. This conclusion is borne out by the behaviour of two cloned DNA fragments, *Sp34* and *Sp88*, which have been shown to occur as RNA transcripts in the early egg and which are ubiquitously represented in the nuclear RNA (Lev *et al.* 1980). Both are present on polysomes in the 16-cell stage but not in the blastula. *Sp34* is found on polysomes in the adult intestine, whereas *Sp88* is not. Thus, there is a continuous nuclear expression of these two genomic clones despite their differential cytoplasmic expression.

We can consider two extreme viewpoints to explain the paradoxical situation that much more of the genome appears to be transcribed than is exported to the cytoplasm in *Strongylocentrotus purpuratus*. One possibility is that a significant proportion is transcribed into RNA which has no regulatory or translatory function, and that the selection of what is useful in an informational sense is made at the time of RNA export from the nucleus into the cytoplasm. This is certainly the case in the oocyte nuclei of amphibians, where giant transcripts, produced by the lampbrush chromosomes, include readthrough products of normal transcription (see § 3.4.4). An alternative possibility is that much of the RNA transcribed is used in a regulatory capacity within the nucleus itself, as has been amply expanded in models in which RNA transcripts are assumed to play a role in the initiation of gene expression.

In the absence of specific developmental mutants, it is difficult to put the important components of the changing RNA population of the sea urchin into a coherent developmental context. How, for example, are we to discern which subset of the large RNA population is critical in making developmental decisions, as opposed to gene products which fluctuate in amount during development because the genes themselves are locked into specific developmental circuits but do not themselves play critical developmental roles?

3.3.4.4 *Vertebrates*

The polysomal RNA complexities in those vertebrates which have been analyzed are summarized in Table 3.13. If we leave aside, for the moment, the values for brain tissue, the recorded complexities are not all that different from those of the invertebrates we have previously discussed. In the two amphibians, *Xenopus laevis* and *Triturus cristatus*, egg complexities are of the order of 30–40 million nucleotides, despite the fact that their genome sizes differ by at least a factor of five. Thus, it is evident that

Table 3.13 RNA sequence complexity in vertebrates.

Species	Genome size (mbp)	% scDNA	Tissue/ stage	Total nRNA (mnt)	Poly (A)$^+$ mRNA (mnt)
Xenopus laevis	3000	75	egg		27–40
			tadpole		30
Triturus cristatus	23 000	47	egg		40
Gallus domesticus	1150	87	liver		20
			oviduct		30
			myofibr..		32
Mus musculus	3360	70	embryo		20
			kidney		20
			liver		10
			brain	500	110–140
Rattus norvegicus	3000		spleen	180	
			thymus	170	
			kidney	230	
			liver	410	
			brain	590	130–140

After Rosbach *et al.* 1974, Axel *et al.* 1976, Hastie & Bishop 1976, Young *et al.* 1976, Chikaraishi *et al.* 1978, Davidson & Britten 1979, Chaudhari & Hahn 1983, Milner & Sutcliffe 1983.

organisms of the same morphological complexity may vary enormously in genome size and yet have an equivalent RNA-transcribing capacity. This is an extremely important point since, from both a metabolic and a developmental context, it reduces to insignificance the enormous variation in genome size found in eukaryotes. *Thus, at the level of metabolism and of development, it is clear that there is no C-value paradox.* In the chicken, which has a genome size of only one-third that of humans, the polysomal RNA complexity is much the same, about 30 million nucleotides. This reinforces the notion that, irrespective of genome size, the transcriptive capacity of mammalian somatic tissues is, in general, much the same as that of other vertebrates elsewhere in the evolutionary spectrum.

The nuclear RNA complexities of several different tissues have also been compared. They vary from 170 to 590 million nucleotides. While the smaller are, in effect, a subset of the larger, they are certainly not identical to them. A small proportion, probably about 5%, is tissue-specific. Even so, the overlap is very substantial, indicating that liver sequences are included in the brain set and kidney sequences are included in the liver set. Thus, the situation is similar to that in the sea urchin where, it will be recalled, while some of the gastrula mRNAs are absent from the adult intestine their nuclear RNA transcripts are, nevertheless, present. In mammals, too, the nuclear RNA transcripts in kidney, liver and brain, for example, appear to be ubiquitous, whereas their mRNA populations are selectively expressed. The one clear exception in all of this is the mammalian brain, where initial estimates of the complexity of whole brain polysomal RNA gave a figure of 225 million nucleotides (Chaudhari & Hahn 1983), which would be equivalent to 150 000 different sequences each 1500 nucleotides long.

More recent studies have used cDNA clones, prepared from poly $(A)^+$ cytoplasmic RNA from rat brain, in conjunction with Northern blot techniques (Milner & Sutcliffe 1983). Of 191 randomly selected clones from a rat brain cDNA library, 47 hybridized to mRNAs expressed in brain but not in liver or kidney, and hence were brain specific. Seventy more hybridized to mRNAs whose expression was not limited to brain, though 41 of these were differently expressed in the three tissues. An additional 41 clones contained copies of mRNAs present in brain, as demonstrated by Southern blotting but below the level of detection employed in experiments using Northern blot hybridizations. These are considered to represent brain-specific mRNAs which, while rare within the brain as a whole, may be relatively abundant in a small number of cell types. The brain-specific mRNA molecules are, on average, twice as long as mature non-specific mRNAs, while rarer mRNAs, in all cases, tend to be larger than more abundant species. Thus, the rarest detectable brain-specific mRNAs have an average size of about 5000 nucleotides, three times the average size of brain mRNA. Since rare brain-specific mRNAs are much larger than abundant species, it is clear that earlier estimates of the number of mRNAs expressed in the brain are too high. A more realistic figure obtained from the recent data is that somewhere between 5000 and 30 000 mRNA species will be required to account for the diversity of brain structure and function, and that between 50 and 90% of these are likely to be brain-specific.

In other mammalian cells, tissues and organs that have been examined, essentially all of the sequence complexity is present in the poly $(A)^+$ fraction of polysomal RNA. In the brain, however, there are basically two distinct, non-overlapping populations of poly $(A)^+$ and poly $(A)^-$ mRNA, present in approximately equal frequency. Moreover, most of the poly $(A)^-$ mRNAs are brain-specific. Since the study referred to above considered only the poly $(A)^+$ fraction of brain mRNA, the upper estimate of 30 000 is likely to be the more accurate. Such a diversity of mRNA molecules within the brain is consistent with the extensive heterogeneity and microdifferentiation of the various cell types that occur there.

The poly $(A)^-$ mRNAs of the brain are largely absent at birth and develop later, whereas the poly $(A)^+$ mRNAs are present at birth. By comparison with most other organs, there is elaborate postnatal development in the brain, which involves myelination of neurons and extensive arborization of dendrites, coupled with synaptogenesis (Fig. 3.23). Correlated with the increase in neuronal connections, there is a substantial increase, over time, in the complexity of poly $(A)^-$ mRNA. The chicken again provides an instructive comparison. It has a poly $(A)^+$ complexity similar to that of man and mouse, although its genome is composed, effectively, of single copy DNA. It will be instructive to determine how the poly $(A)^-$ and brain complexities of the chicken compare to other mammals. If they turn out to be quite different, then there may be a way of rationalizing the difference in terms of brain-specific RNAs.

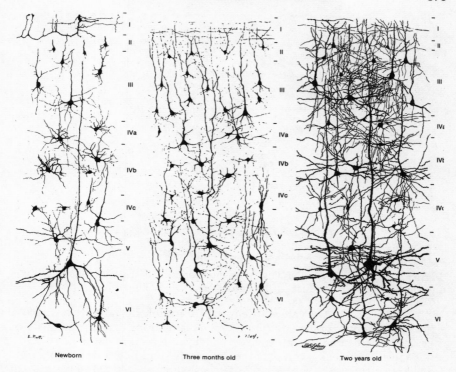

Figure 3.23 Progressive postnatal arborization of the human visual cortex. For the purpose of this comparison the depth of the cortex has been normalized to the same level (data of Conel 1959).

Calculations of transcript numbers from RNA complexity studies suffer from one additional complication. If a significant proportion of the genome consists of pseudogenes, then, when the unique portion of the genome is used in a driver/tracer experiment with excess RNA, stable hybrids should form between the pseudogenes and the RNA made from the normal working genes. Thus, the amount of hybridization will over-estimate the actual copy number of genes involved. What is obviously needed is an estimate of how many pseudogenes there are for the average gene in a vertebrate genome. If there is, on average, one pseudogene per gene, then estimates of somatic gene number will need to be halved. In some instances, at least, there is a vast excess of pseudogenes, as in the case of vertebrate 3-phosphate dehydrogenase (Piechaczyk *et al.* 1984). However, it should be remembered that these will not add to the saturation hybridization value, since they will already have been removed in the repetitive DNA fraction.

3.3.4.5 Developmental capacities of eukaryote genomes

The results obtained from studies on message complexity in eukaryotes show a considerable measure of consistency, and point to four general conclusions regarding the developmental activity of the eukaryote genome:

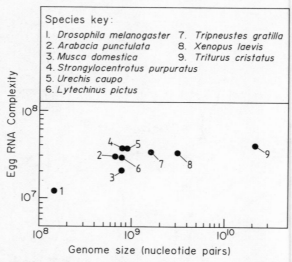

Figure 3.24 Egg RNA complexity (in nucleotides) of selected eukaryotes as a function of genome size (data of Hough-Evans *et al.* 1980).

(a)   When the known values of egg RNA complexity are expressed as a function of genome size (Fig. 3.24), it is obvious that in a majority of species this complexity is largely independent of genome size.

(b)   Most of the eukaryote genome evidently does not code for proteins, since estimates of the number of genes which are active in a tissue or an organism, derived from the complexity of single copy sequences represented in mRNA, fall considerably short of the number expected were all the unique DNA within the genome to function in a coding capacity.

(c)   Since the nuclear RNA complexities of liver and spleen in the rat, for example, are quite different, it is clear that some tissues must have more, and different, nuclear transcripts than others. Thus, there is control of what is transcribed in the nucleus and of what eventually exits into the cytoplasm. We shall be in a better position to determine how many genes behave differentially in this way when we have a representative sample of specific cloned probes for given transcripts, so that their nuclear and cytoplasmic timetables can be defined.

(d)   There is a compelling need to distinguish between those genes that are responsible for important developmental decisions that give rise to morphological diversity, and those whose activity is required either for basic cell functioning or for the specialized functions that characterize specific cell types. Mice and humans, for example, have approximately the same genome size and similar RNA complexities, yet they are morphologically very different. At the moment there is little alternative but to continue characterizing mammalian genomes using recombinant DNA technology. However, in the absence of systematic genetic perturbations of the genome, which might provide clues on genes that are developmentally significant and which can be examined at the DNA level, we are unlikely to be able to directly isolate key developmental genes from the mammalian genomic haystack. However, as the case of the homeo box illustrates, it may be that *Drosophila melanogaster* will provide probes necessary to unlock some aspects of mammalian development.

3.3.5 Differences in heterochromatin content

In terms of the supposed C-value paradox, we have now seen that organisms of comparable morphological complexity, but with widely varying genome sizes, are, nevertheless, also comparable in terms of somatic message complexity, so that the excess DNA is not obviously of metabolic or developmental significance. This applies especially to the class of highly repeated DNA which, in the main, is known not to be transcribed in somatic tissues. The few cases where transcription of heterochromatin has been noted, for example in the germ line of amphibians, can be explained in terms of polymerase readthrough from conventional transcription units (see § 3.4.4). It is important to emphasize that with highly repeated DNA we do not face the same uncertainty in measuring transcription products as we do in the case of conventional genes expressing rare copies, since in the case of highly repeated DNA a large proportion of the genome can be built up from sequences of this type. If, for example, significant transcription was to occur in the kangaroo rat, where more than half the genome consists of simple sequence DNA, it would be unlikely to be missed.

Considerable differences, in fact, exist between different eukaryotes in respect of both the content and the arrangement of highly repeated DNAs within the genome. A particularly striking example of this variation is found in the genus *Drosophila* (Table 3.14). In *Drosophila virilis* the genome is about twice the size of that of *Drosophila melanogaster* (a distant relative), and over 40% of it consists of highly repeated DNA which includes four distinct sequences. In all four, the basic repeat is 7 bp long, and they all occur within the pericentric heterochromatin. Other members of the *Drosophila virilis* group, e.g. *Drosophila ezoana* and *Drosophila littoralis*, have no detectable highly repeated DNA within their genomes, though sequences homologous to the high repeats of *Drosophila virilis* can be detected in the chromocentres of these species using *in situ* hybridization (Cohen & Bowman 1979). Thus, even between closely related species belonging to the same grouping the proportions of highly repeated DNA vary considerably.

Even more striking differences obtain in more distantly related forms. In the Samoan species, *Drosophila nasutoides*, four highly repeated sequences collectively make up some 60% of the genome, and all of them are localized in one giant chromosome (Fig. 3.25). By contrast, in the three Hawaiian species, *Drosophila gymnobasis*, *Drosophila silvarentis* and *Drosophila grimshawi*, approximately 40% of the genome is composed of a single highly repeated sequence. This consists of a 189 bp repeat unit and its variants, and is spread

Table 3.14 Heterochromatin and highly repeated content of *Drosophila* genomes.

Species	% Male hetero-chromatin	Sequence 5'→3'	Highly repeated DNA		
			density¹ (g cm⁻³)	% genome	total %
D. melanogaster	35	AATAT		3.1	
		AATAG		0.23	
		AATAC		0.52	
		AAGAC		2.4	
		AAGAG		5.6	
		AACAA		0.06	21
		AATAAAC		0.23	
		AATAGAC		0.24	
		AAGAGAG		1.5	
		AATAACATAG		2.1	
		359 bp		5.1	
D. simulans	30		1.669	3	22
			1.692	19	
D. virilis	52	ACAAACT	1.671	5	
		ATAAACT	1.688	11	42
		ACAAATT	1.692	18	
			1.704	8	
D. texana	47		1.676	6	
			1.691	25	35
			1.721	4	
D. hydei	30		1.714	4	17
			1.696	13	
D. ezoana	33		no detectable satellite DNA		
D. nasutoides			1.665	28	
			1.669	12	57
			1.682	8	
			1.687	9	
D. gymnobasis			1.690	40	40

After Schweber 1974, Cordeiro-Stone & Lee 1976, Gatti *et al.* 1976, Renkawitz 1978, Wheeler *et al.* 1978, Cohen & Bowman 1979, Miklos & Gill 1981, Lohe & Brutlag 1986.

over all members of the complement, each of which is characterized by a large block of pericentromeric heterochromatin (Miklos & Gill 1981). Thus, different species of *Drosophila* may be characterized by having no detectable highly repeated DNA, or several forms of it (Fig. 3.26). Sequence analysis shows that the minimum repeat unit may vary from 5 to 359 bp within the same species, and from 5 to 189 bp between species. These highly repeated sequences are almost invariably present within the heterochromatic regions of chromosomes, though not all heterochromatin necessarily contains such DNA.

X Y

Figure 3.25 The haploid chromosome set of *Drosophila nasutoides* (2*n* = 8) to illustrate the giant and predominantly C–band positive (solid) nature of the macro element, which is also an isochromosome consisting of two equal and homologous arms (data of Wheeler & Altenberg 1977).

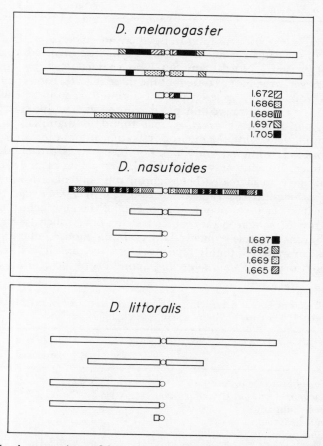

Figure 3.26 A comparison of the satellite DNA sequence content (shaded) of three species of *Drosophila* (data of Peacock *et al.* 1977, Wheeler *et al.* 1978, Cohen & Bowman 1979, Hilliker & Appels 1982).

Table 3.15 Genome variation in the kangaroo rat genus *Dipodomys* (from Hatch *et al.* 1976).

Species/subspecies	2n	Arm no.	C(pg)
D. heermani tularensis	64	94	3.5
D. panamintinus leucogenys	64	96	4.0
D. microps	60	116	4.0
D. merriami merriami	52	100	4.0
D. heermani californicus	52	96	4.3
D. deserti deserti	64	108	4.4
D. agilis perplexus	62	110	4.8
D. spectabilis baileyi	72	94	4.8
D. ordii compactus	74	144	4.9
D. spectabilis spectabilis	72	70	4.9
D. ordii monoensis	72	140	5.5

Comparable variation is found elsewhere in eukaryotes. For example, members of the rodent genus *Dipodomys* (kangaroo rats), which live in the deserts of western North America, are characterized by considerable interspecies variation in the content of the three repeats, consisting of simple sequences from 3 to 10 bp long, present within the genome. The net result (Table 3.15) is that the range of haploid genome sizes within this one genus, 3.5–5.5 pg, is almost as great as that found in mammals as a whole (2.5–5.9 pg). Yet, within kangaroo rats the amount of morphological differentiation is minimal and does not begin to approach that shown between mice, humans or whales. The larger genomes in *Dipodomys* are characterized by an increased heterochromatin content, a feature which is common in mammals (Table 3.16) and, as in *Drosophila*, the highly repeated sequences form major components of the heterochromatin.

In contrast to kangaroo rats, other rodents have much larger repeating units within their simple sequence DNAs. In the mouse, where 10% of the genome is composed of highly repeated DNA, the unit is 234 bp in length. In the laboratory rat, *Rattus norvegicus*, about 3% of the genome consists of

Table 3.16 Variation in heterochromatin content in placental mammals (after Hsu & Arrighi 1971, Hatch *et al.* 1976, Deaven *et al.* 1977).

Species	1C (pg)	%-C band material
Peromyscus crinitus (deer mouse)	3.5	6
Muntiacus muntjak vaginalis (barking deer)	2.5	15
Homo sapiens (human)	3.5	17
Microtus agrestis (vole)	3.2	26
Mus musculus (mouse)	3.5	27
Mesocricetus auratus (golden hamster)	3.5	34
Peromyscus eremicus (deer mouse)	4.7	36
Dipodomys ordii monoensis (kangaroo rat)	5.5	58

highly repeated DNA with a 370 bp monomeric unit, which is itself constructed from underlying 92 and 93 bp sequences. In some species within the genus *Rattus*, however, these two sequences are barely detectable, whereas in others variants of the 92 and 93 bp repeats are found as 185 bp building blocks (Miklos *et al.* 1980). Thus, within rodents the complexities of the basic repeating units vary from 3 to 370 bp, and the amounts of the highly repeated sequences range from less than 1% of the genome to in excess of 50%.

Finally, DNA sequence analysis of highly repeated DNA, both within and between species, reveals that extensive variation also occurs at the sequence level. The categories of variants observed indicate that this non-transcribed component of the genome readily accommodates base changes, deletions, additions and insertions of unrelated DNA. Evidently, eukaryote genomes can tolerate considerable variation in both the amount and the composition of highly repeated sequences without any obvious somatic effects, and such sequences appear to be genuinely and generally dispensable from a metabolic and a developmental point of view.

3.3.6 Differences in coding DNA

We turn now to consider the sorts of genes to be found within the genome with a view to deciding:

(a) whether the kinds and the structure of genes which determine a given grade of morphological complexity do, or do not, differ substantially from those that exist in different morphological systems; and

(b) whether the arrangement of genes within the genome has a functional significance.

3.3.6.1 Multigene families

We begin with the genes responsible for the most fundamental aspects of the cellular machinery, namely the large multigene families.

The rRNA genes. The organization, size and degree of repetition of rDNA in eukaryotes is highly variable (Table 3.17). Considering first the 5S RNA sequences: in both *Saccharomyces cerevisiae* and *Dictyostelium discoideum* the 5S genes are linked to, and interspersed with, the high molecular weight rRNA genes (Maizels 1976), as they are regularly in bacteria. *Saccharomyces cerevisiae*, for example, has 100 rRNA genes per haploid genome, arranged in a tandem array on chromosome 12. Each gene contains 9 kb of DNA and encodes all four species of rRNA. The 25S, 18S and 5.8S are transcribed as a single 35S precursor which is later processed. In a majority of eukaryotes, however, the 5S RNA is encoded by gene loci which are separate from the nucleolus organizing region itself. The average number of 5S sequences per genome varies widely between species, with a range from a minimum of 100 in *Saccharomyces cerevisiae* to a maximum of 300 000 in the urodele *Notophthalmus viridescens*.

Pronounced differences are also evident in the number of 5S DNA sites

Table 3.17 Repetition of 18S + 28S, 5S and tRNA genes in selected eukaryotes (after Long & Dawid 1980).

Species	Number of genes per haploid genome			Haploid genome size (pg)
	18S+28S RNA	5S RNA	tRNA	
Physarum polycephalum	280	690	1050	0.006
Saccharomyces cerevisiae	140	150	360	0.017
Tetrahymena pyriformis	290	780	1450	0.21
Drosophila melanogaster	200	160	750	0.18
Xenopus laevis	450	24 000	6500	3.1
Xenopus borealis	500	9000		3.5
Rattus norvegicus	170	830	6500	3.2
Homo sapiens	200	2000	1310	3.5

within the genome. In man and other primates, as in *Drosophila melanogaster*, there is only one such site. In the genus *Plethodon* there are from one to three 5S DNA sites, according to the species involved. In *Notophthalmus viridescens* there are four, while in *Xenopus laevis* at least 15 of the 18 telomeres in the haploid set carry 5S genes.

Whereas the multiple 5S genes of most species produce only a single category of RNA, *Xenopus laevis* synthesizes at least two, one in somatic cells and the other in oocytes. During embryogenesis, and at later stages, the oocyte 5S genes are silent and only the less abundant somatic 5S RNA genes are expressed. Most of the 24 000 5S genes encode for the oocyte–type 5S RNA and only 450 produce somatic 5S RNA (Miller & Melton 1981).

Turning now to the 18S + 28S system there are quite striking variations here too. The 18S and 28S genes on the X-chromosome of *Drosophila melanogaster* constitute an island in the middle of approximately 16 million bp of highly repeated DNA together with assorted nomadic sequences. Those on the Y are similarly embedded in highly repeated DNA. Each organizer carries about 200 copies of the 18S, 5.8S and 28S genes. Stocks of different origin frequently show significant differences in rDNA content (Spear 1974), but whether this actually signifies differences in the number of rRNA genes, or whether it merely reflects differences in the level to which these genes are amplified within the tissues of the adult in different strains (see § 3.4.2.3) is not clear.

Cloning and sequencing studies reveal that the basic 18S + 28S unit is a mosaic at the DNA level (Fig. 3.27). The 18S coding sequence is both preceded and followed by a transcribed spacer, after which there is a 5.8S coding sequence, a 28 bp AT-rich sequence, a 2S sequence, another transcribed spacer and finally the 28S sequence may, itself, carry an insertion. Individual transcription units within the cluster are separated from one another by material which is not transcribed. Both the 5.8S and the 28S sequences can be interrupted. Most attention has focused on the interruption in the 28S unit, since here there are basically two insert families, termed type I and type II, which have little sequence homology with one another.

The insertions themselves are of variable length, being 0.5–5.0 kb in the case of type I and averaging around 1.5 and 4.0 kb in type II. Both types are inefficiently transcribed. The major transcript of the type I insertion is a 1 kb cytoplasmic RNA molecule. The major type II transcript corresponds to the full length of the insertion, i.e. 3.4 kb, but does not leave the nucleus. The two insertions also differ in their distribution. Approximately half of the 200 28S genes in the X-chromosome carry type I insertions, about 30 have type II insertions, and 70 are devoid of either. On the Y-chromosome, however, only about 30 of the 200 or so 28S genes have insertions, and most of these belong to type II. Additionally, type II insertion sequences are mostly, if not entirely, restricted to the ribosomal locus, whereas approximately 400 kb of sequences homologous to type I insertions occur outside the rDNA locus (Dawid & Botchan 1977). Most of these are in the heterochromatic chromocentre of the polytene system, but a few are found at a euchromatic site (102C) on chromosome 4. Analysis of transcription products reveals that ribosomal genes containing type I insertions are effectively not transcribed, so that in female *Drosophila*, with two X-chromosomes, only about 150 of the 400 or so ribosomal genes are available for potential transcription in diploid tissues (Long & Dawid 1979). Thus, some 60% of the ribosomal genes do not function in transcription in diploid tissues.

As a fundamental component of the metabolic machinery of the cell, the rRNA-producing factory appears, at first sight, inefficiently organized to carry out its function, since it carries an inordinate amount of debris. Fortunately, this debris does not interfere with the functional capabilities of the nucleolus organizing region (NOR) in diploid tissues, while in polytene tissues the normal developmental events involved in polytenization effectively circumvent any problems (Endow & Glover 1979). Thus, only one NOR is selected for amplification in polytene tissues of the female and only the Y-NOR is amplified in male polytene tissues. Additionally, only the uninterrupted repeat is amplified during polytenization. Since interrupted and uninterrupted genes are intermingled at random, this implies either that the unit of amplification is not much larger than a single uninterrupted gene or else that amplification involves an extra-chromosomal mode.

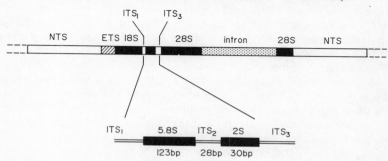

Figure 3.27 Organization of the 18S + 28S gene repeat unit in *Drosophila melanogaster* (after Glover 1981). Note ETS, ITS and NTS represent respectively the externally transcribed, the internally transcribed and the non-transcribed spacer sequences.

This example illustrates four important points:

(a) If a transcription unit is interrupted by a non-homologous piece of DNA at a very large distance from the site of transcription initiation, then transcription does not even begin, despite the fact that the upstream controlling regions are not perturbed.

(b) The type I insertion sequences have some of the characteristics of mobile elements (Dawid *et al.* 1981), and this is consistent with their distribution in other parts of the genome.

(c) Given that in the female, only 150 (40%) of the 400 or so ribosomal genes are available for transcription in diploid tissue, *Drosophila melanogaster*, nevertheless, maintains the 250 non-functional copies in both its diploid somatic tissues and in the germ line, without any apparent embarrassment or difficulty.

(d) It is perfectly possible for the fly to compensate for this large non-transcriptive fraction in polytene tissues by amplifying those rRNA genes that do function in transcription.

In terms of restriction pattern profiles, sequences homologous to the type I insertions of *Drosophila melanogaster* have undergone extensive divergence in six related sibling species (Roiha *et al.* 1983). Moreover, in *Drosophila simulans* and *Drosophila mauritiana*, the two species most closely related to *Drosophila melanogaster*, type I sequences can exist independently of rDNA. The type II sequences, on the other hand, are present in five of the six sibling species and their restriction profile is highly conserved. Even so, type II sequences are completely absent from *Drosophila erecta*.

In *Drosophila simulans* there is no Y-NOR, while in *Drosophila hydei* there are three NORs, one on the X and two on the Y. Electron microscope examination of transcribing versus non-transcribing ribosomal DNA of *Drosophila hydei* reveals two sorts of genes, one of which is longer than the other. The long genes have a 5 kb piece of DNA inserted in them, and are not transcribed in primary spermatocytes (Glätzer 1979). Since these insertions occur predominantly in the rDNA of the X-chromosome, its NOR is effectively transcriptively silent. Finally, in Hawaiian *Drosophila* species the NORs are autosomal.

There is comparable variation in the siting of NORs in vertebrates. *Xenopus laevis* has only one NOR in its haploid set, whereas its fellow amphibian *Notophthalmus viridescens* has three. Some rodents have only one NOR per haploid set and this is sex-linked, but the mouse has NORs close to the centromeres of autosomes 12 and 15–19. In the white handed gibbon there is only one rDNA site in the genome, whereas in man there are NORs on the short arms of five of the autosomes, namely 13, 14, 15, 21 and 22. Thus, in two primates which show marked developmental similarities the nucleolus may occupy a single site or multiple sites.

The variation in the copy number of 18S and 28S rRNA genes per genome is no less extensive (Table 3.18). Variation occurs within, as well as between, species. In a single population of the salamander *Plethodon cinereus*, different individuals range from 330 to 2500 ribosomal genes per genome,

Table 3.18 Numerical variation in the 18S + 28S RNA genes of eukaryotes (from Long & Dawid 1980).

Species		Number of genes
Nematoda		
	Caenorhabitis elegans	55
	Ascaris lumbricoides	300
Insecta		
	Sciara coprophila	45
	Acheta domestica	170
	Bombyx mori	240
	Drosophila melanogaster	200
Echinodermata		
	Lytechinus variegatus	260
Chondrichthyes		
	Squalus acanthias	960
Osteichthyes		
	Tinca tinca	120
	Salmo truta	1080
	Neoceratodus forsteri	4800
Anura		
	Bufo viridis	500
	Bombina variegata	2050
Urodela		
	Plethodon cinereus	2060
	Ambystoma mexicanum	4100
	Notophthalmus viridescens	5800
	Amphiuma means	19 300
Aves		
	Gallus domesticus	200
Mammalia		
	Mus musculus	100
	Cricetulus griseus	250
	Homo sapiens	200

yet all have a single major NOR on chromosome 7 (Macgregor *et al.* 1977). Thus within a population, individuals tolerate a vast excess of rRNA genes at a single site, an excess, moreover, which has no developmental significance.

Andronico *et al.* (1985) have examined four different Italian populations of the urodele *Triturus vulgaris meridionalis*, and have found that the ribosomal RNA gene copy numbers per genome vary from 2000 to 20 000 between individuals. In addition to the single main location of the NOR, there are at least 48 additional ribosomal RNA sites scattered throughout the genome. These are variable in number and distribution between individuals, but are, nevertheless, inherited in a Mendelian fashion.

The variation is not restricted to animal genomes. There is a 2.5-fold range of ribosomal RNA gene numbers between individual plants of *Allium cepa* (Maggini *et al.* 1978), and variation in the number of ribosomal RNA sites between individuals (Schubert & Wobus 1985). There are also claims of variation in the number of ribosomal RNA gene sites between cells of the same individual in *Allium* (Schubert & Wobus 1985) and in *Crotalaria agatiflora* (Leguminosae) (Verma & Raina, 1981). The most spectacular variation in copy number is found between plants of *Vicia faba*, which have from 250 to 24 000 ribosomal RNA genes per haploid genome (Rogers & Bendich 1987, in manuscript). It is obvious, just as in the animal cases, that the vast excess of ribosomal gene number between individual plants has no developmental significance whatsoever.

At the molecular level, the rDNA of *Xenopus laevis* has been particularly well characterized. The entire *Xenopus laevis* unit is roughly 12 kb, whereas man and mouse average 50 and 44 kb respectively. In *Xenopus laevis* the transcribed region contains both gene sequences and spacer regions which are subsequently discarded from the primary transcript. There is also a non-transcribed spacer whose length varies from 3 to 8 kb in different cloned repeats. As is the case for all vertebrates so far examined, including mice and humans, there are no insertions in the 28S gene of the kind that characterizes *Drosophila melanogaster*. Evidently, early in their evolution, the Diptera suffered an insertion into their rRNA genes and have coped with it ever since.

Whereas the base sequence of the transcribed spacer regions remains fairly constant between different species of *Xenopus*, the non-transcribed spacer is heterogeneous between species. Similarly, when non-transcribed spacer sequences are compared between the various sibling species of the *Drosophila melanogaster* subgroup, it is also evident that there has been a replacement of one sequence by another. Thus, for any given species the non-transcribed sequences, themselves, belong to a single family, and the spacer DNA is in fact diagnostic for a species (Coen *et al.* 1982).

We have, so far, dealt only with rRNA genes occurring in tandem arrays. In the crested newt, *Triturus cristatus carnifex*, there are NORs on chromosomes 6 and 9. Additionally, however, there are an unknown number of orphon rDNA sequences, no longer contained within the bosom of their own repetitive families but dispersed as a series of scattered sites, on the long arm of chromosome 1 (Morgan *et al.* 1980).

The histone genes.   The variability in the organization of the histone genes in eukaryotes is also quite remarkable given the overall conservation of sequences in the histone proteins themselves. Thus, the organization of the repeating unit may be very different in different species (Table 3.19).

The histone genes of *Drosophila melanogaster* are arranged, like the rRNA genes, as a single cluster, which includes an array of approximately 100 repeats which exist as 5.0 and 4.8 kb families. Within the histone cluster, the 5.0 kb repeats are predominantly tandemly arranged and they are rarely interspersed with the 4.8 kb repeats. As expected, the odd nomadic element has also been detected within the cluster. Each repeat contains the coding

Table 3.19 Organization of histone genes in animal genomes (from Hentschel & Birnstiel 1981).

Species	Organization of repeat unit	Length of repeat unit (kb)	Approximate repetition frequency
Drosophila melanogaster	←H3 →H4 ←H2A →H2B ←H1	4.8	100
Psammechinus miliaris	→H4 →H2B →H3 →H2A →H1	6.3	300–600
Strongylocentrotus purpuratus	→H4 →H2B →H3 →H2A →H1	6.5	300
Xenopus laevis	→H3 →H4 ←H2A →H2B →H1		20–50
Notophthalmus viridescens	←H4 ←H2A →H2B ←H3 ←H1	9.0	600–800

sequences for all five histone classes, and these are divergently oriented within a cluster. The genes for H2a and H1 are encoded on one strand with those for H3, H4 and H2b on the other (Fig. 3.28).

In situ hybridization indicates that the histone genes are clustered at a single site spanning at least 12 bands at position 39DE on the left arm of chromosome 2. Since the five histone genes are on different strands, and divergently oriented, this implies that one unit is capable of producing five different transcripts. With over 100 of these five transcript-producing units in 12 bands, either there is a large excess of transcription units per band or else the histone cluster must, in some sense, be divided up into 12 functional genetic units. Thus, it seems likely that again there are many more transcription units than there are complementation groups.

Dispersed and solitary members of all five histone gene classes, which are surrounded by flanking sequences quite different from those of the major cluster, have also been identified. These are referred to as orphons (Childs *et al.* 1981), and they exhibit extensive heterogeneity in different individuals. Whether these orphons are physically separated from the cluster is, however, not known, since no *in situ* hybridization data are available for any orphon clone.

In the American spotted newt, *Notophthalmus viridescens*, which has one of the largest known genome sizes (1C = 45 pg), the histone genes have a

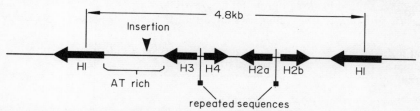

Figure 3.28 Structure of the histone repeat unit of *Drosophila melanogaster*. Arrows indicate direction of transcription (5' → 3'). Two classes of units are known, the one differing from the other by an insertion in the spacer between H1 and H3 (after Lifton *et al.* 1978).

markedly unusual pattern of organization (Stephenson *et al.* 1981). Genes
for the five histone proteins are clustered on a 9 kb segment, and there are
600–800 of these units per genome. The individual 9 kb clusters, however,
are separated by enormous length (50–100 kb) spacer sequences, some, or
all, of which are members of a predominantly centromeric highly repeated
DNA sequence with a 225 bp repeating unit. This contrasts with the
situation in the histone genes of *Drosophila melanogaster*, *Xenopus laevis* and
Mus musculus, which do not contain simple repeated sequences in their
spacers. The highly repeated DNA of *Notophthalmus viridescens* occurs at the
centromeres of all the chromosomes and at two non-centromeric sites on
chromosomes 2 and 6. Thus, somehow, highly repeated centromeric DNA
sequences have become integrated with histone genes at sites far removed
from their centromeric location. Moreover, readthrough of the histone
genes causes this highly repeated DNA to be transcribed by default (see
§ 3.4.4).

Quite a different situation is found in man (Table 3.20). Clones recovered
from a human genomic DNA library show that more than one histone gene
exists within each clone, but that the spacing between such genes fails to
show any ordered periodicity and they are certainly not in regular tandem
arrays. A comparable situation exists in both the mouse (Sittman *et al.* 1981)
and the chicken (Engel & Dodgson 1981), so that the arrangement found in
mammals evidently predated the avian–mammal dichotomy. A human
histone gene cluster (λHHG41), containing H3 and H4 coding sequences,
has been located on the long arm of chromosome 1 (Green *et al.* 1984),
though an earlier report had claimed that the histones of man were localized
specifically to the distal third of the long arm of chromosome 7 (Chandler *et
al.* 1979). The λHHG41 cluster is certainly not present on 7, though this
does not preclude the presence of other histone clusters on 7 or, indeed,
elsewhere in the genome.

Additionally, it is worth drawing attention to the fact that in the
unicellular ciliate *Stylonychia*, in which the macronuclear genome exists as
gene sized pieces, the various histone genes are present as independent
DNA fragments, emphasizing that there certainly is no absolute requirement
for their contiguity (Elsevier *et al.* 1978).

There is also evidence for the occurrence of subtypes of individual
histones, which differ slightly in amino acid sequence, some of which may
appear at different developmental stages or in different tissues. In the sea
urchin embryo, the histone genes encoding the blastula stage proteins, the

Table 3.20 Six types of histone gene containing clones isolated from the human
genome (after Heintz *et al.* 1981, Sierra *et al.* 1982).

H2B H4				
H2B . . . . . . H2A . . . H4				
H3,H4 . . H4 . . . . H3				
H2A,H2B . . . . H3 . . . . . H4				
H4 . . . H3 . . . . H4 . . . . . H2A				
H3 . . H4 . . H2A,H4 . . H3				

α-subtype, are present in several hundred nearly identical tandem repeats, each of which contains one copy of each of the five histone structural genes, separated by non-coding spacer DNA. The structure and topology of the genes that encode the late stage histone proteins is different. Specifically, the H3 and H4 genes are not present in tandem repeats and are not tightly clustered with the H1, H2A and H2B genes, as they are in the early histone genes (Cohen *et al.* 1975). Moreover, the late H3 and H4 genes are divergently transcribed, whereas in the early histone cluster all five histone genes are transcribed off the same strand (Childs *et al.* 1982). Additionally, both tandem early, and dispersed late histone gene families can function independently within the same genome.

Histone orphon genes, have also been identified in sea urchins. Nearly all of them are polymorphic, so that no two individuals have the same set of histone orphons. One such an orphon, which represents a single H2B gene of *Strongylocentrotus purpuratus* embedded in moderately repeated DNA, has its coding region interrupted by a 2.9 kb sequence that has many of the structural characteristics of a transposable element (Hentschel & Birnstiel 1981).

Finally, although all enchinoderms transcribe the evolutionarily highly conserved α-subtype histone genes during embryogenesis, the expression of these genes as maternal mRNAs is restricted to the structurally advanced sea urchins (Raff *et al.* 1984). These maternal mRNAs are, moreover, unique in their time of accumulation during oogenesis, their localization in the egg nucleus and their delayed timing of translation after fertilization.

The number of histone genes also varies considerably in different species. In mice, as in humans, there are about 40, while chickens have some 10–20. *Drosophila melanogaster*, however, has about a hundred, while sea urchins, where several hundred are present, show a greater reiteration of histone genes than any other phylum. There has been an attempt to explain such large differences in histone gene number in terms of the exceptionally large histone requirements during early embryogenesis in organisms which, like sea urchins and *Drosophila melanogaster*, undergo an initial phase of very rapid nuclear division. In mammals, on the other hand, cleavage is characteristically a very slow process. Thus, the mouse embryo takes approximately 3½ days to produce an expanded blastocyst of some 60 cells.

Amphibians, at first sight, appear to provide an exception, since *Xenopus laevis*, with only some 50 histone genes, forms 30 000 cells in the 9 hour period between fertilization and the gastrula stage. Here, however, the embryo solves its problem in a different way, for the histone genes are intensely active throughout oogenesis and give rise to an egg with a high stable histone mRNA content (Woodland 1980). In amphibians, the number of copies of histone genes per genome shows some tendency to increase with increasing C-value (Table 3.21). This, it has been argued, is related to the fact that species with larger genomes require more stored histone and histone mRNA in their eggs to package the larger amounts of DNA per cell at cleavage. However, the increase is most certainly not proportional and there are particularly striking differences between *Ambystoma mexicanum* (38 pg DNA) and *Notophthalmus viridescens* (45 pg DNA) which

Table 3.21 Histone gene copy numbers* in the genomes of amphibians (from Hilder *et al.* 1981).

Species	C-value		Approximate histone gene number	
	Absolute (pg)	Relative	Absolute	Relative
Xenopus laevis	3	1.0	50	1
Triturus cristatus carnifex	23	7.7	640	15
Ambystoma mexicanum	38	12.7	2690	60
Notophthalmus viridescens	45	15.0	700	14

* Based on complementarity to a cloned *Xenopus laevis* histone H4 coding sequence.

have histone gene numbers of 2300 and 700 respectively. Thus, the number of histone genes in *Ambystoma mexicanum* more than compensates for its large genome size. Of course, this assumes that all the copies are functional, which may not be true. Neither do we know what proportion of copies is active at oogenesis.

The tRNA genes. No consistent pattern is evident for the organization of the tRNA genes in eukaryotes. In yeast, most such genes occur widely scattered throughout the genome. Occasionally, two genes encoding different tRNA species lie in close proximity and are then transcribed together as a single unit.

In the *Drosophila melanogaster* genome it has been estimated that there are some 600–750 tRNA genes, which indicates that, on average, there are likely to be 10–13 coding sequences for each tRNA species. Unlike both the rDNA and the histone system, the tRNA genes are not organized into tandem repeating units. They occur in clusters of from 1 to 18 coding sequences, located at more than 30 sites in the genome (Table 3.22 & Fig. 3.29). The precise organization of these clusters appears extraordinarily variable. Identical, or near identical, genes may occur at a single cluster but different tRNA genes are frequently interspersed with them. For example, a cloned 9.3 kb DNA fragment which encodes 1 $tRNA_2^{Arg}$, 3 $tRNA^{Asn}$, 1 $tRNA^{Ile}$ and 3 $tRNA_2^{Lys}$ genes had been partially sequenced (Fig. 3.30). The eight genes are irregularly spaced; one strand contains the coding sequences for five of them (Arg, Ile, Lys, Asn, Asn), and the other strand carries sequences for the remaining three (Lys, Lys, Asn). The genes are separated by DNA regions of varying length. If the various tRNAs in this cluster turn out to have developmentally specific regulation, and hence are transcribed at different times, then there must again be a vast excess of transcription units over polytene bands, since it is highly unlikely that there would be eight bands in 9.3 kb of DNA. Similarly, in region 50AB at least seven genes span a 2.5 kb length of DNA, are irregularly spaced and are distributed on both strands. Evidently, tRNA genes are widely dispersed throughout the genome and, even when clustered, the genes themselves are different. Additionally they may, or may not, be interrupted. Two of the

Table 3.22 The location of known tRNA species in *Drosophila melanogaster* (from Kubli 1984).

Species	Location	Species	Location
$tRNA^{Ala}$	63A, 90C	$tRNA_2^{Met}$	48B5–7, 72F–73A, 83F–84A
$tRNA_2^{Arg}$	42A, 84EF	$tRNA_3^{Met}$	19–20, 46A1–2, 61D1–2, 70F1–2
$tRNA_2^{Asp}$	25D, 29E, 69F–70A		
$tRNA_5^{Asn}$	42A, 59F, 60C, 84F	$tRNA_{2b}^{Ser}$	86A, 88A, 94A6–8
		$tRNA_4^{Ser}$	12DE, 23DE, 56D3–7, 64D
$tRNA_4^{Glu}$	52F, 53A, 56E–57D, 62A	$tRNA_7^{Ser}$	12DE, 23DE, 64D
$tRNA_{GGA}^{Gly}$	58AB, 84C, 90DE		
$tRNA_3^{Gly}$	22AC, 35AC, 56E–57D	$tRNA_3^{Thr}$	47F, 87BC
		$tRNA_4^{Thr}$	93A1–2
$tRNA^{His}$	48F, 49AB, 56E	$tRNA_1^{Tyr}$	19F, 22F, 23A, 28C, 41AB, 42A, 42E, 50C1–4, 56D, 85AB
$tRNA^{Ile}$	42A, 50AB		
		$tRNA_3^{Val}$	70BC
$tRNA^{Leu}$	49F, 50AB	$tRNA_{3a}^{Val}$	64D
$tRNA_2^{Leu}$	44EF, 66B5–8, 79F	$tRNA_{3b}^{Val}$	84D3–4, 90BC, 92B1–9
$tRNA_2^{Lys}$	42A, 42E, 50BC, 56E–57D, 63B	$tRNA_4^{Val}$	56D3–7, 89BC, 90C
$tRNA_5^{Lys}$	29A, 84AB, 87BC, 94AB		

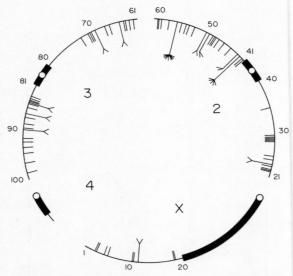

Figure 3.29 Location of tRNA genes in the genome of *Drosophila melanogaster*. Multiple species at the same site denoted by Y (two spp.), ⅄ (five spp.), ⅄ (nine spp.). In addition to identified tRNA species listed in Table 3.22, this figure also includes as yet unidentified species (see Kubli 1984).

genes at 50AB, both assigned to tRNA^{Leu}, contain intervening sequences of 38 and 45 bp, respectively, within the anticodon loop (Robinson & Davidson 1981).

Yet a third mode of distribution occurs in *Xenopus laevis*. Here, the genes encoding seven tRNA species, all having the same polarity, are clustered within a 3.18 kb DNA fragment and in the sequence: Phe, Tyr, Met A, Met B, Asn, Ala, Leu and Lys. This fragment is, however, tandemly repeated some 150 times within the genome, and occurs as a major block of repeats at a single site near the telomere of the long arm of one of the chromosomes in the 13–18 size group (Fostel *et al.* 1984).

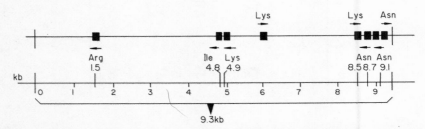

Figure 3.30 The tRNA gene arrangement on the 9.3 kb *Drosophila melanogaster* DNA fragment in pC1T12. Arrows indicate the direction of transcription, $5' \rightarrow 3'$ (data of Hovemann *et al.* 1980).

3.3.6.2 Structural gene systems

A small number of structural gene systems have also been sufficiently characterized to allow for comparisons to be made between organisms of different complexity.

The tubulin genes. Microtubules are filamentous proteins composed principally of heterodimers consisting of one α and one β polypeptide. They form major structural components of cilia, flagella and the spindles on which chromosomes move during cell division, and they play key roles in determining cell shape and cell motility (Weber & Osborn 1985). The proteins involved in microtubule formation are encoded by an important series of highly conserved genes, the tubulins. Despite their conservation, they vary in number in different eukaryotes (Table 3.23). In mammals, however, where they are most numerous, most of the copies are pseudogenes which have a variety of lesions that preclude the translation of a functional polypeptide. Moreover, there is certainly no simple relationship between the number of functional tubulin genes and morphological complexity, since both the chicken and *Drosophila melanogaster* have identical gene numbers.

Using RNAs prepared from a variety of chicken cell lines and tissues, five different mRNAs have been identified which carry β-tubulin sequences. Two of these are encoded by the same gene, so that there are four functional β-tubulin genes. All are very similar in structure, containing three or four small intervening sequences clustered at the 5'-portion of the coding region. A fifth region contains only a 5' fragment of a β-tubulin gene. Three of the

Table 3.23 Tubulin gene numbers in eukaryotes. Where more than one gene is present in the genome they are invariably dispersed in organization (from Cleveland 1983, Cleveland & Sullivan 1985).

Species	Tubulin type	Number of sequences per haploid genome	Number of functional tubulin genes
yeast	α	2	2
	β	1	1
Chlamydomonas	α	2	2
	β	2	2
Drosophila	α	4	4
	β	4	4
chicken	α	4–5	2
	β	6–8	6
rat	α	10–15	2
	β	10–15	3
man	α	15–20	2
	β	15–20	3

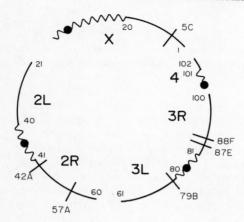

Figure 3.31 Location of the six actin genes within the genome of *Drosophila melanogaster* (data of Fyrberg *et al.* 1980).

five mRNAs produced are exceptionally long, between 3500 and 4000 bases, but, nevertheless, give authentic β-tubulin translation products (Lopata *et al.* 1983). Since yeast has only one α-tubulin gene, it is evident that many of the essential functions of a eukaryote, including meiosis, can be carried out with a single gene copy.

The actin genes. Actin is present in all eukaryote cells as a major component of the cytoskeleton, and it plays a fundamental role in both cell and organism motility. It is not a single protein, but rather consists of several molecular species. Three such species, designated as I, II and III, have been identified in *Drosophila melanogaster*. These are encoded by a family of six genes within the haploid complement (Fig. 3.31). The coding regions of all six are highly homologous, but each contains a single intron whose position varies in the different genes (Fyrberg *et al.* 1980). Actins, like tubulins, are highly conserved proteins. As estimated by Southern blot hybridization, the number of actin genes, like the number of tubulin genes, varies widely in different species (Table 3.24). Also, as in the case of

Table 3.24 Actin gene numbers in eukaryotes (from Fornwald *et al.* 1982, Hightower & Meagher 1986). In some of these cases the number of functional genes has yet to be determined.

Species	Number of sequences per haploid genome
yeast	1
slime mold	17
Drosophila melanogaster	6
sea urchin	11
chicken	4–7
human	20
soybean	6

tubulins, humans carry a much larger number of actin genes than does the chicken, while yeast again satisfies all its needs with a single actin gene.

The collagen genes.   Collagens are a family of major connective tissue proteins (Table 3.25). They share a common triple helical region with a repeating pattern in which every third amino acid is glycine. The various collagens have similar mechanisms of transcription, processing and secretion, but their biosynthesis involves several post-translational modifications of a precursor polypeptide, termed procollagen. This processing occurs extra-cellularly through the mediation of the procollagen peptidases. Vertebrate collagens are encoded by a family of at least 15 genes.

The most abundant and best studied of the vertebrate collagens is type I, a heteropolymer consisting of two α1(I) chains and one α2(I) chain. Genomic clones for the entire chick α2(I) gene have been isolated and analyzed (Wozney *et al*. 1981). The results indicate that 5 kb of coding sequence is contained within a gene at least 38 kb in length, and the coding region is interrupted by 50 introns. It is thus one of the most fractured genes known in eukaryotes, and also contains the smallest known exon, consisting of only 11 bp (Aho *et al*. 1984). This case highlights the more general question of why there are so many introns in some of the mammalian genes. In the case of the chicken α2(I) gene, while individual intron sizes range from less than 100 bp to several thousand, the size of the coding exons is most frequently 54 bp. This suggests that this gene expanded locally by repeated duplication of an ancestral 54 bp sequence, but nevertheless maintained its identity as a single functional unit. Furthermore, since so many introns interrupt what is effectively a monotonous run of Gly-X-Y triplets, it is unlikely that any of these exons have significance in terms of protein domains. This contrasts with the situation in the globins and the immunoglobins, where it has been argued that domains at the DNA level, which lead to the formation of proteins with particular functional domains, have arisen from the juxtaposition of initially separate exons during evolution.

In *Drosophila melanogaster*, by contrast with the vertebrates, there are as yet only two known collagen genes (Monson *et al*. 1982, Natzle *et al*. 1982), one autosomal (25C) and one sex-linked (19EF/20AB). At least one of these, the autosomal gene, encodes a protein of the classic Gly-X-Y repeating unit. This gene has been isolated by screening a *Drosophila melanogaster* genomic library with a chicken pro-α2(I) clone as a hybridization probe. It has only two introns and is otherwise composed of two large coding sequences, which together specify a 6.4 kb RNA species and a protein of some 469 amino acids.

In the nematode worm, *Caenorhabditis elegans*, the genome contains 50 or more collagen genes, the majority of which probably encode cuticle proteins. Two of these genes have been examined in some detail (Kramer *et al*. 1982), and both are small with only about 50 contiguous Gly-X-Y repeats. One is divided into only two exons, the other into three. Unlike the vertebrate situation, none of the exons is 54 bp long and none of the introns occur in the triple helical coding region of the gene.

Table 3.25 Characteristics of the major classes of vertebrate collagen (from Bernfield 1981).

Location	Type	Molecular composition	Form	Tissue
interstitial	I	$[\alpha 1(I)]_2\alpha 2$	large, coarse and strong fibril	skin, bone, tendons and ligaments
	II	$[\alpha 1(II)]_3$	fine fibrils	cartilage and notochord
	III	$[\alpha 1(III)]_3$	reticulin fibrils	foetal skin, blood vessels
cell surface	IV	$[\alpha 1(IV)]_2\ \alpha 2(IV)$	amorphous	basal lamina of placental epithelium and endothelium
	V	disputed	amorphous	basal lamina of muscle cells

In humans, the pro-α2(I) and pro-α1(I) genes, which together constitute a type I collagen, are located on chromosomes 7 and 17 respectively (Huerre *et al.* 1982). Clearly, this difference in location is irrelevant to co-ordinate requirements at the cellular level.

3.3.6.3 *Genome structure and function*

It has become almost a truism to assume that the position of a gene, either on a chromosome or else within a nucleus, is of critical importance to its functional capabilities. This is not borne out by an examination of comparative genome organization, where we find wide variation through-out eukaryotes at every level within the genome. This applies to gene number, gene location and gene structure. The wide range of genome sizes in eukaryotes is paralleled by an equally wide range of patterns of genome organization, which implies that it is imprudent to attach much significance to any one pattern when this is so readily flouted in other cases.

Since many eukaryote genomes contain an excess of families of repeated sequences, it has been assumed that these are, in some sense, implicated in the control of gene expression. A second assumption is that sequence arrangement is important because it determines the pattern of gene regulation, and hence development. As we have seen, there is no hard evidence to support either assumption, and the available evidence contradicts both. In fungi, where there is little excess of non-genic DNA, the control of gene expression evidently cannot operate through moderately repeated DNA. Likewise, the variety of interspersion patterns evident between organisms which must share similar basic developmental pathways, does not support a role for sequence arrangement in controlling gene expression. The available evidence from *Drosophila melanogaster* indicates that a very large proportion of the interspersed long repeats are mobile elements, so that there is no reason to continue to seek a function for gross genome organization in this organism. In vertebrates, the short interspersed repeats are either still nomadic, or have been at some time, and a number of the long repeats have characteristics expected of nomadics. Added to this, reverse transcription is now established as a major mode of genetic reorganization in species as diverse as yeast, *Drosophila* and humans. Baltimore (1985) estimates that as much as 10% of the genome, in some species, may have arisen by the reverse transcription of RNA into DNA. Thus, gross genome organization in vertebrates too, as determined by reassociation data, probably has little functional significance. If there are important regulatory circuits mediated by interspersed repeats, they have yet to be demonstrated. On balance, it seems much more likely that, in all eukaryotes, gene activity will turn out to be specified by those immediate regions of the genome in which a given gene or gene family is located, a conclusion which is strongly supported by the effects of reinserting genes into the *Drosophila melanogaster* genome using P element mediated germ line transposition.

Some of the structural and numerical variation present in eukaryote genomes makes little sense when thought of in conventional terms. However, if the genome is profligate, in both its replication and transcription processes, it is much easier to understand why *Xenopus laevis* has 24 000 copies of the 5S gene when its sister species *Xenopus borealis* has 9000; why such widely different levels of rDNA redundancy exist within eukaryotes; and why gene families responsible for the production of conserved proteins, like histones, show such different numbers of members and different patterns of organization. Ribosome production, for example, requires the co-ordinated expression of genes encoding over 70 different proteins, in addition to that of the genes specifying the ribosomal RNAs. In several mammals, including man, mouse and hamster, individual r-protein genes are encoded by multigene families, with 7–20 genes for each family and a total of about 750 genes in all. This multiplicity of r-protein genes in mammals contrasts sharply with the situation in yeast, *Drosophila melanogaster* and amphibians, where there are only one or two copies of each r-protein gene (Monk *et al.* 1981). Again, there is no apparent rationale for such a marked difference when thought of in conventional terms.

Some of these observed differences in genome organization may, therefore, simply constitute evolutionary debris, which the genome is both able to accommodate and tolerate. This is particularly clear in the case of the localized highly repetitive sequences, which show such dramatic differences in amount and whose addition, or deletion, from a genome has no measurable effect on the somatic phenotype. Pseudogenes represent a further clear case where we can be confident of the source of excess DNA and its lack of function in a metabolic and developmental sense. The same is true for the large DNA difference between two virtually identical species of *Drosophila*, namely *Drosophila melanogaster* and *Drosophila simulans*, where this difference can be accounted for almost exclusively by a difference in the content of mobile elements. It is evident that the eukaryote genome is a much more flexible and fluid entity than has previously been assumed, and that not all, or even a majority, of its components need be of functional utility in a somatic sense.

One does not, however, need to go to comparisons *between* species in order to visualize variation in nuclear DNA amounts. In the Asteraceae, differences in nuclear DNA contents of about 20% occur between plants of *Microseris* species having the same chromosome number, $2n = 18$, and growing side by side in the wild (Price *et al.* 1983). In pocket gophers of the genus *Thomomys* the between-individual nuclear DNA variation can be as high as 35% (Sherwood & Patton 1982). In North American cyprinid fish the variation between individuals with the same chromosome number, $2n = 50$, can be as high as 13.5% (Gold & Price 1985).

The observed variation in genome organization goes some way to accommodating the striking differences in genome size found in eukaryotes. A substantial portion of this size variation can now be accounted for in terms of DNA sequences which have no transcriptive activity within the soma. The non-nuclear components of the cell, namely the mitochondrial and chloroplast genomes, are also very instructive in this regard. While most animal mitochondrial genomes are less than 20 kb in size, yeasts vary from 20 kb to 106 kb. In closely related species of plants, such as *Brassica* species, mitochondrial genome sizes vary only a little, from 190 kb to 220 kb, yet in four cucurbit species the variation is from 350 kb to 2540 kb (reviewed by Bendich 1985). The C-value paradox is, thus, also apparent for mitochondrial genomes. However, the number of mitochondrial translation products is approximately constant, at 20 or so. Thus, there are roughly the same number of protein coding genes among plants and animals, with huge differences in mitochondrial genome sizes – from 14 kb in *Ascaris* (Wolstenholme, personal communication) to 2540 kb in musk melon (Bendich 1985). Thus, although the mitochondrial genome itself varies enormously in size, its informational capacity is largely constant. Furthermore, there is no correlation between the total organellar volume and mitochondrial genome size in four cucurbit species which have a sevenfold difference in mitochondrial DNA genome size (Bendich & Gauriloff 1984).

A similar situation is apparent for chloroplast genomes. In Algae, the chloroplast genome sizes vary from 85 kb to approximately 2000 kb, with variation among four species of *Chlamydomonas* being from 195 to 292 kb. In angiosperms, the variation is between 120 kb and 217 kb (reviewed in Palmer 1985). Although the overall variation is at least from 85 kb to 2000 kb, the same spectrum of genes is coded for by these genomes. Besides the ribosomal RNA genes and the tRNAs, there are approximately 25 protein coding genes. Furthermore, about 20 ribosomal proteins may also be encoded by chloroplast DNA sequences.

Thus, chloroplast and mitochondrial genomes are alike in having a near constant informational capacity in the face of highly variable genome sizes. Clearly, here too, there is no C-value paradox at the level of information production.

In terms of the variation in organellar gene structure and organization, there are again many similarities to the variation in nuclear genomes (reviewed in Palmer 1985). In *Euglena*, the chloroplast genes are interrupted by a total of at least 50 introns, whereas the broad bean has only six introns, or so. While *Euglena* chloroplast genes are frequently interrupted, most angiosperm chloroplast genes have no introns at all.

Furthermore, while the chloroplast gene order in most angiosperms is pretty much conserved, some groups do not obey this generalization. The genomes of pea and mung bean are highly scrambled, differing by more than a dozen inversions and complex rearrangements. Clearly, chloroplast function is not at all perturbed by such extensive genomic scrambling.

The mitochondrial genomes of the mustards are particularly rearranged. *Brassica hirta* and *Brassica campestris* are similar in mitochondrial genome size (208 kb) but differ by at least ten inversions (Palmer & Herbon 1987). Interestingly, the chloroplast genomes in these two species are totally colinear.

Whether or not such sequences perform any function other than the purely transcriptive is unknown. If they are simply 'selfish' or 'ignorant' (see § 4.2.1.3), then clearly there is little need to consider them any further. In some cases, too, not an inconsiderable portion of a genome may reflect the duplication of coding sequences, a situation which, as we shall see, sometimes depends on a wholesale polyploidization of the genome (see § 4.5.1.1).

It is clear that basic metabolic genes are held in common between eukaryotes. Thus, all of them obviously require tRNAs and rRNAs, histones, actins and tubulins. Additionally, many of the loci involved are strikingly similar in composition. Equally, the enzymic machinery used to effect DNA and RNA manipulations by eukaryotes is held in common and, indeed, is basically available even in bacteria, indicating its early evolution in biological systems. Likewise, major metabolic steps, such as photon capture and the subsequent initiation of light-dependent biochemical events in the cell, also evolved early. So, too, did the calcium channels involved in modulating neural activity, which have been found to be universally present from single-celled organisms to mammals. Furthermore, peptides related to known hormones, such as insulin, ACTH and β-endorphin in mammals, are purportedly present in unicellular eukaryotes (Le Roith *et al.* 1980, 1982) and in invertebrate nervous systems. Thus, the fundamental molecular requirements for the production of a neuroendocrine system appear to predate its anatomical development.

In prokaryotes, genes that respond to a single signal or that encode the enzymes involved in a single metabolic pathway, are often linked as an operon which is transcribed into a single polycistronic mRNA. Eukaryotes do not appear to use such polycistronic messengers. Thus, the genes coding for the enzymes catalyzing successive steps of at least five human metabolic pathways most commonly map to different chromosomes (Table 3.26). The one exception concerns four of the 14 enzymes involved in glycolysis, which are encoded by genes on chromosome 12, three of them on the same arm. Fig. 3.32 illustrates yet a further pattern of variation, this time a comparison of the location of presumed homologous X-linked mutations in six species of *Drosophila*. If these genes are genuinely homologous, and this has not yet been determined using cloned probes, then they are evidently able to function at a variety of sites within the X-chromosome and in a variety of positions relative to other characteristically X-linked genes.

In *Drosophila melanogaster* there are two principal forms in which genes may be distributed within the genome. In some few cases genes with a similar function are clustered. This is particularly evident for those sequences exhibiting extensive repetition. This includes the genes encoding the structural RNAs needed for the protein synthesizing machinery of the cell, as well as the histone genes, both of which are arranged in the form of tandem repeats (Fig. 3.33). Such clustering has been generally assumed to provide a rational basis for co-ordinating transcription. This certainly occurs in the rDNA cluster. However, some clustering may occur by default, having arisen as a consequence of local DNA duplication. This may explain, for example, the cluster of the four larval cuticle protein genes located at 44D.

By contrast, with the exception of the small number of gene complexes that have been identified in *Drosophila melanogaster* (Bithorax and Antennapedia) and which are probably the product of duplication, most of the structural genes code for single monofunctional polypeptides, and these, in general, are widely separated from genes of related function. Thus, the genes affecting segmentation of the early embryo, which are fundamental to

Table 3.26 The location of genes controlling enzymes involved in five human metabolic pathways (from McKusick 1980).

Enzyme	Chromosome location (p = short arm; q = long arm)
Galactose metabolism	
galactose-4-epimerase	1p
galactokinase	17q
galactose-1-phosphate uridyltransferase	9p
The urea cycle	
argininosuccinate synthetase	9
argininosuccinate lyase	7
ornithine transcarbamylase	X
Tricarboxylic acid cycle	
aconitase, mitochondrial	22q
aconitase, soluble	9q
isocitrate dehydrogenase, mitochondrial	15q
isocitrate dehydrogenase, soluble	2q
fumerase	1q
malate dehydrogenase, mitochondrial	7
malate dehydrogenase, soluble	2p
citrate synthase, mitochondrial	12
Glycolysis	
Entry to glycolytic sequence	
hexokinase	10
phosphoglucomutase-1	1p
phosphoglucomutase-2	4
phosphoglucomutase-3	6q
mannosephosphate isomerase	15q
First stage of glycolysis	
glucose phosphate isomerase	19
triosephosphate isomerase 1 & 2	12p
Second stage of glycolysis	
glyceraldehyde-3-phosphate dehydrogenase	12p
phosphoglycerate kinase	Xq
enolase-1	1p
enolase-2	12
lactate dehydrogenase-A	11p
lactate dehydrogenase-B	12p
Phosphogluconate pathway (pentose phosphate pathway; hexose monophosphate shunt)	
glucose-6-phosphate dehydrogenase	Xq
6-phosphogluconate dehydrogenase	1p
glyceraldehyde-3-phosphate dehydrogenase	12p

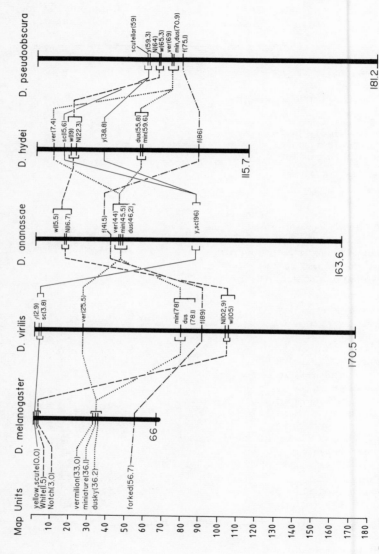

Figure 3.32 Location of X-linked genes in five species of *Drosophila*. In the case of *Drosophila pseudoobscura*, which has a metacentric X, only X_R is shown (data of Sturtevant & Novitski 1941, Alexander 1976, Hess 1976).

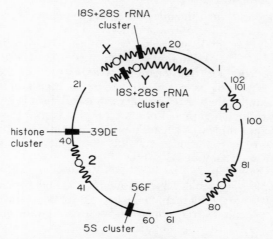

Figure 3.33 Major multigene clusters in the genome of *Drosophila melanogaster*.

many developmental pathways, are widely scattered in the genome (see Table 2.3). So, too, are the genes involved in the ommochrome biochemical pathway. Likewise, the six actin genes are well dispersed (see Fig. 3.31).

Two suggestive cases for the occurrence of what may be developmentally significant gene clusters have been documented in *Drosophila melanogaster*. The first of these occurs near the dopa decarboxylase structural gene (*Ddc*). Here 18 genes, 13 of which affect the formation, sclerotization or pigmentation of the cuticle (Wright *et al.* 1984), occur in an interval of 8–12 polytene bands. However, the precise role of each gene at the biochemical level is unknown and, as the authors themselves point out, it is premature to conclude that a functional cluster exists. The recent work of Eveleth & Marsh (1987), however, has revealed excellent homology between the *Ddc* gene and the closely linked α-methyl dopa hypersensitive (*amd*) gene. Furthermore, two other genes on either side of *Ddc* also show structural similarities to it on the basis of low stringency cDNA hybridization experiments. Thus, the *Ddc* region certainly contains a cluster of structurally and functionally related genes and these, again, appear to have arisen by duplication.

The second case concerns the location of contractile protein genes within the 88F subdivision of chromosome 3. Here, a 20 kb segment located 140 kb proximal to the actin 88F gene contains several myofibrillar contractile protein genes whose products, like those of actin 88F, accumulate only in the indirect flight muscles of the thorax (Karlik *et al.* 1984). However, the five antigenically related myofibrillar proteins encoded in this 20 kb segment are probably the products of only two transcribed regions, which may well represent the products of a tandem duplication, though this leaves unexplained the linkage of these genes to actin 88F. This linkage certainly does not appear to facilitate the co-ordinated expression of the loci concerned, since germ line transformation experiments demonstrate that a 7.2 kb fragment, which includes actin 88F, expresses itself normally when introduced into novel genomic locations.

There are two experimental approaches we can use to determine whether clustering is indeed of crucial functional significance. On the one hand, we can split a tandem array and then examine whether each of the products behaves normally. We can also split an individual repeat unit within a tandem cluster. If, under these circumstances, transcription is undisturbed, as we have seen it to be when the Bithorax Complex is split (Struhl 1984), then it seems fair to conclude that clustering is nothing more than the simple historical outcome of gene duplication and of the inevitable constraints which follow from such a mode of origin.

Karpen & Laird (1986) have provided data on precisely this point, by introducing a single ribosomal RNA gene into a euchromatic site of *Drosophila melanogaster* by P element mediated germ line transformation. This single rRNA gene is transcribed and forms a mini–nucleolus in polytene nuclei. Clearly, there is no strict requirement for either tandem repetition or for heterochromatic sequestering in order to ensure normal rDNA function.

It has been suggested that a distinctive kind of cluster involves the different genes concerned with specific modes of morphogenesis within individual cell lineages. Thus, three of the α-tubulin genes, two of the actin genes and the gene for the production of tropomyosin, all of which are implicated in muscle cell differentiation, are located within the interval 84–88, while two additional genes which are co-ordinately expressed in differentiating muscle cells have also been mapped to 87B. One needs to remember, however, that this is an enormous interval, at least 4 million bp and thus the size of an entire *Escherichia coli* genome. Consequently, such an association may be purely fortuitous.

A majority of the genes so far tested work reasonably well when inserted elsewhere into the genome by P element mediated germ line transformation, so that there is certainly no substantive evidence for the existence of large developmental domains. Thus, the three subunits of larval serum protein 2 are not closely linked, despite the fact that they are co-ordinately transcribed. On balance, therefore, it would appear that the position of any one gene relative to others within the euchromatin is unimportant for gene function.

3.4   GENE DOSAGE RELATIONSHIPS

It is a truism that all developmental programmes require particular gene products to be available at particular times. Genes differ, however, in how much of their product is required to produce a normal phenotype. For example, in the case of the xanthine dehydrogenase gene (Xdh), 1% of the amount normally produced appears perfectly adequate (Chovnick *et al.* 1977). The protein components of muscle, such as myosin, on the other hand, are blatantly inadequate when their level of production falls to 50%. Such genes need, therefore, to operate close to maximum efficiency.

Since the time of action of a gene product is also likely to be limited, there may be an additional constraint on the time available to a gene to produce an adequate amount of product. Such a temporal constraint will be minimal in a gene whose product is required only at a low level of efficiency compared to one where the requirement is for 100% efficiency. This means that if the required time limit is reduced by even a small amount, then genes of the latter category will be markedly affected in their performance.

For the fundamental metabolic systems that underlie all developmental circuits, the amount of gene product is of particular developmental significance. Without an adequate supply of ribosomes, histones, tRNAs and major enzymes no cell can function efficiently. Thus, the total deletion of ribosomal genes is lethal in early embryogenesis. On the other hand, X0 individuals of *Drosophila melanogaster*, with only half the diploid number of ribosomal transcription units, complete morphological development, despite the fact that 60% of the rRNA genes in the X-chromosome of *Drosophila melanogaster* are non-functional. How then is this achieved?

3.4.1 Compensation and magnification

Mutant individuals of *Drosophila melanogaster* carrying a reduced number of rRNA genes normally exhibit a bobbed (*bb*) phenotype, characterized by a reduced growth rate, a thinner cuticle and smaller bristles. If, however, an X-chromosome with a normal complement of rRNA genes is partnered by an X or Y that is either partially (X*bb* and Y*bb*) or wholly (X0 and XX^{N0-}) deficient for its ribosomal genes, than the rDNA content of such an individual increases during its development and the fly does not express a bobbed phenotype. This phenomenon, termed rDNA compensation, was discovered by Tartof, and initially interpreted in terms of an actual increase in the number of rRNA genes in the nucleolus organizing region (NOR) of the normal X-chromosome (Tartof 1973a). The measurements on which this interpretation was based came, however, from adult flies.

Both the larva and the adult of *Drosophila melanogaster* are, as we have seen, a mosaic of cells with varying levels of ploidy. Somatic cells can be diploid, polytene or endopolyploid, and the proportion of cells belonging to these categories is not known. From measurements involving whole individuals, whether adults or larvae, it is difficult to decide whether there has, indeed, been an increase in the actual number of rRNA genes per NOR, or whether there has been an amplification of rDNA in non-diploid tissues. Less ambiguous data can be obtained from measurements made on defined cell populations which are either predominantly diploid (imaginal discs, brain cells, germ cells), predominantly polytene (salivary glands) or endopolyploid (thoracic musculature), and by using rDNAs that have restriction site differences, so that it is possible to identify the precise source of the rDNA being measured.

From the recombinant DNA data that now exists, it is evident that the rDNA of *Drosophila melanogaster* displays a number of novel features:

(a) X0 cells from normally diploid XX tissues contain only half the

Table 3.27 Percent hybridization of rRNA to the DNA of larval diploid (brain and imaginal discs) and polytene (salivary gland) tissues in two strains of *Drosophila melanogaster* (from Spear & Gall 1973).

| | % rDNA | | | |
| | Diploid | | Polytene | |
Strain	XX	X0	XX	X0
Oregon R	0.469 ± 0.016	0.264 ± 0.021	0.078 ± 0.002	0.074 ± 0.007
Urbana S	0.365 ± 0.005	0.236 ± 0.006	0.085 ± 0.011	0.080 ± 0.013

normal amount of rDNA. Evidently, in these cells there is no rDNA compensation. The same level of rDNA is, however, present in polytene tissues of both X0 and XX flies (Table 3.27).

(b)   In XY polytene nuclei only the NOR of the Y-chromosome replicates. It is, therefore, possible to account for the uniform rDNA levels of X0 and XX polytene tissues without postulating compensation. All that is required is that, in XX individuals too, only one NOR is polytenized. That this is indeed the case has been confirmed by Southern blot comparisons of diploid and polytene tissues of interstrain hybrids in which the two X-chromosomes are characterized by distinctive *Eco*RI restriction patterns. Under these circumstances, whereas DNA from diploid XX tissues shows rDNA patterns representative of both X chromosomes, polytene tissues show a pattern characteristic of the X with the highest number of rRNA genes, irrespective of whether this is contributed by the female or the male parent. This phenomenon, referred to as nucleolar dominance, accounts for the polytenization of rDNA of only one of the NORs in both wild type XY cells, as well as in interstrain hybrids with XY, XX or XXY cells (Endow 1980, 1983).

(c)   Ribosomal genes are, however, under-replicated in polytene tissues, relative to euchromatic genes. The exact number of rounds of replication that rRNA genes undergo is not known with certainty, since the precise number of genes within an organizer that polytenize is unknown. This is because there is differential replication even within an organizer, and only genes without insertions are replicated. It has, in the past, been assumed that rRNA genes undergo the equivalent of only 6–7 rounds of replication, whereas euchromatic sequences replicate ten times.

The question that remains to be answered is whether the rDNA increase measured in whole flies occurs exclusively as a result of differential replication of rDNA in polytene and endomitotically growing somatic tissues. There are two additional data sets which have a bearing on this question.

Drosophila hydei has three NORs in its diploid set, one in the X and two in the Y. It is, however, possible to construct a series of translocation chromosomes which carry variant numbers of NORs. Thus, attached-X

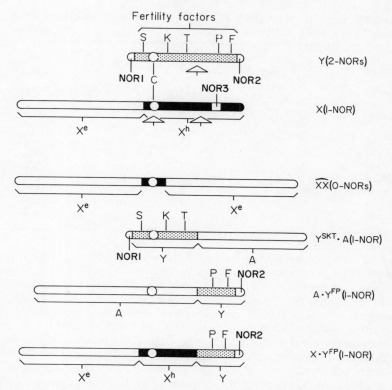

Figure 3.34 Variation in the number of nucleolar organizer regions (NORs) in wild type Y- and X-chromosomes of *Drosophila hydei*, and in experimentally produced structural variants (data of Kunz & Schäfer 1976).

($\widehat{XX}$) chromosomes can be constructed which completely lack the X-NOR, while two reciprocal Y-autosome translocations each carry one of the two Y-NORs (Fig. 3.34). Finally, two X.Y translocation chromosomes with Y fragments comparable to those in the Y-autosome translocations also each carry one of the two Y-NORs.

In wild type individuals, the single X-organizer includes about 520 rRNA genes, while each of the Y-organizers has 200. Consequently, $\widehat{XX}$Y females are expected to have an rDNA equivalent to 400 rRNA genes. The rDNA level, as measured by filter hybridization of DNA isolated from adult flies, however, gives a value in excess of this (Table 3.28). An even greater excess obtains in flies where the attached X is combined with a Y.A homologue, and a still higher value is found in individuals with two complementary Y fragments.

When the rDNA content of salivary glands from individuals carrying different combinations of NORs is compared, however, it is clear that, as in polytene tissues of *Drosophila melanogaster*, the same rDNA level is present irrespective of the number of organizers represented (Table 3.29). This is so

Table 3.28 Percent DNA hybridized with 28S rRNA in adult flies from different stocks of *Drosophila hydei* (from Kunz & Schäfer 1976).

Genotype	Number of nucleolar organizers			% rDNA	rDNA cistron no.
	X	Y	Total		
XX	2	0	2	0.47	1036
XY	1	2	3	0.44	910
$\widehat{X}X$Y	0	2	2	0.26	532
$\widehat{X}X/Y^{SKT}$A	0	1	1	0.32	630
X.Y^{FP}/Y^{SKT}A	0	2	2	0.45	835
X/Y^{SKT}A	1	1	2	0.43	841
$\widehat{X}X/A.Y^{FP}$	0	1	1	0.27	521
$Y^{SKT}X/A.Y^{FP}$	0	2	2	0.35	653
X/A.Y^{FP}	1	1	2	0.40	788

despite a threefold under-replication of rDNA in polytene chromosomes, relative to that of euchromatin. The rDNA content of both $\widehat{X}X/Y^{FP}$ diploid cells and endotetraploid $\widehat{X}X/Y^{SKT}$ and $\widehat{X}X/Y^{FP}$ thoracic muscle cells evidently exceeds, by a considerable margin, the value expected from the presence of only one Y-NOR. In the case of the thoracic musculature, it is conceivable that this is a consequence of the endomitotic DNA replication that operates to produce their tetraploid state. This is not likely to be the case in larval brain cells, since the great majority of these are diploid. The rDNA increase in these suggests that a different mechanism, termed rDNA over-replication, may operate in *Drosophila hydei*. If this is the case, then this mechanism must act differentially on the two individual Y-organizers. Thus, in the brain cells of $\widehat{X}X/Y^{SKT}$ individuals the rDNA value is unchanged and is one-half that of the wild type Y-rDNA.

The second data set which suggests that a genuine compensation mechanism may operate quite independently of nucleolar dominance and differential rDNA replication concerns the behaviour of a special locus present in the penultimate region of the distal X-heterochromatin. This locus, termed compensatory responder (cr^+), acts in *trans* to sense the presence or absence of an homologous locus in its partner X- or Y-chromosome. When cr^+ is absent from its partner it somehow leads to an increase in the rDNA content of the NOR with which it is contiguous (Procunier & Tartof 1978). Precisely how cr^+ works is not known, but deletion of the locus has no effect on nucleolar dominance. The mal^{12} deficiency chromosome, a derivative form of the scute[8] inversion ($In(1)sc^8$), carries a deletion that begins within the NOR and then extends to the euchromatic sequences normally located in the proximal region of the X-chromosome (Fig. 3.35). It thus includes a partial deficiency of rDNA and lacks cr^+. In hybrids which combine an X-chromosome from the Oregon-R stock (X^{OR}) with one carrying a mal^{12} deficiency ($Df(1)mal^{12}$), characterized by a weak bobbed allele, it has been found that the mal^{12} rDNA pattern predominates in polytene cells. Since X^{OR} is normally recessive to sc^8 in

Table 3.29 The rDNA content of diploid larval (brain), polytene larval (salivary gland) and endotetraploid adult (thoracic muscle) tissues of individuals of *Drosophila hydei* with different genotypes (from Grimm & Kunz 1980).

| Genotype | Number of nucleolar organizers | | | % Hybridization with 28S rRNA | | |
	X	Y	total	brain (diploid)	salivary gland (polytene)	thoracic muscle (endotetraploid)
XX	2	0	2	0.418 ± 0.015	0.056 ± 0.003	0.467
XY	1	2	3	0.391 ± 0.007	0.052 ± 0.004	—
XX̂/Y^{SKT}	0	1	1	0.097 ± 0.003	0.065 ± 0.005	0.323 ± 0.016
XX̂/Y^{FP}	0	1	1	0.334 ± 0.019	0.066 ± 0.005	0.340 ± 0.009

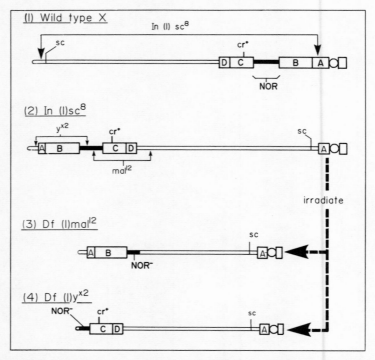

Figure 3.35 Chromosome inversions and deletions associated with the hetero-chromatic regions of the X-chromosome of *Drosophila melanogaster* covered by the mal^{12} and y^{x2} deletions. (Adapted from Durica & Krider 1978. Details of these various chromosomes can be found in Schalet & Lefevre 1976).

respect of nucleolar expression, it is clear that the deletion of cr^+ has no effect on dominance. Nevertheless, these hybrids appear to undergo ribosomal gene compensation (Endow 1983).

From this it has been concluded that partial deletion of genes from the dominant organizer leads to a release of dominance, so that both organizers undergo polytenization. There is, however, an alternative explanation. In both the mal^{12} and the y^{x2} deficiencies where the nucleolar organizers are themselves partially deficient they are also inverted relative to their normal location. Their relocation may simply mean that both are now behaving in a manner equivalent to the single organizer of XO individuals. Consequently, both undergo polytenization.

A bobbed mutation in an X-chromosome (Xbb) can be stably maintained against a Y carrying normal rDNA content (Ybb^+). However, when such an Xbb chromosome is maintained over several generations with a Y that is deficient in rRNA genes (Ybb^-), then the rDNA content of that X reverts to wild type and there is a loss of the bobbed phenotype. This phenomenon has been termed rDNA magnification and, unlike compensation, it is a heritable change. Earlier reports imply that a stepwise increase in the rDNA level of the Xbb chromosome occurs in successive generations during magnification. A more recent study, which examined the restriction cleavage pattern of rDNA from an X-chromosome undergoing magnification, reports the production of extremely bobbed male progeny in the first generation of magnification whose rDNA, nevertheless, exceeded that of wild type flies. This then reduced to a wild type level in subsequent generations with loss of the bobbed phenotype (de Cicco & Glover 1983). This means that although F_1 flies undergoing magnification actually contain sufficient rRNA genes to express a wild type phenotype, they do not do so. This, in turn, implies that the extra rDNA produced is non-functional and, significantly, both insertion$^+$ and insertion$^-$ rDNA units amplify during magnification. Additionally, tandemly arranged type 1 insertions, which interrupt about 60% of the rDNA units of the X-chromosome, also flank the heterochromatin on either side of the NOR and these, too, amplify transiently in the first generation of magnification.

Thus, at least three types of disproportionate rDNA production have been described in *Drosophila melanogaster* – genuine compensation, which appears to involve a direct increase in the number of rRNA genes within an organizer; dominance, which involves differential polytenization of an organizer, or at least parts of it; and magnification. It has been reported that circular DNA molecules can be identified in testis cells of males undergoing magnification, though these are not present in wild type testes (Graziani *et al.* 1977). If extra rDNA copies are being produced by extrachromosomal replication, they must, of course, be subsequently re-integrated into the germ line to be stably inherited.

Single unequal exchanges in a ring chromosome lead to the meiotic loss of the products of exchange; Tartof (1973b) reasoned that it should be possible to use a bobbed ring X-chromosome (X*bb*) to determine the basis of magnification. If magnification could occur independently of unequal exchange, it should be possible to detect it in the progeny of individuals carrying such a ring X. Since it was not, he concluded that magnification must result from unequal exchange.

That unequal exchange may indeed be involved in a magnification-like process is supported by observations on *Saccharomyces cerevisiae*, a species in which unequal exchange is known to occur regularly. In this species there are some 140 rRNA genes per haploid genome, with about 60% of them residing in chromosome 1. When a diploid strain of yeast is constructed and then made monosomic for a chromosome 1 initially deficient for 25% of its rRNA genes, its rDNA content increases over some 300 generations until it reaches a stable diploid level. While, strictly speaking, it is not meaningful to distinguish between compensation and magnification in a unicell, at least part of the increase in rDNA was transmitted through two meiotic cycles into viable haploids, which indicates the increase is indeed heritable (Kalback & Halvorson 1977).

There are also indications that both 5S (Procunier & Tartof 1975) and histone (Chernyshev *et al.* 1980) DNAs are able in some way to increase in quantity in genomes carrying deficiencies for these loci in one homologue. However, since both studies relied on analyses of adult flies it is not as yet meaningful to comment on the mechanisms involved.

3.4.2 Selective gene amplification

The conclusion from the studies on compensation and magnification is that sensing systems must exist within cells to detect the under–representation of tandem arrays. The precise response which particular cells then make will vary according to the replicative machinery operative within that cell type. Given that such mechanisms can operate under abnormal circumstances to increase the number of genes within a genome, do equivalent mechanisms exist in normal developmental sequences?

3.4.2.1 *The chorion genes of* Drosophila melanogaster

In *Drosophila melanogaster* the proteins of the egg shell, or chorion, are synthesized and secreted by the follicular epithelial cells which surround the oocyte. Oogenesis lasts for only two days, and the six major, and many more minor, chorion proteins are all synthesized over a brief five hour period at the end of oogenesis. A developmentally regulated amplification of the chorion genes occurs to boost transcription at this time (Kafatos *et al.* 1985, Kalfayan *et al.* 1985).

Prior to actual chorion gene expression, the follicle cells undergo several rounds of DNA replication in the absence of cell division, reaching a DNA level of 16C by stage 8 of oocyte development. Further increases take place but exact doubling is no longer observed (Fig. 3.36), suggesting the possibility of differential DNA amplification. At the time of onset of

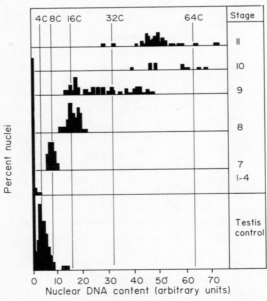

Figure 3.36 The nuclear content of follicle cells of *Drosophila melanogaster* at different developmental stages of the oocyte, compared to testis controls (data of Mahowald *et al.* 1979).

Table 3.30 The chorion gene system of *Drosophila melanogaster* (after Thireos *et al.* 1980, Spradling 1981).

locus	A1	A2	B1	B2	C1	C2
protein	C15	C16	C18	C19	C36	C38
gene	*S15-1*	*S16-1*	*S18-1*	*S19-1*	*S36-1*	*S38-1*
map site	66D		66D		7F	7F

chorion gene transcription in late oocytes the follicle cell nuclei contain about 45C of DNA.

The chorion genes (Table 3.30) are located in two major clusters, each of two genes. The coding sequences for *S38-1* and *S36-1* are on the X-chromosome at 7F, while those for *S18-1* and *S15-1* are on chromosome 3 at 66D. In both clusters, the two genes are separated by only 1–2 kb of DNA. Late in oogenesis these two clusters of chorion genes, and the flanking DNA that surrounds them, undergo a specific amplification, prior to the onset of mRNA synthesis. Amplification occurs by the repeated initiation of bidirectional replication at a specific origin located within each cluster. This results in the differential replication of a chromosome domain some 80–100 kb in length, and so extends well beyond the 12–20 kb, centrally located, chorion gene cluster itself. The sequences flanking the cluster amplify to a lesser extent than the chorion genes themselves, giving rise to 30–40 kb gradients of amplification with no apparent discrete termination sites (Spradling 1981).

While the sequences within a cluster are amplified to a similar extent, the 14–16 fold increase of the X-chromosome cluster is significantly less than the 60-fold increase of the cluster on chromosome 3. Thus, the chorion system provides us with an example where DNA templates are amplified and subsequently transcribed in one class of cells, the follicle cells, and the protein produced then transported to a quite different cell, the oocyte.

Two features of this amplification event are of particular interest. The first relates to the control of amplification within a cluster. The ocelliless (*oc*) mutation in *Drosophila melanogaster* is associated with a small inversion (*In(1)7F-8A*). In such a mutant the 40 kb of DNA lying distal to the 7F breakpoint completely fails to amplify, while the proximal 50 kb, now located at 8A, amplifies to a reduced extent. Consequently the chromosome region which remains at 7F undergoes no amplification, whereas the 7F sequences that have been placed in the 8A region, along with the putative replication origin, amplify in a manner reminiscent of that which obtains in the wild type chromosome (Fig. 3.37).

Using P element mediated germ line transformation, DNA segments derived from the chorion cluster at 66D have been inserted into new sites within the genome. Most of the fragments tested were not amplified, but a 3.75 kb sequence was identified which contained all the information necessary to specify amplification (de Cicco & Spradling 1984). Transposons containing this fragment all showed correct tissue and temporal specificity, and adjacent non-chorion sequences also underwent amplification. Even so, the levels of amplification obtained varied widely and only in one line was it

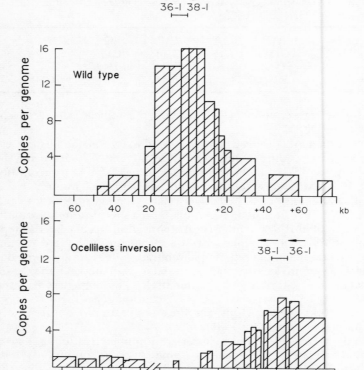

Figure 3.37 The effect of ocelliless inversion (*In[1]oc*) on the pattern of amplification of the X–chromosome chorion gene region (36–1, 38–1) in *Drosophila melanogaster* (data of Spradling & Mahowald 1981).

normal. Attempts to induce amplification using subfragments of the 3.75 kb sequence all proved unsuccessful. There are three possible explanations to account for this. First, that the residual flanking DNA from the P element has an effect on the efficiency of amplification. Secondly, that the fragment which has been cloned contains most, but not all, the signals necessary for efficient amplification. Thirdly, that the DNA landscape into which the vector has been inserted plays a role in determining the extent of amplification. Taken together with the behaviour of the ocelliless mutant, it seems most likely that the surrounding chromatin can have some influence on the level of amplification.

While this system provides evidence for a specific *cis*-acting element responsible for controlling amplification, it also emphasizes that gene 'regulation' may involve a great deal more than initiation of gene expression, for there is both spatial and temporal co-ordination of a series of events involving template amplification at the correct time, to the correct level and in the correct tissue. This is then followed by transcription at the correct time, with subsequent export of the protein product across cell membranes and into another cell type. All these events form part of a programme that takes only a few hours.

A second point of interest relates to the fact that gene *S15-1* produces a polyadenylated RNA early in oogenesis and prior to formation of the C15 protein. This RNA exits into the cytoplasm but does not associate with polysomes and is subsequently degraded. Two hours later, the same gene is again transcribed but this time the transcript becomes associated with polysomes and is translated subsequently to yield C15 chorion protein (Thireos *et al.* 1980). Thus, the same gene can be transcribed at two different times but yields a protein product on only one of the two occasions. This pattern of behaviour suggests that the onset of amplification and transcription are coupled in some way. Whether the early RNA, or its synthesis, is in fact functionally related to the amplification event, or whether the gene is simply locked into an historical regulatory circuit from which it is unable to dissociate itself, is not clear. Whatever the correct explanation, this example cautions us against searching for a function for every gene transcript, or even assuming that every transcript necessarily has a function on each occasion that it is made.

Two X-linked, recessive female sterile mutations, K451 at 12A6,7–12D3 and K1214 at 5D5,6–6D12, are known which cause underproduction of all six major chorion proteins, so that chorion structure is severely disrupted. This results from a substantial reduction in the level of chorion gene amplification (Orr *et al.* 1984). Although X-linked, the mutant loci are located far from the chorion cluster at 7F1–2. Moreover, their effects are most pronounced on the third chromosome cluster. It is not known whether these mutations impinge directly on the amplification process, or whether they affect a non-specific event which subsequently influences the amplification sub-routine.

3.4.2.2   *The DNA puffs of* Rhynchosciara americana

Individual polytene bands of dipterans are subject to metabolically induced variations in the form of a local decondensation, a phenomenon known as puffing (Ashburner & Berendes 1978). In *Drosophila melanogaster* the major sequence of puffing occupies a period of development extending from approximately 10 hours before pupation until 2 hours after pupation. Prior to this, only some ten bands are known to form prominent puffs and these are continuously active for some 30 hours. During the 12 hours of intense puffing, however, the number of bands involved increases dramatically to in excess of 125 (Ashburner 1970). Each band puffs at a definite time after the initiation of the sequence, and is active for a characteristic period before finally regressing. Circumstantial evidence initially led to the hypothesis that puffs are sites of intense transcription. Even now, the number of puffs for which we know the eventual gene product is small. In *Drosophila melanogaster* this includes the puffs responsible for the production of certain of the heat shock proteins (McKenzie *et al.* 1975), as well as those for the glue proteins (Velissariou & Ashburner 1980) which serve to attach the puparium to the substrate and which are synthesized by, and then stored within, the salivary gland until a few hours before puparium production.

A very distinctive series of puffs are formed in the larval salivary gland nuclei of the sciarid, *Rhynchosciara americana*, during the last cycle of polytenization in late fourth instar larvae. Here, puffing involves a process of DNA amplification, so that after the puffs regress there is more DNA in the bands involved than in their neighbours. By using a library of cloned cDNAs, synthesized from poly (A)$^+$ RNA isolated from salivary glands at a time when the DNA puffs are active, it has been possible to identify two genes, located respectively at DNA puff sites C3 and C8, which encode 1.25 kb and 1.95 kb mRNA molecules. Both these genes undergo a 16-fold amplification during RNA puffing (Glover *et al.* 1982). Here, again, we have an example of a developmentally regulated phenomenon presumably aimed at providing a large amount of a specific product, in this instance the peptides required for the production of the communal cocoon, over a short time period. It, thus, parallels the 60-fold amplification of the *S15* and *S18* chorion protein genes.

3.4.2.3   rDNA amplification

Tissue-specific gene amplification of the kind exemplified by the chorion genes does not appear to be a common mechanism for selectively enhancing gene expression, even in the case of egg shell formation. Thus, the chorion of the silkmoth *Bombyx mori*, which is also secreted by the follicle cells, is produced by over 100 linked genes which are not somatically amplified (Kafatos *et al.* 1977). There is, however, one clear case where selective amplification is of very general occurrence, and this involves the rRNA genes of animal oocytes.

Somatic cells of urodeles usually contain only one nucleolus, or at best a small number of nucleoli, per haploid chromosome set. Growing oocytes, however, are characterized by many hundreds of nucleoli per germinal vesicle (Table 3.31), but these are not attached to chromosomes. Although extrachromosomal, they are, nevertheless, analogous to the nucleoli of somatic cells and each is an autonomous site for rRNA synthesis, since they contain circular copies of rDNA. These extra copies of rDNA are in fact synthesized by, and detached from, the NORs during the early stages of oogenesis as closed circles. They continue to synthesize rRNA until the egg is mature and ready for ovulation (Table 3.32). These extrachromosomal copies of rDNA are used for rRNA synthesis only during oogenesis, and are rendered non-functional, and discarded into the cytoplasm, when the germinal vesicle breaks down to form the first meiotic spindle.

In *Xenopus laevis* the number of extrachromosomal nucleoli varies from 500 to 2500 in oocytes of the same stage and from the same female. Consequently, the actual rDNA content of one circle ranges from 0.7×10^{-2} to 1.5×10^{-2} pg, corresponding to about 500–11 000 rDNA cistrons (Thiebaud 1979). In *Xenopus laevis*, a weak amplification of rDNA occurs in oogonia but this is lost at the onset of meiotic prophase and the

Table 3.31 rDNA amplification in oocytes of urodele amphibians (after Brown & Dawid 1968, Macgregor & Pino 1982).

	Soma		Germinal vesicle		Number of nucleoli per germinal vesicle
	DNA (pg)				
	4C total nuclear	4C rDNA	rDNA	extrachromosomal DNA	
Flectonotus pygmaeus	6.8	0.014	5.3	7175	1000
Xenopus laevis	12.6	0.16	5.5	30	1000
Siredon mexicanum	140	0.16	8.5		
Triturus viridescens	178	0.16			
Necturus maculosus	380	0.08	5.3	30	600

Table 3.32 RNA content of *Xenopus laevis* oocytes (from Rosbash & Ford 1974).

Stage	μg RNA per oocyte	%		
		28S+18S	5S	4SRNA
early previtellogenic	0.04	8	45	47
late previtellogenic	0.07	49	31	20
early vitellogenic	1.20	80	11	9
mid vitellogenic	3.50	92	5	3
mature oocyte	4.30	94	3	2.5

Table 3.33 DNA amplification in invertebrate oocytes.

Species	% extra DNA
Acheta domesticus (cricket)	27
Tipula oleracea (cranefly)	59
Dytiscus marginicollis (water beetle)	90
Panagrellus silusiae (nematode)	90

After Lima-de-Faria & Moses 1966, Kato 1968, Cave 1973, Pasternak & Haight 1975.

major synthesis occurs in very young oocytes at pachytene (compare with the amplification of chorion genes in *Drosophila melanogaster*). Since the rDNA content of germinal vesicles isolated from uninucleolate (*1-nu*) mutant females is the same as that of normal females, although the somatic nuclei of *1-nu* heterozygotes can be shown to have lost half their rDNA, the control of rDNA amplification in the oocyte must also be sensitive to its final content in the germinal vesicle.

The tailed frog, *Ascaphus truei*, has an unusual pattern of rDNA amplification. In the last three oogonial mitoses all daughter nuclei remain in the same cell, so that the oocyte is an 8-nucleate syncytium at the onset of meiosis and remains so until late oogenesis, when seven of the nuclei degenerate. All eight nuclei behave synchronously in the synthesis of up to 5 μg of extrachromosomal rDNA per nucleus during and immediately after pachytene (Macgregor & Kezer 1970). *Ascaphus truei* thus increases its rDNA in two ways. First, there is an 8-fold multiplication as a consequence of the formation of the 8-nucleate oocyte. This is then followed by a further synthesis involving rDNA amplification at pachytene in each of the eight nuclei within the oocyte. However, notice that, despite the presence of eight nuclei in the oocyte, there is nothing unusual about the final size of the mature oocyte or of the tadpole it eventually produces.

Large extrachromosomal rDNA-containing bodies have also been described in the oocytes of a number of invertebrate species (Table 3.33). In all these cases, the increase is achieved by the transcriptive activity of extrachromosomal segments of rDNA that accumulate as single bodies following repeated replication during early oogenesis. This form of selective amplification can yield amounts of rDNA within a single oocyte which are 1000–5000 times the rDNA content of the relevant haploid genome.

3.4.2.4 *Somatic endoploidy*

Somatic endoploidy arises from two or more rounds of DNA replication without any breakdown of the nuclear membrane, so that mitosis is either aborted, with no spindle forming (endomitosis), or else completely omitted, so that the nucleus remains in an interphase state throughout the entire operation (endoreduplication). In either event, though the products of replication separate they remain within a common nucleus which, in consequence, increases both in DNA content and in size. In *Drosophila melanogaster*, a majority of the cells in both larvae and adults are endoploid. As in the case of polytenization, where, however, the nucleus is in a permanent prophase stage, different tissues tend to attain characteristic, though not necessarily uniform, DNA levels, with the result that they are usually mosaics of cells with different levels of endoploidy. Also, as in polytenization, some, though not all, endocycles are associated with a differential replication of euchromatin and heterochromatin. Endoploidy, like polyteny, leads to the production of an enhanced number of DNA templates in specialized cell types. Indeed, the highest levels of endoploidy, like those of polyteny, are found in glands or other cells showing intense synthetic activity.

These amplification events are not peculiar to *Drosophila*. There is a form of nuclear DNA amplification during development of the triploid endosperm in maize. Here, amplification yields very different nuclear DNA levels in different strains of maize (Knowles & Phillips 1985). Although the mechanisms of amplification are unknown, they may be akin to polytenization, since polytene chromosomes have been found in cotyledons, suspensor cells, synergids and antipodals of various plant species.

A series of giant neurons in the abdominal ganglion of *Aplysia californica* is known to modulate cardiovascular activity. Each of the neurons devotes a substantial fraction of its transcriptive capacity to the biosynthesis of a single polyprotein product, which acts as a precursor for the neuropeptides used as extracellular messengers by the cells which control the contractile activity of the circulatory system. These giant neurons have a diameter of about 250 μm and their genomes are highly polyploid, containing up to 10 000 times the haploid DNA content, presumably to support their metabolic activity (Nambu *et al.* 1983).

It is true that, at first sight, endoploidy appears to be a rather indiscriminate means of increasing the transcriptive capacity of a cell. Its efficacy, however, can be judged by the fact that the highly endoploid posterior silk gland nuclei of *Bombyx mori* are able to produce some 10^5 molecules of silk fibroin within about 4 days, despite the fact that the DNA template for the corresponding mRNA is present only once in the haploid genome (Suzuki *et al.* 1972). Likewise, in insects two main types of ovary occur, known respectively as panoistic and meroistic. Meroistic follicles are accompanied by a cluster of nurse cells, mitotic siblings of the oocyte itself, which are trophic in function, highly polyploid and connected to the oocyte by fine cytoplasmic strands. Such nurse cells are totally absent in panoistic follicles. Large amounts of RNA are produced by the highly endoploid nurse cells, which in effect contain genomes working in support of the growing oocyte. The advantage they bestow can be gauged by the difference in time required to complete oogenesis in the meroistic ovary of the blowfly, 6 days, compared to the 100 days it takes in the case of the panoistic ovary of *Acheta domestica*, where the oocyte nucleus itself is responsible for its own synthetic activity (Bier *et al.* 1969).

3.4.3   Dosage compensation

In diploid organisms with separate sexes but which lack morphologically distinguishable sex chromosomes, the activity of both copies of each gene is required for normal development. In such cases, if fertilization fails, the progeny are haploid and inviable. Here, either a single segregating recessive lethal locus will cause embryonic death or else the first gene that fails to make an adequate amount of product shuts the developmental system down.

Where differentiated sex chromosomes are present, there will, inevitably, be a dosage problem, since genes on the X-chromosome will be present in different numbers in the two sexes. Here then, it becomes necessary to compensate for this difference in dosage so that the total amount of gene product remains constant regardless of the number of X-chromosomes present. In fact, the expression of genes appears to have been carefully balanced, so that two copies of each autosomal gene and one or two copies of the dosage compensated X genes yield the correct amount of gene product.

3.4.3.1   Female X-inactivation

In humans, transcription is suppressed in the greater part of one of the two X-chromosomes in female somatic cells. It was originally proposed that the entire X was inactivated, but there is now good evidence that the steroid sulphatase (STS) locus and the Xg blood group locus, both of which occur on the short arm of the X, escape inactivation (Lyon 1986). Associated with this form of regulation is the heterochromatization of the inactive X, which consequently appears in female somatic interphase nuclei as the sex chromatin body.

Inactivation is initiated in early embryonic cells. The X-chromosome contributed by the sperm (X^P) is most certainly inactive at the time of fertilization. The maternal X is presumed to be active at this time. By the 8–16 cell stage of the embryo both X^P and X^M are active in individual blastomeres, and both remain active until the blastocyst stage (Fig. 3.38). At this time, the embryonic ectoderm, which gives rise to the cells of the foetus, is distinguishable from the primary endoderm which surrounds it and which gives rise to the extra-embryonic cell types. Whereas both X-chromosomes initially remain active in the embryonic ectoderm, X-inactivation has already occurred in the primary endoderm. By 6 days gestation, however, X-inactivation has also occurred in a majority of the cells of the embryonic ectoderm. In this case the inactivation is random with respect to the parental origin of the X, whereas in the cells derived from the trophectoderm, which are responsible for the production of the extra-embryonic membranes, the paternal X is preferentially inactivated. Once established, the pattern of inactivation becomes part of the somatic heredity of the cell in which it occurred. The clonal perpetuation of random inactivation thus leads to a mosaicism of expression, within the soma, of heterozygous X-linked genes which lie within the inactivated region.

Germ cell progenitors present in the embryonic ectoderm undergo

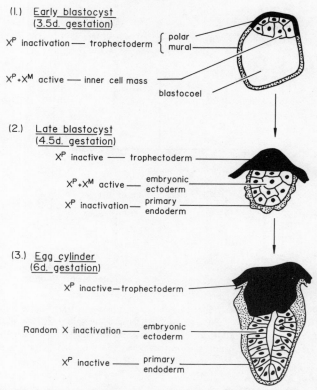

Figure 3.38 X-chromosome behaviour in early stages of the development of the female mouse embryo (data of Martin 1982).

differentiation and X-inactivation later than do other cells. Reactivation of the inactivated X coincides with the entry of oocytes into meiotic prophase at 12–14 days gestation. Thus, there are four distinct occasions at which X-inactivation occurs in placental embryogenesis – in the trophectoderm, in the primary endoderm, in the embryonic ectoderm and in the differentiation of the germ line. Only one of these is reversible, namely that which takes place within the germ line. Here there is a reactivation of the inactive X during the differentiation of oogonia into oocytes (Martin 1982).

Although the fact of inactivation is well known, its control is not. Specifically, one would like to know whether the structure of the DNA itself has been altered during inactivation. One approach to this problem is to examine whether an inactive X can function in DNA-mediated gene transfer of the kind which is possible in cell cultures.

Mus castaneus is known to carry a distinctive electrophoretic variant of the X-linked gene encoding the enzyme, hypoxanthinephosphoribosyl transferase (HPRT). By crossing a male carrying this variant (*Hprt^a*) with a female of *Mus musculus* heterozygous for an X-autosome translocation [*T(X;16)*] and

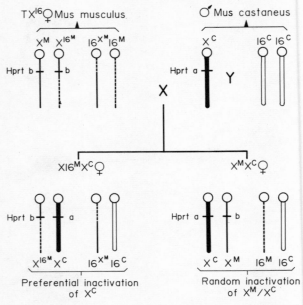

Figure 3.39 X-inactivation behaviour of the hypoxanthinephosphoribosyl transferase (HPRT) gene in the F_1 hybrid between a female of *Mus musculus* (M), homozygous for the *Hprt^b* allele and heterozygous for an X-16 translocation, and a male of *Mus castaneus* (C), hemizygous for *Hprt^a* (data of Chapman *et al.* 1982).

homozygous for the normal allele of HPRT (*Hprt^b*) it is possible to obtain females of two kinds (Fig. 3.39):

(a) Progeny with two normal X-chromosomes, one from *Mus musculus* the other from *Mus castaneus*. In such females inactivation is random and expression of the variant form of HPRT is mosaic.

(b) Progeny with one translocated X from *Mus musculus* and a normal X from *Mus castaneus*. In females of this type it is known that only those genes residing on the translocated X are expressed in somatic tissues. The normal X is inactive in over 95% of the somatic cells. Consequently, such females are not mosaic for gene expression and the variant form of HPRT, though present, is not expressed.

DNA extracted from female progeny carrying a translocated X^{16} chromosome was therefore used in a DNA-mediated gene transfer system to determine whether the inactive X was efficient in effecting HPRT transformation in V79 hamster cells. Sib female progeny with two normal X-chromosomes provided a convenient control. With DNA extracted from them one expects both HPRT variants to be functional in transformation.

Fifty-nine transformant lines were tested with DNA extracted from brain, liver and kidney of three adult *T(X;16)* females heterozygous for *Hprt^{a/b}*. Fifty-eight of these expressed only *Hprt^b*. Twenty-four of these were induced with DNA from a female that had successfully transmitted

Hprt^a to three of her five progeny, indicating that she did indeed carry an intact *Hprt^a* allele. By contrast, in DNA from brain, liver and kidney of two XX female sibs, both HPRT alleles proved capable of eliciting transfer. Therefore, this experiment suggests that the DNA of the inactive X-chromosome may indeed be different from that of the active X.

5-Azacytidine is an analog of cytidine known to inhibit methylation of DNA. By treating mouse–human hybrid cells containing an inactive human X with 5-Azacytidine, it has been possible to demonstrate a reactivation of the human HPRT locus, and some of the clones in which this occurred also showed a reactivation of other X-linked enzymes (Venolia *et al.* 1982). While this suggests that DNA methylation plays a role in the maintenance of X-inactivation, it does not indicate whether methylation is directly responsible for the initiation of inactivation or is concerned in some way with maintaining an inactive state.

By comparing the extent of specific antibody binding to the X-chromosomes of normal females, it has been shown that anti-5-methylcytidine binds to a small extent to the arms of all chromosomes, including both Xs. Since the two X-chromosomes do not differ from one another or from the autosomes in respect of such antibody binding, it is evident that X-inactivation is certainly not associated with a cytologically detectable increase in DNA methylation (Miller *et al.* 1982).

3.4.3.2 Male X-compensation

A very different kind of regulatory mechanism, compensating for differences in the number of X-linked genes in males and females, is found in *Drosophila melanogaster*. Here, in contrast to mammals, both X-chromosomes function in transcription in all somatic cells of the female, so that females heterozygous for X-linked recessive markers do not exhibit mosaicism. A visible sex difference in the morphology of the X-chromosome is evident in the larval salivary gland nuclei of the male, where the single X approaches the width of the two somatically paired X-chromosomes of the female, despite the fact it has only half the DNA content of this pair. Additionally, in both polytene and non-polytene tissues, the activity of X-linked genes is in some sense compensated, so that total activity is constant regardless of the number of X-chromosomes present, and is also directed to match that of the autosomes, so that it is directly dependent on their number too (Table 3.34).

Table 3.34 Activity of X-linked genes in individuals of *Drosophila melanogaster* with differing X-chromosome:autosome ratios (from Stewart & Merriam 1980).

Chromosome constitution	Activity per gene copy
diploid male (X:2A)	1.0
diploid female (2X:2A)	0.5
metafemale (3X:2A)	0.33
metamale (X:3A)	1.25
triploid intersex (2X:3A)	0.75
triploid female (3X:3A)	0.5

The precise molecular basis of this system of dosage compensation is not known. Since, as yet, there has been no direct demonstration of any differential synthesis of mRNA in males and females, it is not possible to determine whether compensation occurs at the level of the number of transcripts produced. The relatively non-specific technique of uridine autoradiography suggests, however, that the transcription of X-linked genes is directly regulated. Similarly, although X-linked enzyme activity levels are equivalent in males and females, there has been no direct demonstration of equality in the number of enzyme molecules synthesized. Neither is it possible to rely on phenotypic end products as a sensible measure of transcription in cases where only phenotype is used. Most of these result from long chains of biochemical reactions controlled by numerous genes. They are, thus, too remote from the act of transcription to serve as a sensible measure of specific gene regulation.

The relevant genetic data we do have concerning dosage compensation in *Drosophila melanogaster* are of five kinds:

(a) The male-specific lethal mutations (*msl-1*, *msl-2* and *mle*) all located at discrete sites on chromosome 2, lead to a 40% reduction in the activity of X-linked enzymes in male larvae homozygous for these mutations but leave autosomal enzyme levels unaffected. Autoradiographic monitoring of RNA synthesis in larval polytene chromosomes of homozygous *mle* males indicates reduced incorporation of labelled nucleotides, consistent with the notion that the maleless gene, and

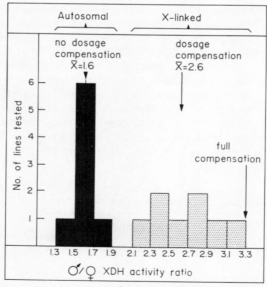

Figure 3.40 The distribution of male to female XDH-specific activity in eight lines containing a rosy gene on the X-chromosome of *Drosophila melanogaster* and eight lines containing an autosomal rosy gene (data of Spradling & Rubin 1983).

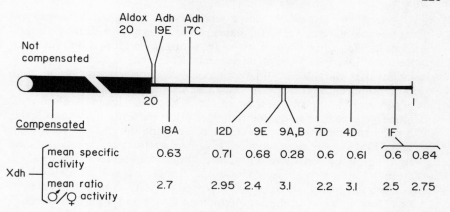

Figure 3.41 Behaviour of autosomal loci inserted into the euchromatin of the X-chromosome of *Drosophila melanogaster* by P element mediated germ line transformation (data of Roehrdanz *et al.* 1977, Roberts & Evans-Roberts, 1979, Goldberg *et al.* 1983, Scholnick *et al.* 1983, Spradling & Rubin 1983).

     presumably also the two other male-specific lethal loci, function in the control of X-chromosome transcription (Belote & Lucchesi 1980).

(b)   The rosy gene, normally located on 3R, shows dosage compensation when inserted into the X-chromosome by P element mediated germ line transfer, involving enhanced enzymatic activity of XDH (Fig 3.40). The average male/female ratio of activity in eight autosomal insertion lines is 1.6, compared to a value of 1.7 in an Oregon-R line containing active rosy genes at their normal autosomal location. By contrast, the equivalent activity ratio in an X-linked insertion line ranged from 1.9 to 3.2 with a mean of 2.6. That is, there is an increase in the absolute specific activity in such males. The same was true for a single insertion of the dopa decarboxylase gene (Scholnick *et al.* 1983). From these experiments, it is clear that the mechanism responsible for dosage compensation in *Drosophila melanogaster* is capable of spreading to rosy insertions in each of a series of 7–8 kb autosomal fragments at a variety of loci in the X (Fig. 3.41). This implies that either there are multiple sites located along the X which mediate hyperactivation, or else that most or all of the X behaves as a single unit in respect of hyperactivation. One of the strains used carried a transposon comprising four rosy genes, apparently derived from a tetramer of the injected plasmid. In this strain, with a tandem insertion, the mean XDH activity (3.3) was evidently proportional to the gene dosage. This argues that differential *Xdh* activity is indeed a direct result of differential transcription.

(c)   In wild type stocks, the X-linked *Sgs-4* glue protein gene normally shows dosage compensation, though there is some variability in different strains of *Drosophila melanogaster*. Transformation analysis has been carried out using a 4.9 kb *Sgs-4* clone, containing 2.6 kb of 5′ and

1.5 kb of 3′ sequences, integrated into a P element transposon. Autosomally located transformants produce between 1.3 and 1.9 times as much *Sgs-4* protein as females (Krumm *et al.* 1984). This suggests that the DNA sequences responsible for dosage compensation are included in the 4.9 kb DNA fragment used for transformation, and thus supports the conclusion that there are multiple sites within the X which mediate dosage compensation. The white locus of *Drosophila melanogaster* is dosage compensated, as are mutations which map at the distal end of the locus. A P element carrying 11.7 kb of wild type white DNA continued to exhibit dosage compensation when transduced to a variety of autosomal locations. The actual pigment ratios varied from 1.5 to 1.9 with an average of 1.5 (Hazelrigg *et al.* 1984). The 11.7 kb fragment used contains approximately 3 kb of sequences upstream to the 5′ end of the white transcript and 3 kb of sequences distal to the 3′ end of the transcript. Thus, the white locus, like *Sgs-4*, is also under the control of its own specific dosage compensation site. Likewise, when a small segment covering some 35 bands of the X-chromosome and carrying the structural gene for 6-phosphogluconate dehydrogenase (6PGD) is inserted into chromosome 3, the gene continues to exhibit dosage compensation. Similar results have been obtained for tryptophan pyrrolase activity using a different X-3 translocation.

(d)   *Adh* loci inserted into the X by equivalent P element mediated germ line transfer are not compensated in males (Goldberg *et al.* 1983). Here, the transduced *Adh* gene is expressed in both males and females at comparable levels which cannot be distinguished from the wild type. An additional autosomal locus which is not compensated when transferred to the X by a translocation is *Aldox* (Roehrdanz *et al.* 1977). In this case, however, an autosome segment of some 12 bands was involved. Finally, at least two normally X-linked loci are known which do not show dosage compensation, namely the white-eosin (w^e) mutation, which maps at the proximal end of the white locus, and the α-chain gene of larval serum protein-1 (Roberts & Evan-Roberts 1979).

(e)   A fusion of autosomal and X-arms has occurred repeatedly in the phylogeny of *Drosophila*. Thus, only arms 3R and 4, as defined in the *Drosophila melanogaster* karyotype, never appear as part of the X-chromosome in other species. *Drosophila pseudoobscura*, for example, has a fusion involving the X and the long arm of 3, and here both heat shock and enzyme loci on the right arm of the metacentric X, which is equivalent to arm 3L of *Drosophila melanogaster*, have been found to be dosage compensated (Abraham & Lucchessi 1974). Evidently, when an autosomal arm becomes part of the X-chromosome, as a consequence of fusion, so that it is represented only once in the male genome, then the activity of its genes is appropriately increased.

The occurrence of dosage compensation raises the more general question of how much of any genome is haplo-insufficient. In contrast to X-linked loci, it has generally been assumed that autosomal loci are dose dependent. That is, the amount of product synthesized by an autosomal gene is directly proportional to the number of structural gene templates available for transcription. For example, a gene located at 92A-B in *Drosophila melanogaster*, which encodes the major species of opsin, is known to influence the rhodopsin content of the retina. In flies with only one copy of this locus, instead of two, the normal rhodopsin content of the major class of photoreceptors is reduced. Three doses of this region increase the rhodopsin content (Scavarda *et al.* 1983).

In *Drosophila melanogaster* some of the genes responsible for the production of contractile proteins (actin, myosin, tropomyosin and troponin) are clustered in the genome. Thus, one of the actin genes and two tropomyosin genes map to 88F on the right arm of chromosome 3. This site is near, or within, a cluster of seven dominant flight muscle mutations which interfere with the assembly of myofibrils in the indirect flight muscle. The *Drosophila melanogaster* myosin heavy chain gene, located at 36B, is also close to, or within, a second cluster of three dominant flightless mutations. Flies heterozygous for deficiencies between 36A1–2 and 36C1–2 are flightless and have disrupted myofibrils in their indirect flight muscles, indicating that this region, which contains the myosin heavy chain gene, is haplo-insufficient for indirect muscle flight function (Bernstein *et al.* 1983). In such flies, the myosin heavy chain content is reduced by some 43–75% indicating that the amount of this protein is dependent on the dosage of the 36AB region. In a more general examination of the effects of segmental aneuploidy in the *Drosophila melanogaster* genome, it has been shown that there are at least 57 dosage-sensitive loci, a majority of which show haplo-abnormal effects (Lindsley *et al.* 1972). The most common class of such mutations are the Minutes which, when hemizygous, lead to a reduced developmental rate, variable, though generally low, viability and small bristles. These are present on all chromosomes, with seven on the X, 17 on chromosome 2, six on chromosome 3 and one on chromosome 4. There is also at least one triplo lethal locus (*Tpl*).

Moscoso del Prado & Garcia–Bellido (1984) draw attention to an interesting situation in the Achaete-Scute Complex (AS-C) of *Drosophila melanogaster*. This complex contains a group of genetic functions involved in the differentiation of the chaetae and sensillae of the cuticular sensory organs. Mutations of hairy (*h*) and extramacrochaetae (*emc*), although recessive in euploid flies, become dominant and haplo-insufficient in the presence of extra doses of $As\text{-}C^+$. Evidently, h^+ and emc^+ code for repressors of $AS\text{-}C^+$ that interact with the achaete and scute regions of the complex.

Despite the existence of numerous haplo-insufficient loci, phenotypic expression of the diploid genome of *Drosophila melanogaster* is relatively insensitive to small changes in gene dosage. Thus, heterozygosity for deletions or duplications passes unnoticed if it involves only a few genes. One notable exception is the Beadex (*Bx*) locus, alleles of which mimic gene duplications in their phenotypic expression, as well as in their interaction with *Bx* deletions and suppressors of Beadex (Lifschytz & Green 1979).

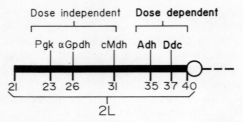

Figure 3.42 Dose-dependent behaviour of five loci in individuals of *Drosophila melanogaster* trisomic for 2L (data of Devlin *et al.* 1982).

Although females are normal with three doses of Bx^+, they are mutant with four and manifest extreme wing scalloping with five. Wing scalloping by Beadex mutations is more extreme in males than females, a consequence of the fact that the gene is dosage compensated. Even so, gene dosage is of critical importance for normal development, since the genome of *Drosophila melanogaster* is able to compensate for varying dosages of individual genes, and still give rise to phenotypic normalcy, only within extremely narrow limits.

By using enzyme assays performed on third instar larval individuals trisomic for the left arm of chromosome 2, three enzyme loci – *Pgk*, α*Gpdh* and *cMdh* – exhibit levels of activity expected of a diploid. Two rather more proximally located loci – *Adh* and *Ddc* – are dose dependent (Fig. 3.42). It is unfortunate that this study was carried out on whole larvae, since a majority of larval tissues are either polytene or endopolyploid, and it is impossible, therefore, to decide to what extent the compensation demonstrated is a function of these states. An examination of this system using diploid tissues certainly seems worthwhile, since it is not obvious why autosomal loci should be compensated.

3.4.4 Default transcription

During oocyte development there is an accumulation of organelles and molecules required for the rapid phases of cell division and cell movement that characterize early embryogenesis. In amphibians, this depends on the formation of a system of giant lampbrush chromosomes. These are diplotene bivalents which persist in a highly extended and diffuse state for up to six months, or more, according to the species. Each of the four chromatids within each bivalent give rise to a series of lateral looplike projections at numerous points along their length. The total number of loops varies according to the size of the genome, as does the length of the loops themselves. The loops go through a reversible phase of extension, during which they are transcriptively active, followed by retraction (Macgregor 1980). During the extended phase the transcriptive events are geared to the production of massive amounts of ribosomes, tRNAs and histones, as well as to stable mRNA molecules. Additionally, the mature egg contains a very large and structurally more complex population of RNA transcripts, which resemble unprocessed or partially processed mRNA precursors. The net result is that there is something approaching a 100-fold increase in the transcriptive capacity of the genome. Moreover, since the average loop in the newts *Triturus cristatus* and *Notophthalmus viridescens* is some 50 μm long, and hence contains many hundreds of kilobase pairs (1 μm ≡ 3 kb), the average size of transcription units in lampbrush chromosomes is considerably greater than in somatic cells. This implies that either initiation or termination, or both processes, are regulated differently in oocytes and somatic cells. Additionally, in at least some cases, the lateral loops may not include structural genes. Thus, related species of plethodontid salamanders, which have the same chromosome number, have very different genome sizes, differences which are largely attributable to their content of middle repeated DNA (see § 4.5.1.3). Species with higher DNA values also have more loop pairs (Macgregor *et al.* 1976). Since the increased middle repeat content of the larger genomes is apparently scattered throughout the genome, leaving the relative dimensions of the mitotic chromosomes unchanged, it follows that some of the loops must consist of middle repeated sequences. Evidently, therefore, there must be a widespread transcription of many sequences, both repetitive and non-repetitive within the lampbrush system (Jamrich *et al.* 1983).

The Satellite 1 DNA, which forms long spacers between the histone repeat units of *Notophthalmus viridescens*, occurs predominantly at the centromeres of all the chromosomes, as well as at the two non-centromeric histone sites located subterminally on chromosomes 2 and 6, close to the sphere loci of the equivalent lampbrush chromosomes in this species. These sites give rise to a series of very large lampbrush loops which produce long transcripts containing both histone and Satellite 1 sequences (Diaz *et al.* 1981). Since the units coding for H1, H3, H2A and H4 in *Notophthalmus viridescens* are located on one strand, with those for H2B on the other, it follows that both strands are transcribed, though in different loops or parts of loops. Presumably, transcription begins at a histone gene promoter and fails to terminate at the end of the histone unit, so that it continues without interruption into the adjacent satellite DNA by simple readthrough.

A general failure of termination signals in oocytes might explain the unusually long transcription products which tend to characterize lampbrush systems. In *Xenopus laevis*, a number of repetitive sequences are known to be transcribed by oocyte lampbrush chromosomes, and in each case both DNA strands are involved. One of these, a long repeated sequence, designated *1723*, has the properties of a transposable element with some 8500 interspersed copies per genome, though only about 2% of these are involved in transcription (Kay *et al.* 1984). The heterologous size of the *1723* RNAs, their location in the cell nucleus, their symmetrical transcription and their parallel accumulation with total poly (A)$^+$ RNA in ovaries, embryos and liver, all suggest that these elements are transcribed by readthrough from flanking regions.

This raises the more general issue of how much of any RNA population is relevant in a developmental sense, and to what extent junk RNA may exist as a result of such default transcription. About 70% of the cytoplasmic poly (A)$^+$ RNA in both *Xenopus laevis* and *Strongylocentrotus purpuratus* oocytes consists of message sequences covalently linked to sequences transcribed from repetitive gene families. This RNA, too, could well represent the post transcriptional products of segments transcribed by default. Thus, it is possible to show that interspersed poly $(A)^+$ RNA of *Xenopus laevis* oocytes, which contains repetitive sequence elements distributed within regions transcribed from single copy sequences, are not translated (Richter *et al.* 1984).

3.4.5 Genetic balance and development

Sexual reproduction involves two events – meiosis and fertilization. Either, or both, of these may fail. If fertilization fails, the progeny are haploid and, as a result, few, feeble and infertile. It is true that most fungi favour a haploid condition of the genome in their vegetative phase, and here, by contrast, it is diploidy which proves to be unstable. In filamentous fungi, for example, the haploid multinucleate mycelia of two different strains grown in close proximity will fuse to form a loose association in which the nuclei of the two strains cohabit a common cytoplasm. Such an association is termed a heterokaryon, since it contains a mixture of haploid nuclei of different origins. Some fungi (e.g. *Neurospora*) have only a heterokaryotic phase to their life cycle. In others (e.g. *Aspergillus*) a small proportion of the conidiospores produced by a heterokaryon are diploid, combining the genomes of the two original haploid strains, and these give rise to a diploid mycelium. This diploid phase is not stable and breaks down, by successive chromosome loss, to restore a haploid condition. In a majority of eukaryotes, however, there is a dominant diploid phase which does not survive well in a haploid condition. The one apparent exception to the inviability and infertility of haploids is the haplodiploidy, involving the formation of haploid males by parthenogenetic eggs produced in diploid females, which is regularly found in the Hymenoptera (ants, bees and wasps). Here, however, a majority of the male tissues are in reality not haploid but endopolyploid. Added to this, fertility is assured, despite the haploid set, by the replacement of meiosis with a single mitotic division in the male germ line.

In the Jimson weed, *Datura stramonium*, naturally occurring triploid individuals ($2n = 3x = 36$, where $2n$ represents the diplophase of the life cycle and $3x$ the number of chromosome sets present in the diplophase) differ only slightly in appearance from normal diploids ($2n = 2x = 24$). By crossing triploids and diploids in the laboratory, it is possible to produce primary trisomics in which one of the chromosomes within the diploid set is present in an extra dose ($2n = 2x+1 = 25$). Since there are twelve different chromosomes in the haploid set, there are twelve different possible primary trisomes. Each of these has a distinctive morphology, differing both from each other, and from the normal diploid, in a whole range of characters affecting all parts of the plant (Blakeslee 1934). Individual trisomic types can, therefore, be recognized by differences in the form of leaves, flowers, stigmas and seed capsules. They also show consistent differences in their internal anatomy.

This striking contrast in the character of the trisomics in comparison with the triploid plant has conventionally been explained by the principle of genetic balance. This implies that a complete haploid genome, or whole multiples of it, is an internally balanced system capable of sustaining normal, or near normal, development. Gains and losses involving only part of a genome, especially when they involve predominantly euchromatic chromosomes or chromosome segments, on the other hand, lead to imbalance and modified, often defective, development. What applies to *Datura stramonium* holds also for a range of angiosperm plants and only in four species – *Zea mays*, *Petunia* hybrids, *Clarkia unguiculata* and *Collinsia heterophylla* – is there any overt tolerance of trisomy (Rick 1971). For example, in *Clarkia unguiculata*, it is possible to produce plants with chromosome numbers ranging from diploid ($2n = 2x = 18$) to triploid ($2n = 3x = 27$) levels in the laboratory, though those with numbers of 24 and above are difficult to obtain and show considerable imbalance (Vasek 1956). Notice, however, that in none of these cases which show aneuploid tolerance are naturally occurring trisomics ever found. They are known only in experiment or, in the case of *Petunia*, under cultivation.

A similar variability in respect of aneuploid tolerance is known in animals. In the newt, *Pleurodeles waltlii*, experimentally produced trisomics for at least six of the twelve members of the haploid set are known to survive to adulthood, and three of these, involving chromosomes 8, 10 and 11, are even fertile (Guillemin 1980). By contrast, in man, despite the fact that 50% of all spontaneous human chromosome abnormalities are trisomics, only three autosomal trisomes are known which survive to birth and all of them exhibit gross developmental abnormalities (Table 3.35). The only human trisomic that is not only near normal but also fertile is the triple-X female, and this situation can be accommodated in terms of the X-inactivation which operates in the development of the female soma. The same principle of balance that affects somatic development of humans may also operate within the germ line. Thus, female germ cells require the presence of two X-chromosomes for normal survival, whereas male germ cells die if they have more than one X-chromosome. This suggests that X-dosage is important for maintaining germ cells, and that sterility in X0 women and XXY men can be related to poor germ cell survival.

Mice, contrary to humans, do not possess a single autosome whose imbalance is compatible with postnatal development. In the case of trisomes for chromosomes 14 or 19 of the mouse, development may proceed to full term and such embryos are sometimes born alive but they do not survive more than a few hours after birth. Trisomy for most murine chromosomes causes substantial damage during major organogenesis, especially of the neural tube and the brain, leading to early resorption of affected embryos (Baranov 1980). In the chicken, too, it is rare that an embryo with an abnormal chromosome complement survives beyond five days incubation and rarer still that it hatches, though some 2–5% of 4182 embryos sampled at four days have been found to carry chromosome abnormalities (Table 3.36).

The situation in birds and mammals differs from that in amphibians and

Table 3.35 Phenotypic consequences of polysomy in Humans (from Mange & Mange 1980).

Condition	Phenotype	Fertility
Autosomes		
trisomy 13 (Patau's syndrome)	gross abnormalities in many organs; mean survival time 2⅓–3 months	
trisomy 18 (Edward's syndrome)	multiple skeletal defects; mean survival time 2½ months	
trisomy 21 (Down's syndrome)	severe mental retardation with defects in many organs; 50% die within first year	males invariably sterile; females with ovarian defects
Sex chromosomes		
triple-X females	many appear normal, some show tendency to mental retardation	many are fertile
poly-X females (XXXX, XXXXX)	all institutionalized with severe mental retardation	?
Kleinefelter males (XXY)	poorly developed male secondary sexual characters; extreme limb development; often decreased mental capacity	sterile
Kleinefelter variants (XXXY, XXYY, XXXXY, XXXYY)	several mental retardation	sterile

fish, where polyploids survive and reproduce. Indeed, in amphibians inviable diploid hybrid embryos can sometimes be rescued by the addition, either spontaneously or experimentally, of a second haploid set of maternal chromosomes. For example, although hybrid offspring of the cross *Rana clamitans* ♀ × *Rana catesbeiana* ♂ are both viable and fertile, the offspring of the reciprocal cross arrest at gastrulation. A blastopore is formed but fails to close. The abnormal morphogenesis generated by this early lesion precludes the continued differentiation of otherwise normal cells. An egg of *Rana catesbeiana* developing as a gynogenetic haploid, on the other hand, is capable of gastrulating normally, as does a triploid which combines 2 sets of *Rana catesbeiana* chromosomes with one of *Rana clamitans* (Elinson & Briedis 1981). In the higher vertebrates, polyploid genomes are not balanced and do not complete development. The few cases of human triploids and tetraploids which have survived beyond a few days after birth have all proven to be mosaics with some diploid tissues. Thus, an abnormal chromosome constitution acts as a dominant lethal in the embryogenesis of birds and mammals.

The whole issue of genetic balance may well revolve around the dosage

Table 3.36 Types and frequencies of spontaneous chromosome abnormalities in a sample of (a) 4182 four-day-old chicken embryos (after Bloom 1972) and (b) 1926 human abortuses (after Chandley 1981).

(a) Chicken

Item	Diploid	Haploid mosaics		Triploid				Tetraploid		Trisomic	Total abnormal
		A/2A	A/2A/3A	ZWW	ZZW	ZZZ	?	ZZWW	ZZZZ		
number	4079	51	8	15	7	9	2	1	1	9	103
		59		33				2			
% of all abnormalities	—	57.3		32.0				1.9		8.7	
% of total sample	97.5	1.4		0.8				0.1		0.2	2.5

(b) Human

Item	Monosomics 45 (X0)	Mosaics	Triploid	Tetraploid	Autosomal trisomics	Other	Total
number	383	38	318	121	981	85	1926
% of all abnormalities	19.9	2.0	16.5	6.3	50.9	4.4	100.0

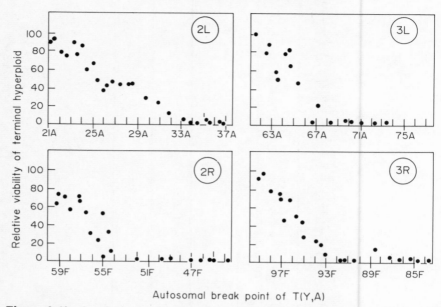

Figure 3.43 Survival of terminal hyperploids for the four major autosomal arms of *Drosophila melanogaster*, as a function of the length of the trisomic region involved in numbered polytene subdivisions (data of Lindsley *et al.* 1972).

sensitivity of individual loci within a genome. As we have seen earlier, two major dose sensitive phenomena are known in *Drosophila melanogaster*, both involving the male X-chromosome. One is sex determination, the other is dosage compensation. Additionally, there are some loci that are haplo-insufficient and at least one, *Tpl*, is also triplo-lethal. Thus, we see that in extreme forms the dosage of individual genes can lead directly to inviability. Overall there are at least 100 dosage-sensitive loci in *Drosophila melanogaster* (Lindsley *et al.* 1972) and these appear to be reasonably evenly distributed throughout the genome, though the left arm of chromosome 2 certainly contains fewer dose-sensitive genes (Fig. 3.43). Hence, the imbalance of trisomies can be explained by the cumulative effect of individual genes which are dose sensitive, rather than by involving undefined properties of entire chromosomes.

3.5   THE DEVELOPMENTAL DILEMMA

Development is a sequential programme of gene activity involving a complex of integrated gene circuits. In such a hierarchy of interacting systems it is self evident that regulation of the programme requires control of the timing of events within individual subprogrammes, and especially of the critical binary choices which operate within that subprogramme. Equally, individual subprogrammes have to be integrated with one another.

While the basic features of the transcriptional and translational control of eukaryotic genes are now clear enough, what is lacking is the ability to transpose this information into those developmental events which lead to cell movement, cell–cell interaction and the sequential activation of other genes or gene circuits. What is important for analyzing developmental regulation is to define how genes are turned on at the right time in the correct developmental circuit and how their protein products contribute to the continuity of that circuit, or else lead to the inception of new circuits.

As far as individual genes are concerned, if a manipulable genetic system is unavailable, one molecular way of distinguishing those that are likely to play key roles in particular subprogrammes would be to take time transects during a period of particular developmental significance, with a view to discovering which of the RNAs then being produced are also formed at other non-specific times. By initially discounting these, it should be possible to define developmentally specific loci and to identify their protein products. Angerer & Davidson (1984) report some initial attempts in this direction but, as yet, this approach is in its infancy. What has been done is to identify genes which are assumed to be of developmental significance, based on their time of action, and then clone them. Unfortunately, there is no assurance that a gene so cloned is of genuine developmental significance. It may simply be locked into a particular circuit from which it is unable to escape. The *Drosophila melanogaster* nomadic element B104, for example, when present, shows extensive transcription at the preblastoderm stage (Scherer *et al.* 1982). Yet its transcription is clearly irrelevant to the developmental programme. What is required to confirm its developmental significance is to delete the gene in question from the genome and then see if the deletion produces an effect on development.

A practicable alternative is simply to extrapolate from phenotype to genotype using a conventional genetic approach. This has been successfully developed by Garcia-Bellido, using deficiencies to produce an amorphic condition for a given gene (Ripoll & Garcia-Bellido 1979, Garcia-Bellido & Robbins 1983). Since most small deficiencies in *Drosophila melanogaster* contain at least one gene which is indispensable to the zygote, the approach uses clones of cells induced by mitotic recombination. Regrettably, such an approach cannot be applied to mammals.

As far as defining gene circuits is concerned, two theoretically different approaches are possible. First, one can enquire whether genes under equivalent systems of transcriptional regulation are likely to be involved in a common circuit. This approach has not, so far, proven to be very productive in the studies that have been made on gene transcription. Nor is this surprising, since the upstream signals responsible for the transcription of individual genes may simply reflect the evolutionary history of the particular landscape in which a given gene happens to reside.

An alternative, and potentially more productive, approach is to identify the metabolic pathways involved in defined gene circuits, with a view to determining the protein intermediates involved in the sequence and their influence on it. For example, two main types of pigment are carried in the eyes of *Drosophila melanogaster*, the ommochromes and the pterins. Each of

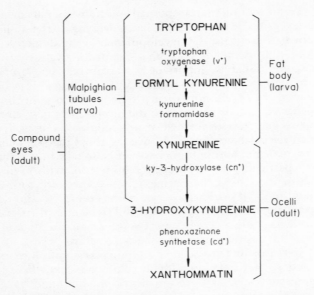

Figure 3.44   The ommochrome pathway of *Drosophila melanogaster*.

these is produced by a different set of genes. The metabolic pathway leading to ommochrome production is reasonably well defined (Fig. 3.44). Eye pigment precursors begin to be formed in the larval fat body and this leads to the secretion of kynurenine into the haemolymph at the pupal stage. This is subsequently taken up by the imaginal discs responsible for the formation of the compound eyes and the ocelli of the adult, and then transformed into xanthommatin which is the ommochrome component of the primary pigment cells. Three of the genes involved in this pathway are known, so that from their isolation and an examination of their modes of regulation it should be possible to discover how they are integrated within this specific subprogramme.

While the definition of individual gene circuits is a necessary first step, the question of the integration of different circuits is a further major issue that needs to be addressed. Thus, the deposition of xanthommatin within the pigment cells of the eye has to be co-ordinated with the developmental circuit that leads to the formation of these cells. Both pterins and ommochromes become attached to protein granules, and it is in this form that they are deposited in the eye cells. For example, the mutation *white* prevents the attachment of pigment to these granules so that such mutants appear white eyed.

In the absence of detailed information on the developmental circuitry of eukaryotes, it is worthwhile examining briefly what is known about such circuits in prokaryotes which pose problems which are similar in principle, if morphogenetically much simpler, to those found in eukaryotes. T4 phage, for example, has a genome size of 1.7×10^5 nucleotides, which implies a total coding capacity of between 150–200 genes. Excluding those which are likely to control non-developmental functions, it is probable that between 50 and 100 of these are required to direct viral morphogenesis (Wood 1973). In T4, morphogenesis involves three initially independent events, namely production of the head, production of the tail and assembly of the tail fibres. These are followed by two further events – attachment of heads to tails and attachment of tail fibres to fibreless particles.

This assembly line is characterized by four general properties:

(a) With but few exceptions the order of individual steps is fixed and invariant.

(b) This order is governed by gene product interaction and not by the level of gene expression. Thus, all the proteins involved in phage morphogenesis are synthesized simultaneously during the latter half of the infectious cycle, and temporal control of protein production is not used to order the steps.

(c) There is no evidence of feedback control of gene expression by structural end products. It is the actual transcription patterns of phage genes which impose built-in controls on the relative levels of particular proteins.

(d) The most obvious class of gene products are self-assembling proteins, and the combination of the three independently assembled components of the phage particle involves a self-assembly mechanism which requires no information other than that present in the subunits themselves.

Thus, the ordering of the later stages of morphogenesis in T4 results from gene product interaction rather than from the actual level of gene expression. While one can reasonably anticipate that equivalent principles operate in eukaryotes too, little attention has been paid to the question of self-assembly systems though systems of this sort are known to be involved in the development of the cytoskeleton in both Protozoa (Sonneborn 1967) and Metazoa (Borisy et al. 1984).

The principles of self-assembly have been gleaned largely from viral proteins where, as in the case of the T4 programme, the assembly line itself is very much 'hard wired' into the amino acid sequence of the proteins involved. There may well be more flexibility in eukaryotes, and especially in the spontaneous self-assembly of microtubules, intermediate filaments and actin filaments, which associate from subunits to yield polymers that are held together by non-covalent interactions. Kirschner & Mitchison (1986), for example, have pointed out that spindle morphogenesis is quite variable, far more so, in fact, than can be attributed to thermal fluctuations that yield variation in the structure of intermediates. In this sense, as was apparent in the case of the chicken–yeast chimeric β-tubulin gene we referred to earlier (see § 1.2.2), spindle morphogenesis is evidently far sloppier and less demanding than phage assembly.

Now that we have completed an examination of the mode of action of the genome in development, we turn in the remaining sections of this book to consider the extent to which evolutionary events can be interpreted in terms of modifications in the developmental system.

4

Genome change and evolutionary change

'The past consists of not what actually happened but what men believe happened.'

Gerald White Jackson

The word 'evolution', which literally means *unfolding*, describes the changes of form which have taken place in living organisms over time. As such, it necessarily involves internal and external morphological modifications, resulting in the production of new forms of life from pre-existing forms. The essential problem in evolution, therefore, is to account for such changes. Any permanent change in morphology must, of necessity, involve a change in development, so that evolution can only be satisfactorily defined in terms of the changes in the developmental programmes that have ultimately been responsible for producing different morphologies. In relation to such changes we can ask six critical questions:

Q.1 How is morphological novelty generated? Is it simply that the same genes, or else divergent duplicates of them, have been assembled into different developmental circuits, or has there been a genuine evolution of new genes with new functions?

Q.2 What do changes at the molecular level tell us about morphological change?

Q.3 Do chromosome changes play any role in producing morphological change?

Q.4 Do evolutionary changes depend on large-scale genomic reorganization?

Q.5 Is speciation in any way involved with the inception of major morphological change?

4.1 THE BASIS OF EVOLUTIONARY CHANGE

As Thoday (1975) has emphasized, the theory of evolution includes two, quite distinct, components. The first is the concept of evolution which postulates that the diversity of the living world is a consequence of modification, over time, of differing lines related by descent. The second component deals with the actual mechanisms of change that have been responsible for such modifications.

The concept of evolution follows logically from the facts concerning the history of life on Earth and is ably supported from a study of comparative morphology. The most direct evidence concerning the kinds of morphological change that have taken place during the course of evolution comes from the study of the fossilized remains of organisms that lived in the past. Unfortunately, this tells us little about the mechanisms of change. It is for this reason that evolution must rank as the conceptually weakest and most speculative of the major biological disciplines since, by its very nature, it deals with events retrospectively. Indeed, the tragedy is that we know so much descriptively, yet understand so little mechanistically, about evolutionary change.

What, then, can be learned about morphological change from a molecular dissection of the genome, bearing in mind that the development of morphology is still relatively poorly understood at the molecular level? There is, as we have seen, abundant evidence from comparative studies of the genome for changes in all of its components. What biological significance can we attach to this?

From the previous sections of this book it should be apparent that many of the components of the eukaryote genome have little to do with the everyday working life of the cell or with the processes of development. This includes the non-coding DNA sequences, which vary in amount and sequence both within and between species. It also includes the transcriptively inactive pseudogenes, at least some of the intergenic spacers and a whole range of mobile elements, many of which are transcribed. Given that these components matter little in metabolism or in development, we now need to examine whether they play a role in evolution, or whether they are simply unavoidable by-products of the molecular mechanisms which govern the replication, recombination and transcription of the coding sequences. Certainly, the evolution of eukaryotic species has sometimes been accompanied by massive quantitative changes in respect of the non-coding fraction of the genome. Do such changes, then, have anything meaningful to tell us about phenotypic change, or are they simply an inevitable, if accidental, outcome of enzymic events that go on within the cell?

The answer to this question depends on being able to pin-point those genomic components which are likely to be prime movers in producing morphological change. We readily accept that, in principle, mobile elements might reorganize a genome in such a way that genes in one regulatory circuit could be transferred into another. The nub of the problem, however, is **to actually demonstrate that the evolution of**

morphological novelty has been more dependent on this category of change than on any other.

The conventional wisdom in dealing with evolution reverses this approach. It takes the known differences in morphology and then proceeds to speculate as to how such differences might have arisen. Consider, for example, the often quoted case of the evolution of the long neck of the giraffe. The conventional solution assumes that the specialized cervical features of this animal have arisen by natural selection because of the survival advantage enjoyed by long necked individuals with access to the foliage on tall trees. This, however, is not a 'solution' at all, for the essence of the problem is to discover how the differential growth of the neck during development was initiated. One approach would be to dissect the gene circuits which regulate the development of the neck, relative to other parts of the body, with a view to predicting how these might be perturbed and the morphological consequences which are likely to arise from their perturbation.

Few biologists have attempted this approach. The net result is that the mechanical basis of morphological change continues to be ignored. The appeal to natural selection in the case of the giraffe's neck is misplaced, since selection was certainly not responsible for the origin of the morphological change whatever role it subsequently may have played in its ultimate fate. Evolutionists, however, have been so preoccupied with determining whether a character is, or is not, adaptive and whether it has 'arisen' as a result of selection, that little attention has been focussed on the developmental genesis of that character. To illustrate this, we now turn to consider how in the past, and indeed even still today, biologists have attempted to deal with the question of evolutionary mechanism.

Here there are three issues to consider. First, the significance of the observed molecular changes that go on within the genome. Secondly, the assumption that changes in genome size have played a role in directing evolutionary change. Thirdly, the argument that genomic change provides a basis for speciation.

4.2   STABILITY AND CHANGE IN THE GENOME

Evolution, conventionally, is defined and analyzed in terms of sustained change in the phenotype. At the molecular level, therefore, it is concerned with those mechanisms that produce heritable changes in the germ line genome which lead to modified patterns of development, and the subsequent fate of such changes within populations. What, then, do we know of events at the molecular level that have important consequences for changes in phenotype within populations, and which can be expected to be mirrored in the events we study as evolution?

It should be apparent that the differences in the content and location of nomadic elements found between different strains of the same species contrasts with the relatively invariant nature of the karyotype and the location of coding genes within it. While within a species the genome appears relatively stable, both in respect of the linkage relationships between its structural genes and their phenotypic end products, it is, nevertheless, in a continuous state of flux in respect of its content of repeated sequences. Thus, at the molecular level, changes have evidently occurred which have spread throughout the population, some of which, at least, have no overt phenotypic effects. How then has this occurred and what consequences does it have for evolution?

As we have already emphasized, most contemporary evolutionists do not study the mechanisms responsible for morphological change. Their interest is in mechanisms which lead to the differential fixation of different morphs in different populations or different species. Their overriding concern is with mechanisms of spread and fixation of new morphs rather than the underlying genetic and developmental changes which lead to the production of these morphs.

We begin, therefore, by considering mechanisms of spread and fixation in order to bring to the reader's attention the important distinction between the role of these mechanisms in evolution compared with the critical developmental mechanisms responsible for actually generating novelty.

4.2.1 The spread and fixation of genome change

There are three current, and conceptually very different, mechanisms for explaining the spread and fixation of germ line variants within populations; natural selection, neutral drift and molecular drive. At the molecular level the consequence of all three is that one particular DNA sequence achieves a frequency advantage over another and eventually replaces it.

4.2.1.1   Natural selection

This is an umbrella term for the interactions that go on between an organism and its environment. Selection operates through two aspects of the phenotype, namely differential survival (viability) and differential reproduction (fertility). This means that selection can only act on gene products that, directly or indirectly, translate into aspects of morphology, physiology or behaviour which in some way differentially affect either the viability or the fertility of an organism. However, molecules which replicate differentially within a cell will obviously compete with each other, and this has led to the concept of intragenomic selection. Such intragenomic selection is in fact intragenomic competition in respect of replication.

Evolution, by definition, implies change with continuity. It does not imply adaptive change though, equally, such change is not excluded. Likewise, selection does not imply adaptation though it, too, may lead to it. Despite this, it has been commonly assumed that most evolutionary changes have been moulded by natural selection as an adaptive response to a given set of environmental circumstances. Not all genomic change can, however, be understood in terms of adaptation. Thus, DNA with no phenotypic expression can hardly be expected to change in an adaptive manner. Moreover, despite persistent statements to the contrary, there is as yet no convincing evidence that the increases and decreases which have evidently occurred in the size of the genome have played any major role in the evolution of eukaryotes (see § 4.5). Added to this, it is extraordinarily difficult to decide which characters of an organism are adaptive and which are selectively neutral (Policansky 1982). For this reason, we do not know how much of the genetic variation present within populations, or how many of the genetic differences found between species, actually result from natural selection (Lewontin 1978). Effects rarely specify their causes, particularly when these effects are judged retrospectively, as they must be in dealing with evolution. Clearly, living things are in some sense adapted to their environment. If they were not they could not possibly survive. But such adaptation is neither necessarily perfect nor precise. The biological imperfections of humans, for example, are both manifold and manifest (Medawar 1957). The human eye is beset with chromatic and spherical aberrations, and while the upright gait of humans undoubtedly has had profound evolutionary advantages, it is not without its mechanical shortcomings. Obviously, adaptations do not need to be perfect to be of advantage. All that is required is that they provide an organism with the ability to compete successfully with other organisms, either of their own kind or of a different kind. Additionally, it is not, and has not been, a requirement of selection that all characteristics of an organism are adaptive, or that selection is responsible for all the changes that go on in populations. Furthermore, selection can vary in intensity, as well as being periodic, simply due to the episodic nature of environmental changes during long evolutionary time spans.

For purely practical reasons, past approaches to the analysis of evolutionary change have concentrated on the causes and significance of the allelic frequency differences which can be demonstrated to exist within and between populations. Indeed, conventional selection theory is predicated on changes in the allelic frequency of single copy genes. We now know that a large number of the gene systems found in eukaryotes are not of this kind. Rather, they consist of multigene families of varying sizes, whose component members produce similar end products of fundamental importance to the final phenotype. In view of the widespread occurrence of these multigene families, it is clear that any theory based simply on changes of gene frequency at individual loci is inadequate for dealing with the evolutionary dynamics of the eukaryote genome. When there are multiple copies of a gene within a genome and a variant arises in one of their number, selection will be unable to distinguish this variant unless its effect overrides, or in some other way influences, that of a majority of the other members of the family to which it belongs. Consequently, selection can only monitor the adequacy of the family as a whole (Ohta 1983).

4.2.1.2 Neutral drift

Most amino acids are specified by two or more synonymous triplet codons which usually differ only at their third position. Nucleotide substitutions at the unconstrained third position leave the amino acid unaffected. It is commonly assumed that the redundant nature of the genetic code means that a change from one redundant sequence to another will be equally effective in promoting the survival and reproduction of the individual. For example, in amino acid sites 24–34 in histone IV of the sea urchins *Lytechinus pictus* and *Strongylocentrotus purpuratus* there are five synonymous differences in third position nucleotides in the coding triplets, which leave the actual amino acid sequence unaffected (Table 4.1). Of course, if two transcripts specifying the same protein have distinct secondary or tertiary structures, they might still produce different phenotypic effects.

In cases where a DNA variant is selectively equivalent to its progenitor sequence, its fate is left to chance. Under these circumstances, its frequency is expected to fluctuate fortuitously over time until eventually, by chance, it either becomes fixed or lost. In the former event, a change will have occurred within the genome that is random relative to the environment. Given that eukaryote genomes have the replicative capacity to generate tandem arrays of virtually any sequence, then some, at least, of the sequence changes in highly repeated DNA, for example, can probably be explained by neutral drift, though others, as we shall see in a moment, are probably a consequence of molecular drive. Some of the dysfunctional pseudogenes, those products of duplication which have accumulated mutations precluding their correct transcription, processing or translation, may also have become fixed within a population by drift. Notice, however, that the variant in question need not necessarily be functionless, but simply that it is no less effective in promoting the survival and reproduction of the individual.

The past decade has witnessed two major arguments concerning mechanisms of evolution. The first relates to selection vs. neutrality (Kimura 1979, 1983, 1986) and the second to gradualism vs. punctualism (Hoffman 1982). The neutralist argument is not, and should not be considered as, opposed to the concept of selection. Thus, in no sense does it invalidate the constraints which selection may impose on genes or their products in either leading to the removal of the unfit or the possibility of producing adaptive change. Few now doubt that some evolutionary changes have resulted from selection while others are a consequence of neutral drift. Comparisons of amino acid sequences between different species of *Drosophila* provide an example where substitutions appear to have been neutral. For example, by dissociaton and reassociation of protein subunits it has been possible to construct heterospecific dimers of acid phosphatase in which the two subunits combined are from different species of *Drosophila*. Fifteen such dimers gave activity estimates which were identical to each other and to five homospecific enzymes, despite the fact that the two subunits of the experimentally constructed dimers had very different amino acid sequences (MacIntyre & Dean 1978). The amino acid substitutions that have occurred within the different species appear, therefore, to be neutral.

A second argument concerns gradualism vs. punctualism. The data relevant to this argument have been discussed in detail by Hoffman (1982) who endorses the view, long familiar to palaeontologists, that there are varying speeds in evolution. We can be confident, therefore, that some evolutionary changes have occurred gradually, while others have been abrupt. The extreme view of punctuated evolution assumes that most

Table 4.1 Nucelotide substitutions in the mRNAs coding for sequences 24–33 of the histone IV protein in two species of sea urchin (from Kimura 1979, 1983).

mRNA	Amino acid sequence									
	Asp	Asn	Ile	Gln	Gly	Ile	Thr	Lys	Pro	Ala
Lytechinus pictus	GAU	AAC	AUC	CAA	GG**A**	AU**A**	AC**U**	AA**A**	CC**G**	GCA
Strongylocentrotus purpuratus	GAU	AAC	AUC	CAA	GG**C**	AU**C**	AC**C**	AA**G**	CC**U**	GCA

evolutionary change occurs at the time of speciation, with stasis between speciation events (Stanley 1979). The important question, however, is not whether punctualism or gradualism are mutually exclusive, but rather whether one has resulted in quantitatively more significant or qualitatively different forms of morphological change. In other words, what do the modes of evolution, as inferred by palaeontologists, actually signify in terms of the developmental events which underlie morphological evolution? The controversies which have raged over selection vs. neutrality, and punctualism vs. gradualism have, in fact, been little more than side tracks that have diverted attention from the really substantive issue. In reality, in most cases we lack the basic data required to decide between neutrality and selection on the one hand, and between gradualism and punctualism on the other.

The neutralist–selectionist debate, for example, has been referred to by Levin (1984) as the two camp syndrome: 'Positions (almost always two for any given issue) are fiercely defended even when they are not mutually exclusive. New ideas are liable to attack for no reasons other than their real or even apparent, violation of orthodoxy.'

4.2.1.3 Molecular drive

Natural selection and neutral drift, either alone or in combination, have, in the past, been considered to be the dominant processes responsible for the spread and fixation of evolutionary changes. In the case of selection, a variant can only increase in frequency if the individuals which carry it are reproductively more successful than those that lack it. Drift requires no such assumption.

A third mode of change – referred to as molecular drive – which applies especially to multigene families, has recently been defined. This type of change depends upon the fact that the molecular events which go on within the genome have created an environment in which DNA sequences capable of promoting their own amplification and dispersion will inevitably arise (Dover 1982, 1986a,b). This, in turn, has endowed the genome with built-in mechanisms for recurrent DNA changes. Paramount among these is the concerted evolution of repeated sequence families. Such concerted change refers to the tendency of a family of repeated sequences to alter in unison, so that the homogeneity of members of a given family is greater within a species than is the sequence homogeneity found between homologous families of repeats in related species.

Some sort of homogenizing mechanism is required, for example, to reconcile the relative sequence uniformity of spacer DNA in tandemly repeated gene families within a species, with the marked divergence of homologous spacers between related species. Thus, while the rRNA genes are conserved between species, their spacers undergo concerted change (Coen et al. 1982). It is highly improbable that selection would be able to maintain sequence homogeneity, either in a large number of sequences with identical function or in non-coding regions. Rather, it suggests that the type sequence of a given repeat family is fixed not by selection or drift but rather as a consequence of processes which lead to cohesive population change, so that a given variant can increase more or less simultaneously in many individuals within a population in a concerted fashion. The genetic cohesion of a population is maintained throughout molecular drive because the rate at which sexual reproduction randomizes chromosomes between generations is much faster than the rate at which a new mutation spreads between chromosomes by turnover mechanisms (Ohta & Dover 1984). The term, molecular drive, thus incorporates all processes which lead to cohesive change of the kind required to explain concerted evolution, namely transposition, unequal crossing over, replication slippage, conversion and amplification (see § 1.2.3).

Despite the hard data on sequence flux in different multisequence systems, both the concept and the implied consequences of molecular drive have proven an anathema to many evolutionists, presumably because they believe that existing mechanisms are capable of accounting for the observed facts. Much of the existing evolutionary dogma is still predicted on the 'beads on a string' concept of genes and chromosomes, with single genes and selection playing key roles in generating morphological change. Dover (1982, 1986a,b), however, has highlighted turnover in multigene and multisequence families as a prime mover in evolution. This transition from conventional neo-Darwinian gene theory to multigene turnover appears to be a difficult conceptual bridge for many to cross, when compared with the relative security of selectionist or neutralist viewpoints. The data, however, are impossible to ignore. The observed patterns of variation in multigene and multisequence families are such that family members are more alike within a species than between related species.

Molecular drive is the *process* which explains these patterns of concerted evolution, which are now well documented in the histones, the immuno-globulins, the histocompatability genes, the globins, the ribosomal RNA sequences, the small nuclear RNA families and the satellite sequences (Dover & Flavell 1984, Strachan *et al.* 1985, Dover & Tautz 1986). Indeed, in terms of the establishment of evolutionary novelties, molecular drive now ranks in importance with both selection and neutrality. The recombinant DNA data base is expanding so rapidly that there will soon be more relevant information at the sequence level which impinges on molecular drive than on either of the two other members of this trinity.

The case of the ribosomal gene promoters is a particularly trenchant example. In many organisms, and especially in insects and amphibians, there are clear interspecific incompatibilities between the rDNA of one species and the RNA polymerase I complex of a related species. These incompatibilities stem from the rapid divergence of polymerase I genes and their associated co-factors. It is quite probable that the continual homo-genization of newly arising variants in the rDNA promoter regions forces the polymerase I systems to co-evolve with them (Dover & Flavell 1984). As a result, alleles of polymerase I may undergo frequency alterations because they are either more or less efficient in their transcriptional capabilities relative to the slowly altering promoter sites. It is, therefore, easy to envisage the generation of species specificities by this means, and to appreciate how rDNA sequences and their interacting proteins may be causally linked.

It has been possible to test some of the predictions of molecular drive by using the non-genic satellite DNA sequences of *Drosophila*. Strachan *et al.* (1985) have sequenced a large number of clones from the 359 bp 1.688 satellite of seven related species, and examined the patterns of variation within and between them. This study has revealed all transition stages in the spread of variant repeats, and has shown that molecular drive is not only more rapid than mutation but occurs at different rates for different satellite DNA families.

Molecular drive thus forces us to think about how organisms can survive continuous cohesive genomic changes, particularly in terms of the developmental expression of genes whose products are tied in any way to the operation of multigene or multisequence families. It is now not beyond the realm of possibility that major evolutionary changes have stemmed from within, and that it has been the internal genomic readjustments generated by molecular drive that have played a role in generating evolutionary novelty. The dilemma is that we have scant data for the molecular basis of any form of morphological change.

One example does, however, reinforce the possibility that turnover in multisequence families can make a significant contribution to the generation of evolutionary novelty. The egg shell of the silk moth, *Bombyx mori*, is generated by the products of several hundred members of the chorion multigene families (Kafatos 1983). DNA sequencing studies confirm the spread of variants within and between subfamilies. Thus, gradual and cohesive alterations in the egg shell genes may well impose structural and functional constraints on the egg itself. This example also highlights a basic problem involved in gene families with hundreds of members that produce a common protein product, namely how, in the first instance, could any newly arisen mutant members of that family give rise to an effect on phenotype that would be detected by selection?

The question that again needs to be answered is, how much of the plasticity generated by molecular drive is relevant to morphological change? For example, it has been argued that the 'origin' of the long ears of bat-eared foxes may well have been a consequence of molecular drive operating on a multigene family (Dover 1984), an argument that can be extended to many adaptive morphological features (Dover 1986a,b). While this is an important and novel argument, relative to an 'origin' by selection, it does not address what we consider to be the crucial evolutionary problem. This problem is concerned not with how this character spread through the species, which is what most evolutionary arguments address, and the importance of which is self-evident, but rather what embryonic alterations were involved and how these were determined. Did they involve multigene families, single gene changes or modifications in developmental circuits?

A further consequence of molecular drive, which is only now becoming appreciated, is that concerted changes which lead to morphological novelty may provide a basis for organisms to select new niches in which the new forms are well adapted. Under such circumstances, the good fit between organism and environment results from the selection of the environment by the organism rather than the selection of the organism by the environment. Because the concept of selection is based more on logic than on observation, it becomes critical to distinguish between these two modes of adaptation.

4.2.2   Genome turnover

Classical genetics assumed that gene copies were exchanged and assorted largely by the random sexual processes of meiosis and fertilization, coupled, more rarely, with structural chromosome mutations. Molecular drive mechanisms, however, imply that genomic turnover is a much more common event than was formerly assumed, that it leads to an uncoupling of the components of the genome from the chromosomes as mechanical units, and that it includes DNA of at least two kinds. First, ignorant DNA, which is replicated by the cellular machinery largely independently of its sequence content. Secondly, selfish DNA, which has a component of sequence dependence, since sequences compete with each other on the basis of their respective replicative abilities (Doolittle & Sapienza 1980, Orgel & Crick 1980). Mobile elements, for example, are essentially selfish because of their capacity for duplicative transmission.

While bulk hybridization data provide crude average estimates of the divergence of repeated sequence families between genomes, the use of individual cloned probes offers the only objective approach. Thus, a clone from one species can be used to challenge the genome of another and so give a measure of the relative frequency of that clone in both genomes. Table 4.2 summarizes measurements of the frequency changes for 12 families of cloned repeats in the two sea urchins *Strongylocentrotus purpuratus* and *Strongylocentrotus franciscanus*. These data indicate that there can be large, up to 20-fold, frequency differences, in individual repeats between species.

Table 4.2   Frequency changes in twelve families of cloned repeats between the sea urchins *Strongylocentrotus purpuratus* (Sp) and *S. franciscanus* (Sf) (from Britten 1982).

Clone	Repeat length class	Frequency ratio Sp/Sf
CSp2108	long	0.8
CSp2099		1.3
CSp2090		1.5
CSp2096		2.3
CSp2111		2.7
CSp2085		2.8
CSp2136	short	3.0
CSp2109B	short	4.0
CSp2007	long + short	9.0
CSp2101	short	13.0
CSp2109A	short	20.0
CSp2133A	long	21.0

They also demonstrate that different repeat families are simultaneously expanding or contracting their membership, and in this case without any major alteration in genome size. The nomadic middle repeat sequences of different species of *Drosophila* also fluctuate widely in frequency (see § 3.2.1), and a comparable variation can be demonstrated between the repeat families of related plant species (Flavell 1982).

Molecular drive challenges natural selection as a means of changing both multigene and repetitive families. To what extent it challenges selection as a prime mover in evolution is less clear. As has already been stressed, some multigene and nongenic sequences do, indeed, make a contribution to the phenotype. What remains to be established is the relative proportion of multigene, gene and nongenic sequences which do have phenotypic effects, versus those which do not (such as the highly repetitive satellite DNA families). What is lacking is the proof that the molecular changes generated by drive actually contribute to major morphological novelty. However, the situation is little better with respect to selection. While there is no reason to deny that selection might play a primary role in the fixation of many evolutionary changes, it is rarely possible to do more than speculate on the events involved. Arguments of this sort are conceptually weak (Gould & Lewontin 1979). Thus, convincing demonstrations of adaptive change are hard to come by, even for structural genes. Despite the very large number of enzyme polymorphisms which are now known to exist, there are remarkably few cases where both population data and biochemical data provide adequate evidence for a selective basis for that polymorphism (MacIntyre 1982). This situation reminds us that even the existence of a functional difference between two variants is not, in itself, adequate evidence for the operation of natural selection. Selection can only be validated by an actual demonstration of differences in survival rates or in fecundity. Equally, sequence conservation does not necessarily imply selection, though most have assumed it to do so.

On balance, there seems little doubt that the widespread occurrence within eukaryote genomes of DNA sequences that lack any phenotypic effect is the outcome of genomic turnover processes, involving selfish, ignorant or simply junk DNA. Indeed, the most parsimonious explanation of the observed variation in genome size and composition is that many DNA sequences make no contribution to the phenotype of an organism, but are able to make additional copies of themselves and so spread throughout the genome. Under such circumstances, the only restrictions they face stem from whatever constraints excess DNA may impose upon the cell or the individual.

4.3   NUCLEOTYPE AND GENOTYPE

Given that a large proportion of the genome is transcriptively silent, it has been argued that the amount of DNA present in a genome may be important for reasons other than RNA transcription (Cavalier-Smith 1978, Nagl 1978), but especially for so-called nucleotypic effects (Bennett 1971).

This argument distinguishes two very broad kinds of control mechanisms within the genome. One is by means of gene transcription. This kind of control is referred to as genotypic, and is the type expected to operate within developmental gene circuits. **The other is independent of the informational content of base sequences and depends solely on the total amount of DNA within the genome, irrespective of how it is produced or what kind of DNA is involved**. This form of control mechanism has specifically been termed nucleotypic (Bennett 1971). **Notice that the nucleotypic argument concerns only DNA mass. It is immaterial whether the DNA in question is coding or non-coding, or whether it is sequence specific or sequence non-specific**. It is an hypothesis based solely on the total amount of DNA in the genome and, on this basis, its predictions are clear cut.

Although the nucleotype argument is frequently referred to, it has not, to our knowledge, received the rigorous analysis required to assess its evolutionary significance. We deal with it, therefore, in some detail. This hypothesis and its predictions can be examined in a number of ways, but the bulk of the evidence in support of it comes from the correlations that have been claimed between genome size and two important biological parameters, namely, metabolic and developmental rates. It has been suggested, therefore, that the DNA content of a genome is simply a consequence of selection for cell size, the rate of cell division, the rate of metabolism and hence developmental time. Before discussing the argument in detail, we need first to consider some of the data on which it is based.

4.3.1   Genome size and cell size

Good positive correlations have been noted between genome size and cell and nuclear volume in the root tip cells of angiosperm plants (Sparrow *et. al.* 1965). Comparable claims have been made for the nucleated red blood cells of amphibians and reptiles (Olmo 1983). The animal data are, however, complicated by the existence of two quite disparate sets of data which exist within amphibians (Table 4.3), and which inevitably cast doubt on the accuracy of the measurements. The most recent of the amphibian studies (Horner & Macgregor 1983) utilizes a system of internal calibration, using latex spheres of known volume comparable with that of the blood cells being measured. We have, therefore, relied on this data set which, because of its internal control, is more likely to approach the real situation.

These correlations are, however, confounded by the fact that cell and nuclear size vary widely between different cell types within the same individual. Thus, in vertebrates, generally, red blood cells rank amongst the smallest, and neurons often amongst the largest of cell types within an individual. In truth, we do not know precisely what does determine the size of either nuclei or cells. While cell size, nuclear size and DNA content are, indeed, in some sense interdependent, especially in a given cell class, the correlations are far less satisfactory when different cell types are considered, so that the causal relationships involved in a change of any or all of these properties are far from clear. There are, presumably, times when the amount of transcription must influence nuclear and cell size, as has been

Table 4.3 Cell and nuclear volumes relative to genome size in nine amphibians as measured by Horner & Macgregor 1983 (H + M) and by Olmo & Morescalchi 1975, 1978 (O + M).

Species	Genome size (pg)		Nuclear volume (μm^3)		Cell volume (μm^3)	
	H + M	O + M	H + M	O + M	H + M	O + M
Xenopus laevis	3	3.1	10	29	250	295
Bombina variegata	6.9	10.5	60	171	650	1102
Plethodon cinereus	22.5	23.0	144	182	1476	1079
Triturus cristatus carnifex	24.5	21.8	210	91	1500	731
Triturus vulgaris	25.0	24.0	176	77	1663	449
Taricha granulosa	35.0	29.5	272	490	2090	1795
Notophthalmus viridescens	38.0	34.8	168	198	2204	1255
Taricha torosa	38.6	28.0	240	241	2232	1518
Taricha rivularis	38.9	30.0	232	119	2520	1828

argued in the case of large oocytes. There is no doubt that cell volume does play an important role in controlling cellular processes. Thus, cell volume is known to influence the sporulation of yeast (Calvert & Dowes 1984). Cells with a volume less than ~26 μm^3 are unable to sporulate. The probability of sporulation then increases with increasing cell size up to a maximum frequency approaching 100% in cells larger than 130 μm^3. However, these changes in cell volume do not involve increases in genome size, and are clearly not nucleotype dependent.

Whereas in artificially-induced polyploids, cytological and biochemical parameters – such as cell and nuclear size on the one hand, and levels of enzyme activity on the other – are strictly proportional to the degree of ploidy, this is not always the case with natural polyploids. Indeed, in cyprinid fish, cell size is unchanged in diploid and tetraploid species, despite the differences in genome size (Schmidtke & Engel 1975). Paralleling this, the amount of enzyme activity in the tetraploids is similar to that found in related diploids (Schmidtke et. al. 1976). These polyploids have, evidently, undergone some form of secondary diploidization. That cell size can undergo a secondary reduction following a ploidy increase, is supported by some old observations on the moss Bryum caespiticium (von Wettstein 1937). Here, a spontaneous diploid strain, which originated in 1926, was found to have a cell volume of 37 800 μm^3, compared to the value of 16 900 μm^3, which characterized cells of the basic haploid gametophyte. When re-examined 11 years later, however, the cell size of the diploid had decreased from 37 800 μm^3 to 18 600 μm^3, although cytologically it had remained diploid.

4.3.2 Genome size and metabolic rate

By growing larger, any object experiences a reduction in relative surface area, even if its shape is unaltered, simply because volume increases more rapidly than surface area. The concept that genome size might be related to metabolism is based on the assumption that, as the size of a cell increases, the relative reduction of its surface area acts as a limiting factor for cell metabolism. Thus, the very low metabolic rate and the sluggish habit of lungfish, it has been pointed out, parallels their unusually large genome size. The hard data supporting this concept is, however, sparse, and the effect considerably less than proportional. In amphibians, there is a suggestion that respiratory rate is inversely correlated to hepatocyte cell volume, which, in turn, is known to be influenced by genome size (Monnickendham & Balls 1973). Additionally, there is an old claim that the rate of carbon dioxide output by amphibian red cells varies inversely with red cell size (Smith 1925). Until substantial and substantive data are available to support these claims, they can be viewed as little more than anecdotal.

4.3.3 Genome size and division cycle time

In diploid angiosperms, a linear relationship has been found to exist between mitotic cycle time and the DNA content of the genome (Bennett 1972). No really comparable data exist for animals. Even in plants, the extent of this increase is, to some extent, dependent on the quality, as well as the quantity, of DNA (Evans *et. al.* 1972). Thus:

(a) for the same amount of DNA, the duration of the mitotic cycle is, on average, four hours longer in dicotyledonous as opposed to mono-cotyledonous plants, though the rate of increase is the same in both groups;

(b) the rate of increase becomes disproportionately high when the extra DNA is supplied in the form of supernumerary chromosomes.

These points argue against a simple nucleotypic effect as, too, does the fact that there may be quite striking differences between the mitotic cycle times in different diploid cell types within an individual. Thus, in plants the duration of the mitotic cycle varies with age of the organ concerned, and it also varies from one organ to another, and from one tissue to another (Barlow 1973). Added to this, the relevance of influences of genome size on cell size and mitotic cycle time takes on a very different complexion when one recalls that many of the tissues of angiosperms are endopolyploid, not diploid, and do not divide by mitosis.

A simple relationship also exists between the duration of meiosis and the DNA content of the genomes in diploid plants, though the meiotic cycle is, in every case, considerably and proportionately longer than the corresponding mitosis (Bennett 1973). Despite this, meiosis in both artificially produced and natural polyploids is much shorter than in the related diploids (Bennett & Smith 1972). Thus, *Triticum monococcum*, *Triticum dicoccum* and *Triticum aestivum*, with 3C contents of 21, 39 and 54 pg respectively, have meiotic cycle times of 42, 30 and 24 hours, which is unexpected on a simple nucleotypic argument. There is the added complication that although the duration of meiosis is inversely correlated with ploidy level in wheats, that of mitosis is not (Table 4.4). Moreover, by using the range of karyotypic variants available in *Triticum aestivum*, it is possible to show that individual chromosomes or chromosome arms differ in their effects on meiotic time. In particular, the short arm of chromosome 5B has a disproportionate effect on slowing down meiosis (Bennett & Smith 1973), which serves as a reminder of the need to disentangle genic effects from pure DNA mass effects when considering the nucleotypic argument.

4.3.4 Genome size and developmental time

The amphibian life cycle includes a phase of metamorphosis, in which an aquatic tadpole is replaced by a semi–terrestrial adult, and two kinds of developmental correlations have been claimed in this group. According to the first of these, species with smaller genomes spend less time in

Table 4.4 Mitotic cycle time in relation to ploidy in two wheat genera (*Aegilops* and *Triticium*) (from Davies and Rees 1975).

Ploidy	Species	2C(pg)	Duration (h)				Total cycle time (h)	Duration meiosis (h)
			G_1	S	G_2	Mitosis		
2x = 14	*Ae. squarrosa*	11	1.7	6.2	3.0	0.54	11.4	
	T. monococcum	15	3.1	6.0	2.5	0.59	12.5	42
4x = 28	*T. timopheevi*	25	1.3	7.5	5.4	0.83	15.0	
	T. dicoccoides	25	2.7	7.6	4.4	0.76	15.5	
	T. dicoccum							30
6x = 42	*T. spelta*	38	6.3	9.3	2.6	0.93	19.7	
	T. aestivum	38	6.2	8.6	3.8	1.12	19.7	24

Table 4.5 Relationship between genome size and hatch time in ten amphibians (Horner & Macgregor 1983).

Species	1C(pg)	Time to hatching (hours after fertilization)
Pyxicephalus adspersus	1.4	48
Xenopus laevis	3.0	66
Rana pipiens	5.3	140
Bombina variegata	6.9	108
Triturus cristatus carnifex	24.5	600
Ambystoma jeffersonianum	28.0	103
Cynops pyrrhogaster	35.0	504
Ambystoma platineum (3x)	42.0	87
Necturus maculosus	83.0	1200

embryogenesis and hatch sooner. The data are, however, inconsistent (Table 4.5). Thus *Triturus cristatus carnifex* and *Ambystoma jeffersonianum*, with not too dissimilar genome sizes, have a sixfold difference in time to hatching, with the lower sized genome showing a much greater hatch time. Added to this, the triploid *Ambystoma platineum* develops faster than its diploid counterpart *Ambystoma jeffersonianum*, despite its higher C-value.

A second correlation concerns the relationship between genome size and the duration of the larval period in anuran amphibians (Goin *et. al.* 1968). Specifically, the larger the genome the longer the time taken from hatching to metamorphosis, and, interestingly, although triploid *Ambystoma platineum* hatches before diploid *Ambystoma jeffersonianum*, the triploid has a longer period of larval development than the diploid. Urodele amphibians, which, in general, have more DNA than frogs, are known to remain as larvae for longer periods, and their metamorphosis entails a much less drastic reconstruction. Some are neotenic, becoming sexually mature in the larval stage, and there has been an attempt to explain this as a consequence of the high levels of DNA present in the genome. The real situation is, however, by no means as simple (Table 4.6). Thus in *Amphiuma means*, which has one of the highest recorded DNA levels, there is a complete metamorphosis, and *Necturus maculosus* is the only species that is completely neotenic. Likewise, an analysis of the length of larval period in plethodontid salamanders reveals no positive association with genome size (Larson 1984).

When we turn to development times in warm-blooded vertebrates the situation is dramatically different. The development of an elephant is spread over two years, while that of a mouse spans a mere three weeks, despite the narrow range of genome sizes found in mammals. Thus, while genome size shows little variation in mammals, developmental time is extraordinarily variable (Table 4.7).

In the sheep blowfly, *Lucilia cuprina*, the embryonic phase of development is about half that of *Drosophila melanogaster*, yet the blowfly has a genome that is at least four times larger. The only case where a deliberate attempt has been made to assess the effect of altering a defined component of the

Table 4.6 DNA values of semilarval (paedogenetic) and permanently larval (neotenic) urodeles (from Olmo & Morescalchi 1975). See also Table 5.1.

Species	2n	1C(pg)
Semilarval (retention of juvenile characters into adult)		
Salamandridae		
Pleurodeles waltlii	24	19.5
Triturus cristatus	24	22.0
Taricha rivularis	22	30.0
Notophthalmus viridescens	22	35.0
Cryptobranchidae		
Andrias japonicus	64/60	46.5
Andrias davidianus	60	50.0
Amphiumidae		
Amphiuma means	28	75.0
Permanently larval (attainment of sexual maturity in larval state)		
Proteidae		
Proteus anguinus	38	48.5
Necturus maculosus	38	82.5
Sirenidae		
Siren intermedia	46	54.0
Siren lacertina	52	57.0

genome on the duration of development within a species is in *Drosophila melanogaster* (see § 3.2.2.1). Here, it will be recalled, substantial increases and decreases in the amount of heterochromatin (± 34%), in otherwise rigidly controlled genetic backgrounds, produce little effect on developmental time (Miklos 1982). These data provide a direct test of the nucleotype hypothesis in terms of developmental time, and demonstrate that the highly repeated component of the genome does not behave in the manner predicted from the hypothesis.

In plants, too, the DNA content of the genome has been assumed to influence the rate and duration of growth. This assumption rests upon two observations:

(a)   Tropical species tend to have low DNA values and rapid growth. Temperate species have high DNA values and lower rates of growth which, it is claimed, represent an adaptation to having to develop at low temperatures when photosynthetic and transport mechanisms operate slowly (Stebbins 1966). That is, large genomes have evolved under circumstances where growth is limited by the effects of low temperature on rates of cell division. However, the temperate–tropical differential is not consistently present within different families of angiosperms. While it holds for the Poaceae and the Liliaceae, it does not hold for other large cosmopolitan families such as the Fabaceae and the Asteraceae (Levin & Funderberg 1979). Additionally, small genomes are found not only in species living in continuously warm tropical environments but also in plants in which growth is restricted to the summer period in temperate, continental and arctic zones.

(b) Ephemeral plants, which are small in size and need to complete their life cycle within a few weeks if they are to survive, also have low DNA contents. Because of the effect which genome size has on the duration of meiosis and sporogenesis, both of which are relatively slow developmental stages, it follows that no plant with a high DNA content can have a short generation time – so species with a high DNA content are obligate perennials. Indeed, even in perennial species with a high DNA content, the duration of meiosis and pollen division may be longer than the season when active growth can occur. As a consequence, such species complete these gametophytic phases of their life cycle outside the season of vegetative growth and within a perennating organ. This applies, for example, to *Endymion non-scriptus*, *Tulipa kaufmannia*, *Trillium erectum* and *Fritillaria aurea*. Contrary to the generalization that annuals tend to have smaller genomes than perennials, no consistent difference was detected in DNA content between perennial and annual species in a recent survey of 162 species of flowering plants (Grime & Mowforth 1982).

In the Anthemidae, a group of angiosperm plants, a given annual can exhibit either a higher or a lower DNA content than a perennial belonging to the same genus (Table 4.8). Moreover, despite the decrease in genome size which occurs in the annual forms of *Artemisia* and *Chamaemelum*, their cell cycles are still longer than those of equivalent perennials. In *Artemisia* the reduced genome size involves a reduction in heterochromatin content, whereas in *Chamaemelum* the heterochromatin content actually increases. Finally, notice also that the nuclear volume differences that occur between genomes of different size are not those predicted by the nucleotype hypothesis.

This example is instructive because it demonstrates unequivocally that there is no simple and consistent relationship between genome size on the

Table 4.7 Gestation time in mammals (from Witschi 1956).

Species	Gestation time (days)
hamster	16
mouse	19
rat	22
bat	50
guinea pig	68
pig	112
sheep	150
man	267
cow	285
fin whale	330
horse	335

Table 4.8 Relationship between life forms (P = perennial, A = annual), genome and cell characteristics in twelve species of Anthemidae (all 2n = 18). (From Nagl & Ehrendorfer 1974).

Species	Life form	2C (pg)	% heterochromatin at interphase	Nuclear volume (μm^3)	Mitotic index	
					shoot	root
Anacylus depressus	P	12.42	0.18	141.4	9.23	11.06
Anacyclus clavatus	A	10.48	0.95	132.6	13.93	14.00
Anacyclus valentinus	A	11.40	0.07	90.0	13.08	21.15
Anacyclus radiatus	A	16.92	0.76	134.9	17.66	16.21
Anthemis tinctoria	P	7.46	0.27	116.5	5.25	7.00
Anthemis austriaca	A	9.63	3.79	89.2	10.04	9.66
Anthemis cota	A	15.78	1.01	111.5	11.70	13.40
Artemisia absinthium	P	7.28	5.05	75.3	6.36	4.68
Artemisia annua	A	4.05	1.57	72.0	12.23	12.67
Chamaemelum nobile	P	10.65	0.02	70.0	8.95	9.24
Chamaemelum fuscatum	A	10.31	0.33	80.1	11.86	11.90
Chamaemelum mixtum	A	7.60	0.39	67.5	10.55	11.22

one hand, and nuclear volume, division cycle time and developmental time on the other hand. *Clearly, it is not simply DNA mass which influences these parameters*, and the real problem is to identify which components of a genome have the most marked effect under particular circumstances.

When we turn to the only experimental data available in plants, the issue is further confounded. In the genus *Lolium* there is a 40% difference in genome size between inbreeding and outbreeding representatives of the genus, with the more specialized inbreeders having larger genomes and larger chromosomes (Table 4.9). The size differences concerned involve both single copy and repetitive DNA fractions, though most of the additional DNA in the larger genomes is moderately repeated. Despite the differences in DNA content, it is possible to hybridize members of the two major groups, and most of the hybrids are fertile giving F_2s, by selfing or intercrossing, as well as backcross progenies with either parent.

In such F_1 hybrids, the amount of DNA, as expected, is intermediate between the parental values. A cytological examination of these hybrids confirms that the differences in genome size do not depend on differences in localized heterochromatin content, since they are characterized by pachytene bivalents with interstitially located loops or overlaps in euchromatic regions. These range in size from 10 to 60% of the total bivalent length, and at first metaphase of meiosis give rise to seven unequal bivalents.

Meiosis in these F_1s also leads to recombination, both within (crossing over) and between (random assortment) the chromosomes of the two parents, and, as expected, generates gametes with DNA contents ranging between haploid extremes of the low and high parents. A comparable range was expressed in the F_2 progeny, showing that the viability of both gametes and progeny was largely unaffected by DNA content. By comparing a series of 19 features, involving continuously varying aspect of plant development and morphology, in parental backcross and F_2 families, it was clear that differences in genome size resulting from DNA segregation in backcross and F_2 progeny (Fig. 4.1) had no demonstrable phenotypic effects in *Lolium* hybrids. In a genotypic sense this is not surprising, since the differences in DNA amounts involve moderately repeated sequences. This example demonstrates unequivocally that a quite massive supplementation

Table 4.9 Genome characteristics in the genus *Lolium* ($2n = 14$). (After Hutchinson *et al.* 1979 & 1980).

Species	2C(pg)	DNA non-repetitive	repetitive	%-repetitive
Outbreeders				
L. perenne	4.16	1.54	2.62	62.9
L. multiflorum	4.31	1.55	2.76	64.0
Inbreeders				
L. loliaceum	5.49	1.70	3.79	69.0
L. remotum	6.04	1.81	4.23	70.0
L. temulentum	6.23	1.87	4.36	70.0

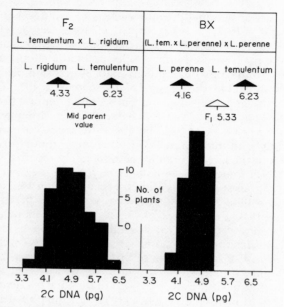

Figure 4.1   DNA distribution in the F$_2$ and backcross (BX) products of interspecies *Lolium* hybrids (data of Hutchinson *et al*. 1979).

of DNA in the hybrid genomes provides no tangible support for any nucleotypic effects on plant development and growth.

The logic of the nucleotypic argument is that natural selection affects nuclear and cell size and that their quantitative variability influences cell metabolism, the rate of cell division and developmental time. A change in genome size, which leads, in turn, to an alteration in cell size, may thus serve to initiate the whole sequence. The extent to which cell size is simply an expression of the total mass of DNA, or whether different classes of DNA make differential contributions to size, is not known.

It is generally assumed that cell size has changed during evolution in response to adaptations to various environments or different modes of life. In the lungfish, for example, the development of tolerance to high concentrations of urea in the blood has been used to explain the parallel increase in both cell size and genome size. However, the amphibian *Scaphiopus holbrookii*, which also accumulates urea, has a low DNA content (Bachmann *et. al*. 1972). Despite inconsistencies of this kind, it has been a convention to assume that all the DNA in a genome is in some sense of adpaptive significance. Many biologists appear reluctant to consider the possibility that not all the DNA in a genome need have adaptive significance, let alone to accept that there might well be disadvantages to a large genome size. Yet salamanders, which parallel lungfish in having exceptionally large genomes, are far less successful than frogs. There are fewer species, they are less widely distributed and show less ecological diversity. On the other hand, the largest family within the order, the

Plethodontidae, and the only one to have successfully invaded the tropics, has the lowest DNA content. One needs, therefore, to exercise considerable caution in accepting arguments which, despite their plausibility, lack either a direct observational basis or indirect experimental evidence.

There are certainly some situations which suggest that DNA mass *per se* may be correlated with parameters which can be assumed to be of selective significance. For example, embryonic development in *Xenopus laevis* begins with a series of rapid cell divisions. These are nearly synchronous for all blastomeres and occur approximately every 35 minutes. This phase terminates abruptly after 12 divisions and is followed by a desynchronization and elongation of the embryonic cell cycles as a result of the inclusion of novel G_1 and G_2 phases coupled with an extension of both S and M. Additionally, the total volume of all nuclei increases exponentially, and at a constant rate, until cleavages 12–14, when it levels off. The timing of this mid–blastula transition in cell cycle behaviour depends on a critical ratio of nucleus to cytoplasm. This can be demonstrated experimentally by varying independently either the initial nuclear content or the initial cytoplasmic content of the fertilized egg (Newport & Kirschner 1982a,b). Thus, embryos of higher ploidy, produced by polyspermy, undergo the mid-blastula transition at correspondingly earlier times than do diploid embryos used as controls. Likewise, if the total DNA content of the fertilized egg is artificially increased by the injection of plasmid (pBR322) DNA, it promotes premature transition. Alternatively, eggs constructed in such a way that the nucleus is present in a cytoplasmic fragment one fourth of the original egg volume, undergo the transition to cell asynchrony two cycles earlier.

These experiments, though suggestive, do not directly address the issue of nucleotype. Indeed, the blastomeres within the same embryo differ widely in volume, yet all undergo the mid–blastula transition within 1 hour of each other. What is needed is a comparison of transition patterns in related species with known differences in genome size and where the differences can be partitioned in terms of the kinds of DNA involved. An obvious place to carry out such a study is in the genus *Xenopus* itself. This genus includes species with ploidy levels ranging from 2x to 12x (Table 4.10), and so provides a unique opportunity to examine nucleotypic effects on the mid-blastula transition.

4.4   GENOME CHANGE AND SPECIATION

In attempting to explain the morphological differences between species, biologists have turned their attention to a variety of other specific differences, in the hope that they might prove to be the causative factors underlying the differences in morphology. In adopting this approach, each new technology that has been developed has provided an opportunity to re-examine the same question in terms of different information.

One of the earliest findings was that related species often differ in either the number or the structure of their chromosomes. In recent years, such

Table 4.10 Genome variation in the genus *Xenopus* (after Thiebaud & Fischberg 1977, Tymowska & Fischberg 1982).

Species	2n	DNA (pg)
X. tropicalis	20	3.55
X. epitropicalis	40	7.10
X. laevis (l. laevis, l. petersi, l. victorianus)		
X. gilli		6.35
X. fraseri	36	
X. borealis		7.10
X. muelleri		7.60
X. clivii		8.45
X. witteri = X. l. bunyoniensis	72	12.57
X. vestitus		12.83
X. ruwenzororiensis	108	16.25

differences have led to the assumption that they are in some way fundamental to both altered morphology (Bush 1981) and to speciation (White 1978). The requirements for substantiating these claims are not difficult to define:

(a)   to demonstrate that the alterations of the genome which accompany the chromosome differences in question actually affect somatic development in such a way as to cause the observed differences in morphology;

(b)   to be able to distinguish between those changes which preceded the origin of the species in question, those which might have arisen subsequent to the speciation event and those which not only accompanied speciation but were responsible for it.

4.4.1   Species differences in mammals

A simple, but very convincing, example, involving a comparison of two species of deer, provides data on both these points. One species has 46 chromosomes in the female, the other only six. Despite this massive disparity in the organization of the genome, there is very little morphological difference between them. This case is therefore worth considering in some detail.

4.4.1.1   Muntjacs

While, in general, diploid species within a genus show a reasonably narrow range of chromosome numbers, there is a particularly striking difference in four of the five species of barking deer that have been studied (Table 4.11). The extremes of this series are found in the Indian and Chinese muntjacs (Wurster & Atkin 1972). These two species, which show such a marked karyotypic dissimilarity, have a pronounced morphological similarity (Fig. 4.2). The two hybridize in captivity and can be reciprocally crossed. The F_1

Table 4.11 Genome variation in barking deer (genus *Muntiacus*). For assessing relative genome size that of man has been taken 1.0. (After Soma *et. al.* 1983, Wurster Hill & Seidel 1985).

Species	Relative DNA Content	Diploid number	
		♀(XX)	♂
M. muntjak (Indian muntjac)			
M. m. vaginalis (red muntjac)	0.73	6	7(XY$_1$Y$_2$)
M. m. muntjak	0.95	8	9(XY$_1$Y$_2$)
M. rooseveltorum		6	7(XY$_1$Y$_2$)
M feae		12,13,14	
M. reevesi (Chinese muntjac)	0.89	46	46(XY)
F$_1$ *M. m. vaginalis* × *M. reevesi*		26	27(XY)

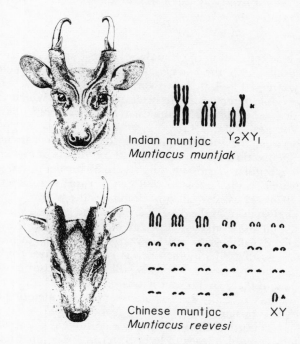

Indian muntjac
Muntiacus muntjak
Y$_2$XY$_1$

Chinese muntjac
Muntiacus reevesi
XY

Figure 4.2 A comparison of the Chinese (*Muntiacus reevesi*) and Indian (*Muntiacus muntjak*) muntjacs (data of Short 1976).

hybrids, as expected, have $2n = 27\sigma$ and $26\female$. This implies that the developmental circuits are, in all essentials, identical within these two species, and that it is irrelevant whether the genes which underlie these circuits are in six or 46 chromosomes. Moreover, a hybrid nucleus, in which one set of genes is contained in three chromosomes and the other in 23, operates no differently from either of the parents in terms of its capacity to produce a normal morphological end product.

Since most species of Cervidae, including the closely related tufted deer, *Elaphodus cephalophus* ($2n = 47$), have a high diploid number, it is assumed that the differences between the Indian and Chinese muntjacs can be explained by a combination of centric and tandem fusions. These have been accompanied by a slight reduction in the DNA content of the genome in the red muntjac, though the amounts of single copy DNA in the two species are almost identical (Schmidtke *et. al.* 1981).

Although somatic development is completely normal in the male F_1 obtained from the cross Chinese $\female$ × Indian σ (Liming *et. al.* 1980), germ line development is not. The seminiferous tubules of the testis are certainly lined with spermatogonia, but very few normal meiocytes occur. Those that are present are at zygotene–pachytene, and none show complete synapsis. Most are in various stages of degeneration and no spermatids or sperm are ever seen. Effectively, meiosis is arrested in prophase I (Liming & Pathak 1981).

4.4.1.2   Equids

A comparable case is found in the Equidae. This is a small family with only seven extant species, all in the one genus *Equus*, which are notable for their divergent karyotypes (Fig 4.3). However, they show a remarkable facility for producing viable interspecific hybrids in captivity, which indicates a considerable degree of genetic similarity between them, despite their divergent karyotypes (Table 4.12).

F_1 hybrids between the wild horse and the domestic horse are fully fertile, despite the single fixed fusion difference between them. In male hybrids of this type, there is regular synapsis giving rise to 30 bivalents, a trivalent and an XY sex bivalent, and the fusion multiple of three segregates regularly (Short *et. al.* 1974). However, all other interspecific hybrid combinations are sterile. This is so even in the case of Grevy's and Grant's zebras, as well as in the horse and the donkey (Chandley *et. al.* 1974). In this latter case there are no cells in diakinesis or any later stage in the testis of the hinny (F_1 horse $\female$ × donkey σ). Meiotic activity is certainly more evident in the mule (F_1 donkey $\female$ × horse σ), but is still poor by comparison with a normal stallion, and there are no spermatids or sperm. The infertility of the male mule and hinny is, thus, directly attributable to a meiotic block at the primary spermatocyte stage.

What neither of these two mammalian cases resolve is the actual cause of the meiotic blockage which leads to hybrid sterility. The probability is that it does not depend on the chromosome differences, but, rather, results from faulty genic interaction, though this has not as yet been established. These cases are, however, instructive in demonstrating that the components of the

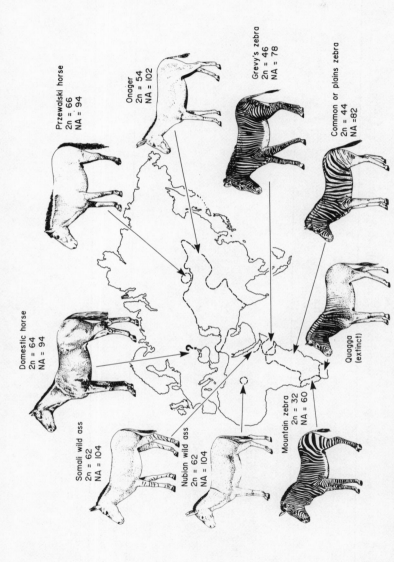

Figure 4.3 Geographical distributions of the wild Equidae, showing their diploid ($2n$) chromosome numbers and the number (NA) of chromosome arms (data of Short 1976).

Table 4.12 The consequences of inter-specific hybridization in the genus *Equus* (from Short *et. al.* 1974).

Species	2n	Equus przewalskii	Equus caballus	Equus asinus	E. hemionus onager	Equus grevyi	Equus burchelli	E. zebra hartmannae
				Fertility of F$_1$ hybrids				
E. przewalskii (wild horse)	66	fertile						
E. caballus (domestic horse)	64	fertile $(2n = 65)$	fertile					
E. asinus (ass, donkey)	62	–	sterile $(2n = 63)$	fertile				
E. hemionus onager (onager)	54	sterile	sterile	sterile	fertile			
E. grevyi (Grevy's zebra)	46	–	sterile $(2n = 55)$	sterile	–	fertile		
E. burchelli (Grant's zebra)	44	sterile	sterile	sterile	–	–	fertile	
E. zebra hartmannae (Hartmann's zebra)	32	–	sterile	sterile	sterile	–	–	fertile

mammalian genome may be arranged very differently in different species without any pronounced effects on somatic development. They, thus, put into clear perspective the incredible range of chromosome numbers found in mammals, with a minimum of six in the female of the red muntjac and a maximum of 92 in the fish-eating rodent, *Anotomys leander* (Table 4.13), despite the relatively narrow range of genome size. These cases also raise the more general question of what role, if any, structural chromosome rearrangements play in speciation. The fact that closely related species do often differ in karyotype suggests that, at the very least, the conditions which lead to speciation may also favour the fixation of chromosome rearrangements, but this says nothing concerning a causal role for these rearrangements in the speciation process itself.

4.4.1.3 Mice

Yet a third instructive situation is to be found in wild populations of the European long-tailed mouse, *Mus musculus* subspecies *brevirostris* and *domestica*. Here, populations containing individuals homozygous for from 1 to 9 whole arm fusions (Fig. 4.4) have been reported. Some of these are surrounded by conventional, $2n = 40$, populations, where all the chromosomes are one-armed, acrocentric, and completely lack two-armed, metacentric members. The sex chromosomes are never involved in the production of such fusion metacentrics and neither, apparently, is the smallest autosome, number 19. By using G-banding (Holmquist *et. al.* 1982) to assess the structure of individual fusion metacentrics, it has been established that each of the other eighteen autosomes (1–18) can occur in many different metacentric combinations (Fig. 4.5), a fact that has been confirmed by the analysis of meiotic configurations in hybrid males. Thus, of the 171 possible arm combinations, at least 57 have been detected in fusion populations. A distinctive feature of this situation, stemming from

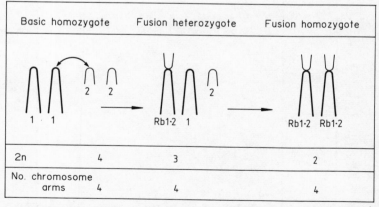

Figure 4.4 Heterozygosity (l/Rbl.2/2) and homozygosity (Rb1.2/Rb1.2) for Robertsonian (Rb) karyotype variation resulting from whole arm fusion between two, non-homologous, one-armed chromosomes, 1 and 2.

Table 4.13 Distribution of diploid chromosome numbers in various orders of mammals (from Arrighi 1974).

Order	Diploid number										Number of species
	0–10	11–20	21–30	31–40	41–50	51–60	61–70	71–80	81–90	91–100	
Marsupialia	1	12	4								17
Insectivora			1	3	3	2	1				10
Dermoptera						2					2
Chiroptera		1	11	9	9						30
Primates				1	24	6	4				35
Edentata					1	2	2				5
Lagomorpha				1	6		1				8
Rodentia		3	14	22	46	21	5			1	112
Cetacea					5						5
Carnivora			1	35	8	1	2	8			55
Pinnipedia				5							5
Tubulidentata		1									1
Proboscidea						2					2
Hydracoidea						1					1
Perissodactyla			3		2	1	3		2		9
Artiodactyla	1			7	12	22	10	1			55
Number of species	2	17	34	83	116	60	28	9	2	1	352

Figure 4.5 Involvement of standard acrocentric autosomes of *Mus musculus* [101 feral (●) and 10 laboratory (○) individuals] in the production of fusion metacentrics (combined data of Gropp & Winking 1981, and Capanna 1982).

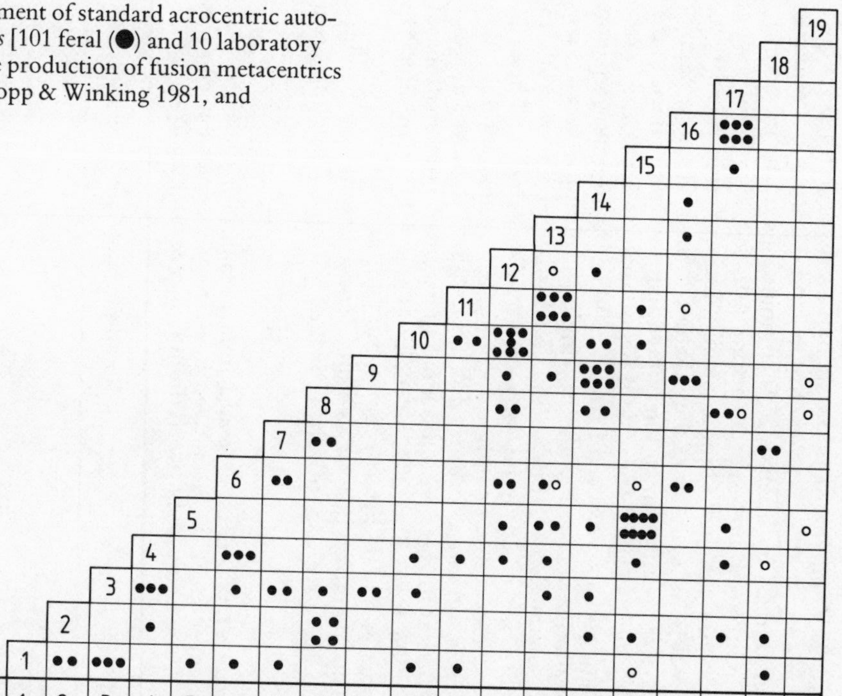

Chromosome number	1	2	3	4	5	6	7	8	9	10	11	12	13	14	15	16	17	18	19
No. times involved in fusion — feral	11	11	15	13	12	12	9	13	13	15	11	15	13	15	13	13	12	4	0
No. times involved in fusion — lab.	1	0	0	1	1	2	0	2	1	0	1	1	2	0	2	1	1	1	3
No. different fusions — feral	8	7	9	9	6	7	5	6	5	7	5	7	7	8	6	5	6	3	0
No. different fusions — lab.	1	0	0	1	1	2	0	2	1	0	1	1	2	0	2	1	1	1	3

the randomness of the fusion pattern, is the fact that different fusion populations, whether located adjacent to one another or placed far apart, have different combinations of metacentrics, with some shared and some distinctive fusions. None of these show any significant morphological differences.

Natural hybrids are known to occur in narrow hybrid zones, which sometimes form in contact areas between fusion and non-fusion populations, and these are characterized by heterozygous combinations of the metacentric and acrocentric chromosomes which distinguish the two hybridizing populations. However, although chromosomally distinct fusion populations are known to co-exist in sympatry, they do not form hybrids in nature, though hybrids between different fusion populations can be obtained in the laboratory. F_1 hybrids, derived from wild fusion male × laboratory non-fusion female matings, show reduced fertility, resulting from malsegregation of the chromosomes associated with the single or multiple metacentric heterozygosity generated by hybridization. Such malsegregation leads to the production of unbalanced, hypo- or hyper-modal, gametes. These, in turn, give rise to monosomic or polysomic embryos, which abort.

The actual rates of chromosome non-disjunction at meiosis show considerable variation in different individual metacentric combinations of feral origin, ranging from 2 to 28% in males (Table 4.14). Considerably higher rates occur in female heterozygotes carrying the same metacentric combination. Mice heterozygous for two metacentric chromosomes which share a common arm give rise to a chain of four multiple (Fig. 4.6). Such

Table 4.14 Malsegregation rates, estimated from metaphase II counts, in male fusion (Rb) heterozygotes of the long tailed house mouse originating either in Switzerland (Bnr) or in central Italy (Rma) (from Gropp & Winking 1981).

Fusion heterozygote	Non-disjunction rate
Rb(10.11)8Bnr/+	2
Rb(1.10)10Bnr/+	
Rb(16.17)Bnr/+	4
Rb(8.12)5Bnr/+	
Rb(6.13)3Rma/+	6.6
Rb(4.12)9Bnr/+	8
Rb(9.14)6Bnr+	10
Rb(1.3)1Bnr/+	14
Rb(10.11)5Rma/+	14.4
Rb(2.18)6Rma/+	16.3
Rb(9.16)9Rma/+	16.6
Rb(3.8)2Rma/+	18
Rb(4.15)4Rma/+	22
Rb(4.6)2Bnr/+	
Rb(5.15)3Bnr/+	28
Rb(11.13)4Bnr/+	

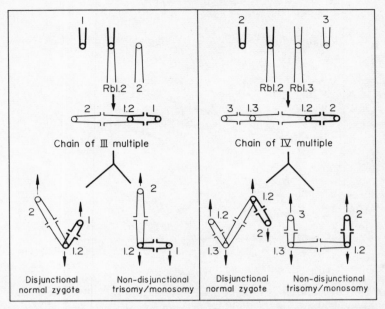

Figure 4.6 Meiotic behaviour of a single fusion heterozygote (l/Rb1.2/2) and a double heterozygote with overlapping (monobrachial) homology (2/Rb1.2/Rb1.3/3).

mice show considerably higher levels of malsegregation, with male rates now exceeding those of females. Moreover, several such double metacentric heterozygotes are male sterile, though female fertile. Polymetacentric hybrids can be obtained by crossing different feral populations. Here, either numerous chains of three, or else larger multiples, are present at meiosis, and the overall non-disjunction rate can rise to still higher values (Fig. 4.7), with females, again, showing more non-disjunction than males carrying the same hybrid combination. Added to this, the presence of large chain or ring multiples may also lead to aspermy, or at least severe oligospermy, as a result of a block in male meiosis.

The data thus indicate that the fusion metacentrics which distinguish different populations are only likely to play a role in building up reproductive isolating barriers in one of two situations. First, where individual metacentrics are introduced into the genetic background of a different fusion population, and especially where they share overlapping arm homology with the existing metacentrics of that population. Secondly, when complex multiples are formed at hybridization. Even under these circumstances, however, reproductive isolation cannot be described as complete (Gropp *et. al.* 1982). Thus, even when numerous structural differences exist between the two chromosome sets, these differences do not produce complete reproductive isolation, and the extent to which the genotype is involved in the infertility produced has yet to be clarified.

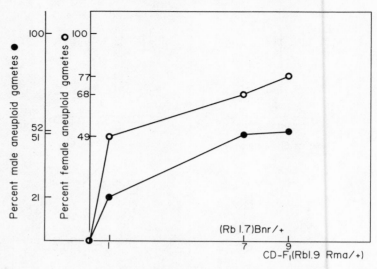

Figure 4.7 Relationship between extent of Robertsonian (Rb) heterozygosity and gametic aneuploidy in *Mus musculus* hybrids (data of Capanna *et al.* 1977).

4.4.1.4 Rats

Considerable karyotypic variation is also present in the genus *Rattus*, where diploid numbers range from 32 to 52 and chromosome arm numbers range from 52 to 83. The relevant data for seven representatives of this genus, all of which have been compared by G-banding, are summarized in Table 4.15. Differences are specially pronounced in the *Rattus sordidus* group, which comprises three species, *sordidus, colletti* and *villosissimus*, and where some 11 different autosomal fusions are involved. These three species can be successfully and reciprocally crossed in all combinations in the laboratory, and at least one of them, *Rattus colletti* × *Rattus villosissimus*, has limited fertility, despite the fact that, at meiosis in the F_1, pairing gives rise to one chain of five and three chains of three chromosomes. Thus, backcross progeny are produced from F_1 hybrids irrespective of whether the hybrid is used as male or female parent, though litter size is reduced to about 70% that found in the parental species.

Thus, none of these cases offer convincing evidence that the fixed fusion differences observed between related mammal species are likely to have played a causative role in the speciation process as a result of the malsegregation they produce in hybrid combinations. Neither do they support the general argument that chromosome rearrangements are common ways in which the regulatory circuits within a genome can be reorganized, so as to lead to novel patterns of development which can account for the observed morphological differences between related species. Indeed, what is so striking in these cases is the overall similarity of phenotype despite the gross differences in karyotype.

Table 4.15 Karyotype differentiation in Australian species of *Rattus* (from Baverstock *et al.* 1983).

Species	2n	(Rb) fusion chromosomes
R. lutreolus	42	Presumed ancestral, no fusions
R. tunneyi	42	Differs from *lutreolus* by 1 peri. inv.
R. fuscipes	38	5.9, 11.13
R. leucopus cooktownensis	36	2.13, 3.10, 4.7
R. leucopus leucopus	34	2.13, 3.10, 4.7, 5.9
R. sordidus	32	2.4, 3.6, 5.7, 8.11, 10.13
R. colletti	42	2.4, 3.10, 5.11, 6.9, 7.8
R. villosissimus	50	2.4, 6.10

4.4.2 Structural changes in the *Drosophila* genome

It has long been held that a diploid number of 12, consisting of five rod pairs and one small dot pair, represents the basic karyotype of the genus *Drosophila* (Patterson & Stone 1952). At the cytological level, departures from this base have resulted from the fusion of whole chromosome arms, from pericentric inversion or from changes in heterochromatin content. The extent of these three types of change vary in the different groups of *Drosophila*. In one cluster of the *Drosophila repleta* group, for example, changes in metaphase configuration can all be explained, predominantly, in terms of either centric fusions, or additions or deletions of heterochromatin (Fig 4.8). Fusions have been particularly prominent, eight distinct fusions having been fixed during the history of six of the species. By contrast, in the 155 species of Hawaiian *Drosophila* that have been analyzed, only 18 have karyotypes altered by fusion. Thus, in the majority of cases in Hawaiian *Drosophila*, speciation has gone on without significant gross chromosome rearrangement. A further 14 species have a large amount of heterochromatin concentrated on the dot 6, so converting it into a rod (Table 4.16), and equivalent situations are common throughout the genus.

While structural chromosome changes provide important clues to the history, and presumed phylogeny, of the species within the genus *Drosophila*, there is no evidence that they have been of any importance to the speciation process itself. *Drosophila mojavensis* and *Drosophila arizonensis*, for example, differ from one another by at least seven inversions, but mate relatively readily and produce fertile offspring and even hybrid swarms (Wasserman 1982). Likewise, fertile hybrids are produced in crosses between *Drosophila virilis* and *Drosophila americana texana*, which are distinguished by a fusion difference (Throckmorton 1982).

The heterochromatin differences that occur are also irrelevant to speciation. When satellite DNA was first discovered it was suggested that gain of a satellite might be expected to create a divergence in chromosome homology which, in turn, would lead to hybrid sterility, and so to speciation. This suggestion, however, totally failed to take into account the extent to which satellite DNA sometimes varied even within a species. It

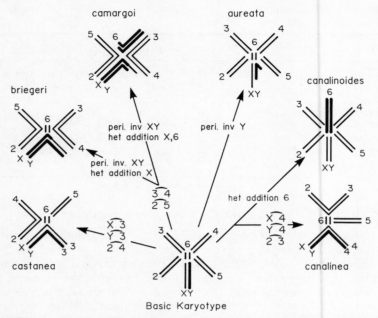

Figure 4.8 Karyotypes of the *castanea*, *canalinea*, *dreyfusi* and *aureata* species groups of *Drosophila*. Heterochromatic regions in thick line (data of Wasserman 1982).

was also unsupported by the fact that in several groups of *Drosophila*, including the *virilis* group, the viability and fertility of interspecies hybrids evidently did not depend on similarities or differences in the satellite DNA content of the parent species involved (reviewed by John & Miklos 1979).

The same is true for the *immigrans* group. Here, *Drosophila nasuta nasuta* ($2n = 8$) and *Drosophila nasuta albomicana* ($2n = 6$) are two chromosome races which are morphologically identical and fully cross fertile (Raganath & Krischnamurthy 1981). The difference in chromosome number between these two races is a consequence of an $\widehat{X3}$ fusion which has reduced the chromosome number from 8 to 6 (Fig. 4.9). There are also differences in the heterochromatin and satellite DNA contents of the two taxa, and their patterns of heterochromatin distribution are strikingly different. In terms of polytene banding, their euchromatin component is effectively cytologically identical (Raganath & Hägele 1982).

There is only one satellite in *D. n. nasuta* (1.664 g/cm^3), whereas in *D. n. albomicana* there are three (1.674, 1.665 & 1.661 g/cm^3 respectively). With the exception of the Y chromosome, the satellite DNAs of both races hybridize to all heterochromatic sites in both races, indicating that there are homologous satellite sequences in the heterochromatin of both taxa. In *D. n. nasuta*, the satellite fraction corresponds to only 7.5% of the genome, whereas in *D. n. albomicana* the three satellites collectively make up 28–30% of the total DNA. Evidently, there has been an amplification of satellite sequences in *D. n. albomicana* in both autosome 4 and the Y, and a

Table 4.16 Karyotype variation in endemic Hawaiian members of the genus *Drosophila* (from Carson & Yoon 1982).

Subgenus	Metaphase configuration				
	5rod 1dot $2n = 12$	6 rod $2n = 12$	1.V 3rod 1dot $2n = 10$	2.V 1rod 1dot $2n = 8$	Total species
Subgenus *Drosophila*					
Drosophila species group	117	10	8	6	141
Antopocerus species group	6	4	–	–	10
Subgenus *Engiscaptomyza*	–	–	4	–	4
Total fusions			18		
Total heterochromatin additions		14			
Total species					155

concomitant reduction in the satellite content of autosomes 2, 3 and the X. Even so, neither of these series of changes in heterochromatin content, nor the single fusion difference which distinguishes the two races, have any effect on either the viability or the fertility of F_1 hybrids.

It is true that in *Drosophila melanogaster* simple reciprocal translocations, in an otherwise normal genotype, will produce male sterility when substantial segments, generally exceeding half the length of a chromosome arm, of the X and one of the two major autosomes (2 & 3) are exchanged experimentally (Lifschytz & Lindsley 1972, Lindsley & Lifschytz 1972). Moreover, the inability to rescue such X/A steriles, by adding a duplication that covers the X breakpoint, implies that the sterility in question is of chromosomal rather than of genic origin, and cannot be simply explained in terms of the inactivation of a gene or genes in the vicinity of the breakpoint. On the other hand, those translocations where the breakpoints are nearly

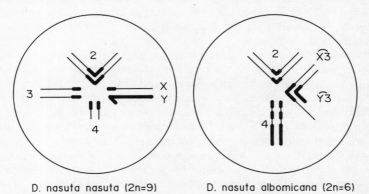

D. nasuta nasuta (2n=9)   D. nasuta albomicana (2n=6)

Figure 4.9 Karyotype relationships between the males of the two chromosome races of *Drosophila nasuta*. Heterochromatic regions in thick line (data of Ranganath & Hägele 1982).

terminal in both X and autosome, or where the X breakpoint occurs within the proximal heterochromatin, turn out to be fertile. Thus, while some induced structural changes in *Drosophila melanogaster* can lead to infertility, there is no convincing evidence that this source of infertility has ever been instrumental in securing reproductive isolation in this genus. Thus, not only is there no evidence from these several examples to indicate any role for structural chromosome change in the speciation process, but there is, again, no evidence to relate chromosome change to morphological change.

4.4.3   Speciation and morphological change

Much attention has recently been focussed on the topic of gradual versus abrupt change. This has led to the claim that there are two quite distinct components to evolution. One represents change within an established species. Such phyletic change is slow. In the other, rates of evolution are, at times, greatly accelerated by rapid divergent speciation. This may occur either because rates of speciation are accelerated, or because an unusually large percentage of speciation events are divergent, following entry into a new environment with a consequent adaptive radiation (Stanley 1979). This latter category of speciational change forms the basis for the theory of punctuated evolution (Gould & Eldredge 1977), which holds that speciation is the focus for accelerated evolution and accounts for most large scale changes. Between such phases of rapid speciation, however, there are long periods of stasis. Consequently, after their origin most species undergo little change before becoming extinct, or else they persist essentially as living fossils.

There is no doubt that the fossil record does reveal that rates of evolution are highly variable, and that cases are known which exhibit the hallmark of punctuated change (Williamson 1981), though, equally, there are also cases which do not conform to the conventional punctuational model (Malmgren *et. al.* 1984). Even these cases are not without their difficulties of interpretation. The only quantifiable changes in morphology that can be recognized in fossils deal either with meristic modifications, often of a relatively minor nature, or else with allometric changes in the relative dimensions of body parts. Speciation, as inferred from hard parts, can, therefore, only be delimited subjectively, since the criteria adopted for defining speciation represent an indirect inference from morphology, on the assumption that this somehow reflects the reproductive capacity of the organism. Indeed, this qualification applies not only to fossil species but to many living ones too. Added to this, the extent to which the assumed morphological constancy of fossil forms may have been associated with sibling speciation of the kind commonly found in living forms is not known. In any event, the question of whether rapid speciation leads to the emergence of morphological novelty is relevant only if speciation is directly related with the production of morphological change.

There is also no doubt that rapid, multiple speciation can, and does, occur. The Hawaiian Islands, the youngest of which is less than one million years old, have all been sites of prodigious recent multiple speciation events in the genus *Drosophila* which, in many cases, have also been associated with morphological change (Carson 1982). Even so, the changes in morphology that have occurred have all been relatively minor and are unlikely to have involved any major developmental reorganization. This example cautions us that the multiplication of species does not, in itself, create major new morphologies. These arise only as a result of altered development. In interpreting the fossil record, therefore, we would need to know how radical were the developmental changes that determined the altered morphologies, and in what manner these changes were associated with the proposed multiple speciation events.

This problem resolves itself into one of genetic cost. How expensive is it, in terms of altering genes or gene circuits, to bring about a major morphological change? A change which appears spectacular might, nevertheless, involve only a minimal change in the genome, especially if that change occurs at a critical point in the developmental programme. Alternatively, it might depend on a major redirection of developmental circuits which involves many genes. It is only by being able to decide between these alternatives that we can really describe a morphological change as major or minor. That is, the distinction needs to be made at the developmental, and not the morphological, level. The crux of the problem is whether developmental changes are major or minor in cost terms, not whether the resultant morphological changes are, or are not, spectacular.

A study of living organisms indicates that speciation may take place in a great many different ways that have little in common except the production of reproductive isolation. They may, or may not, be accompanied by morphological change, and they may, or may not, lead to subsequent morphological change. Morphological change and speciation are clearly distinct genetical and developmental phenomena, so that even when they appear closely correlated it is difficult to determine whether they are causally related and, if so, in what way.

It should now be apparent that little insight into the fundamental processes of evolutionary change can come simply from the continued study of the products of that change. What we need to focus on are not the products of change, but the molecular events that take place within genomes, and the developmental consequences of these events.

All the genetic variation available to an organism depends ultimately on the DNA misdemeanours that go on within the genome. These may arise in a number of ways. The simplest of these are conventional base changes in the coding or flanking sequences of genes. In laboratory populations of *Drosophila melanogaster* many of the known mutations arise as the consequence of the insertion of mobile sequences into, or else close to, genes. Similarly, two or more mobiles can cause large-scale chromosome rearrangements which may, or may not, have significant effects on nearby genes. An inevitable consequence of these molecular findings is that they, too, have been extrapolated as a possible mechanism for producing significant evolutionary change. Thus, a recent claim relates mutational changes arising from the movements of nomadic elements to the induction of speciation. It is to this claim that we turn next.

4.4.4   Hybrid dysgenesis in *Drosophila melanogaster*

The progeny that result from crossing certain strains of *Drosophila melanogaster* are characterized by a syndrome of abnormalities which includes sterility, high rates of both gene mutation and structural chromosome rearrangement, coupled, in some cases, with the occurrence of male recombination in a meiotic system which is conventionally achiasmatic. This syndrome is customarily referred to as hybrid dysgenesis (Bregliano *et. al.* 1980, O'Hare 1985) and is caused by mutator genes (Green 1986) which, in turn, lead to the mobilization of nomadic elements. It should be pointed out, however, that the term 'hybrid dysgenesis' is a rather inappropriate and imprecise one. The phenomena associated with this term are due to mappable entities (MR elements) and not to hybridity *per se*, a point made forcefully by Green (1986).

Two rather distinct systems of interstrain hybrid sterility have been identified to date. In the first of these, transitory sterility occurs in F_1 females obtained from crossing inducer (I) and reactive (R) strains, but only in progeny of the cross (R)♀ × (I)♂. The (I) condition is determined by a transposable element, though the occurrence of transposition requires a reactive (R) cytoplasm (Bucheton *et. al.* 1984). Reactivity is an extra chromosomal state that is transmitted by maternal inheritance and is ultimately dependent upon genotype. Sterile F_1 females lay normal quantities of eggs, but a certain percentage of these do not hatch, since their development stops between the fifth and eighth cleavages. All stocks established from flies recently caught in the wild have been found to be of the (I) type, whatever their geographical origin. (I) strains contain I elements located within the pericentromeric hetrochromatin, in addition to 0–15 copies of the complete I element present on the euchromatic arms. I elements are 5.4 kb mobile sequences which have no sequence homology with P elements and are not members of either the *copia*-like or the FB class of transposable elements. They have two open reading frames which could code for proteins of respectively 429 and 1086 amino acids, and the ends of the I element do not contain any inverted or direct repeats (Finnegan 1986). Thus, I element structure is quite unlike that of other families of nomadic elements, and resembles more the LINE sequences of rodents and primates (Singer 1982b).

In the second system, instability results when a male of a paternal contributing (P) strain from the wild is mated to a female of maternal (M) strain, but not when the reciprocal cross is carried out. (P) strains are characterized by the presence of mobile P elements which are conserved 2.9 kb sequences with terminal inverted repeats of 31 bp (O'Hare & Rubin 1983). While these P elements do not produce instability in (P) strains themselves, they do so when placed in a maternally-derived cytoplasmic background of an (M) strain. Thus, the PM system, like the IR system, involves a combination of chromosomal and cytoplasmic heredity. In contrast to the IR system, however, the PM category causes instability in males as well as females, since it leads to male recombination.

One form of PM instability is the production of a specific category of sterility, gonadal dysgenesis, which leads to gonadal atrophy in both F_1 males and females. This results from an early blockage in the development of the germ cells at temperatures above 24 °C (Engels & Preston 1979). The germ line of *Drosophila melanogaster* is first differentiated some 2 hours after fertilization. At that time, several of the cleavage nuclei migrate into the posterior pole plasm of the egg to form the pole cells, some of which eventually become the germ cells. The early development of the germ line is, thus, dependent on the cytoplasmic environment of the egg, and a failure in this developmental sequence can explain the sterility phenotype observed in hybrid dysgenesis. Other aspects of hybrid instability also appear to be germ line limited, so that they, too, may stem from abnormal interactions in the early development of the germ line.

Many of the gene mutations which appear in (M)♀ × (P)♂ crosses are unstable, and are caused by insertion of P elements which are not represented in the (M) strain and which are normally stable in the (P) strain. When chromosomes carrying P elements are placed in an M-cytotype, the P-elements are derepressed and transpose at high rates. As a result, they induce mutations by inserting into and disrupting genetic loci, or by causing chromosome rearrangements. These mutations are then stable in the P-cytotype but may revert, by excision of the element, in the M-cytotype. Of seven independent mutations at the white locus obtained in F_1 hybrids between (P) and (M) strains, five have been found to result from DNA insertions of, respectively 0.5, 0.6, 1.2 and 1.4 kb. Although heterogeneous both in size and in the pattern of restriction enzyme sites, the four that have been studied in detail proved to be P elements, homologous in sequence (Rubin *et. al.* 1982). These P elements were, as expected, stable in P-cytoplasm but showed reversion rates of 4×10^{-3} in M-cytoplasm. This reversion was accompanied by excision of the P element.

The P element thus functions as a tranposon, and there are 30–50 copies per haploid genome in (P) strains which are completely lacking from (M) strains. All strains of *Drosophila melanogaster* that lack P elements appear to be descendants of gravid females caught in the wild more than 30 years ago, whereas those originating from females caught during the last ten years are usually (P) in type.

Certain crosses between different species of *Drosophila*, also lead to hybrid adults with gonadless sterility reminiscent of interstrain dysgenic sterility which characterizes *Drosophila melanogaster*. Not surprisingly, therefore, hybrid dysgenesis has been suggested as playing a role in speciation within the genus. As yet, however, no known form of hybrid dysgenesis has been found to preclude gene flow between any pair of populations within *Drosophila melanogaster*, let alone between different species within the genus. An extension of the argument, which has yet to be tested, proposes that the combined action of several different dysgenic systems might produce F_1 hybrids incapable of backcrossing to either of their progenitor parents (Rose & Doolittle 1983).

It has also been proposed that P elements may lead to genome reorganization through the induction of chromosome rearrangements, and that this reorganization, in turn, leads to reproductive isolation. Most of the breakpoints induced as a result of hybrid dysgenesis within a strain occur at specific locations rather than being randomly distributed throughout the genome (Engels & Preston 1981). These hot spots have characteristics expected of P elements but not of any other class of mobile elements, so that this form of instability, too, can be interpreted simply as a result of the transposition, excision or other change of inserted P elements. Because of the assay system employed, most of the rearrangements turn out to be inversions where both breakpoints are hot spots, suggesting that these rearrangements result from direct interaction between P elements. The remainder are complex rearrangements. As we have commented earlier, there is no evidence that any of the chromosome rearrangements which distinguish different species of *Drosophila* function in reproductive isolation, so that there is no reason to believe that rearrangements induced by hybrid dysgenesis would lead to speciation.

The possibility that hybrid dysgenesis might in some way contribute to speciation events raises the more general issue of whether any of the molecular turnover processes involving repetitive DNA sequences might be expected to lead to effective reproductive isolation, given that there is evidence to associate changes in DNA sequence families with speciation events. Thus, the major differences in genome structure between related species can sometimes be shown to be due to sequence amplification and rearrangement (Flavell 1982). If, for example, the mechanisms involved in molecular drive were to contribute to a failure of meiotic pairing, or of recombination, in hybrids, then they might well function as contributors to speciation events. While this remains a formal possibility, there are, at present, three data sets which appear at variance with it. First, the genomic differences in question are not restricted to cases of interspecific change, most are also known to occur within as well as between species. Secondly quite massive differences in length and DNA content do not disrupt the effectiveness of meiotic pairing in interspecific hybrids between *Lolium* and *Festuca* on the one hand, or between *Festuca drymeja* and *F. scarioxa* on the other (Rees *et. al.* 1982). Thirdly, substantial quantitative alterations in the highly repeated DNAs of experimentally constructed stocks of *Drosophila melanogaster* have no effect on fertility (John & Miklos 1979, Miklos 1982, 1985). However, if, meiotic pairing sites are subject to turnover mechanisms, then it is possible to envisage how genomic flux could contribute directly to reproductive isolation.

4.5   CHANGES IN GENOME SIZE

We have already documented the extent to which genome size varies both within and between organisms with different grades of morphological complexity. Such differences can arise in different ways. Is there any evolutionary significance to be attached to any of these mechanisms for modifying genome size?

4.5.1 Modes of change

Evolutionary divergence in eukaryotes has commonly involved changes in the size of the genome. Sometimes this represents a gain, and at other times a loss, of DNA, though it is usually impossible to determine the direction of change. Polyploidy is the most blatant and most obvious mechanism leading to an increase in genome size. It is also one of the few where we can be confident about direction.

4.5.1.1 Polyploidy

It is widely acknowledged that polyploidy involving hybridization has played an important role in plant evolution, especially in the pteridophytes and angiosperms. Consequently, in both of these groups, polyploidy contributes substantially to the excessive amounts of DNA present in certain genomes. Until recently it was generally believed that, unlike the situation in plants, polyploidy could not be sustained in sexually reproducing animals, though it was accepted that it could, and did, occur in hermaphrodites, such as flatworms and earthworms, as well as in parthenogenetic populations of both invertebrates and vertebrates. Sexual polyploids, however, are now known in bony fishes (Schultz 1980) and frogs (Bogart 1980). Of the naturally occurring polyploids that have retained sexuality, tetraploids are, without question, the most successful. Such tetraploids are automatically isolated from their diploid progenitors since, to breed true, they must produce diploid gametes and were these to combine with haploid gametes they would necessarily produce triploid progeny. Meiosis in such triploids almost always gives rise to aneuploid gametes with chromosome numbers varying between x and $2x$, or $2x$ and $3x$. Since these represent unbalanced combinations, the triploid is invariably sterile and polyploidy represents an effective mechanism of reproductive isolation.

4.5.1.2 Duplication

It is not always easy to distinguish between extensive duplication and polyploidy as mechanisms for increasing genome size. Thus, lungfish and urodele amphibians are regarded as evolutionary oddities in which the extraordinary size of the genome is assumed to have been achieved exclusively by duplication. It has also been proposed that increases in genome size, from polyploidy, tandem duplication, or both of these processes, occurred in the early phase of chordate evolution, and that changes in some of this, intitially redundant, DNA provided new genetic information required to facilitate subsequent vertebrate radiation (Ohno 1970b). The existence of isoenzymes, multiple molecular forms of enzymes with distinct physiological roles and with different temporal and cellular specifities, for example, is usually attributed to the divergence of duplicated genes. A study of the distribution of isozyme types in five multilocus enzymes, whose genetic bases are well understood, provides evidence in support of the postulate that duplications occurred, in each of these, early in chordate ancestry (Fig. 4.10).

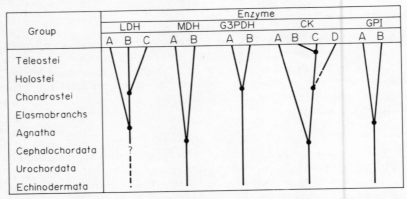

Figure 4.10 Phylogenetic distribution of isozymes for five multilocus enzyme systems in the Chordata. LDH, lactate dehydrogenase; MDH, malate dehydrogenase; G3PDH, glycerol-3-phosphate dehydrogenase; CK, creatine kinase; GPI, glucose phosphate isomerase (data of Fisher *et al.* 1980).

In other instances, duplication has resulted in the formation of families of related proteins, for example, the immunoglobulins and the globins of vertebrates. The latter provide a particularly striking example of divergence following duplication. The initial globin duplication, which initiated the whole sequence, gave rise to the ancestors of contemporary myoglobin and haemoglobin genes. The myoglobin gene has persisted unduplicated, and is present as a single gene in modern mammals (Jeffreys *et. al.* 1984). In the case of the haemoglobin gene, however, there have been a series of duplication events, followed, in some cases, by dispersal and further duplication. Initially, two related globin gene families, α and β, were defined by duplication, and this is the situation currently seen in *Xenopus tropicalis* ($2n = 20$) which has a single gene cluster containing both α and β genes. In *Xenopus laevis* ($2n = 36$), two similar, unlinked clusters have been generated by a tetraploidization of the genome (Jeffreys *et. al.* 1980). Here, the entire globin family consists of 12 members, six of them encoding α-and six encoding β-globins. Eight of these represent larval sequences and only four are expressed in adults (Fig. 4.11). Evidently, there have been additional duplication events within each cluster followed by divergence.

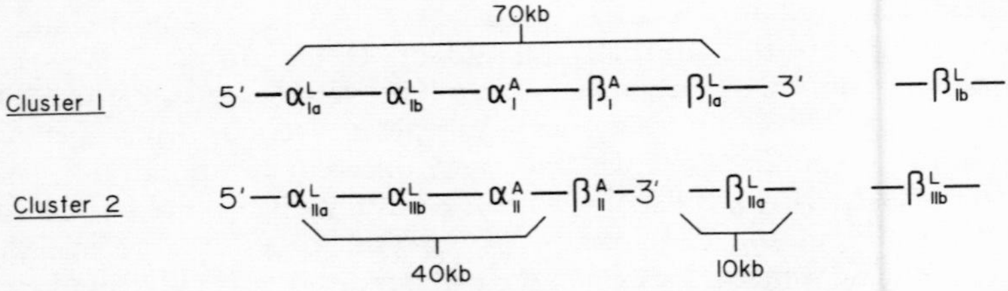

70kb

Cluster 1 $5' - \alpha_{Ia}^{L} - \alpha_{Ib}^{L} - \alpha_{I}^{A} - \beta_{I}^{A} - \beta_{Ia}^{L} - 3'$ $- \beta_{Ib}^{L} -$

Cluster 2 $5' - \alpha_{IIa}^{L} - \alpha_{IIb}^{L} - \alpha_{II}^{A} - \beta_{II}^{A} - 3'$ $- \beta_{IIa}^{L} -$ $- \beta_{IIb}^{L} -$

40kb   10kb

Figure 4.11 Organization of the two globin gene clusters of *Xenopus laevis* (data of Hosbach *et al.* 1983).

Whereas the linkage of α and β genes has been maintained in at least some amphibians, they are no longer linked in birds and mammals. Here the α and β genes have become dispersed into separate chromosomes, so that they are now organized into two unlinked gene clusters. These, by further duplications, have come to contain a variety of regions which specify a range of specialized non-adult and adult haemoglobins whose expression is both developmentally and temporally regulated. Humans, for example, have two unlinked clusters, in both of which the genes are arranged in the order of their expression during development. The α–cluster, located on chromosome 16, includes a ζ gene, expressed during embryogenesis, and two α genes, active in both the foetus and the adult. The β cluster, on chromosome 11, includes a ε gene that functions during embryogenesis, two very similar foetal genes, $^G\gamma$ and $^A\gamma$, and two adult genes, δ and β (Fig. 4.12). Notice that a common feature of these clusters is the occurrence of two immediately adjacent loci that are co-ordinately expressed at a given developmental stage. Pseudogenes, designated by the prefix ψ, are also known in both clusters. The single myoglobin gene of man is not linked with these two haemoglobin clusters and is present on chromosome 22.

An even more complex system is present in the β–globin locus of the goat (Townes *et. al.* 1984). This locus, which extends over 120 kb, includes 12 genes arranged into three homologus 4–gene sets that have evolved by the triplication of an ancestral set of the type ε–ε–ψβ–β (Fig. 4.13). The three β genes located at the ends of each of the 4–gene sets are expressed at different stages of development.

Whereas humans switch directly to adult haemoglobin ($\alpha_2\beta_2$) at birth, goats undergo an additional switch to a juvenile or pre-adult haemoglobin composed of two α and two β^C polypeptides ($\alpha_2\beta^C_2$). At approximately three months of age, this is replaced by adult haemoglobin ($\alpha_2\beta^A_2$). Whether the duplication event has moved sequences into chromosomal domains that are specifically opened up during pre-adult life, or whether mutations in

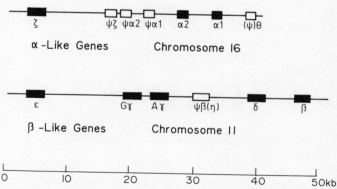

Figure 4.12 Organization of the human α–like and β–like globin gene clusters. Pseudogenes in each cluster are depicted by open boxes (unpublished data of Maniatis).

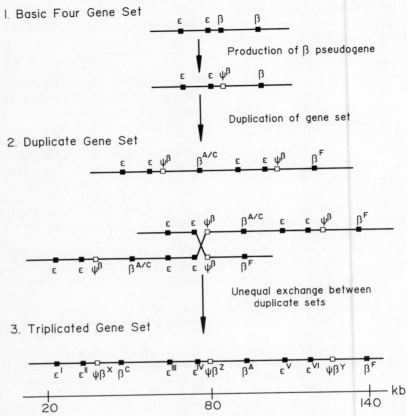

Figure 4.13 Probable mode of origin of the triplicated β-globin gene set of the goat (data of Townes *et al.* 1984).

regulatory sequences have altered the developmental specificity of the gene subsequent to duplication, is not known.

The globin gene system thus provides some clear examples of small multigene families which appear to have arisen as unavoidable products of genomic flux. It illustrates the fact that duplicate genes, however produced, can follow at least three courses. First, they may be used essentially as duplicates, as are the adult α-globin genes. Secondly, they may evolve into related functional genes, as in the case of the embryonic and foetal globin genes. Thirdly, they may be rendered non-functional, as in the case of the globin pseudogenes.

A fourth, and far less common, consequence of gene duplication is the evolution of novel systems of development. In insects, for example, gene duplication has made possible the differential development of serially homologous parts (Raff & Kaufman 1983). Since body segmentation has also been involved in vertebrate evolution, it may be that equivalent events have occurred here, too. The presence in vertebrates of homeo boxes, homologous to those known to occur in the presumed products of duplication in *Drosophila*, provide suggestive evidence on this point.

4.5.1.3 Amplification

Amplification offers an alternative mechanism for increasing genome size. The differences between diploid species generally involve alterations in the amount of both repetitive and non-repetitive DNA, though the pattern differs in different groups. In amphibians, generally, the increase in genome size occurs predominantly in the repeated fraction. In birds, on the other hand, where the bulk of the genome is known to consist of single copy DNA, increases stem largely from this fraction. Turning to individual genera, which offer the most informative comparisons, the situation in the genus *Plethodon* indicates that here changes in genome size occur largely in the middle repeated fraction.

The Plethodontidae is a large family of lungless salamanders in which all but the most so-called 'primitive' genera lay their eggs in moist ground, where they hatch as miniature adults. The genus regarded as the least specialized on morphological grounds, *Pseudotriton* ($2n = 28$), has a genome about the size of *Plethodon cinereus* ($2C = 40$ pg), while the genera believed by taxonomists to be the most advanced – *Oedpina*, *Thorius* and *Chiropterotriton*, all of which have $2n = 26$ and live in central America – have the largest genomes. On this basis it has been argued that, while chromosome number and form have remained little changed within the family, C-values have increased (Mizuno & Macgregor 1974). However, arguments of this sort, which assume that a comparison of present day genome sizes accurately reflects past genome sizes, need to be treated with caution.

There may, as we have seen earlier (§ 4.3.1), sometimes be an unquestioned relationship between increases in genome size and cell size, though to what extent this relationship is due strictly to nucleotype, as opposed to other genetic factors, is unclear. Given that cell size and genome size are related, the only data which meaningfully attempt to compare changes in cell size, and hence indirectly DNA content, over evolutionary time are those obtained from measurements made from the bone tissue of fossil fishes and amphibians (Thomas 1972). Using this approach, cell sizes have been measured in lungfish lineages for which there is reasonably continuous fossil material dating from the inception of the group to recent times. These data have been analyzed in the context of the periods during which the group was known to be undergoing major diversification. One important conclusion from this study is that increase in cell size occurred after the major Devonian diversification of the group (Fig. 4.14). Thus, large cell size is not a primary feature of early lungfish evolution. Additionally, while the general increase in cell size persists until the modern families become recognizable, the rate of increase differs in different lineages and is evidently also independent of major morphological change. If cell size in this group is related reasonably directly to DNA content, then it is clear that the enormous DNA multiplication was largely irrelevant to

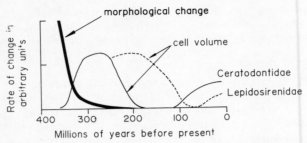

Figure 4.14 Relationship between morphological change and cell volume in the evolutionary history of two families of lungfish (data of Thomson 1972).

the major morphological phase of dipnoan evolution. A more general study of cell size in fossil bone tissue of crossopterygian, lobe-finned, fishes and their presumed descendants, the extinct and living amphibians, suggests that significant increases in cell size may have arisen several times independently within both assemblages. Moreover, as in dipnoans, there is no evidence that increase in cell size preceded evolutionary diversification (Thomson & Muraszko 1978).

As far as the 26 North American species of *Plethodon* are concerned, they have retained the same basic body form and the same diploid number but are distinguishable in terms of their adult body size, their geographical distribution and certain dissimilarities in their proteins. There are also differences in genome size which are particularly pronounced between eastern and western representatives and, correlated with this, species with larger genomes have larger chromosomes. By determining the relative lengths and centromere positions of what, on the basis of length, are assumed to be homologous chromosomes in different species, it has been argued that, despite differences in genome size, the karyotypes of different species are remarkably similar. On the assumption that here, as elsewhere in the family, genome size has increased during evolution (an assumption which has, however, been challenged by Larson 1984), it has been claimed that all the chromosomes of the genome have expanded in a relatively balanced manner (Mizuno & Macgregor 1974).

Let us now examine this point in some detail. The largest part of the genome in all species of *Plethodon* consists of moderately repeated DNA. Closely related species with genomes of similar size have as much as 90% of their moderately repeated sequences in common, at standard stringencies, whereas distantly related species with widely different genome sizes have less than 10% of their moderately repeated sequences in common. Added to this, the large genomes have a larger middle repeat content. Thus, in *Plethodon cinereus*, with a C-value of 20 pg, some 60% of the genome has been claimed to consist of moderately repeated DNA, whereas in *Plethodon vehiculum*, with a C-value of 37 pg, the proportion of repetitive material within the genome is claimed to be 80%. This has been interpreted to mean that expansion of the genome has involved the accumulation, the diversification and the widespread scattering of moderately repeated sequences, leaving the relative dimensions of the chromosome little changed (Macgregor *et. al.* 1976). It should be pointed out, however, that in genomes of this size it can be difficult to estimate proportions of repetitive DNA sequences from reassociation data. Thus, the values of 60% and 80% may, or may not, be significantly different. In addition, when the conditions of stringency are varied, as has been carefully done for four species of vascular plants, the relative values of the repetitive DNA contents between species can change (Bendich & Anderson 1977). It should be noted that the proportion of DNA classified as repetitive by reassociation techniques is based on operational criteria (usually $T_m = 25°C$). How the cell discriminates between repetitive and non-repetitive sequences is another matter altogether. As yet no comparison exists of the sequence components of the various classes of DNA present in the genomes of different species. In view of the finding that much of the moderately repeated sequences in insects and vertebrates consists of nomadic elements, the extent to which genome expansion has mirrored changes in the mobile element population needs to be determined.

The data relating to purely quantitative alterations in genome composition have been taken to indicate that changes in genome size between the species of *Plethodon* can be accounted for by multiple amplifications, followed by dispersion of the products of amplification throughout the length of all the chromosomes. If the genome of plethodonts has indeed grown in such a balanced manner, then one might predict that the positions of individual genes, no less than those of individual centromeres and telomeres, should have persisted unchanged. The only convenient marker applied to this system to date has been the ribosomal RNA probe. When the positions of these genes are compared in seven species of plethodontids, with genome sizes ranging from 20 to 70 pg, it is clear that they do not follow this simple expectation (Table 4.17). They vary in both number and location, and there is no evident relationship between either of these variables and genome size. One explanation for this variation is that the genus has a basic set of potential NOR sites, and that only certain of these are expressed in any one species. However, even when a NOR is located on what is assumed to be the same chromosome in two different species, it is not necessarily found on the same chromosome arm or at the same site on a given arm (Fig. 4.15). Thus, an equally likely alternative is that individual chromosomes expand and contract by large uneven lengths, so that, when the whole karyotype is examined, the newly amplified chromosomes simply take up another position in what appears to be an essentially unaltered series. An evaluation of the manner in which genome expansion or contraction has actually occurred will be facilitated by the use of unique cloned probes that will provide unambiguous labels for each and every chromosome segment.

Regardless of the precise mechanism by which the genome has been expanded, or contracted, it is clear that these changes have had effectively no developmental or morphological significance. Indeed, from what we now know about the properties of the middle repeated component of the genome, this comes as no surprise.

A somewhat similar situation has been reported in the genus *Lathyrus*.

Table 4.17 Location of the main clusters of ribosomal genes in seven species of *Plethodon* (from Macgregor & Sherwood 1979).

Species	1C (pg)	1	2	4	5	6	8	9	14	Total number of sites
					Chromosome number					
P. cinereus	20					SA^I			$\pm$	1–2
P. glutinosus	23		T		LA^T		SA^C			3
P. elongatus	34	$SA^I + LA^{ST}$	T+ST		SA^C	SA^C				4
P. vehiculum	37		SA^{ST}	SA^T				SA^{ST}		3
P. dunni	39					SA^M				1
P. larselli	48								T	1
P. vandykei	69	LA^I			SA^M		LA^T			3

SA, LA = short and long chromosome arms; T = terminal; M = median; I = interstitial; C = centric; ST = subterminal.

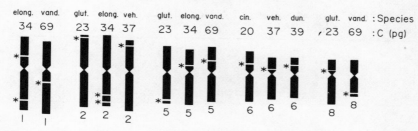

Figure 4.15 Distribution of secondary constrictions in presumed homologues belonging to different species of *Plethodon* (see Table 4.17). Chromosome morphology obtained from relative lengths and centromere positions but not drawn to scale. In terms of absolute size, genomes with larger C-values have larger chromosomes (data of MacGregor & Sherwood 1979).

Species belonging to this group of angiosperm plants are all diploids with $2n = 14$, but show a fourfold range of genome size, with a minimum 2C of 7 pg in *Lathyrus miniatus* and a maximum of 30 pg in *Lathyrus vistitis* (Narayan 1982). The more specialized inbreeding annual species have, on average, significantly smaller genomes than the outbreeding perennials. Since inbreeders must be derived from outbreeders, it has been assumed that there has been a reduction in genome size during the evolution of this genus. But this, of course, also assumes that the progenitor outbreeders had the same genome characteristics as the present-day forms.

Differences in genome size between species of *Lathyrus* (Table 4.18) correlate well with differences in chromosome size, and indicate that the changes that have taken place have involved all the chromosomes with the complement, though this is not to say that all chromosomes are necessarily affected to the same degree. Changes in genome size have involved substantial alterations in the amount of both the single copy and the moderately repeated components, but with only minor changes in the amount of the fast fraction. At the cytological level these changes are paralleled by alterations in both the euchromatic and the heterochromatic components of the genome, though again not equally so (Fig. 4.16). While in *Plethodon*, genome variation does not involve substantial change in the amount of heterochromatin, in *Lathyrus* it does. There are, thus, different ways of changing the size of the genome in terms of its DNA composition (Hutchinson *et. al.* 1980), and the values of the different DNA fractions may fluctuate considerably in all classes, even within a given group. The genomes of amniotes show least interspecific variability. This is particularly true in birds, where there are unusually small repetitive DNA components and where the variation that does occur involves mainly single copy DNA. In angiosperm plants, on the other hand, the bulk of the genome often consists of repetitive DNA.

There seems little doubt that the various sequence categories present within a genome are interchangeable to some extent. If we rely on the evidence obtained by simply comparing related modern species, then there are certainly grounds for concluding that the frequency of some repetitive

Table 4.18 Discontinuous nature of DNA distribution in 21 species of the plant *Lathyrus* (from Narayan 1982).

| Species | 2C-DNA (pg) | | |
	species mean	group mean	Group difference
Group1			
L. miniatus	6.86	6.86	
			3.90
Group 2			
L. angulatus	10.76	10.76	
Group 3			
L. articulatus	12.15		
L. nissolia	12.92		
L. maritimus	13.15		2.98
L. clymenum	13.43		
L. ochrus	13.63		
L. setifolius	14.01	13.74	
L. aphaca	14.04		
L. cicera	14.04		
L. sphaericus	14.18		3.14
L. pratensis	14.72		
L. annus	14.93		
Group 4			
L. sativus	16.78		
		16.88	
L. odoratus	16.96		
			3.63
Group 5			
L. tuberosus	19.52		
L. hirsutus	19.93	20.51	
L. tingitanus	22.08		
			4.22
Group 6			
L. sylvestris	24.65	24.73	
L. latifolius	24.78		
			4.49
Group 7			
L. vistitis	29.22	29.22	

sequences, and the structure of others, has changed since the presumed divergence of the species in question. However, this evidence is not particularly rigorous or decisive.

4.5.2 Genome size, specialization and speciation

It has been postulated that the DNA content of a genome correlates both with specialization and speciation (Hinegardner 1976). In support of the first of these claims, it has been held that unspecialized, and presumed primitive, forms have large genomes, whereas specialized forms have small genomes,

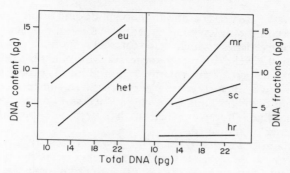

Figure 4.16 A comparison of genome composition in species of *Lathyrus* in terms of euchromatin (eu) and heterochromatin (het) content (left), and in terms of the content of middle repeated (mr), single copy (sc) and highly repeated (hr) DNA (right) (data of Rees & Narayan 1977).

and that this relationship holds among vertebrates in fishes, amphibians and mammals. In reality, however, most organisms turn out to be mosaic, consisting of both specialized and unspecialized features, and much depends on which of these features are chosen for emphasis. For example, dipnoan fishes and paedogenetic urodeles both have very large genomes. Are these to be regarded as unspecialized? The capacity to tolerate physiological uraemia, shown by the Dipnoi, and the persistence of larval characters throughout life, shown by the paedogenetic urodeles, are clearly specialized features. In plethodontid salamanders, the most specialized forms are assumed to have the highest, not the lowest, DNA content. In mammals, the aardvark, which is specialized for a subterranean mode of life, has the largest genome known in that group, with a value 1.7 times that of humans (Benirschke *et. al.* 1970). Likewise in molluscs, the highly specialized cephalopods (squids and octopuses) have much larger genomes than the bivalves and snails (Hinegardner 1973). Not only can the claim that specialized forms have small genomes be challenged in terms of its validity, but it is open to serious criticism in terms of its rationality. The reason offered to explain why specialized forms should be expected to have smaller genomes is that specialization leads to a loss of functions and hence also to the loss of the structures required for those functions. This, in turn, it is held, leads to a loss of DNA (Hinegardner & Rosen 1972).

In mammals, with only a relatively small variation in genome size, development results in such diverse morphological end products as bats, whales, armadillos and monkeys. This makes a mockery of any generalization which argues that specialization leads to the loss of DNA. It is true that specialization sometimes involves regressive change. For example, the eyes of *Platanista indi*, the Indian river dolphin, and *Platanista gangetica*, the gangetic dolphin, begin embryological development in a conventional manner, as do the optic nerves. Later, however, the optic nerves atrophy and in the adults the eyes are rudimentary and have no lenses (Pilleri & Gihr 1971). This means that the gene pathways required for the production of both eye and nerve are present, and functional, despite the fact that the adult is blind. An even more interesting situation is the reappearance in the lynx of dental characteristics which had formerly been assumed to have disappeared in the Felidae during the early Miocene phase of evolution (Kurtén 1963). The genetic information required for the production of these characteristics must, therefore have persisted in an unexpressed form for literally millions of years. Thus, gene circuits necessary for the development of a character which, for some reason, has become superseded may be suppressed rather than deleted.

A second claim holds that the average DNA content of a class or order appears to be inversely correlated with the number of species in that group. In support of this claim it is pointed out that the two most speciose groups of vertebrates, the teleost fishes and birds, both have small genomes. While one cannot dispute the fact that birds are indeed speciose and that birds also have small genomes, the postulate, as it stands, offers no sensible explanation for why these two features should be causally related. Without such an explanation the generalization has little meaning. Of greater interest is the fact that speciation in birds appears to have involved behavioural differences in courtship pattern, and, if interspecific matings are brought about in captivity, the hybrids are viable and often fertile (Gray 1958). If one analyzes taxonomic groupings narrower than classes and orders, there is again no correlation between genome size and species status. Amongst modern mammals, with little significant variation in genome size, rodents are, without question, the most speciose group. There is no reason to believe that their overall genetic composition is in any significant way different from that of primates. Yet rodents are speciose, primates are not.

4.5.3 Supernumerary chromatin

Despite the infrequent occurrence of conventional aneuploids (see § 3.4.5), representative species in a wide variety of eukaryote groups carry extra, or supernumerary, chromosomes, often referred to as B-chromosomes. These constitute a somewhat heterogeneous class which are clearly distinguished from the standard, or A-chromosomes, in both morphology and behaviour, and lack pairing homology with them. Indeed, the absence of meiotic pairing between A- and B-chromosomes is a prerequisite for the survival and independent evolution of the supernumerary. Where B-chromosomes are present within a species their number and frequency varies not only from population to population but also from individual to individual. Though they are often small in size, they, in fact, range from the smallest to the largest member of the complement. They are also often, though not always, either partly or wholly heterochromatic in character. The molecular composition of B-chromosomes has been examined in a number of different plants and animals. In none of them has a B-specific DNA component been identified. Rather, they contain representatives of some, or most, of the nucleotide families present in the A-chromosomes of the species to which they belong (Chilton & McCarthy 1973, Dover 1975, Rimpau & Flavell 1975, Dover & Henderson 1976, Klein & Eckhart 1976, Amos & Dover 1981).

Table 4.19 DNA content of some plant B-chromosomes (from Jones & Rees 1982).

Species	Diploid genome		Diploid genome + Bs		
	2n	1C (pg)	Number of Bs	total DNA (pg)	DNA per B (pg)
Lolium perenne	14	5.0	3	6.0	0.3
Zea mays	20	11.0	34	28.0	0.5
Secale cereale	14	18.9	8	29.3	1.3
Picea glauca	24	38.3	6	44.4	1.0

In a majority of eukaryotes, individuals with or without Bs cannot be distinguished from one another on morphological grounds, though the Bs may add appreciably to the DNA content of the genome (Table 4.19). This implies that Bs are specialized to the extent that they either do not have, or else do not express, structural genes. They do, however, sometimes add to the continuous variation expressed in the phenotype.

Supernumerary chromosomes display an additional property which has undoubtedly played an important role in their establishment within a species, namely, most of them are characterized by one or more self accumulation mechanisms operative within the germ line. These mechanisms lead to B-chromosomes being transmitted to progeny in higher frequencies than expected from the known constitution of the parents and, unchecked, this is expected to lead to a progressive accumulation of supernumeraries within a species. The fact that they do not accumulate *ad nauseam*, however, implies the existence of some counteracting principle. While low numbers of Bs appear to be largely neutral, in higher numbers they almost invariably produce adverse effects on viability and fertility. For example, rye plants with four or more B-chromosomes are virtually sterile. In maize, where the Bs are smaller, at least ten have to be present to produce a noticeable effect on fertility. Animals, on the other hand, are far less tolerant of B-chromosomes and only in cases where they are exceptionally small are more than two or three supernumeraries found in an individual.

The existence of accumulation mechanisms has led to the hypothesis that B-chromosomes have no useful effects of any kind, and are maintained within populations by the efficiency of their accumulation mechanisms and so are leading a virtually parasitic existence within the genome (Östergren 1945). The fact that B-chromosomes reduce either viability or fertility, or both, when present in higher numbers, can be expected to lead to their loss from a population. Thus, the tendency for numerical increase by accumulation will be balanced by such a loss and this maintains Bs within populations at equilibrium frequencies (Nur 1977). If this interpretation is indeed correct, it would appear that supernumerary chromosomes are simply curiosities which the genome is stuck with, and whose effects it can tolerate, at least when they are present in small numbers. In this sense B-chromosomes are no different from other genomic components which have inbuilt mechanisms for expansion and accumulation. They demonstrate, yet again, how tolerant eukaryotic genomes are to excesses of DNA in whatever form.

This applies equally to a second category of supernumerary chromatin to be found in natural populations, and which again varies from individual to individual within a species creating a novel form of chromosome polymorphism, also with no morphological effect. Here additional, usually heterochromatic, segments are present on standard members of the chromosome complement. Such supernumerary segments may exist in either a heterozygous or a homozygous state and, while they can occur in any position within a given chromosome, they are most commonly found either proximal to the centromere or else terminally at the distal extremity of a chromosome arm. Like B-chromosomes, these supernumerary segments lead to variation in the DNA content of the genome between different individuals of the same species which, at the extreme (as in the grasshopper *Atractomorpha similis*), may be as high as 30% (John & King 1983).

4.6   SUMMARY STATEMENT

Our examination of eukaryote genomes has revealed that every genomic component can be subject to change, at literally every level and in almost every conceivable combination. This includes all DNA sequence classes, whether repetitive or unique, mobile or stationary. Indeed, the middle repetitive sequences are evidently in a permanent state of flux, both with regard to their number and their location within genomes. Changes in DNA sequence structure and sequence organization may, or may not, be accompanied by alterations in genome size. Additionally, such changes may occur in the face of a strict conservation of chromosome number, or else there may be wholesale structural and numerical reorganization of the chromosome system. Between these extremes an enormous variety of combinations of genomic and karyotypic changes are possible, but there is no simple, consistent or meaningful relationship between these genomic changes and the principal evolutionary events that have occurred within eukaryotes.

Given time and its own enzymology, a genome can reorganize itself seemingly very subtly, or as in the case of the muntjacs, very dramatically. However, there is, in general, little phenotypic movement as a consequence of either minute or massive restructuring. Thus, no clear correlation exists between the size of the genomic perturbation and the size of the phenotypic perturbation. A single base pair change in an executive gene may well have cataclysmic or totally novel consequences, whereas the addition, deletion or reshuffling of tens of millions of base pairs of satellite DNAs may have no phenotypic consequences of note.

All of the conventional approaches have, so far, been oblique to the central problem of the origin of morphological novelty. There is patently no C-value paradox, and neither nucleotypic nor karyotypic alterations impinge meaningfully on the genesis of morphological changes. Thomson's lungfish example beautifully demonstrates that morphological evolution and gross genomic evolution can be uncoupled. The seductive appeal of characterizing genomes in finer and finer molecular detail with a view to bringing the broad landscape of evolutionary phenomena into sharper focus is a misplaced hope. **It should be remembered that the prime aim is to discover the *origins* of *principal* evolutionary events**. Thus, the initial choice of the genomic compartment to be explored is crucial. In the light of the data of the previous chapters, it is obviously now imprudent to sequence to the very last base those genomic landscapes that breeze in and breeze out of populations. Data so obtained are simply not relevant to issues of how morphological and neuronal novelties arose during the course of evolution. The broader question is how much of any genome is likely to fall into this category? Molecules will neither come to Darwin's aid, nor ours, if the wrong category is chosen for analysis.

It was clear from the data on the two sea urchins, *Strongylocentrotus purpuratus* and *Strongylocentrotus franciscanus*, that certain repetitive sequences fluctuated wildly in frequency between these two genomes. But has this told us anything about the kind of morphological changes that initially gave rise to sea urchins as a developmentally engineered biological entity, different, for example, in their design to arthropods? The answer is no.

The genome sizes of plethodontid salamanders vary enormously in size, yet the organisms themselves are remarkably uniform in terms of biological structure. One could certainly make a large molecular evolutionary investment in such a group and derive phylogenetic trees, calculate evolutionary distances and measure the rates of evolution of various biological components. However, would such studies rapidly bring us any closer to the mechanisms that gave rise to this morphological type, and which distinguish it from its near relatives the frogs? The answer is again a definite no.

What then were the underlying processes which governed the emergence of the major groups of organisms on this Earth? It is to this critical issue that we now turn in the final two chapters of this book.

5

The unsolved problem – the origin of morphological novelty

'What men want is not knowledge but certainty.'

Bertrand Russell

The neo-Darwinian approach to the analysis of evolution is based on a mathematics which reduces the entire process to the study of changes in gene frequencies within populations, assuming that such changes can, by extrapolation, eventually explain all evolution. Here, the organism has, in effect, been discarded and only lip service is paid to the developmental interactions necessary for producing the required changes in phenotype. In the case of microevolutionary changes, involving essentially quantitative alterations within populations through time, this is of little serious consequence. But in respect of macroevolutionary change, involving complex shifts in phenotype of the kind required for the formation of the major eukaryotic taxa, it fails to address the real issue. Since all biological structure is the result of developmental processes, it follows that all phenotypic structural change must originate through changes in development. This, in turn, implies that the only sensible speculations about the origin of the various eukaryotic phyla, classes and orders must be based on developmental considerations. Thus, phylogeny unfolds historically as a sequence of modified ontogenies involving descent with modification.

What our consideration of the conventional approaches to analyzing evolution has shown is that all of them have been peripheral to the central issue of how morphological novelty is generated. Neither the selectionist–neutralist debate nor the punctualist–gradualist controversy concerns itself with the *mechanisms* responsible for changes in morphology. The cases we have considered make it self-evident that most chromosome rearrangements cannot be held responsible for morphogenetic change and, while speciation might sometimes set up the conditions of isolation which lead to subsequent morphological divergence, there is no reason for believing that speciation is itself directly concerned with producing modified morphology. As far as changes in genome size are concerned, there is certainly evidence that gene duplication has played a role in producing developmental variants, such as the various members of the haemoglobin gene cluster but, with the exception of the putative duplication products involved in the gene complexes of *Drosophila*, their morphological effects are of only a minor nature.

The situation is no better when we turn to consider the changes that are revealed by the cataloguing of structural gene changes at the genomic level. Initial studies of evolution at the recombinant DNA level concentrated on structural genes because they were the most accessible – simply because they are high RNA producers. There is, however, no *a priori* reason why abundant RNA producing genes should be those which are important for the developmental circuits involved in morphogenesis. Recall that gene *B104* of *Drosophila* is known to be developmentally specific in its action, in that it is highly expressed during blastoderm formation. It was one of the sequences uncovered in a hunt for blastoderm-specific expression, as a prelude to isolating genes which acted very early in development. It is also an abundant RNA producer. *B104* is, however a mobile element with no known morphological connection of any kind. Presumably, it has become locked into a developmental circuit which is of little consequence to itself or to the organism.

Thus, without an independent criterion for defining key morphological genes, a comparison of high RNA producers is unlikely to do more than provide an overview of the sorts of structural gene changes which the genome may be able to tolerate without any overt morphological effects.

For example, by comparing some of the common proteins of humans and chimpanzees, using a combination of immunological and protein sequencing techniques, it has been found that the two species have identical fibrinopeptides A and B, cytochrome c, haemoglobin chains of the α, β and γ categories and lysozymes. In the case of myoglobin and the δ chain of haemoglobin there is a single amino acid difference, while the sequences of albumins, transferrins and carbonic anhydrases differ only slightly. Electrophoretic comparisons of 44 different structural genes, including those listed above, likewise indicate that the average human protein is more than 99% identical in amino acid sequence to its chimpanzee homologue (King & Wilson 1975). Despite this similarity in protein composition the two species differ in a number of anatomical features, including locomotory (bipedal gait, upright stance), masticatory (teeth and jaws) and cranial (skull and brain) characters. Clearly, none of these differences can be explained in terms of the structural gene products that have been used so far in molecular comparisons between the two species, so the fact that they are so similar in respect of these products is irrelevant to the issue. What these molecular comparisons do emphasize is that changes in these structural genes are unlikely to have anything to do with the production of morphological change. Equally, the existence of living fossils provides evidence that morphology can be maintained unchanged, presumably in the face of considerable molecular change within the genome.

Because G-banding studies indicate that the chromosomes of chimps and humans differ by at least one fusion and ten inversions and translocations, it has been suggested that such structural rearrangements provide a basis for regulatory changes in gene expression, and hence phenotype (King & Wilson 1975). Genome reorganization, consequent upon chromosome rearrangement, is then assumed to establish new developmental programmes which, in turn, may provide a basis for reproductive isolation, and hence speciation.

This argument suffers from serious defects of two kinds. First, simply by comparing 44 common structural proteins and finding no interspecific differences of substance does not require an explanation necessitating regulatory gene involvement to explain the genuine morphological differences that do exist between two species. These differences could just as easily be brought about by other structural genes of developmental significance. If, as appears likely, such genes are few in number, then, in a very biased sample of 44 loci from the entire coding portion of the genome, the probability of isolating a key morphological gene is remote. Secondly, as emphasized by Charlesworth et al. (1982) there is no evidence in eukaryotes for a sharp distinction between structural and regulatory genes of the kind invoked by King & Wilson. Indeed, as pointed out by MacIntyre (1982), no single compelling observation was ever responsible for spotlighting the putatively significant evolutionary role of regulatory genes. Thus, interpretations that invoke gene regulation are purely speculative. Additionally, as Raff & Kaufman (1983) point out, while the distinction between structural and regulatory genes is perfectly clear in prokaryotic systems, apart from the GAL 4 protein of yeast, we know of few gene products in any eukaryote whose function is solely regulatory.

When we have used the term 'regulation' elsewhere in this book this in no sense implies the kind of regulatory control exemplified by prokaryotic systems. Moreover, in the gene network which characterizes bithorax, which itself is a seminal gene controller, it is self-evident that while a number of genes influence the expression of the bithorax locus they, in turn, come under the control of other genes. Where, then, in this sequence of gene interactions do we make the distinction between structural and regulatory loci?

A key question, therefore, is whether the large-scale structure of the genome is in any way significant for morphological development, in which case evolutionary changes involving shifts of morphology might depend on genome reorganization (Gillespie et al. 1982). We have already seen how, in the case of muntjacs, a mammalian genome can be significantly remodelled in terms of genome reorganization and yet yield negligible morphological change. We have also seen how mitochondrial and chloroplast genomes can be very extensively rearranged in closely related species without impairing organellar function or perturbing plant phenotype. However, in spite of such data let us consider chromosome rearrangements in the context of position effect variegation. When cells of Drosophila melanogaster carry a chromosome rearrangement which juxtaposes a euchromatic gene near to a heterochromatic breakpoint, there may be no developmental expression of that gene in a fraction of the cells in which it is normally expressed. Consequently, the organism is mosaic for the activity of the rearranged gene. This position effect variegation has, in the past, led many to assume that large–scale genomic structure is, indeed, important for morphological development. Almost all rearrangements which juxtapose euchromatic loci to a breakpoint in the heterochromatic centric blocks of the autosomes of *Drosophila melanogaster* induce variegation of these loci. This *cis*–inactivation of DNA sequences neighbouring the heterochromatic breakpoint may extend to include other more distant genes, though variegation diminishes with distance from the breakpoint. The heterochromatin of the sex chromosomes, which constitutes by far the major fraction of the total heterochromatin of the genome, is far less frequently involved in variegation (Spofford 1976). Thus, more than half of the rearrangements which juxtapose euchromatin to X or Y heterochromatic breakpoints do not disturb the functioning of the transposed gene loci. Thus, the *Sgs-4* locus, normally located in the X–euchromatin, retains normal expression when inserted into the heterochromatic Y–chromosome (McGinnis *et al.* 1980). Similarly, Lewis (1950) reports that a complete reversal of variegation was obtained when a variegated gene located in the hetero–chromatin of IV was transferred to the distal heterochromatin of the Y. Moreover, the variegation induced by Y breakpoints is often closely localized with little or no spreading effect, and additional Y–chromosomes suppress variegation in other rearrangements, while fewer than the normal number enhances such variegation. Henikoff (1981) showed that when the heat shock sequences at 87C1 are placed next to Y–heterochromatin in a T(Y:3) translocation system, they are fully, or nearly fully, polytenized in salivary gland cells. Thus, the lack of gene expression in this particular situation is not due to a decrease in template number, i.e. to under–replication, but rather to a transcriptional block. Additionally, we wish to point out that where position effect variegation involves non–polytene cells, as it must do in some tissues, under–replication cannot be the cause of the failure of gene expression.

White locus rearrangements associated with variegation have been found to undergo reversion when reinverted by X–ray mutagenesis, despite the fact that at least some of the heterochromatin remains adjoined to the white locus in these revertants (Tartof *et al.* 1984). This implies that, in this case, variegation is initiated at, or near, a site distant to the actual heterochromatic breakpoint, yet still within the heterochromatin. These molecular data of Tartof *et al.* (1984) are particularly relevant, since the three variegating rearrangements allow such clear distinctions to be made between hypotheses. Two of the rearrangements, w^{m4} and w^{mMc}, are next to very similar heterochromatic DNA sequences, yet variegate differently. On the other hand, w^{mMc} and w^{m51b} are adjacent to different DNA sequences, but variegate similarly. Thus, it is likely that the type of flanking heterochromatin is not all that important for position effect variegation. Furthermore, an analysis of revertants of these rearrangements reveals that some hetero–chromatic sequences still persist at the breakpoint, indicating that variegation is not initiated at the heterochromatic junction *per se*, but is propagated from a site some distance from the breakpoint. On these bases, Tartof *et al.* (1984) have argued that the data are consistent with a model in which there are two types of sequences in heterochromatin, termed initiator sequences (i) and terminator sequences (t). Heterochromatinization is initiated at i sequences and terminated at t sequences. If this model is correct, then position effect variegation can be rather easily conceptualized. There can be a number of i to t domains in a block of heterochromatin, and the domain density may vary from chromosome to chromosome. If a rearrangement occurs between i and t sequences, then variegation results. A revertant simply has to remove or inactivate i, some distance from the breakpoint itself, and variegation will be eliminated.

It has also been shown that the extent of variegation of a gene can be a function of histone gene multiplicity. Thus, heterozygous deficiencies which delete one entire histone gene cluster enhance the active expression of variegating genes (Moore *et al.* 1979). Control deficiencies which delete regions adjoining the histone locus, on the other hand, leave the extent of variegation unaffected. A reduction in the number of histone genes does not, of course, necessarily imply a decrease in the amount of cellular histone protein, since histone genes are repeated about 100 times within a single cluster. Indeed, individuals whose genomes contain either single or triple histone clusters are phenotypically normal and viable. Even so, the effect of the loss of a histone cluster on the transcriptional fate of a variegating gene is assumed to depend on a reduction in the coding capacity for histone protein which, in turn, affects the condensation of the variegated locus into a transcriptionally inactive state.

A considerable number of chromosome rearrangements involving whole euchromatic regions have been found to be associated with stable, as opposed to variegated, changes in gene action. In none of these cases, however, is there any evidence that the change is due to a position effect rather than to intragenic mutation (Lewis 1964). Thus, these cases do not provide a sensible basis for arguing that gross genome organization *per se* is developmentally important. The added facts that gene position seems to matter little for gene function provided there is an adequate immediate landscape, and that gene regulation depends predominantly on elements located close to individual transcription units, negate the claim that most chromosome rearrangements can be expected to lead to major developmental changes.

Arguments which involve regulatory changes to explain modifications in morphogenesis owe their popular origin to the original Britten & Davidson model (1969) and its subsequent elaboration (Britten & Davidson 1971, Davidson & Britten 1979). These models are predicated on the unavoidable assumption that there are genomic regulatory networks which functionally relate physically distant genes, and that genes required for different circuits need to have near them elements which allow them to be transcribed at the correct times, correct sites and correct levels within different circuits. Such regulatory elements have, however, proven a great deal more elusive than structural genes. Apart from generalized polymerase recognition sites and enhancer sequences, which can elevate gene expression, we know of few such regulators in eukaryotes, though we accept that they must exist. One clear example occurs in front of the HIS4 gene in yeast. Accepting the validity of interacting gene circuits that are modified during evolution, it is now clear that they will not be defined by continuing to search for common upstream repetitive elements but rather by genetic approaches of the kind pioneered by Nüsslein-Volhard & Wieschaus (1980).

The probability of any chromosome rearrangement altering a major developmental circuit must depend on the number of genes whose expression can be grossly altered to produce a phenotypic effect of consequence. Genes that are haplo-insufficient would be severely affected. These, however, are few in number within the genome as, too, are the developmentally significant switch genes. Added to this, in *Drosophila melanogaster*, at least 50% of all induced breakpoints involved in gross chromosome rearrangements have no overt effects of any kind (Lefevre 1974). Thus, with not more than half the breakpoints having any possible effect and with so few genes available, the gross rearrangement argument really is somewhat contrived. While gene regulation is clearly of the utmost importance for the development of eukaryotes, the appeals so far made to altered gene regulation as a basis for morphological novelty have been unhelpful and largely semantic.

It is instructive to compare the situation in prokaryotes and eukaryotes. In the prokaryote λ phage, development begins with a common early phase involving both transcription and replication. It then switches to either a productive or a lysogenic pathway. Productive development largely depends on the activity of two regulatory proteins, *Cro* and *Q*. *Cro* functions as the major switch protein, since it represses the transcription of the early-acting genes. *Q*, on the other hand, turns on the transcription of the genes involved in the synthesis of head, tail and lysis proteins. In lysogenic development, the transcription characteristic of productive growth is turned off and the λ genome becomes incorporated into the host genome through site-specific recombination (Echols 1980).

We know of no eukaryotic regulatory genes equivalent to those of prokaryotes, such as *Q* and *Cro* in λ, though the homeo box products may turn out to be regulatory in nature.

Moreover, developmental regulation in eukaryotes involves considerably more than just gene regulation, since it operates through four mechanisms, three of which have no counterpart in prokaryotes, namely:

(a) through developmental differences in timing,
(b) through binary switch mechanisms,
(c) through cell interactions, and
(d) through cell position.

It is possible to modify development and morphology through alterations in any or all of these systems of regulation. Regulation also includes systems capable of monitoring cellular conditions throughout the entire life of the individual, and not just during its development, systems which, within limits, are capable of initiating corrective action to maintain the operating state of the organism.

5.1   TIMING ADJUSTMENTS

Alterations in the timing and rate of developmental events fall under the general heading of heterochrony, and include four classes of change (Table 5.1). The best understood of these is neotony. In preparation for terrestrial life, aquatic anuran larvae undergo a complex metamorphosis which involves an elaborate series of changes (Table 5.2). In the central nervous system, and particularly in the spinal cord, there is a large amount of cellular necrosis coupled with further differentiation of the surviving neurons. The alimentary canal and the pancreas are almost entirely remodelled. The pronephros disappears and the skin undergoes successive additions, deletions and modifications of different cell types (White & Nicoll 1981). During metamorphosis the organs essential for the aquatic environment (tail and gills) do not begin to degenerate until the growth and development of organs essential for the terrestrial environment (legs and lungs) are nearly complete. In the later stages of larval development the blood chemistry also changes, coincident with the transition from gill to lung respiration. Thus, whereas in insects metamorphosis depends on setting aside discs of embryonic cells specifically for the larval-adult transformation, in amphibians it involves the production of adult tissues within the larva.

Accompanying these anatomical alterations there are some very fundamental biochemical modifications, two of which, in particular, are worthy of note. The first involves the induction of the urea cycle enzymes in the liver. Whereas ammonia is rapidly lost by diffusion in an aquatic environment, it is extremely toxic to a terrestrial organism and needs to be converted to a non-toxic form before it is excreted from the body. The second important biochemical change involves photopigment conversion. The rods present in the retina of vertebrate eyes contain one of two photopigments – porphyropsin or rhodopsin. The porphyropsin system characterizes most freshwater fish and the sea lamprey, whereas most marine fish and all terrestrial vertebrates possess rhodopsin. In many, though not all, anurans, porphyropsin is found in the eyes of aquatic larvae whereas rhodopsin is the photopigment in the eyes of the terrestrial adults

Table 5.1 Categories of heterochronic timing shifts during development (from Gould 1977).

Category	Somatic development	Germ-line development	Evolutionary result
acceleration	accelerated	unchanged	adult trait appears in juvenile descendants (RECAPITULATION)
hypermorphosis	unchanged	retarded	
paedogenesis	unchanged	accelerated	juvenile trait appears in adult descendants (PAEDOMORPHOSIS)
neoteny	retarded	unchanged	

Table 5.2 Morphological and biochemical changes induced by thyroid hormones during amphibian metamorphosis (from White & Nicoll 1981).

Body region	Pattern of change
skin	Formation of dermal glands, formation of nictitating membrane, degeneration of operculum, sodium transport
nervous system	Growth of cerebellum, growth of mesencephalic V nucleus, growth of hypothalmic nucleus preopticus, increase in retinal rhodopsin
muscle	Degeneration of tail muscle, growth of limb muscle, growth of extrinsic eye muscles
kidney	Pronephric resorption, induction of prolactin receptors
gastro–intestinal tract	Regression of gills, intestinal regression and reorganization, pancreatic reduction and restructuring, induction of urea-cycle and other enzymes in liver
supportive tissues	Degeneration of gill arches, restructuring of head and mouth, calcification of axial and appendicular skeleton

(Wald 1981). Thus, amphibian metamorphosis seems to recapitulate a number of the changes that would have accompanied the movement of vertebrates from water to land.

All these metamorphic changes are under endocrine control, and depend on the interaction of two hormones, thyroid hormone and prolactin, which is produced by the anterior hypophysis. These interact as antagonists. While the thyroid is already functional in the larva, it is overbalanced by prolactin which acts as a larval growth hormone. As the concentration of thyroid hormone rises, it stimulates metamorphosis. This two–hormone interplay is similar to that which operates in insects. Here juvenile hormone (neotenin) maintains larval development, while moulting hormone (20-hydroxyecdysone) stimulates metamorphosis and adult development. In *Drosophila melanogaster* the X-linked mutation, giant (*gt*), results in animals remaining as larvae some 3–5 days longer than normal. During this time they continue to grow and so produce giant larvae. These, in turn, give rise to giant pupae which eclose as giant adults. Giant turns out to be an ecdysteroid-deficient mutant, which is, nevertheless, competent to respond to exogenous moulting hormone. The induced metamorphosis is complete and gives rise to viable, fertile and normal sized adults (Schwartz *et al.* 1984).

Metamorphosis, which is a major event in anurans, may also occur in a partial or limited way in many urodeles and, in some few cases, never occurs. Thus, *Necturus maculosus* and *Proteus anguinus* become sexually mature and breed without reaching an adult stage. Here, the animal produces the thyroid hormone which normally triggers metamorphosis, but the target cells no longer respond to it. The Mexican axolotl, *Ambystoma mexicanum*, is also neotenic, whereas in related species, such as *Ambystoma tigrinum*, metamorphosis occurs in warm climates but not in colder ones. This is because low temperature inhibits thyroxine production and, at the same time, renders tissues insensitive to the hormone (Smith Gill & Berven 1979). By crossing axolotls with metamorphosing individuals of *Ambystoma tigrinum*, Humphrey (1944) was able to demonstrate that the neotenic condition results from recessive homozygosity at a single gene locus (see Tompkins 1978). Whereas most neotenic urodeles do not respond to exogenous thyroid hormone, indicating their insensitivity to it, the axolotl can be induced to metamorphose by a single injection of thyroxine, or even by immersion in thyroxine solution. Here, then, is an example of a major developmental change that is demonstrably under single gene control.

Certain salamanders, such as *Notophthalamus viridescens*, while undergoing a normal metamorphosis, return subsequently to water to breed and live out their lives there. Under these circumstances they undergo a partial, reverse metamorphosis. During this second metamorphosis they redevelop characteristics, such as the tail fin, integument changes and the eye pigment, porphyropsin – which is readily interconvertible with rhodopsin (Fig. 5.1). They also modify their pattern of nitrogenous excretion and increase their output of ammonia. As in the larva-adult transition, this second metamorphic shift is also under the control of both thyroid hormone and prolactin.

Several aberrations of amphibian metamorphosis are also known. Thus, the clawed toad *Xenopus laevis*, metamorphoses completely but never leaves the water. Here, adults possess porphyropsin almost exclusively in their retinas. By contrast, in the terrestrial Bolitoglossini the aquatic larva has been eliminated from the life cycle, but selected juvenile features are retained in development and persist in the adult stage. Thus, neoteny provides suggestive evidence that major evolutionary changes might not be extreme at the genomic level.

By contrast with neoteny, acceleration of somatic development could

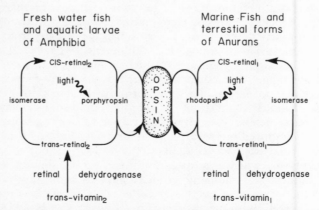

Figure 5.1 Visual pigment systems of lower vertebrates. Each photopigment is a conjugated protein, opsin, which contains either retinal$_1$ or retinal$_2$ as its prosthetic group. These are the aldehydes of vitamins A$_1$ and A$_2$ which cycle respectively with red rhodopsin and purple porphyropsin. There is a transition from porphyropsin synthesis to rhodopsin synthesis in amphibians which undergo metamorphosis (data of Smith-Gill & Carver 1981).

well play a role in bringing about allometric changes, that is changes in the relative size of different body parts. The immense antlers of the Irish elk, for example, are the result of a positive allometry between antler size and increased body size. This is a condition which obtains in all cervids, but it reaches an extreme in the Irish elk where it produces a non-adaptive hypermorphosis (Gould 1977). As body size increases other morphological correlates inevitably appear. A variety of vertebrate lineages all start with a very small body size which subsequently increases. Large terrestrial vertebrates have to develop heavy columniform legs with disproportionately large bones, simply because with an increase in body size the volume of the body, and hence its weight, increases by the third power whereas that of the bones supporting it increases only by the second power. Such columniform legs are thus present in the larger members of a variety of vertebrate orders, including elephants, rhinoceroses, ostriches and dinosaurs (Rensch 1959).

In elephants, the principal changes, apart from the mechanical readjustment of the limbs, have concerned the head. As a rule, in mammals the development of long limbs requires a corresponding lengthening of the neck, as in horses and giraffes. In mastodons and elephants, however, there has been a shortening of the neck and the development of a large head and an elongate proboscis. To what extent these changes can also be expected to follow on from an increase in size is not clear. Neither do we know what factors lead to size increases, nor the changes that are required in a genetic sense to bring them about, though in most general terms the epigenetic mechanisms that trigger heterochrony are often hormonal in nature. In domestic breeds of dogs, allometric changes resulting from genetically controlled endocrine alterations have also radically reshaped the morphology of the adult (Stockard 1941). Heterochrony thus demonstrates that the relative timing of processes can be dissociated without any adverse effects on developmental canalization, and yet still produce considerable morphological change. This is well illustrated by a number of different facets of whale development. In highly specialized forms like the large baleen whales of the genus *Balaenoptera*, for example, development is very retarded in the fore extremities (Müller 1971) which, in the 90 mm foetus, do not show any sign of imminent ossification. Pilleri & Wandeler (1970) have compared the ontogeny of the eye of the fin whale with that of a number of other mammalian species, including the horse, the dog and man. They find that while the whale eye starts out as a typical mammalian structure it undergoes specific changes during development. The ciliary muscle, for example, forms relatively early but is subsequently dispensed with. There is also an earlier differentiation of the retina and the vascular system of the eye, but a later differentiation of the sclera and a later regression of the embryonic blood vessels to the eye. Also, in the fin whale there is a complete set of tooth buds in the eight-month-old foetus whereas these are completely resorbed later when rudiments of baleen appear (Slijber 1976). By contrast, in the adult bottle nose whale, there are only 1–2 teeth in each half of the jaw but a whole set of rudimentary teeth can be found within the gums.

Changes in the temporal patterns of development may also result from heterochronic mutations which, like homeotic mutations, alter cell fates.

The recessive mutation *lin-14* of *Caenorhabditis elegans*, for example, accelerates development by causing cells to express fates normally seen only in descendant cells. The dominant *lin-14* mutation, on the other hand, retards development by altering the fates of cells so that they repeat the lineages of their progenitors (Ambros & Horvitz 1984).

Another possibility concerns timing alterations in developmental circuits. The larvae of *Amphioxus* normally live for several months in the plankton, achieving a length of 4–5 mm, and with some 14–15 primary gill openings, before undergoing metamorphosis. On rare occasions when the larvae do not metamorphose at the conventional stage, and so spend a much longer time in the plankton, they can attain a length of up to 9 mm with as many as 27 primary gill openings (Bone 1972). This example alerts us to the latent potential inherent in the developmental circuits which are allowed to run beyond their conventional time span. Phenocopies, that is environmentally induced phenotypes which mimic mutant phenotypes, similarly demonstrate the potential which developmental circuits show for ready modification.

5.2   BINARY SWITCH MECHANISMS

While we are now in a position to identify at least one set of causes for altered morphology, we are still faced with the difficulty of accounting for those relatively large-scale evolutionary changes that have, in the past, been responsible for what appear to be novel structures or radically new body plans. Here, the crux of the problem is that the types of morphological change that can be studied in extant organisms are relatively trivial, compared with those required to account for macroevolution. Such macroevolutionary changes include the rapid initial formation of the major metazoan phyla, and the development of groups, like whales and lungfishes, which appear in the fossil record with remarkable abruptness. On *a priori* grounds one might have anticipated that changes of this type would have had to involve genes operating early in development.

Arguments that changes in structural genes do not influence macro-evolutionary change rest largely upon comparisons which involve relatively late-acting genes, as in the case of those used in the human–chimpanzee study referred to earlier. The establishment of metameric segmentation in the *Drosophila melanogaster* embryo provides the clearest example of how early-acting switch genes are of critical importance in determining basic body pattern. The mutant forms of these switch genes include, as we have seen, maternal effect lethals that alter the overall polarity of the embryo, zygotic lethals that alter segment number or pattern and homeotic mutations that alter segment identity. Using the known *Drosophila melanogaster* mutations, it is possible to speculatively reconstruct the developmental events involved in the production of an advanced dipteran from a non-segmented ancestor (Fig. 5.2), a macroevolutionary change of some substance.

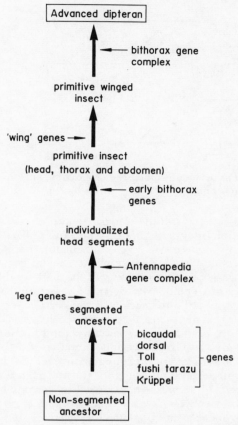

Figure 5.2 Hypothetical reconstruction of the steps involved in the production of an advanced dipteran from a non-segmented ancestor (after Gerhart 1982).

5.3   CELL INTERACTIONS

The problem of the programming of development can be reduced in its essentials to that of defining the prospective areas which give rise to the germ layers of the gastrula. This is the most crucial event in the whole of development, since it is the subsequent transfer of information between cell populations originating from the germ layers that governs cell morphogenesis.

Consider for example, morphogenesis of the skin in amniote vertebrates. The basic configuration on which all developmental programmes in amniotes are constructed involves an epidermal layer and a dermal layer, separated by a basal lamina. In this system, the epidermal cells are attached to the underlying dermal matrix by collagen fibres. The first common step in the programme involves the thickening of the epidermal layer, giving rise to the epidermal placodes. Now the programme divides into a number

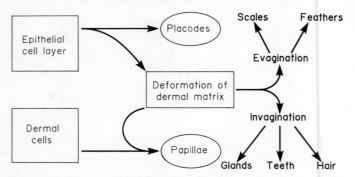

Figure 5.3   The developmental programme of skin in terrestial vertebrates (data of Oster & Alberch 1982).

of alternative pathways. If the epidermal layer evaginates, it leads ultimately to the formation of either scales (reptiles) or feathers (birds). When, on the other hand, the epidermal layer invaginates, then hair and skin glands (mammals) result (Fig. 5.3). The event that initiates the branching of the programme is the pattern of folding of the epidermal layer. This, in turn, is a consequence of an epidermal–dermal interaction (Oster & Alberch 1982). This sequence, like the formation of the germ layers themselves, depends upon co-ordinated geometrical changes in cell layers. To correctly interpret such changes we need to understand how modifications in cell shape are co-ordinated to give rise to the various patterns of deformation in cell layers which underlie subsequent morphogenetic events. Thus, the level at which we now seek understanding is that of the mechanical forces that act on cells and cell layers. While we certainly need to know which gene sets provide the potential for cell interactions, the problem is no longer one of simple gene regulation but rather one of cell mechanics. We see a similar principle at work in the case of the development of the panda's thumb (Gould 1980a).

The panda, unlike ordinary bears, has six digits on its forepaws, which include an opposable thumb. This, however, is simply a much enlarged and elongated radical sesamoid bone, of the type normally found in the wrist. This bone is, in fact, slightly enlarged even in ordinary bears, who also use their forelimbs for manipulating their food, though with nothing approaching the dexterity of the panda. The muscles which operate this extra digit also represent conventional parts of the anatomy whose development has been remodelled for the new function. In effect, the panda's thumb represents a mechanical response to the growth of the sesamoid bone.

Epithelial morphogenesis illustrates yet a further important principle. Namely, that genic alterations which occur before a programme divides into a number of alternative pathways may be amplified, in the sense that small alterations around a threshold point can have major morphological consequences. Altered morphology, thus, emerges as a consequence of modified patterns of interaction between cell populations. These are determined partly by their geometric contiguity and partly by their gene activity. It is for this reason that looking solely at the control of gene activity cannot provide satisfactory answers to developmental problems.

5.4 CELL POSITION

That cell position may play a role in developmental regulation is clear from the fact that the level of activity of a gene within a cell may be dependent on the position which that cell occupies within a given organ. For example, using monoclonal antibodies it is possible to define a cell membrane protein which is distributed in a dorso-ventral gradient within the retina of the chick and other birds. The protein in question, toponimic, can therefore be used as a marker of cell position within the dorso-ventral axis of the retina. Dissociated retinal cells continue to synthesize toponimic in culture, and do so in amounts expected from cells of the corresponding region of the intact retina from which they originated (Nirenberg *et al*. 1983). These results indicate that gene regulation, in this case, must depend on the position of a cell within the gradient, and that cells express different amounts of toponimic purely as a function of their position in the retina.

5.5 THE EVOLUTIONARY DILEMMA

There are three key issues that need to be resolved in order to provide solutions to, as yet, unanswered evolutionary questions. First, which of the components of the eukaryote genome are required for those changes which lead to major patterns of reorganization during development (macro-evolution), as opposed to those which bring about relatively minor developmental changes (microevolution)? In all eukaryote groups there is, as the reader will now be aware, a minimum genome size which is substantially smaller than that shown by many members of that particular group. This minimum genome defines the upper limit of the essential coding capacity required for the production of the grade of organization of the group in question. It does not, however, tell us whether there is a requirement over and above the minimum coding capacity in order to obtain, for example, a particular cell size, or whether the remainder of the genome is simply a combination of evolutionary debris and the accumulated consequences of genomic turnover processes.

There has also been a tendency to assume that increasing complexity necessarily requires an increase in the actual number of genes within a genome. To the extent that the duplication of major genes may lead to small multigene families, which appear to be more extensively represented in more complex organisms, there may be some truth to this proposition, though the case of the tubulins, where many pseudogenes exist, cautions us in this regard. This type of argument, which relates increasing morphological complexity to increasing gene number, is a variant of the novel gene hypothesis, which postulates that new synthetic genes form the basis for major morphological change. One form of this hypothesis is that exons, or genes in pieces, can be reshuffled to give new functional combinations (Gilbert 1979). An alternative possibility is that increasing complexity results not so much from an increase in gene types as from the effects of altered interactions between a relatively constant number of genes. Thus, relatives of most known structural genes seem to occur in a very wide range of organisms, whereas the specific patterns of development in which they are utilized tend to differ in different kinds of organisms.

A second issue is whether the coding component of a genome can itself be further partitioned into a fraction which encodes major developmental genes, as opposed to those genes which, while necessary for endorsing developmental decisions, are not involved in actually making them.

The third and final issue concerns the extent to which molecular changes resulting from drive mechanisms within the genome are instrumental in bringing about developmental and evolutionary changes. One of the significant features of molecular drive, which distinguishes it from both selection and drift, is that a sequence variant may spread not only to all other family members within one genome, but also to all individuals in an interbreeding population. Under a concept such as molecular drive, all individuals within a population are within a narrow variance for a particular series of DNA sequences. If these are control sequences for the timing of gene expression, then entire circuits can be delayed or accelerated. Thus, since all individuals within the population can be affected to the same extent, any developmental alteration takes place in all individuals, and either they all survive as changed forms or else they all become extinct. We have already seen how, in theory, concerted change might impose restrictions of potential evolutionary significance on single copy genes. Since rDNA spacer sequences are known to contain polymerase I promoters, concerted change in those sequences might, via changes in gene frequency, force the polymerase gene to be altered in concert with changes in that multigene family. It has also yet to be generally appreciated that, in theory, molecular drive can lead not only to morphological novelty but, additionally, can provide a genomic basis for some categories of extinction.

If punctualism, as seen in the fossil record, is a valid indicator of the nature of evolutionary change, then very radical developmental reorganization must have occurred in a very short geological time span. The case which perhaps best illustrates this situation for vertebrates is the origin of whales. Even the oldest known fossil, *Pakicetus*, is already very whale-like. Thus, we either have to conclude that the stringent morphological and physiological changes were extremely rapid, or else accept that the fossil record is incomplete and that the evolution of cetaceans occurred at a more gradual rate. Paralleling the case of whales, amphibians, turtles and pterodactyls all appear abruptly in the fossil record (Lull 1940), which suggests that some morphological novelties may well have arisen rapidly and cannot continue to be explained simply by appealing to an incomplete fossil series. Many non-placoderm bony fishes, such as lungfishes, also appear abruptly in the early Devonian. These progenitor lungfish were also more heavily ossified than at any other time in their evolutionary history (Campbell & Marshall 1987). In the light of this, and since many fishes are known from older rocks, it is unlikely that a long line of unpreserved lungfishes has been missed.

In the case of the invertebrates, two major fossil data sets are relevant to the appearance of evolutionary novelties among metazoans. The first concerns the echinoderms, while the second relates to the fauna of the Burgess Shale of British Columbia.

Because their skeletons closely reflect their soft anatomy, echinoderms provide some of the best palaeontological material available for the study of large-scale evolution. As far as the Palaeozoic history of this phylum is concerned, major changes in morphological design, leading to the formation of new classes, appear abruptly and are concentrated in the Cambrian and Ordovician. None of these classes could have resulted from the diversification of pre-existing classes, and the new structural designs were established in two bursts of about 35 million and 55 million years, during the early phase of echinoderm evolution. From a detailed analysis of the data, Campbell & Marshall (1987) conclude that the appearance of such marked morphological diversification cannot be correlated with a phase of adaptive radiation; nor did the establishment of these designs result in any subsequent large evolutionary radiations. They argue that some unusual feature of the primitive echinoderm genome may have allowed it to change in such a way that marked morphological change could take place over relatively short periods of geological time. They find no evidence that such change was initiated by environmental conditions. Apparently, these morphological novelties were only produced during the early stages of echinoderm evolution and in what must have been the primitive echinoderm genome.

The second data base stems from the fossil fauna of the Burgess Shale. This is part of a sedimentary sequence of silts and muds in which the processes of decay that normally lead to the destruction of soft tissues and delicate skeletons have been avoided. Consequently, the fossils in this deposit provide a unique record of the diversity of marine life present during the earliest known phase of metazoan history. The Burgess Shale fauna includes representatives of all the major metazoan phyla. The most abundant and varied, in terms of specimens collected, are the arthropods and, apart from the trilobites, which are invariably the most common fossils present in Cambrian rocks, include thirty or more kinds of essentially soft-bodied, non-trilobite arthropods that cannot be matched with any other early Palaeozoic animal elsewhere in the world. Additionally, it includes a number of miscellaneous forms with a unique anatomy and which cannot be readily accommodated in any of the known metazoan phyla (Conway Morris & Whittington 1985).

We illustrate three of these bizarre forms (Fig. 5.4), namely *Anomalocaris* (Whittington & Briggs 1985), *Opabinia regalis* (Whittington 1975) and *Wiwaxia corrugata* (Conway Morris 1985), all of which have distinctive body plans. *Anomalocaris* and *Opabinia regalis* both appear to be descended from segmented animals from which either arthropods or annelids, or both, were derived, but each represents a quite separate line of descent. *Wiwaxia corrugata* is certainly not an annelid, and may represent an early off-shoot during the evolution of molluscs from turbellarian ancestors. Other organisms, such as *Hallucigenia sparsa*, are so odd that it is not known what to compare them with.

While it is well known that most of the extant animal phyla evolved

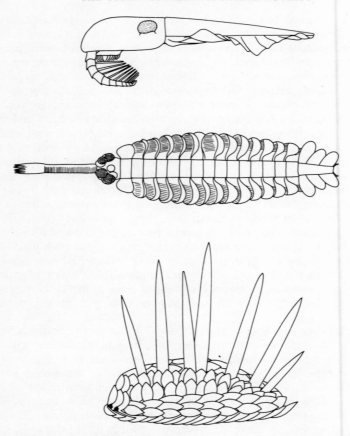

Figure 5.4 Reconstructions of three 'bizarre' fossils from the Burgess Shale of British Columbia. Top, lateral view of *Anomalocaris* (redrawn from Whittington & Briggs 1985). Middle, dorsal view of *Opabinia regalis* (redrawn from Whittington 1975). Bottom, lateral view of *Wiwaxia corrugata* (redrawn from Conway Morris 1985).

during the Cambrian, it would appear that other body plans were also represented at this time, but have not survived to the present. These soft-bodied animals, which are known at intervals throughout the Palaeozoic, have not been seriously considered in discussions relating to the emergence of evolutionary novelty. The bizarre body plans they illustrate are probably bold, though unsuccessful, experiments in metazoan design. The reasons for their lack of continuation are not known. In the case of one major group of Palaeozoic trilobites, Raup (1981) has argued that their extinction could, just as easily, have resulted from 'bad genes' as against 'bad luck', whatever this may mean in genomic terms.

Evolutionarily, this is the crux of our dilemma. We are unable to reproduce nature's own experiments because they have involved a unique combination of environmental and genetic circumstances. How, then, can we glean information about macroevolutionary change? We are unlikely to gain such information from a more detailed examination of the fossil record which, while it may reveal some of the structural changes that have taken place during evolution, can tell us nothing about how these changes came about since, to the palaeontologist, evolution presents itself simply as a succession of adult forms. Neither is it likely to come from a preoccupation with whether the changes were spread through a population by selective or other means, since this does not address the question of the genesis of the changes themselves. The only sensible place to start is in very early embryogenesis. The essential problem here is that we still do not know what combinations of developmental circuits have contributed to the evident adaptations that have arisen during evolution, and the extent to which these circuits involve single genes, multigenes or multisequence families. What is clear is that defining these circuits will be infinitely more tractable in *Drosophila melanogaster* than in any other organism simply because of the power and potential of its genetic system, the ability to define and categorize early-acting genes and the relative simplicity of the developmental end product, namely the fly itself. The essential step of identifying the patterns of cell interactions involved in early development of *Drosophila melanogaster* has already begun. Initially, we perceive this in terms of the molecular fate mapping of specific cells in relation to the defined functions which those cell lineages collectively give rise to. This, in turn, should lead to a more precise molecular definition of the cell properties which underlie critical developmental decisions.

We appreciate that changes in developmental timing may already have obscured the circuits responsible for the original changes. If this were to be the case, then we would be left with an intractable problem in so far as being able to completely resolve how macroevolution has taken place. Even under these circumstances, however, we would most certainly gain increased understanding of developmental circuits, the constraints they impose on morphogenesis and how altered morphologies can be produced. And this, after all, is what evolution is really about.

6

Coda

'Selection may account for the survival of the fittest but fails to
account for the arrival of the fittest.'

Anon.

6.1 FACTS AND CONCLUSIONS

Now that we have completed an examination of the structure and function
of eukaryote genomes it may be helpful to summarize the conclusions we
believe can be drawn from the facts, in relation to the problems of
development and evolution.

Before doing so, we need to define what the molecular analysis of these
two processes can be expected to reveal, and what it cannot reveal. The
periodicities noted by Raup & Sepkoski (1984) for mass extinctions mean
that evolutionary strategies may well have been determined, in part at least,
by external events about which molecular biology can have no input. The
narrow specializations of some organisms to their food sources has also
undoubtedly contributed to extinctions. Clearly, where a food source
declines, for whatever reason, the very specializations which allowed for the
adaptation of the consumer now seal its fate and lead, inevitably, to
extinction.

Much of evolutionary thought had its origins in comparative morphology,
but this phase was effectively nearing the end of its significance early this
century simply because, as Bateson (1922) pointed out, '. . . *it was obvious
that no progress was being made.*' Little has changed in the intervening years.
While we can reconstruct the probable patterns of change and can invent
whatever mechanisms, molecular or otherwise, that appeal to our imagination
to account for these patterns, we will be unable to test them in any
meaningful sense. For example, we might argue that the critical event that
initiated a significant morphological sequence of change, such as the
development of jaws in vertebrates, was the insertion of a nomadic element
into the control sequence of a developmentally important gene. Subsequent
genomic changes, however, may have assumed control of the developmental
network in question, with the mobile sequence having become excised and
allowed to continue its nomadic meddlings elsewhere within the genome.
How will we ever possibly know if such sequences contributed to the

initiation of a programme, when their footsteps have been erased forever from the genomic landscape?

The general problem has been put into perspective by Gehring (1985a,b) who has pointed out that organisms, by necessity, have developed along historical lines, and there is no reason whatsoever to believe that they have been assembled in the most rational and efficient manner. They simply have to work at the time, and reproductively and developmentally, of course, they may be optimized in the course of evolution. Thus, there are aspects of development which will only make sense in terms of evolutionary history. The problem is whether we can ever determine what that history was, and separate the essential basic mechanisms from those which were no more than historical accidents.

In attempting to come to terms with evolutionary problems, we face a clear choice. On the one hand, we can attempt to extrapolate back to what might have been the structure and function of a DNA sequence, or even the entire genome, at an earlier stage in evolution, in much the same way that the early evolutionists did, though using morphological characters. Alternatively, we can attempt to define the extent to which we can experimentally perturb a developmental programme, and the phenotype it gives rise to, using the tools of molecular biology.

In order to put the data we have gleaned about the genome into a developmental and evolutionary perspective, we now return to the sets of questions we posed at the beginning of each of the three major chapters of this book.

(1) How much of the genome is coding; how does the coding component function and how do genes control development?

In the biological sciences, progress is too often defined in terms of new questions rather than new answers. The advent of recombinant DNA technologies has, however, radically altered this situation for they allow us, for the first time, to identify those genomic components that have no effect on development and, hence, cannot be expected to have any major evolutionary consequences. The evidence is perhaps clearest in the case of the highly repeated and tandemly arranged DNA sequences. In some crabs, 30% of the genome is pristine poly (AT) with the odd G or C nucleotide appearing in an otherwise monotonous landscape. Is this the stuff of development and evolution? Surely not. Rather it reflects a replication machinery that is profligate and which leads to variation in the amount of highly repeated DNA between individuals, but which has no influence on phenotype. Indeed, the copepod *Cyclops*, where heterochromatic segments are chopped out of the chromosomes prior to the fifth cleavage division of the egg, demonstrates most effectively that the DNA contained within these segments can play no role in somatic development. Similarly, the repetitive DNA sequences of ciliates are dispensed with during the formation of the macronucleus which controls all vegetative functions in these unicells. Added to this, satellite DNA sequences of amphibians can be transcribed by default from histone gene promoters, so that not only does the genome carry junk DNA but it can also produce junk RNA. Finally, most of the cellular hypotheses which have attempted to relate tandemly repeated sequences to somatic functions have now been rigorously tested, either by experiment or by comparisons of sequence data bases, and found wanting (Singer 1982a, Miklos 1985, Miklos & John 1987). This includes hypotheses dealing with chromosome recognition, chromosome rearrangement, developmental timing and speciation phenomena.

While it is true that some simple sequences are involved with the promotion or enhancement of gene expression, or with gene switching mechanisms, the bulk are not of this kind. Rather, they reflect a replicative machinery that is subject to regular slippage. This is known to occur *in vivo* between short direct repeats, leading to deletions or expansions of particular motifs, and is capable of shuffling old and new motifs at relatively high rates.

The enzymic machinery of the cell is also capable of mobilizing middle repetitive sequences. Furthermore, mobile elements can be inserted into other mobile elements and so may eventually account for the production of clustered scrambled repeats. Alternatively, sequences located in different chromosomes can interconvert each other, as we saw in the case of the serine tRNAs of yeast. Nor, as many have argued, does the cell appear to be in any way embarrassed by the presence of so many useless components and commodities. In *Drosophila melanogaster* where 60% of the 18S+28S rRNA genes are non-functional, the fly continues to replicate them without any overt metabolic cost. Likewise, in *Plethodon cinerus* different individuals in the same population and with identical phenotypes carry from 330 to 2500 ribosomal genes at the same single nucleolus organizing region. In *Triturus vulgaris meridionalis*, ribosomal RNA gene numbers vary from 2000 to 20 000 between individuals, and in *Vicia faba* from 250 to 24 000 between different plants. The chloroplast genomes of different species of algae vary from 85 kb to 2000 kb, and the mitochondrial genomes of four cucurbits species from 350 kb to 2540 kb. Thus, organellar genomes behave in a not too different way to their nuclear counterparts. Eukaryote genomes also suffer from the genomic fallout of non-functional pseudogenes which frequently litter the landscape.

As far as the interspersed repetitive sequences are concerned, it is now clear that most are probably mobiles, or else derelict copies of mobiles. *Drosophila melanogaster* and *Drosophila simulans* differ in their nomadic element contents by the enormous figure of 20 million bp, yet there are only minor morphological differences between them. Tschudi *et al.* (1982) have described an interesting example of the extensive genomic rearrangements that can be wrought by mobile elements. They showed that the mobile element *B104* causes deletions of varying length when it excises from a location within the 5S RNA gene cluster. Moreover, even when 60% of the entire cluster is deleted, and the deletion chromosome then made homozygous, the flies are viable and fertile. This indicates a significant functional redundancy within the cluster, since 70 or so 5S RNA genes, out of a total of about 160, still support normal development. A further potentially interesting consequence of these data is that disparate sequences can be brought into apposition by deletions of the sequences which normally separate them. The ability to cause deletions upon excision is not restricted to either *B104* or to the 5S RNA gene cluster. It occurs for other loci and other nomadics, such as *copia* (Telford & Pirrotta unpublished). If the induction of deletions by mobiles is a general phenomenon then one can expect the *Drosophila melanogaster* genome to be literally 'riddled' with deletions.

It should now be reasonably clear that if we are ever to resolve the problems of development then there is only limited payoff in studying gross genomic architecture. From a developmental viewpoint, chromosomes are evidently nothing more than convenient evolutionary packages, most of whose contents are indeed somatic junk, a fact appreciated by Ohno almost two decades ago. While the insertion of many hundreds or even thousands of repetitive elements have had major effects on the structure of the genome, there is little evidence that they have led to developmental novelty. There can be no doubt that the genomic flux seen in non-coding sequences is an unavoidable by-product of the misdemeanours of DNA and its attendant retinue of enzymes. As a modification of an old adage would have it – 'if it is not biochemically forbidden, it will occur'.

If we turn to the mode of operation of the coding components of the genome, the *Drosophila melanogaster* data gives us every expectation for optimism. Saturation-mutation schemes are now uncovering the interacting genes which operate at the critical, early, stages in the developmental programme (Nüsslein-Volhard & Wieschaus 1980, Nüsslein-Volhard *et al.* 1985). Consequently, for the first time in any eukaryote, we now have molecular analyses of gene groups of this kind. This includes the maternal type effect genes which constitute the dorsal group and which determine the fundamental dorso-ventral gradient on which the ultimate architecture of the embryo is predicated. Still others, like folded gastrulation (*fog*) and twisted gastrulation (*tsg*), control the early morphogenetic movements which follow from the definition of positional co-ordinates (Zusman & Wieschaus 1985), while loci, such as Antennapedia, bithorax, fushi tarazu, Krüppel and engrailed, are involved in the control of body segmentation and compartmentalization. All three of these gene groups appear to involve an extensive network of regulatory interactions. Thus, engrailed, and other selector genes, act as pleiotropic regulators of transcription events whose protein products interact with the DNA near the promoters of a number of target genes. In turn, the expression of engrailed is, itself, regulated by the products of the pair-rule, segmentation, genes.

The most exciting and promising developments are clearly in the area of the regulatory hierarchies which operate between developmentally significant genes. When the interactions between engrailed, fushi tarazu and even-skipped are examined, it can be demonstrated that both *ftz* and *en* expression are changed in a mutant *eve* genotype, and that *eve* expression is quite different in a mutant *en* genotype. Furthermore, the patterns of expression of a number of homeotic genes are also altered in mutant *eve* embryos (Harding *et al.* 1986). Additional insights into the hierarchical interactions between different segmentation genes come from the work of

Howard & Ingham (1986) and Carroll & Scott (1986), who have examined the effect of some gap, pair-rule and segment polarity mutations on the expression of the ftz^+ gene. All four gap mutations tested, Krüppel, knirps, hunchback and giant, have an effect on ftz expression, as do three of the pair-rule mutations – hairy, runt and even-skipped. However, five other pair-rule mutations, odd-skipped, paired, odd-paired, sloppy-paired and engrailed, have no effect. Two segment polarity mutations, hedgehog and patched, are also without effect. As yet, no simple inter-relationships are apparent. However, there is clearly a network of regulatory interactions among segmentation genes (O'Farrell et al. 1985), and it is this network which produces the spatial pattern that gives rise to embryonic form. The next crucial level of analysis will undoubtedly be the study of multiple gene expression in single cells of the embryo.

Thus, gradually, we are at last beginning to discern the precise developmental networks that are involved in particular subprogrammes. Four additional networks which are likely to prove especially important in the future are:

(a)   the loci that interact with bithorax to prescribe body segmentation;
(b)   the genes which collectively influence sex determination and dosage compensation;
(c)   the numerous loci, including Notch, Delta, mastermind, big brain, neuralized, almondex and Enhancer of split, that function in early neurogenesis; and
(d)   the numerous loci involved in synaptic connectivity in the nervous system.

The requirements for specifying development evidently include both temporal information, based on the availability of particular gene products at particular times, and spatial information, which determines the site at which particular gene products will act. In viruses, as we saw earlier, much of morphogenesis involves self-assembly processes, which require no more information than that contained within the constituent macromolecular subunits. While self-assembly is certainly involved in eukaryote development (Kirschner & Mitchison 1986), especially in relation to the production of the cytoskeleton (Lazarides 1984), it is clear that it operates in conjunction with gene determined systems of temporal and spatial control.

(2)   How is gene activity regulated and does the large scale rearrangement of genes in chromosomes have any functional significance in development and evolution?

Gene activity can be regulated in a variety of ways, including transcription, splicing, exit from the nucleus, binding to polysomes, translation and post-translational events. If there is one point that has become obvious, it is that almost every enzymatic quirk that the cell can muster has been exploited for the purpose of gene regulation. There are sequences that enhance gene activity and silencers that diminish it (Brand et al. 1985). There are sequences that control mRNA stability and proteins that produce topological constraints on DNA. It also seems likely that antisense RNA will be found to be regulatory, and it would not be surprising if the transcriptive capabilities of chromatin could be altered at the nucleoskeletal level.

It is also true that some introns provide for flexibility in the expression of individual gene products, by providing variant splicing sites, and so regulate the production of different forms of the same protein in different cells. This includes those genes involved in muscle formation, in the synthesis of hormones and the synthesis of extracellular matrix proteins. The majority of introns within genes, however, have no known function. Some genes also show an obvious correlation between exons and functional protein domains, but most do not show such a correlation. This is true, for example, in the case of the two myosin heavy chain genes of *Caenorhabditis elegans* (Karn *et al.* 1983), where the pattern of intron positions is inconsistent with the proposal that proteins evolve by exon shuffling of discrete structural and functional domains.

While a number of individual loci have now been studied in sufficient detail to define their mode of regulation, most of these studies have, of necessity, concentrated on well-known metabolic genes. What is urgently required is a study of the regulation of regulators, such as Polycomb, since this is likely to provide direct insight into the question of developmental networks.

Bryant *et al.* (1981) have shown how the information of spatial pattern in development can be understood in terms of strictly local interactions, which involve the co-ordination of the activities of embryonic cells in space and time. Some of the genetic approaches that have been adapted to this problem illustrate how the study of gene regulation, when applied to cellular topographies, can prove to be infinitely more rewarding than the conventional monotony of simply describing upstream and downstream landscapes.

For example, runt (*run*) is one of the pair-rule loci required for proper segmentation of the *Drosophila melanogaster* embryo. In its mutant form it produces periodic deletions in the segmentation pattern along the antero-postero axis. When, however, the wild type dosage of runt is increased, an antirunt phenotype forms, in which the deletions in the segmentation pattern are out of phase with those caused by the mutant alleles (Gergen & Wieschaus 1986). Fig. 6.1 illustrates the requirement for runt in the establishment of blastoderm cell positional identity in the A2/A3 region of the future abdomen. The shaded cells require runt$^+$ activity, though the actual level of such activity is different in different cells. Not only is this the case, but the unshaded cells have an obligatory requirement not to express runt$^+$. Thus, for normal segmentation to occur, some blastoderm cells must express runt$^+$ at varying levels, while others must not express runt$^+$, in a geography which we have illustrated as eight-cell widths. As has been pointed out by Gergen & Wieschaus (1986), the observation that extra doses of runt$^+$ cause pattern deletions, and the fact that these deletions are out of phase with those caused by loss of function mutations, suggests that controlling the level of runt$^+$ activity is a critical aspect of positional identity determination in all blastoderm cells. This example also emphasizes

316 CODA

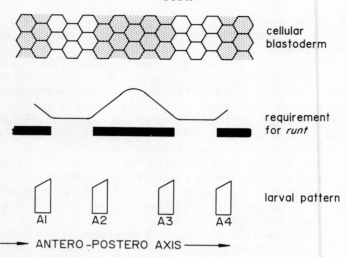

cellular
blastoderm

requirement
for *runt*

larval pattern

A1   A2   A3   A4

⟶ ANTERO-POSTERO AXIS ⟶

Figure 6.1 Diagrammatic representation of blastoderm cells in the A1–A4 abdominal region of the *Drosophila melanogaster* embryo, and their requirements for *runt*⁺ activity. The shaded hexagons represent cells with *runt*⁺ products, while those in which *runt*⁺ is inactive are drawn clear. Within the shaded cells the varying levels of *runt*⁺ products are indicated by the 'requirement for *runt*' curve. Finally, the positions of the denticle bands in segments A1–A4 are shown as 'larval pattern' (data of Gergen & Wieschaus 1986).

that transcript and product analysis need to be carried out on *single adjacent* blastoderm cells in order to glean *the* important biological information relating to regulation.

These experiments with the runt locus, in which it has been inferred that it is gene activity in single blastoderm cells that is so crucial, bear directly on a major problem of embryogenesis. Are morphogens perceived by *groups* of cells or by *individual* cells? The accepted view is that the uniformity in a sheet of cells in the early embryo is converted into discontinuities, termed compartments or polyclones. The decision-making unit is a *group* of cells, in which the blastoderm is represented as a series of discrete segments or para-segments. However, an alternative viewpoint is that *individual* cells themselves are the basic decision-making units (Gergen *et al.* 1986).

The data base for such an alternative hypothesis rests largely on the phenotypic diversity of pair-rule mutants and their combinations. The difficulties with the *group of cells* model show up when embryos are constructed which are homozygous for the mutations even-skipped and odd-skipped. If the individual defects were additive, the resulting double mutant embryos should be deficient for both odd and even numbered denticle bands. In fact this is not the case, and these double mutant embryos have eight partial denticle bands. Thus, if mutation in one pair-rule gene results in loss of function, some pattern can be *regained* when there is loss of function in a second, and different, pair-rule gene.

Furthermore, we saw in the case of the runt gene that when the dosage of the wild type gene product was increased, an antirunt phenotype was observed. This phenotype is closer to that produced by loss of function mutations at other pair-rule loci, such as paired and even-skipped. Thus, overexpressing runt$^+$ is similar to underexpressing paired or even-skipped.

The results of these examples are to be expected if it is the *ratio* of different segmentation gene products which controls the fates of *individual* blastoderm cells. If this turns out to have generality, so that at blastoderm each cell is, in fact, different from its neighbours, then our attention should be directed to the challenging problem of how *individual* cells sense and process morphogen concentrations (Gergen *et al.* 1986).

A further example concerns the pattern of spatial differentiation of chaetae in *Drosophila* (Moscoso del Prado & Garcia-Bellido 1984b). Here, two competing hypotheses have been proposed to explain pattern formation. The first relies on assigning positional values to cells on the basis of their location, with the cells then differentiating according to their interpretation of positional cues. The second proposes that the generation of chaetae pattern results from an interaction between neighbouring cells, and not on positional cues. This implies that when a cell is destined to become a chaeta it creates an inhibition on its neighbours which prevents them from following suit. Moscoso del Prado & Garcia-Bellido provide evidence to support the second hypothesis, and demonstrate that the position of chaetae is, in fact, a function of the level of activity of the Achaete-Scute gene Complex. Thus, here too, it is gene activity in a restricted cellular geography that appears crucial for developmental pattern.

Both these examples remind us that it is at the single cell level that we must first examine developmental systems, if we are to obtain meaningful information on pattern formation. We saw previously how the loci hairy and extramacrochaetae both interacted with the Achaete-Scute Complex, and how gene dosage of all three loci was critical for correct developmental expression (Moscoso del Prado & Garcia-Bellido 1984a). Recall, also, how in *Caenorhabditis elegans*, mutations at the *lin-12* locus behave as binary switches in a flip-flop mechanism for determining alternative cell fates in development (Greenwald *et al.* 1983).

Some gene systems do not appear to have such tight constraints on dosage activity. In the dopa decarboxylase (*Ddc*) system of *Drosophila*, Marsh *et al.* (1985) have shown that, with up to the equivalent of ten copies of the *Ddc* locus fully active within the genome, flies are still viable, fertile and phenotypically normal (Fig. 6.2).

Drosophila has not only provided a number of single gene systems for studying how gene activity is regulated, but, additionally, has been in the vanguard for identifying networks of interacting genes. As yet, however, almost no molecular inroads have been made into the so-called polygenic systems. It is worth pointing out that a Gaussian distribution for the expression of a given character may result from as few as two alleles at a locus, provided that there are environmental effects and that the heritability of the character in question is low. For cases of genetic effects without an environmental component, as few as five interacting loci will also yield a Gaussian distribution for the character they affect. Polygenic systems have,

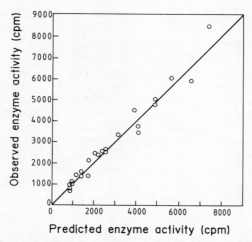

Figure 6.2 Effect of gene dosage on dopa decarboxylase activity (1000 cpm = 50 pmoles dopa) in different strains and developmental stages of *Drosophila melanogaster* carrying up to 10 copies of the *Ddc* gene. These multiple copies result from multiple insertion events generated by P element mediated transduction. The perturbation of the normal ratio of gene products resulting from the presence of extra functional copies of the *Ddc* locus has no detectable phenotypic consequences (data of Marsh *et al.* 1985).

so far, been set aside, mainly because of the current concentration on more tractable gene systems, but ultimately they, too, will need to be categorized in order to determine whether they are, or are not, of importance in the generation of evolutionary novelty.

As far as gene transcription and genomic topography is concerned, the P element vectors, engineered by Rubin & Spradling, indicate that given a modest flanking landscape of a few thousand bases, most genes seem to function quite happily almost anywhere in the euchromatic portion of the genome. We may reasonably anticipate, therefore, that their actual location within a given genome is simply a reflection of their past history of movement, either as a result of structural rearrangement of the chromosomes themselves or else by other means. This argument is reinforced by the fact that whereas the ciliate *Tetrahymena pyriformis* has ten conventional chromosomes in its macronucleus controlling all its vegetative functions, in *Stylonychia* the macronucleus contains a shattered genome consisting of thousands of individual gene-sized pieces. Likewise, the muntjac situation also shows us that massive rearrangements of the genome, this time in the form of structurally rearranged chromosomes, leave metabolism, development and phenotype as undisturbed in this multicellular eukaryote as in the unicellular ciliates with a shattered macronuclear organization. The extensive rearrangements that have occurred in the pea and mung bean chloroplast DNAs, as well as between the mitochondrial DNAs of the closely related species of *Brassica*, remind us that gene order in organellar genomes is also largely irrelevant to gene function.

While functional domains have been inferred to exist within chromosomes, there are data which place severe restrictions on the significance of such domains in respect of individual gene activity. Henikoff *et al.* (1986), for example, have identified a functional cuticle gene located within an intron of the Gart housekeeping locus of *Drosophila melanogaster*. While the Gart locus, which codes for purine pathway activities, is expressed throughout development, the intercalated cuticle gene has a very restricted expression during the production of the pre-pupa. The calmodulin housekeeping gene also has an intercalated and developmentally regulated gene within an intron (Tobin *et al.* 1986). Yet a further example is found in the case of gene 1, which lies within the small heat shock gene cluster (*hsp* 22, 23, 26 and 27) but is not activated by either heat shock or by hormone treatment (Ireland *et al.* 1982). Here, again, the domain of transcriptional regulation is the individual gene and not the entire landscape, which contains both the four *hsp* genes and the intercalated gene 1.

There is, however, one *caveat* to be noted in mammalian systems. Here, it appears that housekeeping genes, in general, replicate early in S phase. Tissue-specific genes also replicate early in S if they are to be expressed in a given cell type, but replicate late in S if they are to remain transcriptively inactive (Goldman *et al.* 1984). On this basis, it has been argued that there are two distinct functional 'genomes', with different patterns of replication, within the mammalian nucleus. These have been termed the housekeeping and ontogenetic 'genomes' by these authors, and they have been postulated to occur, respectively, in the Giemsa-stained light and dark bands of the mammalian chromosome system (Holmquist *et al.* 1982). If this genomic segregation also applies to invertebrates, then one might expect that the replication patterns of the 'genes within genes' described for *Drosophila melanogaster* should differ from that of the landscapes in which they are intercalated.

(3) How is major morphological novelty generated and what do changes at the molecular level tell us about morphological change. How is speciation related to morphological change?

While we still do not know how major morphological novelty is generated, we can, at least, suggest where we need not look for an explanation. Clearly the non-coding portion of the genome can be discarded, as can most of the interspersed nomadic elements, their immobilized derivatives and the pseudogenes. Genome size, too, is unlikely to provide any sensible prospect for significant evolutionary change in multicellular forms, though it may have played a role in the evolution of unicells. This is especially clear in cases where marked differences in genome size between related species do not contribute to differences in phenotype. The obvious tolerance of experimentally produced *Lolium* hybrids to changes in genome size, but with no overt accompanying phenotypic effect, is especially compelling. Equally striking are cases where genome sizes vary enormously between vertebrates of similar morphology. Thus, some puffer fish have genomes only one three-hundred-and-fiftieth the size of genomes of other bony fish,

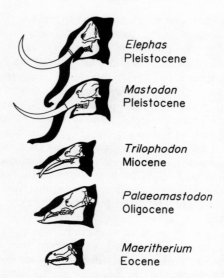

Elephas
Pleistocene

Mastodon
Pleistocene

Trilophodon
Miocene

Palaeomastodon
Oligocene

Maeritherium
Eocene

Figure 6.3 Evolution of the head in mastodons and elephants to show progressive development of tusks and trunk (after Lull 1940).

and only twice that of *Drosophila melanogaster*. Evidently, such a tiny teleost genome encodes all the necessary information for basic vertebrate morphological design. Similarly in plants, *Arabidopsis thaliana* with a genome size one thousand times smaller than the largest angiosperm genome, and half that of *Drosophila melanogaster*, has all the necessary coding capacity for angiosperm characteristics. Nucleotypic effects, too, provide little prospect either for significant evolutionary change or for the production of novelty. Thomson's telling tale of lungfish evolution reveals that while major morphological diversification of the group occurred in the Devonian, changes in cell size, and by inference genome size, occurred much later.

The reconstructed fossil sequences of groups such as the Proboscidea, shown in Fig. 6.3, certainly suggest a gradual and progressive pattern of morphological change, consistent with slow genetic change, though we can say nothing about its underlying basis. Here, what appear as quite novel morphologies in terms of the extremes of the series, are connected by clear intermediates. This is certainly not true of other groups. Thus, amphibians, turtles and pterodactyls, like the whales, all appear abruptly in the fossil record (Lull 1940). Not only are pterodactyls fully formed when they first appear, but their subsequent morphoplogical changes mainly involve increase in size, loss of teeth and modifications of the shoulder girdle. Likewise, the early fossil amphibians, such as *Loxomma* and *Pholidogaster*, are clearly not transitional forms, since they, too, are fully developed. Lungfish, which also appear suddenly, were more heavily ossified when they first appeared than at any other time in their evolutionary history, so that it is simply not credible to explain their abrupt appearance in terms of an incomplete fossil record (Campbell & Marshall 1987). If, as appears increasingly likely, major morphological change did underlie such a seemingly rapid appearance of novelty, then we will need to account for it developmentally. The key question here is whether a major morphological change requires a major genetic change. As yet, we do not know the precise developmental basis for any major change in morphology, other than that evident in amphibian metamorphosis, though the case of the alternative modes of epidermal development analyzed by Oster & Alberch (1982) may provide important clues. In particular, we are unable to decide whether morphological novelty involved change at a single locus, a newly created locus constructed from the shuffling of different gene segments, whether it involved a multilocus or multisequence system, or whether it involved a novel interaction between previously unintegrated circuits. We are also unable to decide whether it involved selection, drift, drive or a combination of these processes.

We keep being drawn back to the problem of genetic cost and developmental cost. What does it cost, in terms of genes, to elicit what one would term a large phenotypic change? Wessels (1982) has drawn attention to face shape changes in the mouse, and how they can be altered by the behaviour of mesenchymal cells. When the head of the mouse embryo elongates to form a snout, the mesenchymal cell density decreases and snout tissue volume increases. In certain mutants, where the face remains flat, the mutant mesenchymal cells actually clump together. This aggregation prevents tissue expansion to form a snout. Thus, it is conceivable that such a 'simple' developmental change, with a correspondingly 'simple' underlying genetic basis, may underlly face shape modification in many vertebrates. Hence, a rather simple programmable event in a cell population can have quite significant phenotypic consequences. In a similar fashion, cell death may provide boundaries between two groups of cells. In the chicken, cell death breaks up mesenchymal cell condensations in the forelimb, so that radius and ulna are separated from each other. In certain mutants, cell death does not occur in this zone and only a single bone arises in the forearms. Likewise, widespread target sampling, coupled with selective axon elimination, may have played an important role in the evolution of neural pathways within the vertebrate central nervous system (Katz *et al.* 1983). In invertebrates, programmed cell death is best described in *Caenorhabditis elegans*, where approximately one cell in eight of all cells formed is doomed to die. These cellular extinctions are predictable and are rarely a consequence of interaction with surrounding cells. Most of this elimination occurs in the nervous system and, unlike the situation in vertebrates, cannot be explained in terms of excessive cell production, followed by the culling of those that do not find synaptic connections (Sulston 1983).

Maynard-Smith (1983) has pointed out that population geneticists can 'explain' the pattern of evolution in so far as it is 'punctuational', by assuming that, most of the time, selection is stabilizing, leading to stasis, but occasionally directional, leading to punctuated change. While we agree that this is indeed an 'explanation', it remains both insufficient and inadequate to us because it does not address the more fundamental question of the developmental mechanisms which underlie the production of morphological novelty, on the one hand, and the lack of morphological change over long time periods, on the other hand. Selection acting on conventional genetic variability might, indeed, produce morphological changes at a rate which, in geological terms, would appear instantaneous in the fossil record. However, there is no reason why, in theory, drift or molecular drive should not be able to do exactly the same.

As we have previously emphasized, the possibility of interpreting events of the past is only likely to become a reality when we have a better idea of how to perturb developmental networks, and so uncover the range of permissible phenotypes which follow such perturbation. Even then, it is hardly conceivable that we will be able to *prove* that a given evolutionary sequence was the result of selection, drift or molecular drive, whether acting alone or in combination. Dover has emphasized that a critical factor in the evolution of biological novelty could have been molecular drive involving multigene or multisequence families, though he accepts that such drive would be expected to interact with selection and drift.

J. H. Campbell (1983) points out that essentially all of evolutionary theory is based on the 'traditional, passive, sacred, stereotyped, holistic, Mendelian, single copy determinant'. At least some complex characters, like the egg chorion of the silk moth, are unquestionably based on multigene families, and these need to be seriously evaluated in the context of neo-Darwinian theory.

While, theoretically, nomadic elements or chromosome rearrangements could have played a role in the origin of morphological novelty, there is no convincing evidence that either have done so. While genomic shuffling of one kind or another has certainly given rise to novel functional proteins, many of these are either relatively recent products of evolution, or else are concerned with similar processes in dissimilar organisms (Doolittle 1985). Consequently, it is unlikely that most of these could have contributed meaningfully to major evolutionary modifications. As far as gross chromosomal changes and morphological diversity is concerned, both the primates and the carnivores include representatives of highly stable and highly unstable genomes in terms of chromosome organization, using the human map as a standard. Within the primates, the gibbons and the New World monkey, *Aotus trivirgatus*, exhibit very diverse karyotypes both within and between genera, while the remaining primates show levels of extreme conservatism. Similarly, in the carnivores, the Felidae are highly conserved whereas the Canidae and Ursidae are dramatically rearranged karyotypically (O'Brien *et al.* 1985). Thus, within two major orders of placental mammals, where within each order there is reasonable morpho-logical conservatism, the karyotype can be very conserved or extensively shuffled. The other side of the coin is seen in the comparison of linkage group associations between cat and human (Fig. 6.4) which show the most extensive syntenic homologies known to exist between two mammalian orders, despite their marked phenotypic differences.

Since only a very small proportion of the genome codes for protein products, the hunt must be for the key genes that influence major developmental decisions, as opposed to their more conventional housekeeping

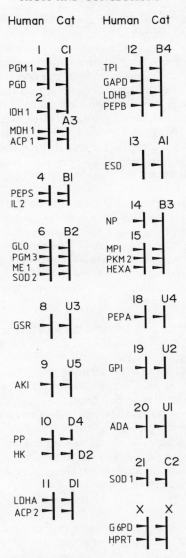

Figure 6.4   A comparison of the known syntenic human and feline linkage groups. The gene symbols employed are those in standard use for human enzyme loci (data of O'Brien *et al.* 1985).

colleagues. How, then, can such key genes be rapidly identified and, having been identified, how do we unravel the developmental networks to which they belong? One thing is certain, it is the invertebrates, such as *Drosophila melanogaster* and *Caenorhabditis elegans*, which must provide the vanguard for this assault. When one considers the evident dearth of relevant mammalian data on such a basic issue as developmental morphogens, it is clear that we can only hope to realistically evaluate vertebrate development from the clues provided by the invertebrates.

A persistent theme in evolutionary studies is the desire to identify mechanisms of speciation, and to determine which components of the genome, if any, were causative factors in the speciation process. Literally any genomic component, whether coding or non-coding, as well as gross genomic reorganization, has at some time been proposed to be a prime mover in speciation events. As we stressed earlier, what is absolutely crucial in discussions of this sort is to be able to distinguish those changes which were causative in species formation from those which either preceded, or else occurred subsequent to, the event. This is the crux of the problem and, to our knowledge, most data bases are incapable of deciding between these alternatives.

We, for example, have analyzed data sets dealing with gross genomic reorganization in muntjacs, equids, rodents and Diptera, and find no evidence whatsoever in these cases for major chromosome reorganization providing a prime mover in speciation phenomena. Nor do we find any convincing case for accepting the mobile elements of a genome as causative factors in speciation events. We stress that this does not mean that a chromosomal rearrangement, or a mobile induced event, could not ever have led to speciation. The difficulty, however, lies in demonstrating that this was the case.

Consider, for example, two species, A and B, that are assumed to have diverged on the basis of differentiation in their behavioural genes. How do we actually show this to be the case, other than by arguing convincingly? If we compare the two genomes, to determine what components in them have changed, we are likely to find many differences. Consequently, we could just as easily argue that the initial causative event was a change in the regulatory activities of a particular tRNA, or the turnover of a ribosomal spacer sequence. How do we distinguish between all the changes we can discern and those which actually led to speciation itself? The short answer is that we cannot, and we are unlikely ever to be able to do so with certainty. Many of the hypotheses which relate to speciation are impossible to falsify, but just as impossible to prove.

A much more important unresolved question is how organisms have achieved such radically different phenotypes in the light of what we now know about their genomes. Thus, a unicellular ciliate, a multicellular fly and probably a multicellular angiosperm all appear to have about 12 000 genes, many of which must be held in common to ensure conventional metabolism. This prompts us to ask, how many novel genes or multigene families were involved in the evolution of multicellularity, and the consequences of multicellularity, and to what extent existing genes or gene products were simply utilized in different combinations. Lima-de-Faria (1980) concluded that many genes which are present in multicellular eukaryotes are also represented in bacteria and unicellular eukaryotes. Consequently, he believes that metazoan evolution may have largely been concerned with 'sorting out' genes and gene combinations already present in unicells. Certainly, as we have seen elsewhere in this book, there are some striking examples of DNA sequence conservation across the entire phyletic spectrum of living things. In reality, however, we have no idea of how many genes or multigene families were involved in producing a multicellular state. Neither do we know whether it involved novel genes or novel multigene families, or whether it resulted from existing proteins being used in different combinations. There are some 2500 known genes in *Escherichia coli*, and the genome is large enough to accommodate up to 1500 more. By contrast, the genome of *Drosophila melanogaster*, which is approximately 40 times that of *Escherichia coli*, contains about 12 000 genes. If we consider only those genes that produce enzymes, then, as Doolittle (1985) points out, over 2000 structurally distinct enzymes have been identified on the basis of the chemical reactions they catalyze, and many of these must be common to most organisms. Since randomly generated proteins do not exhibit a biological function, Doolittle argues that a new functional protein is much more likely to come from a modification of an existing protein. He, therefore, believes that the greater majority of extant proteins must be descended from a small number of archetypes, and that gene duplication served as the primary mechanism of protein evolution. This process has undoubtedly led directly to the production of multisequence systems, like the haemoglobins, where variant forms of the same protein are produced. This is true also of the duplication that has led to the formation of two functional cytochrome c genes in *Drosophila melanogaster*. These two intron-less genes are found within a 4 kb DNA fragment at position 36A in the genome, and encode proteins having the same function but exhibiting totally different patterns of activity. The expression of one is relatively high and is stage and tissue specific, while the expression of the other is relatively low and is not stage or tissue specific. Thus, these two genes, encoding effectively the same protein product, have been sequestered into quite different developmental circuits (Limbach & Wu 1985).

Indirectly, duplication may well have played a role in producing new proteins with new functions. Thus, several enzyme systems which differ greatly in function, are known to include a common domain of about 70 amino acids which functions as a binding site for mononucleotides, and which shares the ability to bind coenzymes which include a mononucleotide in their structure. Here, as Doolittle points out, the common domain may well have had its origin in an archetypal protein which has, subsequently, been incorporated through duplication into new enzyme systems during the course of evolution. Alternatively, the continuing reassembly of two or more useful domains into new combinations, coupled with duplication events, has sometimes led to the production of an extended family of novel proteins (Fig. 6.5).

Of course, we need to bear in mind that, as cases of atavism remind us, many pathways of development have persisted within the genomic repertoire, despite the fact that their end products are not functional. There is, for example, a case on record of a humpback whale, *Megaptera*, which externally carried two tapered cylinders, approximately four feet in length, which included a femur, tibia and vestiges of a tarsus and metatarsus. Other, less striking, cases include the presence of foetal hair in elephants and

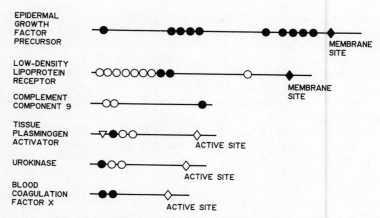

Figure 6.5 An extended family of disparate vertebrate proteins produced evolutionarily by the shuffling of common multiple domains. Five such domains are involved, each represented by a different geometrical symbol. Comparable domains in different proteins share a homologous, though not identical amino acid sequence (data of Doolittle 1985).

whales, and of vestigial teeth buried in the gums of embryonic whale-bone whales. How are these suppressed developmental programmes still maintained within the genome? Is it that many of the genes used in such programmes are also used in other developmental pathways? Were this indeed to be the case, it would imply that the executive genes, which canalize a particular pathway, do so as a result of a threshold effect as, for example, in the case of the *lin-12* locus of *Caenorhabditis elegans* (Greenwald *et al.* 1983). If this was not the case, one might have expected the unused executives to have become functionless. However, if genes of this type were part of a multigene family, then continuous homogenization events, particularly if they were molecularly driven, could cohesively maintain whole developmental networks, particularly if, in an extreme case, only one member of the multigene family is necessary for function.

If some of the key developmental switches involve multigene or multisequence families, then, like duplications, they will be buffered from mutagenic assault, since to become effective a newly arisen mutant will need first to spread through such a family. It is also possible that if multigene or multisequence switch families change cohesively through time, then key developmental pathways could be either formed or deleted when a family went to fixation.

It will surely not have escaped the reader's attention that all of the developmentally significant loci which we have discussed in *Drosophila melanogaster* involve either single or complex loci but not multigene families. This situation may, however, be more apparent than real, since usually only mutations in single complementation groups can actually be isolated by the mutagenesis screens used to identify genomic components with developmental roles. Multigene families are rarely, if ever, represented because it is not possible to mutagenize all their members simultaneously. Consequently, some parts of the *Drosophila melanogaster* genome may well have been excluded from evaluation, and we have no idea, as yet, of how crucial this defect may prove to be. The abnormal oocyte (*abo*) system may well be one such example. This maternal effect causes a reduction in egg to adult survival which can be ameliorated by paternally inherited sequences localized in heterochromatic regions on the Y, the X, and the second chromosome. These heterochromatic regions are denoted ABO-Y, ABO-X and ABO-2. Any one of these ABO regions is dispensable, since its function is adequately performed by the others. Phenotypic effects, however, require the simultaneous deletion of most, or all, of these ABO regions (Pimpinelli *et al.* 1986).

Mammals, on the other hand, have most recently been examined using recombinant DNA techniques which principally isolate genes that produce abundant RNAs. Consequently, the bias here has been for multigene families or high RNA producers, and it is the unique, low RNA producing, genes that have been excluded.

It is also possible that some genetic functions may be difficult to characterize because of shunts or redundancies. One such an example is the α-glycerophosphate dehydrogenase system of *Drosophila melanogaster*. O'Brien & Shimada (1974) isolated 'null' alleles at the α-*Gpdh* locus and found that although homozygous flies were viable they were, nevertheless, flightless. After 25 generations, however, a homozygous null stock regained the ability to fly and was quite viable and fertile, despite the continued absence of measurable activity. Since the α-glycerophosphate cycle is a major circuit involved in energy production, it is indeed surprising that abolition of the activity of the α-*Gpdh* locus does not result in inviability. Clearly, there must be a functional redundancy, involving different biochemical routes to the same end product, which acts as a biological buffer system in this instance.

Equivalent buffer systems operate in at least 11 other metabolic genes in which 'null' alleles also abolish activity (Table 6.1); yet all are viable as homozygotes despite the absence of critical functions (O'Brien 1973). In a number of these systems, the resolution of detection of enzyme activity for the 'nulls' was 0.1% of wild type levels, and in the case of acid phosphatase and xanthine dehydrogenase the amount of cross-reacting material proved negligible. Furthermore, double 'null' mutants were constructed for alkaline and acid phosphatases, and this combination yielded morphologically normal, viable and fertile organisms. Thus, genes which code for enzymes that, *a priori*, may appear critical for normal metabolism can, nevertheless, be dispensed with, and the genome possesses a remarkable capacity to compensate for what, at first sight, appears to be a serious metabolic lesion. Additionally, 6-phosphogluconate dehydrogenase (6PGD) and glucose-6-phosphate dehydrogenase (G6PD) are two enzymes involved in the pentose phosphate shunt. A null activity mutant allele of 6PGD in *Drosophila melanogaster* is lethal, suggesting that the proper functioning of the pentose shunt pathway is necessary for the survival of the organism. That this is not so is apparent from the fact that a mutant of G6PD which has no measurable

Table 6.1 Buffered genes in *Drosophila melanogaster*. At each of these eleven loci, null alleles are all homozygous viable so that while these silent alleles eliminate their respective gene products they do not appear to be lethal. The sensitivity of the assay employed detects at least 5% normal enzyme levels though in the case of *Acph*-1, rosy, *Adh* and α-*Gpdh*-1, analytical enzyme assays, with a sensitivity near 0.1% of wild type enzyme levels, also failed to detect trace activity in 'null' homozygotes. (From O'Brien 1973.)

Locus	Product	Number of null alleles
Est-C (3R)	esterase-C	1
Est-6 (3L)	esterase-6	1
Aph (3L)	alkaline phosphatase	1
Acph-1 (3R)	acid phosphatase	15
rosy (3R)	xanthine dehydrogenase	79
Aldox (3R)	aldehyde oxidase	2
Adh (2L)	alcohol dehydrogenase	14
Idh (3L)	isocitrate dehydrogenase	2
α-*Gpdh*-1 (2L)	α-glycerophosphate dehydrogenase	4
vermillion (X)	tryptophan pyrrolase	10
cinnabar (2R)	kynurenine hydroxylase	3

activity is, nevertheless, both viable and fertile. Moreover, double mutant combinations, lacking significant activity of both enzymes, lead to the suppression of the lethal character of the 6PGD allele, and so also produce viable and fertile flies (Hughes & Lucchesi 1977). This example provides yet another case of a system that is evidently buffered, though its dispensability obviously depends on the specific position of the block introduced into the pathway.

Ohno (1985) has reviewed the vertebrate data which show that some of the most abundant proteins are totally dispensable for the viability and well being of the organism. He has also challenged the popular belief that any gene which is under strict developmental control is indispensable to the organism. In so doing, he has pointed out something that has been a part of *Drosophila* folklore for decades, namely that the indispensability of a gene can only be assessed by a homozygous deficiency (or by loss of function mutations) at the locus.

The total absence of serum albumin, which is a major transport protein for steroid hormones and cholesterol, has no deleterious consequences in teleost fishes, rats, dogs or humans. The *Slp* gene in the mouse, although under strict developmental control, produces a defunct form of C_4 complement. Multiple IgCH homozygous deletions in the human immunoglobulin heavy chain cluster are also apparently quite harmless (Migone *et al*. 1984). The complement-fixing activity of IgG-2 can be replaced by IgG-1 or IgG-3, and the functions of IgA-1 can be performed by IgA-2. Furthermore, lysozyme is dispensable in the duck and the rabbit. Finally, fibrinopeptide B, the aminoterminal portion of the β-subunit of the plasma glycoprotein fibrinogen, has suffered all forms of random mutations, including amino acid substitutions, insertions and deletions. In fact, Ohno has stressed that fibrinopeptide B is the fastest changing amino acid sequence known.

These cases not only highlight the existence of buffered pathways, but point to the very real possibility of functionless proteins. In the case of fibrinopeptide B, there are similarities to the spectrum of mutational perturbations that have occurred in satellite sequences without any overt phenotypic effects. There is every reason to believe that some functionless protein coding sequences will exist in eukaryote genomes, a subset of which will be locked into a number of strictly regulated developmental circuits. If they do not compete with wild type enzymes for substrates and coenzymes, they will be harmless to the cell and the organism. In fact, it may be many millions of years before such developmentally regulated defunct genes, producing defunct protein products, can be silenced by mutation (Ohno 1985).

How many gene systems then are involved in this inbuilt redundancy? This is a very important question because it has a direct bearing on our notions of gene 'importance'. How do we recognize behavioural or neuroanatomical gene systems of importance? How do we devise paradigms for their isolation? Even in the most sophisticated systems, the common phenotypes are based on lethal, morphological, enzymatic, behavioural, meiotic, sterile and neuroanatomical types. These various categorizations, however, usually depend on the method of isolation, and all must select against multigene, multisequence or multicircuit systems. Thus, the isolation of some mutations may well be extremely difficult. Moreover, even if one finds a single mutant allele that yields a fancy phenotype, this is no guarantee that the gene in question is of real developmental significance. The only way to gauge this is to examine the phenotypes produced by amorphic alleles, or else to produce loss–of–function via deficiencies; no mean feats in mammalian systems.

J. H. Campbell (1983) has argued that some complex phenotypes in mammals are based on multigene families. It is worth considering, therefore, whether a substantial part of the higher order genomic bureaucracy might consist of such families. To the extent that this might indeed be so, the genome would obviously be buffered against rapid changes, since any mutation arising in such a family would need to spread through that family before any significant perturbation could occur. This has some interesting theoretical consequences. Earlier, we drew attention to the tremendous diversification of phyla that occurred during the Cambrian, when major metazoan body plans were formulated. If genomes at this time contained relatively few developmental gene circuits, relative to present day standards, and did not contain many buffering characteristics, such as multigene families or metabolic and developmental shunts, then major morphological innovations may have been relatively easy to introduce. Once the genome began to increase in size through gene duplication and the diversification of multigene families, however, it may have been more and more difficult to radically alter a particular developmental programme. As the genome became more complex and more redundant, developmental gene circuits might be expected to have become more inflexible, so that subsequent evolutionary novelty would be harder to produce. This could explain why the major metazoan phyla were sorted out in the Precambrian and Cambrian, and why morphological novelties on that scale have not appeared since. Rather, evolution has involved progressively finer tuning of a more and more recalcitrant genome.

6.2 FUTURE PROSPECTS

In the past, most attention to evolutionary problems has been focussed on morphological aspects of the organism. In evolutionary terms, however, it seems to us that the nervous system offers much more prospect for future study, since it was through behavioural evolution that the major advances leading to primates were consummated. The molecular information now available also suggests that it would be more profitable to turn our attention to the nervous system. Not only is it the first system to form in the embryo, but, in mammals, it appears that much of the RNA complexity lies in the brain. Thus, between 5000 and 30 000 genes appear to have an operational base there (Milner & Sutcliffe 1983). This creates quite a complication in the context of conventional views on protein evolution (see, for example, Doolittle 1985). It implies that during vertebrate evolution almost as many 'brain' genes have appeared in mammals as already existed for determining metabolic and morphological diversity. We can anticipate that, in terms of neuronal cell types, the mammal brain will probably be found to be almost as diverse as all the remaining cell types of the body put together. Within the confines of vertebrate morphological constraints, there appears to have been a veritable explosion of protein evolution, largely confined to wiring circuits within the brain. This, in turn, suggests that the major steps in vertebrate evolution were neural, so we need to de-emphasize the more spectacular morphological changes which have commanded most attention in the past.

This leads us, yet again, to the issue of cost effectiveness at the genetic level. Clearly, setting up and maintaining a sophisticated mammalian nervous system is costly in terms of gene number. Katz *et al.* (1983) have emphasized that increases in the complexity of the synaptic neuropile of vertebrates may occur in one of two ways. Either as a consequence of local expansion, or else through the establishment of new connections between previously unconnected areas. Local enlargements of specific areas, such as have occurred in the cerebellum, tectum and cerebrum, are, in general, due to increases in the number of local circuit neurons, whereas different local areas of the central nervous system are connected by a distinct class of projection neurons. While many of the major axon tracts of projection neurons are conserved in vertebrates, a few have appeared without apparent precursors. One such a tract is the (dorsal) corpus callosum, which is absent from monotremes and marsupials but present in placental mammals. In a manner analogous to the developing grasshopper system, most projection tracts follow pre-existing substrate pathways that guide growing axons during development. An important problem is to determine how new tracts arise, and what it costs genetically to set up a new tract. The glial sling constitutes such a novel pathway. It represents a population of glial cells, which normally migrate to the midline of the foetal brain before the axons arrive and then serve to guide axons to the opposite hemisphere. This glial structure is malformed in many mice of the inbred strain BALB/c, consequently, the corpus callosum fails to develop. While deficiency of the corpus callosum is completely recessive, it does not exhibit single locus Mendelian inheritance, and so is clearly not monogenic (Wahlsten 1982). The acallosal mutation is able to selectively delete the glial sling while leaving the callosal axons intact. If, as appears likely, the sling is a developmentally separate entity which is used by other axons as a 'neural highway', then it may have been one of the critical innovations responsible for the evolution of the corpus callosum.

This case has some similarities to the small optic lobes system of *Drosophila melanogaster*, where mutant forms of specific genes delete complementary subsets of neurons in the anterior optic tract (Fishbach & Lyly-Hünerberg 1983). The brain is obviously assembled in a modular form through time, and it may well be that identifying subsets of neural highways molecularly is initially a productive way to proceed.

There is every indication that, not only did the evolution of the nervous system involve a far more significant sequence of events than the morphological systems with which most evolutionists have been preoccupied, but, additionally, that its evolution in turn directed many of the morphological changes that have occurred. Gans & Northcutt (1983), for example, point out that, in many respects, the vertebrate head is really an addition to a protochordate body, and that in the evolution of vertebrates it was primarily the epidermal nerve plexus which was elaborated.

While one can be overcome by the sheer complexity of the morphological changes that have occurred during evolution, they pale in comparison with the amazing complexity of brain structure and function, especially that of mammals. Here again, however, the only sensible approach to the development of such evident complexity, is to concentrate less on the most complicated systems and use a compromise approach involving different organisms (see *Trends in Neurosciences*, June 1985). Thus, it is only in the simpler multicellular forms that we can call upon the required combination of genetics, suitable recombinant DNA technology and neuroanatomy, since, if we isolate a gene which we believe to be of neurological significance, how do we demonstrate it? It is, in fact, only in *Drosophila melanogaster* that we can easily use mosaic analysis to identify the focus of action of that gene. It is not much use finding an anatomical fault in the retina of the eye, for example, if we are unable to determine whether the mutant gene which leads to that fault exerts its primary effect there. However, by constructing flies in which the head is mutant and the remainder of the body normal, or vice versa, we can resolve this problem.

Since all nervous systems are constructed out of similar components, the neurons, the essence of neurogenesis is to determine what information is used to tell neurons where and when to go and what to do. The central problem of neurobiology is, therefore, how do neurons make synaptic connections and how do they distinguish between appropriate and inappropriate connections?

The work on grasshoppers offers an inroad into the molecular neuro-anatomy of simple neurological pathways (Doe & Goodman 1985). For lineage studies we can use *Caenorhabditis elegans* and then hopefully launch genomic raids into flies and mammals, though we appreciate that this avenue is not without its limitations. The neurogenetic approach to the 10^5 neuron *Drosophila melanogaster* system outlined in 1973 by Benzer, and recently reviewed by Hall (1985), has already begun to go molecular, with the fusion of classical neurobiology, recombinant DNA techniques and the use of monoclonal antibodies (see, for example, *50th Cold Spring Harbor Symposia* 1985; *Trends in Neurosciences*, vol. 8 1985; *The First European Meeting* of Drosophila *Neurogenetics* 1987; the *Molecular Neurobiology of* Drosophila *Meeting, Cold Spring Harbor* 1986).

In relation to the question of memory in humans we have only somewhat simple-minded theories on hypothetically long-lived proteins at synapses. Our aim is to find long-lasting changes in synaptic chemistry. In *Drosophila* there is at least one mutant, amnesiac, whose phenotype is described and whose chromosomal location is known to within two polytene bands. When more alleles of this locus have been isolated, it may well be worthwhile initiating a full-scale molecular investigation of this gene.

As far as brain function is concerned, we have little choice, as yet, but to continue to characterize neuronal components in the reasonable expectation that these are likely to be highly conserved between vertebrates and invertebrates. For example, Venter *et al.* (1984) have used monoclonal antibody cross-reactivity to demonstrate that muscarinic cholinergic receptors isolated from *Drosophila melanogaster* heads and rat and human brain are highly conserved. Recall that nearly half of the monoclonal antibodies made from the brain of *Drosophila melanogaster* cross-react to the human brain (Miller & Benzer 1983). As these authors point out '. . . it may be feasible to transfer a selected gene from human to fly in order to study its function.'

Young *et al.* (1985) have shown by direct DNA sequence comparisons that the *per* gene of *Drosophila melanogaster*, which plays such a pivotal role in the function of the biological clock, is also represented by a conserved sequence in the mouse. Furthermore, as we saw earlier, some *Drosophila* neurogenic proteins are homologous to the epidermal growth factor of mammals, and the *Drosophila* and vertebrate sodium channel subunits are highly conserved.

With the nervous system, we really are late molecular starters relative to the rest of the organism, simply because we know so little about the basic molecules that even impinge on structure, namely novel neuropeptides such as neurotransmitters, neuroregulators and neurohormones and neural cell adhesion molecules. We are only just now nibbling away at channels and pumps. For example, we do not even know how many types of calcium channels exist in neurons. Since calcium entry into a cell is basic to understanding processes such as transmitter release, spike initiation and rhythmic firing, this is part of the essential data that we need to uncover first in coming to grips with structure.

As has been so appropriately expressed by a number of people, while neurobiologists continue to raise questions for which molecular answers have not even been sought, molecular biologists have arrived at answers for which one still has to formulate the appropriate questions. This is symptomatic of exciting work to come. We are mining the genome faster than we can analyze the products on the surface. The use of monoclonal antibodies is already identifying previously unappreciated relationships among neurons, and it is likely that the molecular description of neuroanatomical networks will probably also outstrip the behavioural or functional correlates that go with them.

However, as we well appreciate, except in the very simplest organisms nervous systems are not hard wired. A major challenge, which is really beyond meaningful discussion at the present time, is to investigate the molecular basis of how synapses modify their functions by increases or decreases in number; in effect, how they turn over. How are old junctions sloughed off and new ones reconnected? Such synaptic modelling is well known in the peripheral nervous system of monkeys, where active turnover of synapses in the ciliary muscle has a half life of about 18 days. The mammalian nervous system thus has an evident capacity to renew its circuitry (Cotman & Nieto–Sampedro 1984).

There is one final and very pressing problem. How are we going to sift the developmentally and neurologically significant genes from the genomic haystacks of invertebrates and vertebrates (Struhl 1982)? The cross raiding of libraries is, as yet, largely unevaluated as a tool for systematically determining functional relationships. Thus, although the human acetyl-cholinesterase gene has been identified by homology to its *Drosophila melanogaster* counterpart (Soreq *et al.* 1985), and although the structure of the muscarinic cholinergic receptor is functionally very similar between *Drosophila melanogaster* and human brain (Venter *et al.* 1984), these are not really the genes we want in order to obtain significant insight into what gave rise to the sorts of novelties we saw, for example, in the acallosal mutant of mouse.

The problem has been put into perspective by Jackson (1986) who has pointed out that 'The . . . gunslingers of modern developmental biology shoot hard, fast and often inaccurately. Any likely gene which raises its head is quickly cloned and analyzed.' However, what could be more frustrating than to clone and sequence a gene, determine its cell and tissue location, and yet be unable to undertake the important next step, namely to determine where it contributes in an unknown developmental hierarchy. This is the fate of many studies, since without the capacity to manipulate gene dosage, or to routinely make null alleles or deficiencies at a locus, or to easily screen for interacting genes, the system will remain hamstrung. Gurdon (1985) has drawn attention to this very difficult problem of determining the functions of gene products, and has raised the spectre of gene unemployment; their products may be known, but this does not yield a route to function.

Unless we can develop and modify an *in vivo* mutagenesis scheme, such as the P element system in *Drosophila melanogaster* (Rubin *et al.* 1985, Laski *et al.* 1986, Rio *et al.* 1986), we are unlikely to ever open up an independent approach to vertebrate genomes. Even if directed *in vivo* mutagenesis via a P element type delivery vehicle is successful, vertebrates still face the same limitations as *Drosophila melanogaster*, namely how to find the key genes involved in significant circuits. This is actually a thornier problem than most like to admit. We can try and use mutant phenotypes, but this will not be very successful for multigene families or, as we saw, for buffered systems with shunts. What, moreover, is the mutant phenotype we wish to find, particularly for connectivity mutants dealing with deep brain functions? If, for example, a thousand particular neurons were deleted by a mutant allele, we may not even realize it if the phenotype is subtle, or if we do not have paradigms to test certain behaviours. Thus, many of these neurological circuits cannot be easily approached either in a molecular or genetic way as yet. We have to wait for detailed neuroanatomical studies before we know what is being perturbed. This area is only now being systematically examined in *Drosophila melanogaster* (Heisenberg & Böhl 1979, Fischbach & Heisenberg 1981, 1984), and at present we have no useful answer to offer to such a question.

6.3   FINAL STATEMENT

As far as the topic of evolution is concerned, molecular biologists have been rather more fortunate than their mathematically minded biological colleagues. By focussing on genome structure and function through recombinant DNA techniques, they have been led to identify developmental systems before attempting to put them into an evolutionary perspective. Consequently, they have been almost totally uninfluenced by the constraints of attempting to 'explain' evolution in mathematical terms. They have also not been sidetracked by mechanistically oblique issues, such as the selectionist-neutralist debate. It was Orgel & Crick (1980) and Doolittle & Sapienza (1980) who broke the fragile nexus between the obsessive hunt for function for every geonomic compartment and the 'solutions' offered in the context of evolutionary phenomena. These authors cautioned against seeking functional and adaptive answers to every piece of genomic flotsam and jetsam. In fact, Doolittle & Sapienza's (1980) brilliant encapsulation of the phenotype paradigm can be very usefully extended. They pointed out that once one is unable to assign a function to a DNA sequence of immediate benefit to an organism, 'evolutionary' explanations are usually resorted to. In the case of mobile elements, at least, once there was a strategy for genomic survival (namely transposition), no other explanations of their existence are necessary. The search for 'other explanations may prove, if not intellectually sterile, ultimately futile . . .' (Doolittle & Sapienza 1980). This futility is very obvious in some areas of evolutionary study, particularly in those dealing with speciation phenomena. Most of the hypotheses attempting to 'explain' evolution in such areas are near impossible to falsify in a pragmatic sense.

It is obvious that the differences which have led to present-day species are due to a plurality of causes, and proving which of these was the major one for any speciation event is a pious hope. The situation has been summed up by Dover (1980): **'Naturally we cannot reconstruct the myriad of events that have gone into shaping the contemporary pattern of relationships of a species group; nor can we realistically test . . . our hypotheses as to the prime determinative forces that have been involved.'**

The extrapolation from the arguments of Orgel, Crick, Doolittle and Sapienza is that there is, quite simply, no need to seek adaptive and functional reasons for every evolutionary phenomenon. In fact, if we turn to phylogenetic reconstructions, and if we are able to derive *all* the phylogenies for *all* extant species of organisms, would we find ourselves significantly closer to the developmental mechanisms which have produced morphological novelties? The answer is assuredly no. Knowing all the moves on the evolutionary chess-board gives little insight into the forces that led to these moves.

As unpalatable as it may seem to many biologists, certain aspects of conventional evolutionary biology have become stalled, and it is futile to pretend that continuing study along the well-worn, mathematically oriented neo-Darwinian pathways will provide significant insights into key evolutionary phenomena. **It is in the molecular analysis of development where significant hope is to be found**. Not only do the molecular analyses of eukaryote genomes confirm the need to unravel development in order to understand evolution, but, of greater significance, they emphasize the need to pay particular attention to the molecular basis of developmental networks and circuits. Specifically, in engineering terms, we need, at the level of cell biology, to understand the boundary conditions of the morphological assembly line in order to be able to predict the array of forms which are likely to follow from modifications in that assembly line.

Although we have concentrated quite deliberately on DNA in this book, we recognize that ultimately this is *not* where the real developmental future lies. That future is with proteins and cells, and this is now becoming a more concrete realization. Rubin (1985b) has argued that the major future conceptual breakthroughs in developmental biology will stem from the molecular analyses of the protein products of developmentally important genes. Edelman (1985) has highlighted the **cell** as the basic unit of control, and the **cell surface** as the nexus of control events. In the developmental hierarchy we are still at the bottom of the information ladder, simply because presently this is the most tractable place to be. As Doolittle (1985) has quite rightly pointed out, while DNA supplies the information, organisms *are* their proteins.

Thus, when we lift our horizons away from 5' and 3' landscapes, we are led to consider much more interesting problems, such as the basis of programmed cell movement. The lamellipodial motion of individual mesodermal cells during embryogenesis, for example, can be contrasted with the ectodermal and endodermal cell populations which move as sheets of cells. How is this social behaviour genetically controlled? We saw in the cases of neurological pathways how cells could follow the previously established route plans laid down by pioneer neurons. Thus, adhesive or chemical gradients must also exist to orient cell movements.

It is when one attempts to examine the molecular basis of morphogenesis, that the irrelevance of neo-Darwinian theory is put into real perspective. This theory can predict what is likely to survive once it has appeared, but it has absolutely nothing whatsoever to say about the origin of change. If our assumption that metazoan phyla underwent most experimentation early in their evolutionary histories because developmental gene hierarchies were less complex and more easily subject to perturbation, is valid, then present-day studies which bear only marginally on developmentally significant circuits are unlikely to be addressing the crucial evolutionary issues. How important, for example, is it to continue to erect and estimate the reliability of phylogenetic trees, particularly one as well known as that familiar quartet of human, chimp, gorilla and orang? Does it really matter which is our closest relative? Is it not more important to discover the underlying molecular mechanisms that have given rise to the different morphologies? Similarly, how critical is it to obtain detailed information on the random fixation of selectively neutral mutations, or the calibrations of molecular clocks, or estimates of the divergence times of species? Have such studies made significant contributions to our molecular understanding of the *genesis* of evolutionary novelties? Readers can judge this for themselves, but we believe that new insights are now urgently required into solving problems of developmental engineering. In our view, we need to put our major efforts into molecular embryology where the quintessence of morphological novelty must obviously lie.

The current position is that biologists are still preoccupied at the level of the genome, though elsewhere in this book we have, at least, attempted to clarify the issues somewhat in determining detritus levels within genomes. The next level for study is that of the protein products of the developmentally important part of the genome, and beyond that the analysis of the cellular properties of these proteins. It is patently obvious that if we are to understand the shape and behaviour of organisms, then our ultimate sights must be set on three-dimensional changes through time. Although this is an admirable goal, it will only be achievable through the detailed examination of tractable two-dimensional systems.

In focussing our attention on cellular phenomena, we will initially need to consider single cells and, in order to understand cell shape and behaviour, to direct attention to the assembly and morphogenesis of the cytoskeleton (see, for example, *Molecular biology of the cytoskeleton*, Borisy et al. 1984). Beyond this, there is the question of the social behaviour of cells and, in terms of morphogenesis in multicells, particular attention will have to be directed to the control of cell shape. As has been discussed by Wessels (1982), there are a limited number of available options, since morphogenesis ultimately depends on some form of cell contact. Cells need to adhere to substrates, in the form of sheets, fibre networks or other cells, in order to undergo division. Thus, there must be significant interplay between the structure of the cytoskeleton and the anchorage capabilities of cells. It is only in terms of the social interplay of cells that we can hope to unravel major morphological shifts of the kind implicit in the feather-hair dichotomy described by Oster & Alberch (1982), the underlying basis of which depends on cell changes which follow on from the evagination or invagination of entire sheets of cells. A direct molecular approach to this specific case is hardly feasible, simply because the system cannot be easily examined in a genetic way. The morphogenetic events involved in the imaginal discs of *Drosophila melanogaster*, however, offer a much more promising system, involving not too disimilar principles. Thus, discs can be isolated in bulk and induced to undergo morphogenetic changes, including evagination, following the application of ecdysteroids, such as 20-hydroxyecdysone (Yund *et al.* 1978). The use of an even more potent ecdysteroid, Ponasterone A, which is 50 times more active than 20-hydroxyecdysone has opened up the quantitative characterization of the mass isolated imaginal disc system in biochemical terms, an approach not previously possible.

There are basically two sequential phases in disc development; a morphogenetic phase in which the shape is altered, and a differentiative phase in which specialized cells are formed. Fristrom & Rickoll (1982) have drawn particular attention to the manner in which tissue shape changes can be achieved, including localized cell division or cell death, cell shape changes and cell movements. The morphogenetic changes that convert simple monolayers into an elongated leg, or the precise array of ommatidia that are formed from the eye disc, are examples of the dramatic changes of cellular events involved in disc differentiation, and both have been analyzed in some detail. Cell division and cell death are of limited importance during leg disc evagination, since total cell number remains approximately the same, and cell movement and rearrangement within the epithelial sheet provides the predominant basis for evagination (Fristrom & Fristrom 1975). In eye discs, on the other hand, it is mainly changes in cell shape that influence ommatidial organization.

There are numerous possible approaches to disc perturbation, though it is well to remember that it will be important to determine the *specificity* of each mutation. For example, decapentaplegic is a mutation which affects imaginal disc development and is thought to be involved in the elaboration of positional information (Spencer *et al.* 1982, Gelbart *et al.* 1985). However, as Struhl (1982) points out, there are far less spectacular possibilities such as localized cell death. Similarly, Szabad & Bryant (1982), in examining various lethal mutations which give rise to a 'discless' phenotype, have shown that these mutations are not imaginal disc specific but cause alterations in cell proliferation in the larval nervous system, the imaginal discs and the gonads. Of course, the basic molecular approaches to the disc system, in terms of morphogenesis, are only in their infancy and deal mainly with proteins that are the result of transcriptional modulations of the system. As Fristrom & Rickoll (1982) point out, the real action is yet to come, via the isolation of cell surface proteins that have a morphogenetic function, and these studies are only just beginning.

We hope it is now obvious to the reader that the time has come to de-emphasize the mathematical approaches to neo-Darwinism as a means of explaining the critical factors of evolution. Indeed, Gould (1980b) has gone even further and comments that 'The modern synthesis, as an exclusive proposition, has broken down on both of its fundamental claims:

extrapolationism (gradual allelic substitution as a model for all evolutionary change) and nearly exclusive reliance on selection leading to adaptation', adding that **neo-Darwinism '. . . as a general proposition, is effectively dead, despite its persistence as a text-book orthodoxy.'** Whether or not one accepts such damning comments, it is clearly time to refocus on the significant aspects of molecular developmental biology, which is the only available key with which to unlock the generation of evolutionary novelty. The next challenging horizons are clearly in protein–protein interactions, self-assembly systems and cell–cell interactions.

Since the 'rules' of development must deal with the protein interactions that guide cellular interactions, these rules can be expected to differ substantially from those dealing with gene regulation. Even so, it is only when subsets of the genome allow a particular subset of proteins to be produced that specific protein interactions, and hence cellular interactions, are able to occur. Thus, the genetic programme constrains the three-dimensional procession of proteins through time, both in a developmental and an evolutionary sense.

References

'To know where we can find a thing is the chief part of learning.'

Persius

Abraham, I & J. C. Lucchessi 1974. Dosage compensation of genes on the left and right arms of the X-chromosome of *Drosophila pseudoobscura* and *D. willistoni*. *Genetics* **78**, 1119–26.

Aho, S., V. Tate & H. Boedtker 1984. Location of the 11 bp exon in the chicken pro α2(I) collagen gene. *Nucleic Acids Research* **12**, 6117–25.

Akam, M. E. 1983. The location of Ultrabithorax transcripts in *Drosophila* tissue sections. *The EMBO Journal* **2**, 2075–84.

Akam, M. E., A. Martinez-Arias, R. Weinzierl & C. D. Wilde 1985. Function and expression of Ultrabithorax in the *Drosophila* embryo. *Cold Spring Harbor Symposia on Quantitative Biology* **50**, 195–200.

Alberch, P. 1982. Developmental constraints in evolutionary processes. In *Evolution and development*, J. T. Bonner (ed.), 313–32. Dahlem Konferenzen. Berlin: Springer.

Alexander, M. L. 1976. The genetics of *Drosophila virilis*. In *The genetics and biology of Drosophila*, vol. 1c., M. Ashburner & E. Novitski (eds), 1365–1427. New York: Academic Press.

Alfageme, C. R., G. T. Rudkin & L. H. Cohen 1980. Isolation, properties and cellular distribution of D1, a chromosomal protein of *Drosophila*. *Chromosoma* **78**, 1–31.

Ali, Z., B. Drees, K. G. Coleman, E. Gustavson, T. L. Karr, L. M. Kuvar, S. J. Poole, W. Soeller, M. P. Weir & T. Kornberg 1985. The engrailed locus of *Drosophila melanogaster*: genetic, developmental and molecular studies. *Cold Spring Harbor Symposia on Quantitative Biology* **50**, 229–32.

Alonso, A., B. Kühn & J. Fischer 1983. An unusual accumulation of repetitive sequences in the rat genome. *Gene* **26**, 303–6.

Ambros, V. & H. R. Horvitz 1984. Heterochronic mutants of the nematode *Caenorhabditis elegans*. *Science* **226**, 409–16.

Ammerman, D., G. Steinbrück, L. von Berger & W. Hennig 1974. The development of the macronucleus in the ciliated protozoan *Stylonychia mytilus*. *Chromosoma* **45**, 401–29.

Amos, A. & G. Dover 1981. The distribution of repetitive DNAs between regular and supernumerary chromosomes in species of *Glossina* (Tsetse): a two-step process in the evolution of supernumeraries. *Chromosoma* **81**, 673–90.

Amstutz, H., P. Munz, W-D. Heyer, U. Lenpold & J. Kohli 1985. Concerted evolution of tRNA genes: intergenic conversion among three unlinked serine tRNA genes in *S. pombe*. *Cell* **40**, 879–86.

Anderson, K. V. 1984. Toll: a key gene in the establishment of polarity in the dorsal-ventral axis of the *Drosophila* embryo. Crete Drosophila Meeting, Kolymbari. (Abst).

Anderson, K. V. & C. Nüsslein-Volhard 1984. Information for the dorsal-ventral pattern of the *Drosophila* embryo is stored as maternal mRNA. *Nature* **311**, 223–7.

Anderson, K. V., G. Jürgens & C. Nüsslein-Volhard 1985. Establishment of dorso-ventral polarity in the *Drosophila* embryo: genetic studies on the role of the *Toll* gene product. *Cell* **42**, 779–89.

Andrews, D. L., L. Millstein, B. A. Hamkalo & J. M. Gottesfeld 1984. Competition between *Xenopus* satellite 1 sequences and Pol III genes for stable transcription. *Nucleic Acids Research* **12**, 7753–69.

Andronico, F., S. De Lucchini, F. Graziani, I. Nardi, R. Batistoni & G. Barsacchi-Pilone 1985. Molecular organization of ribosomal RNA genes clustered at variable chromosomal sites in *Triturus vulgaris meridionalis* (Amphibia Urodela). *Journal of Molecular Biology* **186**, 219–29.

Angerer, R. C. & E. H. Davidson 1984. Molecular indices of cell lineage specification in sea urchin embryo. *Science* **226**, 1153–60.

Arrighi, F. E. 1974. Mammalian chromosomes. In *The cell nucleus*, H. Busch (ed.), 1–32. New York: Academic Press.

Arrigo, A-P., J-L. Darlix, E. W. Khandjian, M. Simon & P. F. Spahr 1985. Characterisation of the prosome from *Drosophila* and its similarity to the cytoplasmic structures formed by the low molecular weight heat shock protein. *The EMBO Journal* **4**, 399–406.

Artavanis-Tsakonas, S., M. A. T. Muskavitch & B. Yedvobnick 1983. Molecular cloning of Notch, a locus affecting neurogenesis in *Drosophila melanogaster*. *Proceedings of the National Academy of Sciences* **80**, 1977–81.

Artavanis-Tsakonas, S., B. Grimwade, D. Hartley, K. M. Johansen, A. Preiss, R. Ramos, K. Wharton & T. Xu 1987. Molecular biology of neurogenesis. In *Molecular Biology of neurogenesis*. 1st European Meeting of *Drosophila* Neurogenetics, Simonswald. *The Journal of Neurogenetics* **4**, 111 (Abst).

Ashburner, M. 1970. Patterns of puffing activity in the salivary gland chromosomes of *Drosophila*. *Chromosoma* **31**, 356–76.

Ashburner, M. & H. D. Berendes 1978. Puffing of polytene chromosomes. In *Genetics and biology of* Drosophila, M. Ashburner & T. R. F. Wright (eds), 315–95. London: Academic Press.

Atkin, N. B., G. Mattinson, W. Becak & S. Ohno 1965. The comparative DNA content of 19 species of placental mammals, reptiles and birds. *Chromosoma* **17**, 1–10.

Awgulewitsch, A., M. J. Utset, C. P. Hart, W. McGinnis & F. H. Ruddle 1986. Spatial restriction in expression of a mouse homeobox locus within the central nervous system. *Nature* **320**, 328–35.

Axel, R., P. Felgelson & G. Schutz 1976. Analysis of the complexity and diversity of mRNA from chicken liver and oviduct. *Cell* **7**, 247–54.

Bachmann, K., O. B. Goin & C. J. Goin 1972. Nuclear DNA amounts in vertebrates. *Brookhaven Symposia in Biology* **23**, 419–50.

Bacon, J. P. & N. J. Strausfeld 1986. The dipteran 'Giant fibre' pathway: neurons and signals. *Journal of Comparative Physiology* **A158**, 529–48.

Baird, D., E. Aceves, R. Wyman, J. Davies, V. Pirrotta & G. L. G. Miklos 1986. Genetics and cloning of Passover, a mutant with altered neural connectivity. In *Molecular neurobiology of* Drosophila, p. 61 (Abst). New York: Cold Spring Harbor Laboratory.

Baker, B. S. & K. A. Ridge 1980. Sex and the single cell. I. On the action of major loci affecting sex determination in *Drosophila melanogaster*. *Genetics* **94**, 383–423.

Baltimore, D. 1985. Retroviruses and retrotransposons: the role of reverse transcription in shaping the eukaryotic genome. *Cell* **40**, 481–2.

Bantock, C. R. 1970. Experiments on chromosome elimination in the gall midge *Mayetiola destructor*. *Journal of Embryology and Experimental Morphology* **24**, 257–86.

Baranov, V. S. 1980. Mice with Robertsonian translocations in experimental Biology and Medicine. *Genetica* **52/53**, 22–32.

Bardwell, J. C. A. & E. A. Craig 1984. Major heat shock gene of *Drosophila* and the *Escherichia coli* heat inducible dnaK gene are homologous. *Proceedings of the National Academy of Sciences* **81**, 848–52.

Bargiello, T. A., F. R. Jackson & M. W. Young 1984. Restoration of circadian behavioural rhythms by gene transfer in *Drosophila*. *Nature* **312**, 752–4.

Barlow, P. W. 1973. Mitotic cycles in root meristems. In *The cell cycle in development and differentiation*, M. Balls & F. S. Billett (eds), 133–65. Cambridge: Cambridge University Press.

Bass, B. L. & T. R. Cech 1984. Specific interaction between the self splicing RNA of *Tetrahymena* and its guanosine substrate: implications for biological catalysis by RNA. *Nature* **308**, 820–6.

Bastiani, M. J., C. Q. Doe, S. L. Helfand & C. S. Goodman 1985. Neuronal specificity and growth cone guidance in grasshopper and *Drosophila* embryos. *Trends in Neuroscience* **8**, 257–66.

Bate, C. M. 1976. Pioneer neurones in an insect embryo. *Nature* **260**, 54–6.

Bateson, W. 1922. Evolutionary faith and modern doubts. *Science* **40**, 55–61.

Bautch, V. L., R. V. Storti, D. Mischke & M. L. Pardue 1982. Organisation and expression of *Drosophila* tropomyosin genes. *Journal of Molecular Biology* **162**, 231–50.

Baverstock, P. R., M. Gelder & A. Jahnke 1983. Chromosome evolution in Australian *Rattus* – G banding and hybrid meiosis. *Genetica* **60**, 93–103.

Beachy, P. A., S. L. Helfand & D. S. Hogness 1985. Segmental distribution of bithorax complex proteins during *Drosophila* development. *Nature* **313**, 545–51.

Beermann, S. 1977. The diminution of heterochromatic chromosome segments in *Cyclops* (Crustacea, Copepoda). *Chromosoma* **60**, 297–343.

Beermann, S. 1984. Circular and linear structures in chromatin diminution of *Cyclops*. *Chromosoma* **89**, 321–8.

Beermann, W. 1952. Chromomerrenkonstanz und spezifische modifikationen der chromosomen-struktur in der entwicklung und organ differenzierung von *C. tentans*. *Chromosoma* **5**, 139–98.

Bell, L., C. Cromiller, P. Hinds, E. Maine, H. Salz, P. Schedl & T. W. Cline 1986. Crete Drosophila Meeting, Kolymbari (Abst).

Belote, J. M. & J. C. Lucchesi 1980. Male-specific lethal mutations of *Drosophila melanogaster*. *Genetics* **96**, 165–86.

Belote, J. & M. McKeown 1986. Molecular cloning and characterization of the transformer locus. Crete Drosophila Meeting, Kolymbari (Abst).

Belote, J. M., A. M. Handler, M. F. Wolfner, K. J. Livak & B. S. Baker 1985. Sex-specific regulation of yolk protein gene expression in *Drosophila*. *Cell* **40**, 339–48.

Belote, J. M., M. B. McKeown, D. J. Andrew, T. N. Scott, M. J. Wolfner & B. S. Baker 1985. Control of sexual differentiation in *Drosophila melanogaster*. *Cold Spring Harbor Symposia on Quantitative Biology* **50**, 605–14.

Bender, W., P. Spierer & D. S. Hogness 1983a. Chromosomal walking and jumping to isolate DNA from the *Ace* and *rosy* loci and the Bithorax complex in *Drosophila melanogaster*. *Journal of Molecular Biology* **168**, 17–33.

Bender, W., M. Akam, F. Karch, P. A. Beachy, M. Peifer, P. Spierer, E. B. Lewis & D. S. Hogness 1983b. Molecular genetics of the bithorax complex in *Drosophila melanogaster*. *Science* **221**, 23–9.

Bender, W., B. Weiffenbach, F. Karch & M. Peifer 1985. Domains of *cis*-interaction in the Bithorax Complex. *Cold Spring Harbor Symposia on Quantitative Biology* **50**, 173–80.

Bendich, A. J. 1985. Plant mitochondrial DNA: Unusual variation on a common theme. In *Genetic flux in plants*, B. Hohn & E. S. Dennis (eds), 111–38. Vienna: Springer.

Bendich, A. J. & R. S. Anderson 1977. Characterization of families of repeated DNA sequences from four vascular plants. *Biochemistry* **16**, 4655–63.

Bendich, A. J. & L. P. Gauriloff 1984. Morphometric analysis of cucurbit mitochondria: the relationship between chondriome volume and DNA content. *Protoplasma* **119**, 1–7.

Benirschke, K., D. H. Wurster, R. J. Low & N. B. Atkin 1970. The chromosome complement of the Aardvark, *Orycteropus afer*. *Chromosoma* **31**, 68–78.

Bennett, D., K. Artzt, T. Magnuson & M. Spiegelman 1977. Developmental interactions studied with experimental teratomas derived from mutants at the T/t locus in the mouse. In *Cell interactions in differentiation*, M. Karkinen-Jääskelainen, L. Saxen & L. Weiss (eds), 389–98. New York: Academic Press.

Bennett, M. D. 1971. The duration of meiosis. *Proceedings of the Royal Society, London* **B178**, 277–99.

Bennett, M. D. 1972. Nuclear DNA content and minimum generation time in herbaceous plants. *Proceedings of the Royal Society, London* **B181**, 109–137.

Bennett, M. D. 1973. The duration of meiosis. In *The cell cycle in development and differentiation*, M. Balls & F. S. Billett (eds), 111–31. Cambridge: Cambridge University Press.

Bennett, M. D. & J. B. Smith 1972. The effects of polyploidy on meiotic duration and pollen development in cereal anthers. *Proceedings of the Royal Society, London* **B181**, 81–107.

Bennett, M. D. & J. B. Smith 1973. Genotypic, nucleotypic and environmental effects on meiotic time in wheat. *Proceedings of the 4th International Wheat Genetics Symposium*, 637–44. Missouri Agricultural Experimental Station, Columbia.

Bennett, M. D., J. B. Smith & J. S. Heslop-Harrison 1982. Nuclear DNA amounts in angiosperms. *Proceedings of the Royal Society, London* **B216**, 179–99.

Bentley, D. & H. Keshishian 1982. Pathfinding by peripheral pioneer neurons in grasshoppers. *Science* **218**, 1082–8.

Benzer, S. 1973. Genetic dissection of behaviour. *Scientific American* **229**, 24–37.

Berendes, H. D. 1963. The salivary gland chromosomes of *Drosophila hydei* Sturtevant. *Chromosoma* **14**, 195–206.

Berendes, H. D. & W. Th. M. Thijssen 1971. Developmental changes in genome activity in *Drosophila lebanonensis* Pipkin. *Chromosoma* **33**, 345–60.

Berg, O. G., R. B. Winter & P. H. Von Hippel 1982. How do genome-regulatory proteins locate their DNA target sites? *Trends in Biochemical Sciences* **7**, 52–5.

Bernfield, M. R. 1981. Organisation and remodelling of the extracellular matrix in morphogenesis. In *Morphogenesis and pattern formation*, T. G. Connelly, L. L. Brinkley & B. M. Carlson (eds), 139–62. New York: Raven Press.

Bernstein, S. I., K. Mogami, J. J. Donady & C. P. Emerson 1983. *Drosophila* muscle myosin heavy chain encoded by a single gene in a cluster of muscle mutations. *Nature* **302**, 393–7.

Bier, K., W. Kunz & D. Ribbert 1969. Insect oogenesis with and without lampbrush chromosomes. *Chromosomes Today* **2**, 107–15.

Biggs, J., L. L. Searles & A. L. Greenleaf 1985. Structure of the eukaryotic transcription apparatus: features of the gene for the largest subunit of *Drosophila* RNA polymerase II. *Cell* **42**, 611–21.

Bingham, P. M., M. G. Kidwell & G. M. Rubin 1982. The molecular basis of P-M hybrid dysgenesis: the role of the P element, a P strain-specific transposon family. *Cell* **29**, 995–1004.

Bird, A. P. 1984. DNA methylation – how important in gene control. *Nature* **307**, 503–4.

Birnstiel, M. L., M. Busslinger & K. Strub 1985. Transcription termination and 3' processing: the end is in site! *Cell* **41**, 349–59.

Bishop, J. M. 1983. Cellular oncogenes and retroviruses. *Annual Review of Biochemistry* **52**, 301–54.

Bishop, J. M. 1985. Viral oncogenes. *Cell* **42**, 23–38.

Bishop, J. M., B. Drees, A. L. Katsen, T. B. Kornberg & M. A. Simon 1985. Proto-oncogenes of *Drosophila melanogaster*. *Cold Spring Harbor Symposia on Quantitative Biology* **50**, 727–31.

Blackburn, E. H. 1984. Telomeres: do the ends justify the means? *Cell* **37**, 7–8.

Blakeslee, A. F. 1934. New Jimson weeds from old chromosomes. *Journal of Heredity* **25**, 81–108.

Bloom, S. E. 1972. Chromosome abnormalities in chicken (*Gallus domesticus*) embryos: types, frequencies and phenotypic effects. *Chromosoma* **37**, 309–26.

Boeke, J. D., D. J. Garfinkel, C. A. Styles & G. R. Fink 1985. *Ty* elements transpose through an RNA intermediate. *Cell* **40**, 491–500.

Bogart, J. P. 1980. Evolutionary implications of polyploidy in Amphibians and Reptiles. In *Polyploidy*, W. H. Lewis (ed.), 341–78. New York: Plenum Press.

Bond, J. F., J. L. Fridovich-Keil, L. Pillus, R. C. Mulligan & F. Solomon 1986. A chicken–yeast chimeric β-tubulin protein is incorporated into mouse microtubules *in vivo*. *Cell* **44**, 461–8.

Bone, Q. 1972. *The origin of chordates*. Oxford Biology Readers **18**, 1–16.

Bonicelli, E., A. Simeone, A. LaVolpe, A. Faiella, V. Fidanza, D. Acampora & L. Scotto 1985. Human cDNA clones containing homeobox sequences. *Cold Spring Harbor Symposia on Quantitative Biology* **50**, 301–6.

Borisy, G. G., D. W. Cleveland & D. B. Murphy 1984. *Molecular biology of the cytoskeleton*. New York: Cold Spring Harbor Laboratory.

Borschenius, S. N., N. A. Belozerskaya, N. A. Merkulova, V. G. Volfson & V. I. Vorob'ev 1978. Genome structure of *Tetrahymena pyriformis*. *Chromosoma* **69**, 275–89.

Borst, P. 1983. Antigenic variation in trypanosomes. In *Mobile genetic elements*, J. A. Shapiro (ed.), 621–59. New York: Academic Press.

Bossy, B., L. M. C. Hall & P. Spierer 1984. Genetic activity along 315 kb of the *Drosophila* genome. *The EMBO Journal* **3**, 2537–41.

Bossy, B., L. M. C. Hall, M. Ballivet & P. Spierer 1986. Molecular genetics of the cholinergic synapse of *Drosophila*. In *Molecular neurobiology of Drosophila*, p. 30 (Abst). New York: Cold Spring Harbor Laboratory.

Bostock, C. J. & C. Tyler-Smith 1982. Changes in genomic DNA with methotrexate-resistant cells. In *Genome evolution*, G. A. Dover & R. B. Flavell (eds), 69–93. London: Academic Press.

Boswell, R. E. & A. P. Mahowald 1985. *Tudor*, a gene required for assembly of the germ plasm in *Drosophila melanogaster*. *Cell* **43**, 97–104.

Boswell, R. E., L. A. Klobutcher & D. M. Prescott 1982. Inverted terminal repeats are added to genes during macronuclear development in *Oxytricha nova*. *Proceedings of the National Academy of Sciences* **79**, 3255–9.

Bourouis, M. & Jarry B. 1983. Vectors containing a prokaryotic dihydrofolate reductase gene transform *Drosophila* cells to methotrexate-resistance. *The EMBO Journal* **2**, 1099–104.

Brand, A. H., L. Breeden, J. Abraham & R. Sternglanz & K. Nasmyth 1985. Characterisation of a 'silencer' in yeast: a DNA sequence with properties opposite to those of a transcriptional enhancer. *Cell* **41**, 41–8.

Brand, M., K. Bremer, A. de la Concha, U. Dietrich, F. Jimenez, E. Knust, G. M. Technau, U. Trepass, H. Vässin, D. Weigel & J. A. Campos-Ortega 1986. Early neurogenesis in *Drosophila melanogaster*. Crete Drosophila Meeting, Kolymbari (Abst).

Bregliano, J. C., G. Picard, A. Bucheton, A. Pelisson, J. M. Lavige & P. L'Heritier 1980. Hybrid dysgenesis in *Drosophila melanogaster*. *Science* **207**, 606–11.

Brennan, M. D., R. G. Rowan & W. J. Dickinson 1984. Introduction of a functional P element into the germ line of *Drosophila hawaiiensis*. *Cell* **38**, 147–51.

Brent, R. & M. Ptashne 1984. A bacterial repressor protein or a yeast transcriptional terminator can block upstream activation of a yeast gene. *Nature* **312**, 612–15.

Britten, R. J. 1982. Genomic alterations in evolution. In *Evolution and development*, J. T. Bonner (ed.), 41–64. Berlin: Springer.

Britten, R. J. & E. H. Davidson 1969. Gene regulation in higher cells: a theory. *Science* **165**, 349–58.

Britten, R. J. & E. H. Davidson 1971. Repetitive and non-repetitive DNA sequences and a speculation on the origins of evolutionary novelty. *Quarterly Review of Biology* **46**, 111–38.

Brown, D. D. 1984. The role of stable complexes that repress and activate eucaryotic genes. *Cell* **37**, 359–65.

Brown, D. D. & I. B. Dawid 1968. Specific gene amplification in oocytes. *Science* **160**, 272–80.

Bryant, P. J. 1978. Pattern formation in imaginal discs. In *Genetics and biology of Drosophila*, Vol. 2c, M. Ashburner & T. R. F. Wright (eds), 229–335. London: Academic Press.

Bryant, S. V., V. French & P. J. Bryant 1981. Distal regeneration and symmetry. *Science* **212**, 993–1002.

Bucheton, A., R. Paro, H. M. Sang, A. Pelisson & D. J. Finnegan 1984. The molecular basis of IR hybrid dysgenesis in *Drosophila melanogaster*: identification, cloning and properties of the I factor. *Cell* **38**, 153–63.

Buchner, E., S. Buchner, M. Heisenberg & A. Hofbauer 1987. Cell-specific staining by immunochemistry in the visual system of *Drosophila*: towards cellular characterization of mutant brain defects. In *Molecular biology of neurogenesis*, 1st European Meeting of *Drosophila* Neurogenetics, Simonswald. *The Journal of Neurogenetics* **4**, 117–18 (Abst).

Bush, G. L. 1981. Stasipatric speciation and rapid evolution in animals. In *Evolution and speciation*, W. R. Atchley & D. S. Woodruff (eds), 201–8. New York: Cambridge University Press.

Calabretta, B., D. L. Robberson, H. A. Barrera-Saldana, T. P. Lambrou & G. F Saunders 1982. Genome instability in a region of human DNA enriched in *Alu* repeat sequences. *Nature* **296**, 219–25.

Calvert, G. R. & I. W. Dowes 1984. Cell size control of development in *Saccharomyces cerevisiae*. *Nature* **312**, 61–3.

Campbell, J. H. 1983. Evolving concepts of multigene families. *Current Topics in Biological and Medical Research* **10**, 401–7.

Campbell, K. S. W. & C. R. Marshall 1987. Rates of evolution among palaeozoic echinoderms. In *Rates of evolution*, K. S. W. Campbell & M. Day (eds), London: Allen & Unwin.

Campos-Ortega, J. A. 1985. Genetics of early neurogenesis in *Drosophila melanogaster*. *Trends in Neuroscience* **8**, 245–50.

Campuzano, S., L. Balcells, R. Villares, L. Carramolino, L. Garcia-Alonso & J. Modolell 1986. Excess function hairy wing mutations caused by *gypsy* and *copia* insertions within structural genes of the achaete-scute locus of *Drosophila*. *Cell* **44**, 303–12.

Capanna, E. 1982. Robertsonian numerical variation in animal speciation: *Mus musculus*, an emblematic model. In *Mechanisms of speciation*, C. Barigozzi, G. Montalenti & M. J. D. White (eds), 155–77. New York: Alan R. Liss.

Capanna, E., M. V. Civitelli & M. Cristaldi, 1977. Chromosomal rearrangement, reproductive isolation and speciation in mammals. The case of *Mus musculus*. *Bolletino di Zoologia* **44**, 213–46.

Carrasco, A. E., W. McGinnis, W. J. Gehring & E. M. de Robertis 1984. Cloning of an *X. laevis* gene expressed during early embryogenesis coding for a peptide region homologous to *Drosophila* homeotic genes. *Cell* **37**, 409–14.

Carroll, D. 1983. Genetic recombination of bacteriophage λ DNA's in *Xenopus* oocytes. *Proceedings of the National Academy of Sciences* **80**, 6902–6.

Carroll, S. B. & M. P. Scott 1986. Zygotically active genes that affect the spatial expression of the *fushi tarazu* segmentation gene during early *Drosophila* embryogenesis. *Cell* **45**, 113–26.

Carson, H. L. 1982. Evolution of *Drosophila* on the newer Hawaiian volcanoes. *Heredity* **48**, 3–25.

Carson, H. L. & J. S. Yoon 1982. Genetics and evolution of Hawaiian *Drosophila*. In *The genetics and biology of* Drosophila, Vol. 3b, M. Ashburner, H. L. Carson & J. N. Thompson (eds), 297–344. London: Academic Press.

Cavalier-Smith, T. 1978. Nuclear volume control by nucleoskeletal DNA, selection for cell volume and cell growth rate, and the solution of the DNA C-value paradox. *Journal of Cell Science* **34**, 247–78.

Cave, M. D. 1973. Synthesis and characterisation of amplified DNA in oocytes of the house cricket, *Acheta domestica* (Orthoptera: Gryllidae). *Chromosoma* **42**, 1–22.

Cech, T. R. 1983. RNA splicing: three themes with variations. *Cell* **34**, 713–16.

Chandler, M. F., L. H. Kedes, R. H. Cohn & J. J. Yunis 1979. Genes coding for histone proteins in man are located on the distal end of the long arm of chromosome 7. *Science* **205**, 908–10.

Chandley, A. C. 1981. The origin of chromosome aberrations in man and their potential for survival and reproduction in the adult human population. *Annales de Genetique* **24**, 5–11.

Chandley, A. C., R. C. Jones, H. M. Dott, W. R. Allen & R. V. Short 1974. Meiosis in interspecific equine hybrids I. The male mule (*Equus asinus* × *E. caballus*) and hinny (*E. caballus* × *E. asinus*). *Cytogenetics and Cell Genetics* **13**, 330–41.

Chapman, V. M., P. G. Kratzer, L. D. Siracusa, B. A. Quarantillo, R. Evans & R. M. Liskay 1982. Evidence for DNA modification in the maintenance of X-chromosome inactivation of adult mouse tissues. *Proceedings of the National Academy of Sciences* **79**, 5357–61.

Charlesworth, B., R. Lande & M. Slatkin 1982. A neo-Darwinian commentary on macroevolution. *Evolution* **36**, 474–98.

Chaudhari, N. & W. L. Hahn 1983. Genetic expression in the developing brain. *Science* **220**, 924–8.

Chen, C.-N., H. Takayasu & R. L. Davis 1986. Molecular characterization of the dunce locus. In *Molecular neurobiology of* Drosophila, p. 37 (Abst). New York: Cold Spring Harbor Laboratory.

Chernyshev, A. I., V. N. Bashkirov, B. A. Leibovitch & R. B. Khesin 1980. Increase in the number of histone genes in case of their deficiency in *Drosophila melanogaster*. *Molecular and General Genetics* **178**, 663–8.

Cherry, J. M. & E. H. Blackburn 1985. The internally located telomeric sequences in the germ line chromosomes of *Tetrahymena* are at the ends of transposon-like elements. *Cell* **43**, 747–58.

Chikaraishi, D. M., S. S. Deeb & N. Sueoka 1978. Sequence complexity of nuclear RNAs in the adult rat tissues. *Cell* **13**, 111–20.

Childs, G., R. Maxon, R. H. Cohn & L. Kedes 1981. Orphons: dispersed genetic elements derived from tandem repetitive genes of eucaryotes. *Cell* **23**, 651–63.

Childs, G., C. Nocente-McGrath, T. Lieber, C. Holt & J. A. Knowles 1982. Sea urchin (*Lytechinus pictus*) late stage histone H3 and H4 genes: characterisation and mapping of a clustered but non tandemly linked multigene family. *Cell* **31**, 383–93.

Chilton, M. D. & B. McCarthy 1973. DNA from maize with and without B chromosomes: a comparative study. *Genetics* **74**, 605–14.

Chovnick, A., W. Gelbart & M. McCarron 1977. Organisation of the rosy locus in *Drosophila melanogaster*. *Cell* **11**, 1–10.

Christie, N. T. & D. M. Skinner 1979. Interspersion of highly repetitive DNA with single copy DNA in the genome of the red crab *Geryon quinquedens*. *Nucleic Acids Research* **6**, 781–96.

Ciliberto, G., G. Raugei, F. Costanzo, L. Dente & R. Cortese 1983. Common and interchangeable elements in the promoters of genes transcribed by RNA polymerase III. *Cell* **32**, 725–33.

Cleveland, D. W. 1983. The tubulins: from DNA to RNA to protein and back again *Cell* **34**, 330–2.

Cleveland, D. W. & K. F. Sullivan 1985. Molecular biology and genetics of tubulin. *Annual Review of Biochemistry* **54**, 331–65.

Cline, T. W. 1979. A male-specific lethal mutation in *Drosophila melanogaster* that transforms sex. *Developmental Biology* **72**, 266–75.

Cline, T. W. 1984. *Early steps in the regulation of* Sxl *for the control of* Drosophila *sex determination*. Crete Drosophila Meeting, Kolymbari (Abst).

Cline, T. W. 1986. Crete Drosophila Meeting, Kolymbari (Abst).

Coen, E. S., T. Strachan & G. A. Dover 1981. The dynamics of concerted evolution of rDNA and histone gene families in *Drosophila*. *Journal of Molecular Biology* **153**, 841–70

Coen, E., T. Strachan & G. Dover 1982. Dynamics of concerted evolution of ribosomal DNA and histone gene families in the *melanogaster* species subgroup of *Drosophila*. *Journal of Molecular Biology* **158**, 17–35.

Cohen, E. H. & S. C. Bowman 1979. Detection and location of three simple sequence DNA's in polytene chromosomes from *virilis* group species of *Drosophila*. *Chromosoma* **73**, 327–55.

Cohen, L. H., K. M. Newrock & A. Zweidler 1975. Stage-specific switches in histone synthesis during embryogenesis of the sea urchin. *Science* **190**, 994–7.

Colberg-Poley, A. M., S. D. Voss & P. Gruss 1985a. Expression of murine genes containing homeobox sequences during visceral and parietal endoderm differentiation of embryonal carcinoma stem cells. *Cold Spring Harbor Symposia on Quantitative Biology* **50**, 285–90.

Colberg-Poley, A. M., S. D. Voss, K. Chowdhury & P. Gruss 1985b. Structural analysis of murine genes containing homeo box sequences and their expression in embryonal carcinoma cells. *Nature* **314**, 713–18.

Colberg-Poley, A. M., S. D. Voss, K. Chowdhury, C. L. Stewart, E. F. Wagner & P. Gruss 1985c. Clustered homeo boxes are differentially expressed during murine development. *Cell* **43**, 39–45.

Collins, M. & M. Rubin 1983. High-frequency precise excision of the *Drosophila* foldback tranposable element. *Nature* **303**, 259–60.

Conel, J. L. 1959. *The postnatal development of the human cerebral cortex*. Cambridge, Mass.: Harvard University Press.

Conner, W. G., R. Hinegardner & K. Bachmann 1972. Nuclear DNA amounts in polychaete annelids. *Experientia* **28**, 1502–4.

Conway Morris, S. 1985. The middle Cambrian metazoan *Wiwaxia corrugata* (matthew) from the Burgess Shale and Ogygopsis Shale, British Columbia, Canada. *Philosophical Transactions of the Royal Society, London* **B307**, 507–82.

Conway Morris, S. & H. B. Whittington 1985. *Fossils of the Burgess Shale: A national treasure in Yoho National Park, British Columbia*. Geological Survey of Canada, Miscellaneous Report no. 43, 1–31.

Cooper, K. W. 1959. Cytogenetic analysis of major heterochromatic elements (especially Xh and Y) in *Drosophila melanogaster* and the theory of 'heterochromatin'. *Chromosoma* **10**, 535–88.

Cordeiro-Stone, M. & C. S. Lee 1976. Studies on the satellite DNAs of *Drosophila nasutoides*: their buoyant densities, melting temperature, reassociation rates and localisations in polytene chromosomes. *Journal of Molecular Biology* **104**, 1–24.

Cotman, C. W. & M. Nieto-Sampedro 1984. Cell biology of synaptic plasticity. *Science* **225**, 1287–94.

Cowman, A. F., C. S. Zuker & G. M. Rubin 1986. An opsin gene expressed in only one photoreceptor cell type of the *Drosophila* eye. *Cell* **44**, 705–10.

Crain, W. R., E. H. Davidson & R. J. Britten 1976a. Contrasting patterns of DNA sequence arrangement in *Apis mellifera* (honeybee) and *Musca domestica* (housefly). *Chromosoma* **59**, 1–12.

Crain, W. R., F. C. Eden, W. R. Pearson, E. H. Davidson & R. J. Britten 1976b. Absence of short period interspersion of repetitive and non-repetitive sequences in the DNA of *Drosophila melanogaster*. *Chromosoma* **56**, 309–26.

Crews, S., J. Thomas & C. Goodman 1987. Molecular analysis of a *Drosophila* gene that affects the development of midline neurons. In *Molecular biology of neurogenesis*, 1st European Meeting of *Drosophila* Neurogenetics, Simonswald. *The Journal of Neurogenetics* **4**, 120–1 (Abst).

Crick, F. H. C. & P. A. Lawrence 1975. Compartments and polyclones in insect development. *Science* **189**, 340–7.

Czihak, G. 1971. Echinoids. In *Experimental embryology of marine and freshwater invertebrates*, G. Reverberi (ed.), 363–506. Amsterdam: North Holland Pub.

Dambley-Chaudière, C. & A. Ghysen 1986. The sense organs in the *Drosophila* larva and their relation to the embryonic pattern of sensory neurons. *Roux's Archives of Developmental Biology* **195**, 222–8.

Daneholt, B. & J. E. Edstrom 1967. The content of DNA in individual polytene chromosomes of *C. tentans*. *Cytogenetics* **6**, 350–6.

Daniels, G. R. & P. L. Deininger 1985. Repeat sequence families derived from mammalian tRNA genes. *Nature* **317**, 819–22.

Darnell, J. E. 1982. Variety in the level of gene control in eukaryotic cells. *Nature* **297**, 365–71.

Darnell, J. E. 1985. RNA. *Scientific American* **254**, 54–64.

Davidson, E. H. & R. J. Britten 1973. Organisation, transcription and regulation in the animal genome. *Quarterly Review of Biology* **48**, 565–613.

Davidson, E. H. & R. J. Britten 1979. Regulation of gene expression: possible role of repetitive sequences. *Science* **204**, 1052–9.

Davidson, E. H., B. R. Hough-Evans & R. J. Britten 1982. Molecular biology of the sea urchin embryo. *Science* **217**, 17–26.

Davidson, E. H., H. T. Jacobs & R. J. Britten 1983. Very short repeats and coordinate induction of genes. *Nature* **301**, 468–70.

Davidson, E. H., G. A. Galau, R. C. Angerer & R. J. Britten 1975. Comparative aspects of DNA organisation in Metazoa. *Chromosoma* **51**, 253–9.

Davidson, E. H., C. N. Flytzanis, J. J. Lee, J. J. Robinson, S. J. Rose & H. M. Sucov 1985. Lineage-specific gene expression in the sea urchin embryo. *Cold Spring Harbor Symposia on Quantitative Biology* **50**, 321–8.

Davidson, J. N. & L. A. Niswander 1983. Partial cDNA sequence to a hamster gene corrects defect in *Escherichia coli* pyr B mutant. *Proceedings of the National Academy of Sciences* **80**, 6897–901.

Davies, J., V. Pirrotta & G. L. G. Miklos 1987. Analysis of the shaking B locus of *Drosophila melanogaster*. In *Molecular biology of neurogenesis*, 1st European Meeting of *Drosophila* Neurogenetics, Simonswald. *The Journal of Neurogenetics* **4**, 123 (Abst).

Davies, P. O. L. & H. Rees 1975. Mitotic cycles in *Triticum species*. *Heredity* **35**, 337–45.

Davis, A. H., G. H. Kidd & C. E. Carter 1979. Chromatin diminution in *Ascaris suum*: two-fold increase of nucleosomal histone to DNA ratios during development. *Biochimica et Biophysica Acta* **565**, 315–25.

Dawid, I. B. & P. Botchan 1977. Sequences homologous to ribosomal insertions occur in the *Drosophila* genome outside the nucleolus organizer. *Proceedings of the National Academy of Sciences* **74**, 4233–7.

Dawid, I. B., E. O. Long, P. P. DiNocera & M. L. Pardue 1981. Ribosomal insertion-like elements in *Drosophila melanogaster* are interspersed with mobile sequences. *Cell* **25**, 399–408.

de Cicco, D. V. & D. M. Glover 1983. Amplification of rDNA and type 1 sequences in *Drosophila* males deficient in rDNA. *Cell* **32**, 1217–25.

de Cicco, D. V. & A. C. Spradling 1984. Localisation of a cis-acting element responsible for the developmentally regulated amplification of *Drosophila* chorion genes. *Cell* **38**, 45–54.

de Couet, H. G., J. Davies, V. Pirrotta & G. L. G. Miklos 1987. Genetic and molecular studies of a flightless mutant of *Drosophila melanogaster*. In *Molecular biology of neurogenesis*, 1st European Meeting of *Drosophila* Neurogenetics, Simonswald. *The Journal of Neurogenetics* **4**, 132–3 (Abst).

de Robertis, E. M., A. Fritz, J. Goetz, G. Martin, I. W. Mattaj, E. Salo, G. D. Smith, C. Wright & R. Zeller 1985. The *Xenopus* homeoboxes. *Cold Spring Harbor Symposia on Quantitative Biology* **50**, 271–5.

Deak, I. I. 1976. Demonstration of sensory neurones in the ectopic cuticle of spineless-aristapedia, a homeotic mutant of *Drosophila*. *Nature* **260**, 252–4.

Deaven, L. L., L. Vidal-Rioja, J. H. Jett & T. C. Hsu 1977. Chromosomes of *Peromyscus* (Rodentia, Cricetidae) VI. The genomic size. *Cytogenetics and Cell Genetics* **19**, 241–9.

Deininger, P. L. & G. R. Daniels 1986. The recent evolution of mammalian repetitive DNA elements. *Trends in Genetics* **2**, 76–80.

Del Rey, F. J., T. F. Donahue & G. R. Fink 1982. *Sigma*, a repetitive element found adjacent to tRNA genes of yeast. *Proceedings of the National Academy of Sciences* **79**, 4138–42.

Deuchar, E. 1975. *Interactions in development*. London: Chapman and Hall.

Devlin, R. H., D. G. Holm & T. A. Grigliatti 1982. Autosomal dosage compensation in *Drosophila melanogaster* strains trisomic for the left arm of chromosome 2. *Proceedings of the National Academy of Sciences* **79**, 1200–4.

Diaz, M. O., G. Barsacchi-Pilone, K. A. Mahon & J. G. Gall 1981. Transcripts from both strands of a satellite DNA occur on lampbrush chromosome loops of the newt *Notophthalmus*. *Cell* **24**, 649–59.

DiBerardino, M. A. 1979. Nuclear and chromosomal behaviour in amphibian nuclear transplants. *International Review of Cytology Supplement* **9**, 129–60.

DiDomenico, B. J., G. E. Bugaisky & S. Lindquist 1982. The heat shock response is self regulated at both the transcriptional and post transcriptional levels. *Cell* **31**, 593–603.

Doe, C. Q. & C. S. Goodman 1985. Neurogenesis in grasshopper and *fushi tarazu Drosophila* embryos. *Cold Spring Harbor Symposia on Quantitative Biology* **50**, 891–903.

Dolecki, G., S. Wannakrairoj, R. Lum, G. Wang, H. D. Riley, R. Carlos, A. Wang & T. Humphreys 1986. Stage-specific expression of a homeo box-containing gene in the non-segmented sea urchin embryo. *The EMBO Journal* **5**, 925–30.

Donelson, J. E. & M. J. Turner 1985. How the trypanosome changes its coat. *Scientific American* **252**, 32–9.

Doolittle, R. F. 1985. Proteins. *Scientific American* **253**, 74–83.

Doolittle, W. F. & C. Sapienza 1980. Selfish genes, the phenotype paradigm and genome evolution. *Nature* **284**, 601–3.

Dover, G. A. 1975. The heterogeneity of B-chromosome DNA: no evidence for a B-chromosome specific repetitive DNA correlated with B-chromosome effects on meiotic pairing in the Triticinae. *Chromosoma* **53**, 153–73.

Dover, G. A. 1980. Problems in the use of DNA for the study of species relationships and the evolutionary significance of genomic differences. In *Chemosystematics: principles and practice*, F. A. Bisby, J. G. Vaughn & C. A. Wright (eds), 241–68. New York: Academic Press.

Dover, G. A. 1982. Molecular drive: a cohesive mode of species evolution. *Nature* **299**, 111–17.

Dover, G. A. 1984. Forza Evolutione. *Kos* **5**, 129–36.

Dover, G. A. 1986a. The spread and success of non-Darwinian novelties. In *Evolutionary processes and theory*, S. Karlin & E. Nevo (eds), 199–237. New York: Academic Press.

Dover, G. A. 1986b. Molecular drive in multigene families: how biological novelties arise, spread and are assimilated. *Trends in Genetics* **2**, 159–65.

Dover, G. A. & R. B. Flavell 1984. Molecular Coevolution: DNA divergence and the maintenance of function. *Cell* **38**, 622–3.

Dover, G. A. & S. A. Henderson 1976. No detectable satellite DNA in supernumerary chromosomes of the grasshopper *Myrmeleotettix*. *Nature* **259**, 57–8.

Dover, G. A. & D. Tautz 1986. Conservation and divergence in multigene families: alternatives to selection and drift. *Philosophical Transactions of the Royal Society, London* **B312**, 275–89.

Dowsett, A. P. 1983. Closely related species of *Drosophila* can contain different libraries of middle repetitive DNA sequences. *Chromosoma* **88**, 104–8.

Dowsett, A. P. & M. W. Young, 1982. Differing levels of dispersed repetitive DNA among closely related species of *Drosophila*. *Proceedings of the National Academy of Sciences* **79**, 4570–4.

Doyle, H. J., K. Harding, T. Hoey & M. Levine 1986. Transcripts encoded by a homeo box gene are restricted to dorsal tissues of *Drosophila* embryos. *Nature* **323**, 76–9.

Dudler, R. & A. A. Travers 1984. Upstream element necessary for optimal function of the hsp70 promoter in transformed flies. *Cell* **38**, 391–8.

Durica, D. S. & H. M. Krider 1978. Studies on the ribosomal RNA cistrons in interspecific *Drosophila* hybrids II. Heterochromatic regions mediating nucleolar dominance. *Genetics* **89**, 37–64.

Dusenberry, R. L. 1975. Characterisation of the genome of *Phycomyces blakesleanus*. *Biochimica et Biophysica Acta* **378**, 363–77.

Dynan, W. S. 1986. Promoters for housekeeping genes. *Trends in Genetics* **2**, 196–7.

Dynan, W. S. & R. Tjian 1985. Control of eukaryotic messenger RNA synthesis by sequence-specific DNA binding proteins. *Nature* **316**, 774–8.

Echols, H. 1980. Bacteriophage λ development. In *The molecular genetics of development* T. Leighton & W. F. Loomis (eds), 1–16. New York: Academic Press.

Edelman, G. M. 1984. Cell adhesion and morphogenesis: the regulator hypothesis. *Proceedings of the National Academy of Sciences* **81**, 1460–4.

Edelman, G. M. 1985. Cell adhesion molecule expression and the regulation of morphogenesis. *Cold Spring Harbor Symposia on Quantitative Biology* **50**, 877–89.

Eden, F. C., J. P. Hendrick & S. S. Gottlieb 1978. Homology of single copy and repeated sequences in chicken, duck, Japanese quail and ostrich DNA. *Biochemistry* **17**, 5113–21.

Edgar, R. S. 1980. The genetics of development in the Nematode *Caenorhabditis elegans*. In *The molecular genetics of development* T. Leighton & W. F. Loomis (eds), 213–35. New York: Academic Press.

Eibel, H., J. Gafner, A. Stotz & P. Phillipsen 1981. Characterisation of the yeast mobile element Tyl. *Cold Spring Harbor Symposia on Quantitative Biology* **45**, 609–18.

Eigel, A. & H. Feldman 1982. Tyl and delta elements occur adjacent to several tRNA genes in yeast. *The EMBO Journal* **1**, 1245–50.

Eissenberg, J. C., I. L. Cartwright, G. H. Thomas & S. C. R. Elgin 1985. Selected topics in chromatin structure. *Annual Review of Genetics* **19**, 485–536.

Elder, R. T., E. Y. Loh & R. W. Davis 1983. RNA from the yeast transposable element Tyl has both ends in the direct repeats, a structure similar to retrovirus RNA. *Proceedings of the National Academy of Sciences* **80**, 2432–6.

Elgin, S. C. R. 1981. DNAase 1-hypersensitive sites of chromatin. *Cell* **27**, 413–15.

Elinson, R. P. & A. Briedis 1981. Triploidy permits survival of an inviable amphibian hybrid. *Developmental Genetics* **2**, 357–67.

Ellis, H. M. & H. R. Horvitz 1986. Genetic control of programmed cell death in the nematode *Caenorhabditis elegans*. *Cell* **44**, 817–29.

Elsevier, S., H. J. Lipps & G. Steinbrück 1978. Histone genes in macronuclear DNA of the ciliate *Stylonychia mytilus*. *Chromosoma* **69**, 291–306.

Emmons, S. W., K. S. Ruan, A. Levitt & L. Yesner 1985. Regulation of Tc 1 transposable elements in *Caenorhabditis elegans*. *Cold Spring Harbor Symposia on Quantitative Biology* **50**, 313–20.

Emori, Y., T. Shiba, S. Kanaya, S. Inouye, S. Yuki & K. Saigo 1985. The nucleotide sequences of copia and copia-like related RNA in *Drosophila* virus-like particles. *Nature* **315**, 773–6.

Endow, S. A. 1980. On ribosomal gene compensation in *Drosophila*. *Cell* **22**, 149–55.

Endow, S. A. 1983. Nucleolar dominance in polytene cells of *Drosophila*. *Proceedings of the National Academy of Sciences* **80**, 4427–31.

Endow, S. A. & D. M. Glover 1979. Differential replication of ribosomal gene repeats in polytene nuclei of *Drosophila*. *Cell* **17**, 597–605.

Engel, J. D. & J. B. Dodgson 1981. Histone genes are clustered but not tandemly repeated in the chicken genome. *Proceedings of the National Academy of Sciences* **75**, 2856–60.

Engels, W. R. & C. R. Preston 1979. Hybrid dysgenesis in *Drosophila melanogaster*: the biology of female and male sterility. *Genetics* **92**, 161–74.

Engels, W. R. & C. R. Preston 1981. Identifying P factors in *Drosophila* by means of chromosome breakage hot spots. *Cell* **26**, 421–8

Epplen, J. T., J. R. McCarrey, S. Sutou & S. Ohno 1982. Base sequence of a cloned snake W-chromosome DNA fragment and identification of a male specific putative mRNA in the mouse. *Proceedings of the National Academy of Sciences* **79**, 3798–802.

Evans, G. M., H. Rees, C. L. Snell & S. Sun 1972. The relationship between nuclear DNA amount and the duration of the mitotic cycle. *Chromosomes Today* **3**, 24–31.

Eveleth, D. D. & J. L. Marsh 1986. Sequence and expression of the *Cc* gene, a member of the dopa decarboxylase gene cluster of *Drosophila*: possible translational regulation. *Nucleic Acids Research* **14**, 6169–83.

Eveleth, D. D. & J. L. Marsh 1987. Evidence for evolutionary duplication of genes in the dopa decarboxylase region of *Drosophila*, in manuscript.

Feldman, D., L. G. Töfkés, P. A. Stathis, S. C. Miller, W. Kurz & D. Harvey 1984. Identification of 17β-estradiol as the estrogenic substance in *Saccharomyces cerevisiae*. *Proceedings of the National Academy of Sciences* **81**, 4722–6.

Ferrus, A., S. Llamazares, J. L. de la Pompa & R. Yuste 1987. On the genetic organization of the Shaker complex of *Drosophila*. In *Molecular biology of neurogenesis*, 1st European Meeting of *Drosophila* Neurogenetics, Simonswald. *The Journal of Neurogenetics* **4**, 125–6 (Abst).

Finnegan, D. 1986. Crete Drosophila Meeting, Kolymbari (Abst).

Finnegan, D. J., B. H. Will, A. A. Bayev, A. M. Bowcock & L. Brown 1982. Transposable DNA sequences in eukaryotes. In *Genome evolution*, G. A. Dover & R. B. Flavell (eds), 29–40. London: Academic Press.

Firtel, R. A., & K. Kindle 1975. Structural organisation of the genome of the cellular slime mold *Dictyostelium discoideum*: interspersion of repetitive and single copy DNA sequences. *Cell* **5**, 401–11.

Firtel, R. A. K. Kindle & M. P. Huxley 1976. Structural organisation and processing of the genetic transcript in the cellular slime mold *Dictyostelium discoideum*. *Federation Proceedings* **35**, 13–22.

Fischbach, K. F. & M. Heisenberg, 1981. Structural brain mutant of *Drosophila melanogaster* with reduced cell number in the medulla cortex and with normal optomotor yaw response. *Proceedings of the National Academy of Sciences* **78**, 1105–9.

Fischbach, K. F. & M. Heisenberg 1984. Neurogenetics and behaviour in Insects. *Journal of Experimental Zoology* **112**, 65–93.

Fischbach, K. F. & I. Lyly-Hünerberg 1983. Genetic dissection of the anterior optic tract of *Drosophila melanogaster*. *Cell tissue Research* **231**, 551–63.

Fischbach, K. F., U. Boschert, F. Barleben, B. Houbè & T. Rau 1987. New alleles of structural brain mutants of *Drosophila melanogaster* derived from a dysgenic cross. In *Molecular biology of neurogenesis*, 1st European Meeting of *Drosophila* Neurogenetics, Simonswald. *The Journal of Neurogenetics* **4**, 126–8 (Abst).

Fisher, S. E., J. B. Shaklee, S. D. Ferris & G. S. Whitt 1980. Evolution of five multilocus isozyme systems in chordates. *Genetica* **52/53**, 73–85.

Fjose, A., W. J. McGinnis & W. J. Gehring 1985. Isolation of a homeo box-containing gene from engrailed region of *Drosophila* and the spatial distribution of its transcript. *Nature* **313**, 284–9.

Flavell, R. B. 1982. Sequence amplification, deletion and rearrangement: major sources of variation during species divergence. In *Genome evolution*, G. A. Dover & R. B. Flavell (eds), 301–23. London: Academic Press.

Fornwald, J. A., G. Kuncio, I. Peng & C. P. Ordahl 1982. The complete nucleotide sequence of the chick α-actin gene and its evolutionary relationship to the actin gene family. *Nucleic Acids Research* **10**, 3861–76.

Fostel, J., S. Narayanswami, B. Hamkalo, S. G. Clarkson & M. L. Pardue 1984. Chromosomal location of a major tRNA gene cluster of *Xenopus laevis*. *Chromosoma* **90**, 254–60.

Foster, K. W., J. Saranak, N. Patel, G. Zarilli, M. Okabe, T. Kline & K. Nakanishi 1984. A rhodopsin is the functional photoreceptor in the unicellular eukaryote *Chlamydomonas*. *Nature* **311**, 756–9.

Frei, E., D. Bopp, M. Burri, S. Baumgartner, J-E. Edström & M. Noll 1985. Isolation and structural analysis of the extra sex combs gene of *Drosophila*. *Cold Spring Harbor Symposia on Quantitative Biology* **50**, 127–34.

Freund, R. & M. Meselson 1984. Long terminal repeat nucleotide sequence and specific insertion of the gypsy transposon. *Proceedings of the National Academy of Sciences* **81**, 4462–4.

Frischauf, A–M. 1985. The T/t complex of the mouse. *Trends in Genetics* **1**, 100–3.

Fristrom, D. K. & J. W. Fristrom 1975. The mechanism of evagination of imaginal discs of *Drosophila melanogaster*. General considerations. *Developmental Biology* **43**, 1–23.

Fristrom, D. K. & W. L. Rickoll 1982. The morphogenesis of imaginal discs of *Drosophila*. In *Insect ultrastructure*, Vol. 1, R. C. King & H. Akai (eds), 247–77. New York: Plenum Press.

Fritton, H. P., T. Igo-Kemenes, J. Nowock, U. Strech-Jurk, M. Theisen & A. E. Sippel 1984. Alternative sets of DNase I hypersensitive sites characterise the various functional sites of the chicken lysozyme gene. *Nature* **311** 163–5.

Frommer, M., J. Prosser & P. C. Vincent 1984. Human satellite I sequences include a male specific 2.47 kb tandemly repeated unit containing one *Alu* family member per repeat. *Nucleic Acids Research* **12**, 2887–900.

Fruscoloni, P., G. R. Al-Atia & M. Jacobs-Lorena 1983. Translational regulation of a specific gene during oogenesis and embryogenesis of *Drosophila*. *Proceedings of the National Academy of Sciences* **80**, 3359–63.

Fukuzawa, H., T. Kohchi, H Shirai, K. Ohyama, K. Umesono, H. Inokuchi & H. Ozeki 1986. Coding sequences for chloroplast ribosmal protein S12 from the liverwort *Marchantia polymorpha* are separated far apart on the different DNA strands. *Federation of European Biochemical Societies Letters* **198**, 11–15.

Fullilove, S. L., A. G. Jacobson & F. R. Turner 1978. Embryonic development: descriptive. In *The genetics and biology of* Drosophila, Vol. 2c, M. Ashburner & T. R. F. Wright (eds), 106–227. London: Academic Press.

Fyrberg, E. A., K. L. Kindle, N. Davidson & A. Soda 1980. The actin genes of *Drosophila*: a dispersed multigene family. *Cell* **19**, 365–78.

Galau, G. A., M. E. Chamberlain, B. R. Hough, R. J. Britten & E. H. Davidson 1976a. Evolution of repetitive and non-repetitive DNA. In *Molecular evolution*, F. J. Ayala (ed.), 202–24. Mass: Sinauer Assoc.

Galau, G. A., W. H. Klein, M. M. Davis, B. J. Wold, R. J. Britten & E. H. Davidson 1976b. Structural gene sets active in embryos and adult tissues of the sea urchin. *Cell* **7**, 487–505.

Ganetzky, B., K. Loughney & R. Kreber 1986. Genetic and molecular analysis of the *para* locus. In *Molecular neurobiology of* Drosophila, p. 25 (Abst). New York: Cold Spring Harbor Laboratory.

Gans, C. & R. G. Northcutt 1983. Neural crest and the origin of vertebrates: a new head. *Science* **220**, 268–74.

Garcia-Bellido, A. & L. G. Robbins 1983. Viability of female germ-line cells homozygous for zygotic lethals in *Drosophila melanogaster*. *Genetics* **103**, 235–47.

Garcia-Bellido, A., P. A. Lawrence & G. Morata 1979. Compartmentalisation in animal development. *Scientific American* **241**, 102–10.

Garcia-Bellido, A., P. Ripoll & G. Morata 1973. Developmental compartmentalization of the wing disk of *Drosophila*. *Nature New Biology* **245**, 251–3.

Garfinkel, D. J., J. D. Boeke & G. R. Fink 1985. Ty element transposition: reverse transcriptase and virus-like particles. *Cell* **42**, 507–17.

Gatti, M., S. Pimpinelli & G. Santini 1976. Characterisation of *Drosophila* heterochromatin I. Staining and decondensation with Hoechst 33258 and quinacrine. *Chromosoma* **57**, 351–75.

Gauger, A., R. G. Fehon & G. Schubiger 1985. Preferential binding of imaginal disc cells to embryonic segments of *Drosophila*. *Nature* **313**, 395–7.

Gausz, J., L. M. C. Hall, A. Spierer & P. Spierer 1986. Molecular genetics of the rosy-Ace region of *Drosophila melanogaster*. *Genetics* **112**, 65–78.

Gebhard, W. & H. G. Zachau 1983. Simple DNA sequences and dispersed repetitive elements in the vicinity of mouse immunoglobulin K light chain genes. *Journal of Molecular Biology* **170**, 567–73.

Gebhard, W., T. Meitinger, J. Hochtl & H. G. Zachau 1982. A new family of interspersed repetitive DNA sequences in the mouse genome. *Journal of Molecular Biology* **157**, 453–71.

Gerhart, J. C. 1982. The cellular basis of morphogenetic change. In *Evolution and development*, J. T. Bonner (ed.), 87–114. Dahlem Konferenzen. Berlin: Springer.

Gehring, W. J. 1978. Imaginal discs: determination. In *Genetics and biology of Drosophila*, Vol. 2c, M. Ashburner & T. R. F. Wright (eds), 511–54. London: Academic Press.

Gehring, W. 1985a. Homeotic genes, the homeo box and the genetic control of development. *Cold Spring Harbor Symposia on Quantitative Biology* **50**, 243–51.

Gehring, W. J. 1985b. The molecular basis of development. *Scientific American* **253**, 136–46.

Gelbart, N. M., V. F. Irish, R. D. St Johnston, F. M. Hoffmann, R. K. Blackman, D. Segal, L. M. Posakony & R. Grimaila 1985. The Decapentaplegic Gene Complex in *Drosophila melanogaster*. *Cold Spring Harbor Symposia on Quantitative Biology* **50**, 119–25.

Gerber-Huber, S., F. E. B. May, B. R. Westley, B. K. Feiber, H. A. Hosbach, A-C. Andres & G. U. Ryffel 1983. In contrast to other *Xenopus* genes the estrogen-inducible vitellogenin genes are expressed when totally methylated. *Cell* **33**, 43–51.

Gergen, J. P. & E. Wieschaus 1986. Dosage requirements for *runt* in the segmentation of *Drosophila* embryos. *Cell* **45**, 289–99.

Gergen, J. P., D. Coulter & E. Wieschaus 1986. Segmental pattern and blastoderm cell identities. In *Gametogenesis and the early embryo*, J. G. Gall (ed.), 195–220. New York: Alan R. Liss.

Gershenson, S. 1940. The nature of the so-called genetically inert parts of chromosomes. *Proceedings of the Academy of Sciences USSR* 1–116. (Ukrainian with English summary).

Ghysen, A., L.-Y. Jan & Y.-N. Jan 1985. Segmental determination in *Drosophila* central nervous system. *Cell* **40**, 943–8.

Ghysen, A., C. Dambly-Chaudière, E. Aceves, L-Y. Jan & Y-N. Jan 1986. Sensory neurons and peripheral pathways in *Drosophila* embryos. *Roux's Archives of Developmental Biology* **195**, 281–9.

Gilbert, D., J. Hirsch & T. R. F. Wright 1984. Molecular mapping of a gene cluster flanking the *Drosophila* dopa decarboxylase gene. *Genetics* **106**, 679–94.

Gilbert, W. 1979. Introns and exons: playgrounds of evolution. In *Eukaryotic gene regulation. ICN-UCLA Symposium on Molecular and Cellular Biology* **14**, R. Axel, T. Maniatis & C. F. Fox (eds), 1–12. New York: Academic Press.

Gillespie, D., L. Donehower & D. Strayer 1982. Evolution of primate DNA organisation. In *Genome evolution*, G. A Dover & R. B. Flavell (eds), 113–33. London: Academic Press.

Glätzer, K. H. 1979. Lengths of transcribed rDNA repeating units in spermatocytes of *Drosophila hydei*: only genes without an intervening sequence are expressed. *Chromosoma* **75**, 161–75.

Glover, C. V. C., E. R. Shelton & D. L. Brutlag 1983. Purification and characterisation of a Type II casein kinase from *Drosophila melanogaster*. *Journal of Biological Chemistry* **258**, 3258–65.

Glover, D. M. 1981. The rDNA of *Drosophila melanogaster*. *Cell* **26**, 297–8.

Glover, D. M., A. Zaha, A. J. Stocker, R. V. Santelli, M. T. Pueyo, S. M. deToledo & F. J. S. Lara 1982. Gene amplification in *Rhynchosciara* salivary gland chromosomes. *Proceedings of the National Academy of Sciences* **79**, 2947–51.

Goday, C. & S. Pimpinelli 1984. Chromosome organisation and heterochromatin elimination in *Parascaris canis*. *Science* **224**, 411–13.

Goin, O. B., C. J. Goin & K. Bachmann 1968. DNA and amphibian life history. *Copeia*, 532–40.

Gold, J. R. & H. J. Price 1985. Genome size variation among north american minnows (Cyprinidae) I. Distribution of the variation in five species. *Heredity* **54**, 297–305.

Goldberg, D. A., J. W. Posakony & T. Maniatis 1983. Correct developmental expression of a cloned alcohol dehydrogenase gene transduced into the *Drosophila* germ line. *Cell* **34**, 59–73.

Golden, J. W., S. J. Robinson & R. Haselkorn 1985. Rearrangement of nitrogen fixation genes during heterocyst differentiation in the agrobacterium *Anabaena*. *Nature* **314**, 419–23.

Goldman, M. A., G. P. Holmquist, M. C. Gray, L. A. Caston & A. Nag 1984. Replication timing of genes and middle repetitive sequences. *Science* **224**, 686–92.

Goodman, C. S., & M. J. Bastiani 1984. How embryonic nerve cells recognise one another. *Scientific American* **251**, 50–8.

Goodman, C. S., M. J. Bastiani, C. Q. Doe, S. duLac, S. L. Helfand, J. Kuwada & J. B. Thomas 1984. Cell recognition during development. *Science* **225**, 1271–9.

Gorovsky, M. A. 1980. Genome organisation and reorganisation in *Tetrahymena*. *Annual Review of Genetics* **14**, 203–39.

Götz, K. G. 1987. Relapse to 'preprogrammed' visual flight-control in a muscular subsystem of the *Drosophila* mutant 'small optic lobes'. In *Molecular biology of neurogenesis*, 1st European Meeting of *Drosophila* Neurogenetics, Simonswald. *The Journal of Neurogenetics* **4**, 153–5 (Abst).

Gould, S. J. 1977. *Ontogeny and phylogeny*. Cambridge, Mass.: Bellknap Press.

Gould, S. J. 1980a. *The Panda's thumb*. New York: Norton.

Gould, S. J. 1980b. Is a new and general theory of evolution emerging? *Palaeobiology* **6**, 119–30.

Gould, S. J. & N. Eldredge 1977. Punctuated equilibria: the tempo and mode of evolution reconsidered. *Palaeobiology* **3**, 115–51.

Gould, S. J. & R. C. Lewontin 1979. The spandrels of San Marco and the Panglossian paradigm: a critique of the adaptationist programme. *Proceedings of the Royal Society, London* **B205**, 581–98.

Gray, A. P. 1958. *Bird hybrids: a checklist with bibliography*. Technical Communication 13, Commonwealth Agricultural Bureau, Farnham, England.

Graziani, F., R. Caizzi & S. Gargano 1977. Circular ribosomal DNA during ribosomal magnification in *Drosophila melanogaster*. *Journal of Molecular Biology* **112**, 49–63.

Green, L., R. van Antwerpen, J. Stein, G. Stein, P. Tripputi, B. Emanuel, J. Selden & C. Croce 1984. A major human histone gene cluster on the long arm of chromosome 1. *Science* **226**, 838–40.

Green, M. M. 1986. Genetic instability in *Drosophila melanogaster*: the genetics of an MR element that makes complete P insertion mutations. *Proceedings of the National Academy of Sciences* **83**, 1036–40.

Greenwald, I. S. 1985. *lin-12*, a nematode homeotic gene is homologous to a set of mammalian proteins that includes epidermal growth factor. *Cell* **43**, 583–90.

Greenwald, I. S., P. W. Sternberg & H. R. Horvitz 1983. The *lin-12* locus specifies cell fates in *Caenorhabditis elegans*. *Cell* **34**, 435–44.

Griffin-Shea, R., G. Thireos & F. C. Kafatos 1982. Organisation of a cluster of four chorion genes in *Drosophila* and its relationship to developmental expression and amplification. *Developmental Biology* **91**, 325–36.

Grimaldi, G., C. Queen & M. F. Singer 1981. Interspersed repeated sequences in the African green monkey genome that are homologous to the human *Alu* family. *Nucleic Acids Research* **9**, 5553–68.

Grime, J. P. & M. A. Mowforth 1982. Variation in genome size – an ecological interpretation. *Nature* **299**, 151–3.

Grimm, C. & W. Kunz 1980. Disproportionate rDNA replication does occur in diploid tissue of *Drosophila hydei*. *Molecular and General Genetics* **180**, 23–6.

Gropp, A. & H. Winking 1981. Robertsonian translocations: cytology, meiosis, segregation patterns and biological consequences of heterozygosity. *Symposia of the Zoological Society, London* **47**, 141–81.

Gropp, A., H. Winking, C. Redi, E. Capanna, J. Britton-Davidian & G. Noack 1982. Robertsonian karyotype variation in wild house mice from Rhaeto-Lombardia. *Cytogenetics and Cell Genetics* **34**, 67–77.

Gros, D. B., B. J. Nicholson & J. P. Revel 1983. Comparative analysis of the gap junction protein from rat heart and liver: is there a tissue specificity of gap junctions. *Cell* **35**, 539–49.

Guillemin, C. 1980. Effects phénotypiques de six trisomies et de deux doubles trisomies chez *Pleurodeles walthii* (Amphibien, Urodele). *Annales de Genetique* **23**, 5–11.

Gundelfinger, E. D., E. Krause, M. Melli & B. Dobberstein 1983. The organization of the 7SL RNA in the signal recognition particle. *Nucleic Acids Research* **11**, 7363–74.

Gurdon, J. B. 1985. Introductory comments. *Cold Spring Harbor Symposia on Quantitative Biology* **50**, 1–10.

Hadorn, E. 1978. Transdetermination. In *Genetics and biology of* Drosophila, Vol. 2c, M. Ashburner & T. R. F. Wright (eds), 555–617. London: Academic Press.

Haenlin, M., H. Steller, V. Pirrotta & E. Mohler 1985. A 43 kilobase cosmid P transposon rescues the fs(1)K10 morphogenetic locus and three adjacent *Drosophila* developmental mutants. *Cell* **40**, 826–37.

Hafen, E., A. Kuroiwa & W. J. Gehring 1984. Spatial segmentation of transcripts from the segmentation gene *fushi tarazu* during *Drosophila* embryonic development. *Cell* **37**, 833–41.

Hafen, E., M. Levine, R. L. Garber & W. J. Gehring 1983. An improved *in situ* hybridisation method for the detection of cellular RNAs in *Drosophila* tissue sections and its application for localising transcripts of the homeotic *Antennapedia* gene complex. *The EMBO Journal* **2**, 617–23.

Hall, A. K. 1983. Stem cell is a stem cell is a stem cell. *Cell* **33**, 11–12.

Hall, J. C. 1977. Portions of the central nervous system controlling reproductive behaviour in *Drosophila melanogaster*. *Behavioral Genetics* **7**, 291–312.

Hall, J. C. 1979. Control of male reproductive behaviour by the central nervous system of *Drosophila*: dissection of courtship pathway by genetic mosaics. *Genetics* **92**, 437–57.

Hall, J. C. 1982. Genetics of the nervous system in *Drosophila*. *Quarterly Review of Biophysics* **15**, 223–479.

Hall, J. C. 1985. Genetic analysis of behaviour in insects. *Comprehensive Insect Physiology, Biochemistry and Pharmocology* **9**, 287–373.

Hall, J. C. & R. J. Greenspan 1979. Genetic analysis of *Drosophila* neurobiology. *Annual Review of Genetics* **13**, 27–195.

Hall, L. M. C., P. J. Mason & P. Spierer 1984. Transcripts, genes and bands in 315 000 base-pairs of *Drosophila* DNA. *Journal of Molecular Biology* **169**, 83–96.

Harding, K., C. Rushlow, M. Frasch & M. Levine 1986. Cross-regulatory interactions among homeo box genes. In *Molecular neurobiology of* Drosophila, p. 90 (Abst). New York, Cold Spring Harbor Laboratory.

Hardman, N. 1986. Structure and function of repetitive DNA in eukaryotes. *Biochemical Journal* **234**, 1–11.

Hartenstein, V. & J. A. Campos-Ortega 1985. Fate-mapping in wild-type *Drosophila melanogaster* I. The spatio-temporal pattern of embryonic cell divisions. *Roux's Archives of Developmental Biology* **194**, 181–95.

Hartenstein, V., G. M. Technau & J. A. Campos-Ortega 1985. Fate-mapping in wild-type *Drosophila melanogaster* III. A fate map of the blastoderm. *Roux's Archives of Developmental Biology* **194**, 181–95.

Hartshorne, T. A., H. Blumberg & E. T. Young 1986. Sequence homology of the yeast regulatory protein ADR1 with *Xenopus* transcription factor. *Nature* **320**, 283–7.

Hastie, N. D. & J. O. Bishop 1976. The expression of three abundance classes of messenger RNA in mouse tissues. *Cell* **9**, 761–74.

Hatch, F. T., A. J Bodner, J. A. Mazrimas & D. H. Moore 1976. Satellite DNA and cytogenetic evolution. DNA quantity, satellite DNA and karyotypic variations in Kangaroo rats (genus *Dipodomys*). *Chromosoma* **58**, 155–68.

Haynes, S. R., T. P. Toomey, L. Leinward & W. R. Jelinek 1981. The Chinese hamster *Alu*-equivalent sequence: a conserved highly repetitious, interspersed deoxyribonucleic sequence in mammals has a structure suggestive of a transposable element. *Molecular Cellular Biology* **1**, 573–83.

Hazelrigg, T., R. Levis & G. M. Rubin 1984. Transformation of white locus DNA in *Drosophila*: dosage compensation, *zeste* interaction and position effects. *Cell* **36**, 469–81.

Hedgecock, E. M. 1985. Cell lineage mutants in the nematode *Caenorhabditis elegans*. *Trends in Neuroscience* **8**, 288–93.

Heintz, N., M. Zernik & R. G. Roeder 1981. The structure of the human histone genes: clustered but not tandemly repeated. *Cell* **24**, 661–8.

Heisenberg, M. & K. Böhl 1979. Isolation of anatomical brain mutants of *Drosophila* by histological means. *Zeitschrift für Natürfurschung* **34**, 143–7.

Henikoff, S. 1981. Position-effect variegation and chromosome structure of a heat shock puff in *Drosophila*. *Chromosoma* **83**, 381–93.

Henikoff, S. & C. E. Furlong 1983. Sequence of a *Drosophila* DNA segment that functions in *Saccharomyces cerevisiae* and its regulation by a yeast promoter. *Nucleic Acids Research* **11**, 789–800.

Henikoff, S., M. A. Keene, K. Fechtel & J. A. Tristrom 1986. Gene within a gene: nested Drosophila genes encode unrelated proteins on opposite DNA strands. *Cell* **44**, 33–42.

Henikoff, S., K. Tatchell, B. D. Hall & K.A. Nasmyth 1981. Isolation of a gene from *Drosophila* by complementation in yeast. *Nature* **289**, 33–7.

Hentschel, C. C. & M. L. Birnstiel 1981. The organisation and expression of histone gene families. *Cell* **25**, 301–13.

Hermans-Borgmeyer, I., E. Sawruk, D. Zopf, E. Gundelfinger & H. Betz 1986. Characterization of the mRNA and the gene of a putative neuronal nicotinic acetylcholine receptor protein from *Drosophila melanogaster*. In *Molecular neurobiology of* Drosophila, p. 29 (Abst). New York: Cold Spring Harbor Laboratory.

Herrick, G., S. Cartinhour, D. Dawson, D. Ang, R. Sheets, A. Lee & K. Williams 1985. Mobile repeats bounded by C_4A_4 telomeric repeats in *Oxytricha fallax*. *Cell* **43**, 759–68.

Hess, O. 1976. Genetics of *Drosophila hydei* Sturtevant. In *The genetics and biology of* Drosophila, Vol. 1c, M. Ashburner & E. Novitski (eds), 1343–63. New York: Academic Press.

Hickey, D. A. 1982. Selfish DNA: a sexually-transmitted nuclear parasite. *Genetics* **101**, 519–31.

Hightower, R. C. & R. B. Meagher 1986. The molecular evolution of actin. *Genetics* **114**, 315–32.

Hilder, V. A., R. N. Livesey, P. C. Turner & M. T. Vlad 1981. Histone gene number in relation to C-value in amphibians. *Nucleic Acids Research* **9**, 5737–46.

Hilliker, A. J. & R. Appels 1982. Pleiotropic effects associated with the deletion of heterochromatin surrounding rDNA on the X chromosome of *Drosophila*. *Chromosoma* **86**, 469–90.

Hilliker, A. J., R. Appels & A. Schalet 1980. The genetic analysis of *Drosophila melanogaster* heterochromatin. *Cell* **21**, 607–19.

Hinegardner, R. 1973. Cellular DNA content of the Mollusca. *Comparative Biochemistry and Physiology* **47A**, 447–60.

Hinegardner, R. 1974. Cellular DNA content of the Echinodermata. *Comparative Biochemistry and Physiology* **49B**, 219–26.

Hinegardner, R. 1976. Evolution of genome size. In *Molecular evolution*, A. J. Ayala (ed.), 179–99. Sunderland, Mass: Sinauer Assoc.

Hinegardner, R. & D. E. Rosen 1972. Cellular DNA content and the evolution of teleostean fishes. *American Naturalist* **106**, 621–44.

Ho, R. K. & C. S. Goodman 1982. Peripheral pathways are pioneered by an array of central and peripheral neurones in grasshopper embryos. *Nature* **297**, 404–6.

Hodgkin, J. 1983. Two types of sex determination in a nematode. *Nature* **304**, 267–8.

Hoey, T., H. J. Doyle, K. Harding, C. Weeden & M. Levine 1986. Homeo box gene expression in anterior and posterior regions of the *Drosophila* embryo. *Proceedings of the National Academy of Sciences* **83**, 4809–13.

Hoffman, A. 1982. Punctuated versus gradual mode of evolution. *Evolutionary Biology* **15**, 411–36.

Hoffmann, F. M., L. D. Fresco, H. H. Falk & B-Z. Shilo 1983. Nucleotide sequences of the *Drosophila src* and *abl* homologs: conservation and variability in the *src* family oncogenes. *Cell* **35**, 393–401.

Hoffman, S., C. M. Chuong & G. M. Edelman 1984. Evolutionary conservation of key structures and binding functions of neural cell adhesion molecules. *Proceedings of the National Academy of Sciences* **81**, 6881–5.

Hoffman-Falk, H., P. Einat, B-Z. Shilo & F. M. Hoffman 1983. *Drosophila melanogaster* DNA clones homologous to vertebrate oncogenes: evidence for a common ancestor to the *src* and *abl* cellular genes. *Cell* **32**, 589–98.

Hogness, D. S., H. D. Lipshitz, P. A. Beachy, D. A. Peattie, R. B. Saint, M. Goldschmidt-Clermont, P. J. Harte, E. R. Gavis & S. L. Helfand 1985. Regulation and products of the *Ubx* domain of the Bithorax Complex. *Cold Spring Harbor Symposia on Quantitative Biology* **50**, 181–94.

Holland, P. W. H. & B. L. M. Hogan 1986. Phylogenetic distribution of *Antennapedia*-like homoeo boxes. *Nature* **321**, 251–3.

Holmquist, G., M. Gray, T. Porter & J. Jordan 1982. Characterization of Giemsa dark- and light-band DNA. *Cell* **31**, 121–9.

Homyk, T., J. Szidonya & D. T. Suzuki 1980. Behavioural mutants of *Drosophila melanogaster* III. Isolation and mapping of mutations by direct visual observation of behavioural phenotypes. *Molecular and General Genetics* **177**, 553–65.

Horner, H. A. & H. C. Macgregor 1983. C value and cell volume: their significance in the evolution and development of amphibians. *Journal of Cell Science* **63**, 135–46.

Horvitz, H. R., P. W. Sternberg, I. S. Greenwald, W. Fixsen & M. Elliot 1983. Mutations that affect neural cell lineages and cell fates during the development of the nematode *Caenorhabditis elegans*. *Cold Spring Harbor Symposia on Quantitative Biology* **48**, 453–63.

Hosbach, H. A., T. Wyler & R. Weber 1983. The *Xenopus laevis* globin gene family: chromosomal arrangement and gene structure. *Cell* **32**, 45–53.

Hotta, Y. & S. Benzer 1973. Mapping of behaviour in *Drosophila* mosaics. In *Genetic mechanisms of development*, F. H. Ruddle (ed.), 129–67. 31st Symposium of the Society for Developmental Biology. New York: Academic Press.

Hough-Evans, B. R., M. Jacobs-Lorena, M. R. Cummings, R. J. Britten & E. H. Davidson 1980. Complexity of RNA in eggs of *Drosophila melanogaster* and *Musca domestica*. *Genetics* **95**, 81–94.

Hovemann, B. & R. Galler 1982. Vitellogenin in *Drosophila melanogaster*: a comparison of the YPI and YPII genes and their transcription products. *Nucleic Acids Research* **10**, 2261–74.

Hovemann, B., S. Sharp, H. Yamada & D. Soll 1980. Analysis of a *Drosophila* tRNA gene cluster. *Cell* **19**, 889–95.

Howard, K. & P. Ingham 1986. Regulatory interactions between the segmentation genes fushi tarazu, hairy and engrailed in the *Drosophila* blastoderm. *Cell* **44**, 949–57.

Hsieh, T. & D. L. Brutlag 1979. A protein that preferentially binds *Drosophila* satellite DNA. *Proceedings of the National Academy of Sciences* **76**, 726–30.

Hsu, T. C. & F. E. Arrighi 1971. Distribution of constitutive heterochromatin in mammalian chromosomes. *Chromosoma* **34**, 243–53.

Hudson, K., C. Hashimoto & K. Anderson 1986. Molecular and genetic studies of the *Toll* gene. Crete Drosophila Meeting, Kolymbari (Abst).

Hudspeth, M. E. S., W. E. Timberlake & R. B. Goldberg 1977. DNA seqeuence organisation in the water mold *Achlya*. *Proceedings of the National Academy of Sciences* **74**, 4332–6.

Huerre, C., C. Junien, D. Weil, M-L. Chiu, M. Morabito, N. V. Cong, J. C. Myers, C. Foubert, M-S. Gross, D. J. Prockop, A. Boué, J-C. Kaplan, A. de la Chapelle & F. Ramirez 1982. Human type 1 procollagen genes are located on different chromosomes. *Proceedings of the National Academy of Sciences* **79**, 6627–30.

Hughes, M. B. & J. C. Lucchesi 1977. Genetic rescue of a lethal 'null' activity allele of 6-phosphogluconate dehydrogenase in *Drosophila melanogaster*. *Science* **196**, 1114–15.

Humphrey, R. R. 1944. The functional capacities of heteroplasia gonadal grafts in the mexican axolotl and some hybrid offspring of grafted animals. *American Journal of Anatomy* **75**, 263–87.

Hunkapiller, M., S. Kent, M. Caruthers, W. Dryer, J. Firca, C. Giffin, S. Horvath, T. Hunkapiller, P. Tempst & L. Hood 1984. A microchemical facility for the analysis and synthesis of genes and proteins. *Nature* **310** 105–11.

Hutchinson, J., R. K. J. Narayan & H. Rees 1980. Constraints upon the composition of supplementary DNA. *Chromosoma* **78**, 137–45.

Hutchinson, J., H. Rees & A. G. Seal 1979. An assay of the activity of supplementary DNA in *Lolium*. *Heredity* **43**, 411–21.

Hynes, R. 1985. Molecular biology of fibronectin. *Annual Review of Cell Biology* **1**, 67–90.

Hynes, R. 1986. Fibronectins. *Scientific American* **254**, 32–41.

Ikeda, K. & W. D. Kaplan 1970. Patterned neural activity of a mutant *Drosophila melanogaster*. *Proceedings of the National Academy of Sciences* **66**, 765–72.

Ingham, P. W. 1984. A gene that regulates the bithorax complex differentially in larval and adult cells of *Drosophila*. *Cell* **37**, 815–23.

Ingham, P. W. 1985. Genetic control of the spatial pattern of selector gene expression in *Drosophila*. *Cold Spring Harbor Symposia on Quantitative Biology* **50**, 201–8.

Ingham, P. W., K. R. Howard & D. Ish-Horowicz 1985a. Transcription pattern of the *Drosophila* segmentation gene hairy. *Nature* **318**, 439–45.

Ingham, P. W., A. Martin-Arias, P. A. Lawrence & K. Howard 1985b. Expression of engrailed in the parasegment of *Drosophila*. *Nature* **317**, 634–6.

Ingles, C. J., J. Biggs, J. K-C. Wong, J. R. Weeks & A. L. Greenleaf 1983. Identification of a structural gene for a RNA polymerase II polypeptide in *Drosophila melanogaster* and mammalian species. *Proceedings of the National Academy of Sciences* **80**, 3396–400.

Inouye, S., S. Yuki & K. Saigo 1984. Sequence-specific insertion of the *Drosophila* transposable genetic element 17.6. *Nature* **310**, 332–3.

Ireland, R. C., E. Berger, K. Sirotkin, M. A. Yund, D. Osterbur & J. Fristrom 1982. Ecdysterone induces the transcription of four heat shock genes in *Drosophila* S3 cells and imaginal discs. *Developmental Biology* **93**, 498–507.

Ish-Horowicz, D., K. R. Howard, S. M. Pinchin & P. W. Ingham 1985. Molecular and genetic analysis of the hairy locus in *Drosophila*. *Cold Spring Harbor Symposia on Quantitative Biology* **50**, 135–44.

Ising, G. & K. Black 1981. Derivation-dependent distribution of insertion sites for a *Drosophila* transposon. *Cold Spring Harbor Symposia on Quantitative Biology* **45**, 527–44.

Jack, J. 1986. Molecular genetics of the cut locus. *27th Annual* Drosophila *Conference*. Asilomar, California (Abst).

Jäckle, H., U. B. Rosenberg, A. Preiss, E. Seifert, D. Knipple, A. Kienlin & R. Lehmann 1985. Molecular analysis of *Krüppel*, a segmentation gene of *Drosophila melanogaster*. *Cold Spring Harbor Symposia on Quantitative Biology* **50**, 465–73.

Jackson, F. R., T. A. Bargiello, S-H Yun & M. W. Young 1986. Product of *per* locus of *Drosophila* shares homology with proteoglycans. *Nature* **320**, 185–8.

Jackson, I. 1986. Solid foundation for developmental biology. *Trends in Genetics* **2**, 193.

Jackson, I. J., P. Schofield & B. A. Hogan 1985. Mouse homeo box gene is expressed during embryogenesis and in adult kidney. *Nature* **317**, 745–8.

Jamrich, M., R. Warrior, R. Steele & J. G. Gall 1983. Transcription of repetitive sequences on *Xenopus* lampbrush chromosomes. *Proceedings of the National Academy of Sciences* **80**, 3364–7.

Jan, L-Y. & Y-N. Jan 1982. Antibodies to horseradish-peroxidase as specific neuronal markers in *Drosophila* and in grasshopper embryos. *Proceedings of the National Academy of Science* **72**, 2700–4.

Jan, L-Y., D. M. Papazian, L. Timpe, P. O'Farrell & Y-N. Jan 1985. Application of *Drosophila* molecular genetics in the study of neural function – studies of the Shaker locus for a potassium channel. *Trends in Neurosciences* **8**, 234–8.

Jan, Y-N., R. Bodmer, L-Y. Jan, E. Grell, A. Ghysen & C. Dambly-Chaudière 1986. Mutations affecting neurogenesis, neuronal identity and pathfinding in *Drosophila melanogaster*. In *Molecular neurobiology of* Drosophila, p. 6 (Abst). New York: Cold Spring Harbor Laboratory.

Jeffrey, W. R. 1985. Specification of cell fate by cytoplasmic determinants in ascidian embryos. *Cell* **41**, 11–12.

Jeffreys, A. J., V. Wilson & S. L. Thein 1985. Hypervariable 'mini satellite' regions in human DNA. *Nature* **314**, 67–73.

Jeffreys, A. J., V. Wilson, D. Wood & J.P. Simons 1980. Linkage of adult α and β-globin genes in *Xenopus laevis* and gene duplication by tetraploidisation. *Cell* **21**, 555–64.

Jeffreys, A. J., V. Wilson, A. Blanchetot, P. Weller, A. Geurts von Kessel, N. Spurr, E. Solomon & P. Goodfellow 1984. The human myoglobin gene: a third displaced globin locus in the human genome. *Nucleic Acids Research* **12**, 3235–43.

John, B. & M. King 1983. Population cytogenetics of *Atractomorpha similis* I. C-band variation. *Chromosoma* **88**, 57–68.

John, B. & G. L. G. Miklos 1979. Functional aspects of satellite DNA and heterochromatin. *International Review of Cytology* **58**, 1–113.

Joho, R. & I. L. Weissman 1980. V-J joining of immunoglobulin κ-genes only occurs on one homologous chromosome. *Nature* **284**, 179–81.

Jones, E. A. & H. R. Woodland 1986. Development of the ectoderm in *Xenopus*: tissue specification and the role of cell association and division. *Cell* **44**, 345–55.

Jones, K., H. Steller & G. Rubin 1986. Molecular characterization of two paralytic mutations. In *Molecular neurobiology of* Drosophila, p. 76 (Abst). New York: Cold Spring Harbor Laboratory.

Jones, R. N. & H. Rees 1982. *B-chromosomes*. London: Academic Press.

Jost, E. & M. Mameli 1972. DNA content in nine species of Nematocera with special reference to the sibling species of the *Anopheles maculipennis* group and the *Culex pipiens* group. *Chromosoma* **37**, 201–8.

Joyner, A., C. Hauser, T. Kornberg, R. Tjian & G. Martin 1985. Structure and expression of two classes of mammalian homeo-box containing genes. *Cold Spring Harbor Symposia on Quantitative Biology* **50**, 291–300.

Jürgens, G. 1985. A group of genes controlling the spatial expression of the bithorax complex in *Drosophila*. *Nature* **316**, 153–5.

Jürgens, G., E. Wieschaus, C. Nüsslein-Volhard & H. Kluding 1984. Mutations affecting the pattern of the larval cuticle in *Drosophila melanogaster* II. Zygotic loci in the third chromosome. *Roux's Archives of Developmental Biology* **193**, 283–95.

Kafatos, F. C. 1983. Structure, evolution and developmental expression of the chorion multigene families in silkmoths and *Drosophila*. In *Gene structure and regulation in development*, S. Subtelny & F. C. Kafatos (eds), 33–61. New York: Alan R. Liss.

Kafatos, F. C., S. A. Mitsialis, N. Spoerel, B. Mariani, J. R. Lingappa & C. Delidakis 1985. Studies on the developmentally regulated expression and amplification of insect chorion genes. *Cold Spring Harbor Sysmposia on Quantitative Biology* **50**, 537–47.

Kafatos, F. C., J. C. Regier, G. D. Mazur, M. R. Nadel, H. M. Blau, W. H. Petri, A. R. Wyman, R. E. Gelinas, P. B. Moore, M. Paul, A. Efstratiadis, T. N. Vournakis, M. R. Goldsmith, J. R. Hunsley, B. Baker, J. Nardi & M. Koehler

1977. The egg shell of insects: differentiation-specific proteins and the control of their synthesis and accumulation during development. In *Biochemical differentiation in insects*, W. Beermann (ed.), 45–145. New York Springer.

Kalback, D. B. & H. O. Halvorson 1977. Magnification of genes coding for ribosomal RNA in *Saccharomyces cerevisiae*. *Proceedings of the National Academy of Sciences* **74**, 1177–80.

Kalfayan L., J. Levine, T. Orr-Weaver, S. Parker, B. Wakimoto, D. de Cicco & A. Spradling 1985. Localization of sequences regulating *Drosophila* chorion gene amplification and expression. *Cold Spring Harbor Symposia on Quantitative Biology* **50** 527–35.

Kandel, E. 1983. Neurobiology and molecular biology: the second encounter. *Cold Spring Harbor Symposia on Quantitative Biology* **48**, 891–908.

Kao-Huang, Y., A. Revzin, A. P. Butler, P. O'Conner, D. W. Noble & P. H. Von Hippel 1977. Non specific DNA binding of genome-regulating proteins as a biological control mechanism: measurement of DNA-bound *Escherichia coli* lac repressor *in vivo*. *Proceedings of the National Academy of Sciences* **74**, 4228–32.

Karch, F., B. Weiffenbach, M. Peifer, M. Bender, I. Duncan, S. Calniker, M. Crosby & E. B. Lewis 1985. The abdominal region of the Bithorax Complex. *Cell* **43**, 81–96.

Karlik, C. C., J. W. Mahaffey, M. D. Coutu & E. A. Fyrberg 1984. Organisation of contractile protein genes within the 88F subdivision of the *Drosophila melanogaster* third chromosome. *Cell* **37**, 469–81.

Karn, J., S. Brenner & L. Barnett 1983. Protein structural domains in the *Caenorhabditis elegans* unc-54 myosin heavy chain gene are not separated by introns. *Proceedings of the National Academy of Sciences* **80**, 4253–7.

Karpen, G. & C. Laird 1986. Autonomous function of a single rDNA cistron. *27th Annual* Drosophila *Conference*. Asilomar, California (Abst).

Kato, K. 1968. Cytochemistry and fine structure of elimination chromatin in Dytiscidae. *Experimental cell Research* **52**, 507–22.

Katz, M. J., R. J. Lasek & J. Silver 1983. Ontophyletics of the nervous system: development of the corpus callosum and evolution of axon tracts. *Proceedings of the National Academy of Sciences* **80**, 5936–48.

Kay, B. K., M. Jamrich & I. B. Dawid 1984. Transcription of a long interspersed highly repeated DNA element in *Xenopus laevis*. *Developmental Biology* **105**, 518–25.

Kelly, L. E. 1983. An altered electroretinogram transient associated with an unusual jump response in a mutant of *Drosophila*. *Cellular and Molecular Neurobiology* **3**, 143–9.

Kimura, M. 1979. The neutral theory of molecular evolution. *Scientific American* **241**, 94–104.

Kimura, M. 1983. *The neutral theory of molecular evolution*. London: Cambridge University Press.

Kimura, M. 1986. DNA and the neutral theory. *Philosophical Transactions of the Royal Society London* **B312**, 343–54.

King, D. C. & R. J. Wyman 1980. Anatomy of the giant fibre pathway in *Drosophila* I. Three thoracic components of the pathway. *Journal of Neurocytology* **9**, 753–70.

King, M. C. & A. C. Wilson 1975. Evolution at two levels in humans and chimpanzees. *Science* **188** 107–16.

Kirschner, M. & T. Mitchison 1986. Beyond self-assembly: from microtubules to morphogenesis. *Cell* **45**, 329–42.

Klein A. S. & R. A. Eckhart 1976. The DNAs of the A and B chromosomes of the mealy bug, *Pseudococcus obscurus*. *Chromosoma* **57** 333–40.

Knipple, D. C., E. Seifert, U. B. Rosenberg, A. Preiss & H. Jäckle 1985. Spatial and temporal patterns of Krüppel gene expression in early *Drosophila* embryos. *Nature* **317**, 40–4.

Knowles, R. V. & R. L. Phillips 1985. DNA amplification patterns in maize endosperm nuclei during kernel development. *Proceedings of the National Academy of Sciences* **82**, 7010–4.

Knust, E., A. Ziemer, K. Tietze & J. A. Campos-Ortega 1987a. Genetic and molecular analysis of the neurogenic locus Enhancer of split. In *Molecular biology of neurogenesis*, 1st European Meeting of Drosophila Neurogenetics, Simonswald. *The Journal of Neurogenetics* **4**, 146–7 (Abst).

Knust, E., U. Dietrich, D. Weigel, U. Tepass, H. Vässin, K. Bremer & J. A. Campos-Ortega 1987b. Molecular genetics of neurogenic genes in *Drosophila*. In *Molecular biology of neurogenesis*, 1st European Meeting of *Drosophila* Neurogenetics, Simonswald. *The Journal of Neurogenetics* **4**, 145–6 (Abst).

Knust, E., H. Vässin, U. Dietrich, K. A. Bremer, U. Wetter, A. Escherich & J. A. Campos-Ortega 1984. *Genes affecting early neurogenesis in* Drosophila melanogaster. Crete Drosophila Meeting, Kolymbari (Abst).

Kominami, R., M. Muramatsu & K. Moriwaki 1983. A mouse type 2 *Alu* sequence (M2) is mobile in the genome. *Nature* **301**, 87–9.

Konarska, M. M. R. A. Padgett & P. A. Sharp 1985. Trans splicing of mRNA precursors *in vitro*. *Cell* **42**, 165–71.

Kongsuwan, K., Q. Yu, A. Vincent, M. C. Frisardi, M. Rosbash, J. A. Lengyel & J. Merriam 1985. A *Drosophila* minute gene encodes a ribosomal protein. *Nature* **317**, 555–8.

Kornberg, T. 1982. Compartments in the abdomen of *Drosophila* and the role of the engrailed locus. *Developmental Biology* **86**, 363–72.

Kornblihth, A. R., K. Umezawa, K. Vibe-Pederson & F. E. Baralle 1985. Primary structure of human fibronectin: differential splicing may generate at least 10 polypeptides from a single gene. *The EMBO Journal* **4**, 1755–9.

Kourilsky, P. 1986. Molecular mechanisms for gene conversion in higher cells. *Trends in Genetics* **2**, 60–3.

Kovaleva, V. G & I. B. Raikov 1978. Diminution and re-synthesis of DNA during development and senescence of the 'diploid' macronuclei of the ciliate *Trachelonema sulcata* (Gymnostomata Karyorelictida). *Chromosoma* **67**, 177–92.

Kramer, J. M., G. N. Cox & D. Hirsh 1982. Comparisons of the complete sequence of two collagen genes from *Caenorhabditis elegans*. *Cell* **30**, 599–606.

Krumm, A., A. Hofman, G. Roth & G. Korge 1984. Regulation of the secretion protein gene *Sgs-4*. Crete Drosophila Meeting, Kolymbari (Abst).

Kubli, E. 1984. *Drosophila melanogaster* (fruit fly), transfer RNA and -RNA *in situ* hybridization data. In *Genetic maps*, S. J. O'Brien (ed.), 300–3. New York: Cold Spring Harbor Laboratory.

Kuner, J. M., M. Nakanishi, Z. Ali, B. Drees, E. Gustavson, J. Theis, L. Kauvar, T. Kornberg & P. H. O'Farrell 1985. Molecular cloning of engrailed: a gene involved in the development of pattern in *Drosophila melanogaster*. *Cell* **42**, 309–16.

Kunz, W. & U. Schäfer 1976. Variations in the number of Y chromosomal rRNA genes in *Drosophila hydei*. *Genetics* **82**, 25–34.

Kuroiwa, A., E. Hafen & W. J. Gehring 1984. Cloning and transcriptional analysis of the segmentation gene *fushi-tarazu* of *Drosophila*. *Cell* **37**, 825–31.

Kurtén, B. 1963. Return of a lost structure in the evolution of the felid dentition. *Societas Scientiatum Fennica Arsbok* **26**, 3–11.

Kyriacou, C. P. & J. C. Hall 1987. Genetic and Molecular analysis of behavioural rhythms in *Drosophila*. In *Molecular biology of neurogenesis*, 1st European Meeting of *Drosophila* Neurogenetics, Simonswald. *The Journal of Neurogenetics* **4**, 147–8 (Abst).

Laird, C. D. 1973. DNA of *Drosophila* chromosomes. *Annual Review of Genetics* **7**, 177–204.

Lansman, R. A., S. N. Stacey, T. A. Grigliatti & H. W. Brock 1985. Sequences homologous to the P mobile element of *Drosophila melanogaster* are widely distributed in the subgenus Sophophora. *Nature* **318**, 561–3.

Larson, A. 1984. Neontological inferences of evolutionary pattern and process in the salamander family Plethodontidae. *Evolutionary biology* **17**, 119–217.

Laski, F. A., D. C. Rio & G. M. Rubin 1986. The tissue specificity of *Drosophila* P element transcription is regulated at the level of mRNA splicing. *Cell* **44**, 7–19.

Laughon, A. & M. P. Scott 1984. Sequence of a *Drosophila* segmentation gene: protein structure homology with DNA binding proteins. *Nature* **310**, 25–31.

Laughon, A., S. B. Carroll, F. A. Storfer, P. D. Riley & M. P. Scott 1985. Common properties of proteins encoded by the Antennapedia Complex genes of *Drosophila melanogaster*. *Cold Spring Harbor Symposia on Quantitative Biology* **50**, 253–62.

Lawrence, P. A. & G. Morata 1976. Compartments in the wing of *Drosophila*: a study of the engrailed gene. *Developmental Biology* **50**, 321–37.

Lawrence, P. A. & G. Struhl 1983. Different requirements for homeotic genes in the soma and germ line of *Drosophila*. *Cell* **35**, 27–34.

Lazarides, E. 1984. Assembly and morphogenesis of the avian erythrocyte. In *Molecular biology of the cytoskeleton*, G. G. Borisy, D. W. Cleveland & D. B. Murphy (eds), 131–50. New York: Cold Spring Harbor Laboratory.

Le Douarin, N. M. 1984. Cell migrations in embryos. *Cell* **38**, 353–60.

Le Roith, D., J. Shiloach, J. Roth & M. A. Lesniak 1980. Evolutionary origins of vertebrate hormones: substances similar to mammalian insulins are native to unicellular eukaryotes. *Proceedings of the National Academy of Sciences* **77**, 6184–8.

Le Roith, D., A. S. Liotta, J. Roth, J. Shiloach, M. E. Lewis, C. B. Pert & D. T. Krieger 1982. Corticotropin and β-endorphin-like materials are native to unicellular organisms. *Proceedings of the National Academy of Sciences* **79**, 2086–90.

Leder, P. 1982. Genetics of antibody diversity. *Scientific American* **246**, 72–83.

Lee, M. G. S., C. Loomis & N. J. Cowan 1984. Sequence of an expressed human β tubulin gene containing 10 *Alu* family members. *Nucleic Acids Research* **12**, 5823–36.

Lefevre, G. Jr 1984. The one-band one-gene hypothesis: evidence from a cytogenetic analysis of mutant and non-mutant rearrangement breakpoints in *Drosophila melanogaster*. *Cold Spring Harbor Symposia on Quantitative Biology* **38**, 591–9.

Lefevre, G. Jr 1976. A photographic representation and interpretation of the polytene chromosomes of *Drosophila melanogaster* salivary glands. In *The genetics and biology of* Drosophila, Vol. 1a, M. Ashburner & E. Novitski (eds), 31–66. New York: Academic Press.

Lefevre, G. 1981. The distribution of randomly recovered X-ray induced sex-linked genetic effects in *Drosophila melanogaster*. *Genetics* **99**, 461–80.

Lefevre, G. & W. Watkins 1986. The question of the total gene number in *Drosophila melanogaster*. *Genetics* **113**, 869–95.

Lehmann, R. & C. Nüsslein-Volhard 1984. The development of the *Drosophila* abdomen: localisation and region-specific requirement of *oskar*-dependent cytoplasmic activity. Crete Drosophila Meeting, Kolymbari (Abst).

Lehmann, R. & C. Nüsslein-Volhard 1986. The knirps group of segmentation genes. Crete Drosophila Meeting, Kolymbari (Abst).

Lehmann, R., U. Dietrich, F. Jiménez & J. A. Campos-Ortega 1981. Mutations of early neurogenesis in *Drosophila*. *Wilhelm Roux's Archives* **190**, 226–9.

Leuders, K. K. & E. L. Kuff 1980. Intracisternal A-particle genes: identification in the genome of *Mus musculus* and comparison of multiple isolates from a mouse gene library. *Proceedings of the National Academy of Sciences* **77**, 3571–5.

Leutwiler, L. S., B. R. Hough-Evans & E. M. Meyerowitz 1984. The DNA of *Arabidopsis thaliana. Molecular and General Genetics* **194**, 15–23.

Lev, Z., T. L. Thomas, A. S. Lee, R. C. Angerer, R. J. Britten & E. H. Davidson 1980. Developmental expression of two cloned sequences coding for rare sea urchin embryo messages. *Developmental Biology* **76**, 322–40.

Levin, B. R. 1984. Science as a way of knowing – molecular evolution. *American Zoologist* **24**, 451–64.

Levin, D. A. & S. W. Funderberg 1979. Genome size in angiosperms: temperate versus tropical species. *American Naturalist* **114**, 784–95.

Levine, M., G. M. Rubin & R. Tjian 1984. Human DNA sequences homologous to a protein coding region conserved between homeotic genes of *Drosophila. Cell* **38**, 667–73.

Levine, M., K. Harding, C. Wedeen, H. Doyle, T. Hoey & H. Radomska 1985. Expression of the homeobox gene family in *Drosophila. Cold Spring Harbor Symposia on Quantitative Biology* **50**, 209–22.

Levine, R. B. & J. W. Truman 1982. Metamorphosis of the insect nervous system: changes in morphology and synaptic interactions of identified neurones. *Nature* **299**, 250–2.

Levinger, L. & A. Varshavsky 1982. Protein Dl preferentially binds A+T-rich DNA *in vitro* and is a component of *Drosophila* nucleosomes containing A+T-rich satellite DNA. *Proceedings of the National Academy of Sciences* **79**, 7152–6.

Levinson, G., J. L. Marsh, J. T. Epplen & G. A. Gutman 1985. Cross-hybridizing snake satellite, *Drosophila*, and mouse DNA sequences may have arisen independently. *Molecular Biology and Evolution* **2**, 494–504.

Levis, R., M. Collins & G. M. Rubin 1982. FB elements are the common basis for the instability of the w^{DZL} and w^c *Drosophila* mutations. *Cell* **30**, 551–65.

Levis, R. & G. M. Rubin 1982. An unstable w^{DZL} mutation of *Drosophila* is caused by a 13 kilobase insertion that is imprecisely excised in phenotypic revertants. *Cell* **30**, 543–50.

Levy, L. S. & J. E. Manning 1981. Messenger RNA sequence complexity and homology in developmental stages of *Drosophila. Developmental Biology* **85**, 141–9.

Lewin, R. 1984. Why is development so illogical. *Science* **224**, 1327–9.

Lewis, E. B. 1950. The phenomenon of position effect. *Advances in Genetics* **3**, 73–115.

Lewis, E. B. 1964. Genetic control and regulation of developmental pathways. In *Role of chromosomes in development*, M. Locke (ed.), 231–52. New York: Academic Press.

Lewis, E. B. 1978. A gene complex controlling segmentation in *Drosophila. Nature* **276**, 565–70.

Lewis, E. B. 1985. Regulation of the genes of the bithorax complex in *Drosophila. Cold Spring Harbor Symposia on Quantitative Biology* **50**, 155–64.

Lewontin, R. C. 1978. Adaptation. *Scientific American* **239**, 212–30.

Lewontin, R. C. 1982. Prospectives, perspectives and retrospectives. *Palaeobiology* **8**, 309–13.

Lifschytz, E. & M. M. Green 1979. Genetic identification of dominant overproducing mutations: the Beadex gene. *Molecular and General Genetics* **171**, 153–9.

Lifschytz, E. & D. L. Lindsley 1972. The role of X-chromosome inactivation during spermatogenesis. *Proceedings of the National Academy of Sciences* **69**, 182–6.

Lifton, R. P., M. L. Goldberg, R. W. Karp & D. S. Hogness 1978. The organisation of the histone genes in *Drosophila melanogaster*: functional and evolutionary implications. *Cold Spring Harbor Symposia on Quantitative Biology* **42**, 1047–51.

Lima-de-Faria, A. 1980. How to produce a human with 3 chromosomes and 1000 primary genes. *Hereditas* **93**, 47–73.

Lima-de-Faria, A. & M. J. Moses 1966. Ultrastructure and cytochemistry of metabolic DNA in *Tipula*. *Journal of Cell Biology* **30**, 177–92.

Limbach, K. J. & R. Wu 1985. Characterisation of two *Drosophila melanogaster* cytochrome C genes and their transcripts. *Nucleic Acids Research* **13**, 631–44.

Liming, S. & S. Pathak 1981. Gametogenesis in a male Indian muntjac × Chinese muntjac hybrid. *Cytogenetics and Cell Genetics* **30**, 152–6.

Liming, S., Y. Yingying & D. Xingsheng 1980. Comparative cytogenetic studies on the red muntjac, Chinese muntjac and their F_1 hybrids. *Cytogenetics and Cell Genetics* **26**, 22–7.

Lin, S., & A. D. Riggs 1975. The general affinity of lac repressor for *E. coli* DNA: implications for gene regulation in procaryotes and eucaryotes. *Cell* **4**, 107–11.

Lindsley, D. L. & E. Lifschytz 1972. The genetic control of spermatogenesis in *Drosophila*. In *The genetics of the spermatozoan*, R. A. Beatty & S. Gluecksohn-Waelsch (eds), 203–20. Edinburgh.

Lindsley, D. L., L. Sandler, B. S. Baker, A. T. C. Carpenter, R. E. Dennell, J. E. Hall, P. A. Jacobs, G. L. G. Miklos, B. K. Davis, R. C. Gethmann, R. W. Hardy, A. Hessler, S. M. Miller, H. Nozawa, D. S. Parry & M. Gould-Somero 1972. Segmental aneuploidy and the genetic gross structure of the *Drosophila* genome. *Genetics* **71**, 157–84.

Livingstone, M. S., P. P. Sziber & N. G. Quinn 1984. Loss of calcium/calmodulin responsiveness in adenylate cyclase of *rutabaga*, a *Drosophila* learning mutant. *Cell* **37**, 205–15.

Livnen, E., L. Glazer, D. Segal, J. Schlessinger & B-Z. Shilo 1985. The *Drosophila* EGF receptor gene homolog: conservation of both hormone binding and kinase domains. *Cell* **40**, 599–607.

Lohe, A. R. & D. L. Brutlag 1986. Multiplicity of satellite DNA sequences in *Drosophila melanogaster*. *Proceedings of the National Academy of Sciences* **83**, 696–700.

Long, E. O. & I. B. Dawid 1979. Expression of ribosomal DNA insertions in *Drosophila melanogaster*. *Cell* **18**, 1165–96.

Long, E. O. & I. B. Dawid 1980. Repeated genes in eukaryotes. *Annual Review of Biochemistry* **49**, 727–64.

Lopata, M. A., J. C. Havercroft, L. T. Chow & D. W. Cleveland 1983. Four unique genes required for β tubulin expression in vertebrates. *Cell* **32**, 713–24.

Lucchesi, J. C. & T. Skripsky 1981. The link between dosage compensation and sex differentiation in *Drosophila melanogaster*. *Chromosoma* **82**, 217–27.

Lull, R. S. 1940. *Organic evolution*. New York: Macmillan.

Lynch, G. & M. Baudry 1984. The biochemistry of memory: a new and specific hypothesis. *Science* **224**, 1057–63.

Lyon, M. F. 1986. X chromosomes and dosage compensation. *Nature* **320**, 313.

Macgregor, H. C. 1980. Recent developments in the study of lampbrush chromosomes. *Heredity* **44**, 3–35.

Macgregor, H. C. & J. Kezer 1970. Gene amplification in oocytes with eight germinal vesicles from the tailed frog *Ascaphus truei* Stejneger. *Chromosoma* **29**, 189–206.

Macgregor, H. C. & E. M. Pino 1982. Ribosomal gene amplification in multinucleate oocytes of the egg brooding hylid frog *Flectonotus pygmaes*. *Chromosoma* **85**, 475–88.

Macgregor, H. C. & S. Sherwood 1979. The nucleolus organisers of *Plethodon* and *Aneides* located by *in situ* nucleic acid hybridisation with *Xenopus* ^{3}H-ribosomal RNA. *Chromosoma* **72**, 271–80.

Macgregor, H. C., S. Mizuno & M. Vlad 1976. Chromosomes and DNA sequences in salamanders. *Chromosomes Today* **5**, 331–9.

Macgregor, H. C., M. Vlad & L. Barnet 1977. An investigation of some problems concerning nucleolus organisers in salamanders. *Chromosoma* **59**, 283–99.

MacIntyre, R. J. 1982. Regulatory genes and adaptation. *Evolutionary Biology* **15**, 247–85.

MacIntyre, R. J. & M. R. Dean 1978. Evolution and phosphatase-1 in the genus *Drosophila* as estimated by subunit hybridisation: interspecific tests. *Journal of Molecular Evolution* **12**, 143–71.

Maggini, F., P. Barsanti & T. Marazia 1978. Individual variation of the nucleolus organizer regions in *Allium cepa* and *A. sativum*. *Chromosoma* **66**, 173–83.

Mahowald, A. P. & P. A. Hardy 1985. Genetics of *Drosophila* embryogenesis. *Annual Review of Genetics* **19**, 147–77.

Mahowald, A. P., C. D. Allis, K. M. Karrer, E. M. Underwood & G. L. Waring 1979. Germ plasm and pole cells of *Drosophila*. In *Determinants of spatial organisation*, S. Subtelny & I. R. Konisberg (eds), 127–46. 37th Symposium of the Society for Developmental Biology. New York: Academic Press.

Mahowald, A. P., J. H. Caulton, M. K. Edwards & A. D. Floyd 1979. Loss of centrioles and polyploidisation in follicle cells of *Drosophila*. *Experimental Cell Research* **118**, 404–10.

Maine, E. M., H. K. Salz, T. W. Cline & P. Schedl 1985. The sex lethal gene of *Drosophila*: DNA alterations with sex specific lethal mutations. *Cell* **43**, 521–9.

Maine, E. M., H. K. Salz, P. Schedl & T. W. Cline 1985. Sex lethal, a link between sex determination and sexual differentiation in *Drosophila melanogaster*. *Cold Spring Harbor Symposia on Quantitative Biology* **50**, 595–604.

Maizels, N. 1976. *Dictyostelium* 17S, 25S and 5S rRNA lie within a 38 000 base pair repeated unit. *Cell* **9**, 431–8.

Malmgren, B. A., W. A. Berggren & G. P. Lohmann 1984. Species formation through punctuated gradualism in planktonic Foraminifera. *Science* **225**, 317–19.

Mange, A. P. & E. J. Mange 1980. *Genetics: human aspects*. Philadelphia: Saunders College.

Maniatis, T., E. F. Fritsch, J. Laver & R. M. Lonvac 1980. The molecular genetics of human hemoglobins. *Annual Review of Genetics* **14**, 145–78.

Marrack, P. & J. Kappler 1986. The T cell and its receptor. *Scientific American* **254**, 28–37.

Marsh, J. L., P. D. L. Gibbs & P. M. Timmons 1985. Developmental control of transduced dopa decarboxylase genes in *D. melanogaster*. *Molecular and General Genetics* **198**, 393–403.

Martin, G. R. 1982. X-chromosome inactivation in mammals. *Cell* **29**, 721–4.

Martinez-Arias, A. & P. A. Lawrence 1985. Parasegments and compartments in the *Drosophila* embryo. *Nature* **313**, 639–42.

Martini, G., D. Toniolo, T. Vulliamy, L. Luzzatto, R. Dono, G. Viglietto, M. D'Urso & M. G. Persico 1986. Structural analysis of the X-linked gene encoding human glucose 6-phosphate dehydrogenase. *The EMBO Journal* **5**, 1849–55.

Marx, J. L. 1983. Immunoglobulin genes have enhancers. *Science* **221**, 735–7.

Matsumoto, H., M. Ozaki, K. Isono, L. Randall, M. Florean, D. Larrivee, Q. Pye & W. L. Pak 1986. Nina: class of mutations affecting photoreceptor-specific gene functions. In *Molecular neurobiology of Drosophila*, p. 16 (Abst). New York: Cold Spring Harbor Laboratory.

Maynard-Smith, J. 1983. The genetics of stasis and punctuation. *Annual Review of Genetics* **17**, 11–25.

McClay, D. & G. M. Wessel 1985. The surface of the sea urchin embryo at gastrulation: a molecular mosaic. *Trends in Genetics* **1**, 12–16.

McGinnis, W. 1985. Homeo box sequences of the Antennapedia class are conserved only in higher animal genomes. *Cold Spring Harbor Symposia on Quantitative Biology* **50**,263–70.

McGinnis, W., J. Farell Jr & S. K. Beckendorf 1980. Molecular limits on the size of a genetic locus in *Drosophila melanogaster*. *Proceedings of the National Academy of Sciences* **77**, 7367–71.

McGinnis, W., A. W. Shermoen, J. Heemskerk & S. K. Beckendorf 1983. DNA sequence changes in an upstream DNAase I-hypersensitive region are correlated with reduced gene expression. *Proceedings of the National Academy of Sciences* **80**, 1063–7.

McGinnis, W., R. L. Garber, J. Wirz, A. Kuroiwa & W. J. Gehring 1984a. A homologous protein-coding sequence in *Drosophila* homeotic genes and its conservation in other Metazoans. *Cell* **37**, 403–8.

McGinnis, W., M. S. Levine, E. Hafen, A. Kuroiwa & W. J. Gehring 1984b. A conserved DNA sequence in homeotic genes of the *Drosophila* Antennapedia and bithorax complexes. *Nature* **308**, 428–33.

McKenzie, S. L., S. Henikoff & M. Meselson 1975. Localisation of RNA from heat induced polysomes at puff sites in *Drosophila melanogaster*. *Proceedings of the National Academy of Sciences* **72**, 1117–21.

McKeon, C., H. Ohkubo, I. Pastan & B. de Crombrugghe 1982. Unusual methylation pattern of the $\alpha2(1)$ collagen gene. *Cell* **29**, 203–10.

McKusick, V. A. 1980. The anatomy of the human genome. *Journal of Heredity* **71**, 370–91.

Medawar, P. 1957. The Imperfections of Man. In *The uniqueness of the individual*, 122–33. London: Methuen.

Medford, R. M., H. T. Nguyen, A. T. Destree, E. Summers & B. Nadel-Ginard 1984. A novel mechanism of alternative RNA splicing for the developmentally regulated generation of Troponin T isoforms from a single gene. *Cell* **38**, 409–21.

Melton, D. W., C. McElwan, A. B. McKie & A. M. Reid 1986. Expression of the mouse HPRT gene: deletion analysis of the promoter region of an X chromosome linked housekeeping gene. *Cell* **44**, 319–28.

Meyer, G. F. & H. J. Lipps 1984. Electron microscopy of surface spread polytene chromosomes of *Drosophila* and *Stylonychia*. *Chromosoma* **89**, 107–10.

Meyerowitz, E. M. & D. S. Hogness 1982. Molecular organisation of a *Drosophila* puff site that responds to ecdysone. *Cell* **28**, 165–76.

Meyerowitz, E. M., L. Crosby, T. Crowley, M. Garfinkel, C. Martin & P. Mathers 1984. Puffing and the 68C glue gene cluster. Crete Drosophila Meeting, Kolymbari (Abst).

Mignone, N., S. Olivero, G. De Lange, D. L. Delacroix, D. Boschis, F. Altruda, L. Silengo, M. De Marchi & A. O. Carbonara 1984. Multiple gene deletion within the human immunoglobulin heavy chain cluster. *Proceedings of the National Academy of Sciences* **81**, 5811–5.

Miklos, G. L. G. 1982. Sequencing and manipulating highly repeated DNA. In *Genome evolution*, G. A. Dover & R. B. Flavell (eds), 41–68. London: Academic Press.

Miklos, G. L. G. 1985. Localised highly repetitive DNA sequences in vertebrate and invertebrate genomes. In *Molecular evolutionary genetics*, R. J. MacIntyre (ed.), 241–321. New York: Plenum Press.

Miklos, G. L. G. & A. C. Gill 1981. The DNA sequences of cloned complex satellite DNAs from Hawaiian *Drosophila* and their bearing on satellite DNA sequence conservation. *Chromosoma* **82**, 409–27.

Miklos, G. L. G. & B. John 1987. From genome to phenotype. In *Rates of evolution*, K. S. W. Campbell & M. Day (eds), 263–82. London: Allen & Unwin.

Miklos, G. L. G., D. A. Willcocks & P. R. Baverstock 1980. Restriction endonuclease and molecular analyses of three rat genomes with special reference to chromosome rearrangement and speciation problems. *Chromosoma* **76**, 339–63.

Miklos, G. L. G., M. J. Healy, P. Pain, A. J. Howells & R. J. Russell 1984. Molecular and genetic studies on the euchromatin-heterochromatin transition region of the X chromosome of *Drosophila melanogaster* I. A cloned entry point near the uncoordinated (*unc*) locus. *Chromosoma* **89**, 218–27.

Miklos, G. L. G., L. E. Kelly, P. E. Coombe, C. Leeds & G. Lefevre 1987a. Localisation of the genes shaking-B, small optic lobes, sluggish-A, stoned and stress sensitive-C to a well defined region on the X-chromosome of *Drosophila melanogaster*. *The Journal of Neurogenetics* **4**, 1–19.

Miklos, G. L. G., L. E. Kelly, P. E. Coombe, J. Davies, G. De Couet, V. Pirrotta, M. Yamamoto, A. Schalet & G. Lefevre 1987b. Genetic and molecular studies of genes affecting behaviour in the maroonlike to Suppressor of forked region of *Drosophila melanogaster*. In *Molecular biology of neurogenesis*, 1st European Meeting of *Drosophila* Neurogenetics, Simonswald. *The Journal of Neurogenetics* **4**, 148–9 (Abst).

Miller, C. A. & S. Benzer 1983. Monoclonal antibody cross-reactions between *Drosophila* and human brain. *Proceedings of the National Academy of Sciences* **80**, 7641–5.

Miller, D. A., E. Okamoto, B. F. Erlanger & O. J. Miller 1982. Is DNA methylation responsible for mammalian X-chromosome inactivation? *Cytogenetics and Cell Genetics* **33**, 345–9.

Miller, J. R. & D. A. Melton 1981. A transcriptionally active pseudogene in *Xenopus laevis* oocyte 5S DNA. *Cell* **24**, 829–35.

Milner, R. J. & J. G. Sutcliffe 1983. Gene expression in rat brain. *Nucleic Acids Research* **11**, 5497–520.

Mitsialis, S. A. & F. C. Kafatos 1985. Regulatory elements controlling gene expression are conserved between flies and moths. *Nature* **317**, 453–6.

Mizuno, S. & H. C. Macgregor 1974. Chromosomes, DNA sequences, and evolution in salamanders of the genus *Plethodon*. *Chromosoma* **48**, 239–96.

Mohler, J. & E. F. Wieschaus 1985. Bicaudal mutations of *Drosophila melanogaster*: alteration of blastoderm cell fate. *Cold Spring Harbor Symposia on Quantitative Biology* **50**, 105–11.

Monk, R. J., O. Meyuhas & R. P. Perry 1981. Mammals have multiple genes for individual ribosomal proteins. *Cell* **24**, 301–6.

Monnickendham, M. A. & M. Balls 1973. The relationship between cell sizes, respiratory rates and survival of amphibian tissues in long term organ cultures. *Comparative Biochemistry and Physiology* **44A**, 871–80.

Monson, J. M., J. Natzle, J. Friedman & B. J. McCarthy 1982. Expression and novel structure of a collagen gene in *Drosophila*. *Proceedings of the National Academy of Sciences* **79**, 1761–5.

Montell, C., K. Jones, E. Hafen & G. Rubin 1985. Rescue of the *Drosophila* phototransduction mutation *trp* by germ line transformation. *Science* **230**, 1040–3.

Montell, C., C. Zuker, K. Jones & G. Rubin 1986. Molecular identification and characterization of genes required for phototransduction. Crete Drosophila Meeting, Kolymbari (Abst).

Montiel, J. F., C. J. Norbury, M. F. Tulte, M. J. Dobson, J. S. Mills, A. J. Kingsman & S. M. Kingsman 1984. Characterisation of human chromosomal DNA sequences which replicate autonomously in *Saccharomyces cerevisiae*. *Nucleic Acids Research* **12**, 1049–68.

Moore, G. D., J. D. Procunier, D. P. Cross & T. A. Grigliatti 1979. Histone gene deficiencies and position-effect variegation in *Drosophila*. *Nature* **282**, 312–14.

Morata, G. & P. A. Lawrence 1977. Homeotic genes, compartments and cell determination in *Drosophila*. *Nature* **265**, 211–16.

Morescalchi, A. & E. Olmo 1982. Single-copy DNA and vertebrate phylogeny. *Cytogenetics and Cell Genetics* **34**, 93–101.

Morgan, B. & J. Hirsh 1986. Tissue-specific splicing of the *Ddc* transcript. *27th Annual* Drosophila *Conference*. Asilomar, California (Abst).

Morgan, G. T., H. C. MacGregor & A. Colman 1980. Multiple ribosomal gene sites revealed by *in situ* hybridisation of *Xenopus* rDNA to *Triturus* lampbrush chromosomes. *Chromosoma* **80**, 309–30.

Moscoso del Prado, J. & A. Garcia-Bellido 1984a. Genetic regulation of the Achaete-scute complex of *Drosophila melanogaster*. *Roux's Archives of Developmental Biology* **193**, 242–5.

Moscoso del Prado, J. & A. Garcia-Bellido 1984b. Cell interactions in the generation of chaete pattern in *Drosophila*. *Roux's Archives of Developmental Biology* **193**, 246–51.

Müller, F. 1971. Comparative embryological investigations based on a *Balaenoptera physalis* of 90 mm. length. Pp. 203–215 In *Investigations on Cetacea*, Vol. 3, G. Pilleri (ed.), 203–15. Berne, Switzerland.

Müller, M. M., A. L. Carrasco & E. M. de Robertis 1984. A homeo-box containing gene expressed during oogenesis in *Xenopus*. *Cell* **39**, 157–62.

Müller-Holtkamp, F., D. C. Knipple, E. Seifert & H. Jäckle 1985. An early role of maternal mRNA in establishing the dorsoventral pattern in pelle mutant *Drosophila* embryos. *Developmental Biology* **110**, 238–46.

Munz, P., H. Amstutz, J. Kohli & U. Leupold 1982. Recombination between dispersed serine tRNA genes in *Schizosaccharomyces pombe*. *Nature* **300**, 225–31.

Nagasawa, H., H. Katoaka, A. Isogai, S. Tamura, A. Suzuki, H. Ishizaki, A. Mizoguchi, Y. Fujiwara & A. Suzuki 1984. Amino-terminal amino acid sequence of the silkworm prothoracicotropic hormone: homology with insulin. *Science* **226**, 1344–5.

Nagl, W. 1978. *Endopolyploidy and polyteny in differentiation and evolution*. Amsterdam: Elsevier North Holland Biomedical Press.

Nagl, W. & F. Ehrendorfer 1974. DNA content, heterochromatin, mitotic index and growth in perennial and annual Anthemidae (Asteraceae). *Plant Systematics and Evolution* **123**, 35–54.

Nagylaki, T. & T. D. Petes 1982. Intrachromosomal gene conversion and the maintenance of sequence homogeneity among repeated genes. *Genetics* **100**, 315–37.

Nambu, J. R., R. Taussig, A. C. Mahon & R. H. Scheller 1983. Gene isolation with cDNA probes from identified *Aplysia* neurons; neuropeptide modulations of cardiovascular physiology. *Cell* **35**, 47–56.

Narayan, R. K. J. 1982. Discontinuous DNA variation in the evolution of plant species. The genus *Lathyrus*. *Evolution* **36**, 877–91.

Nasmyth, K. A. & K. Tatchell 1980. The structure of transposable yeast mating type loci. *Cell* **19**, 753–64.

Natzle, J. E., J. M. Monson & B. J. McCarthy 1982. Cytogenetic location and expression of collagen-like genes in *Drosophila*. *Nature* **296**, 368–71.

Neuman-Silberberg, F. S., E. Schejter, F. M. Hofmann & B-Z. Shilo 1984. The *Drosophila ras* oncogenes: structure and nucleotide sequence. *Cell* **37**, 1027–33.

Newport, J. & M. Kirschner 1982a. A major developmental transition in early *Xenopus* embryos I. Characterisation and timing of cellular changes at the midblastula stage. *Cell* **30**, 675–86.

Newport, J. & M. Kirschner 1982b. A major developmental transition in early *Xenopus* embryos II. Control of the onset of transcription. *Cell* **30**, 686–96.

Nirenberg, M., S. Wilson, H. Higashida, A. Rotter, K. Krueger, N. Busis, R. Ray, J. G. Kenimer & M. Adler 1983. Modulation of synapse formation by cyclic adenosine monophosphate. *Science* **222**, 794–9.

North, G. 1984. How to make a fruitfly. *Nature* **311**, 214–16.

Nöthiger, R. 1972. The larval development of imaginal discs. *Results and Problems in Cell Differentiation* **5**, 1–34.

Nur, U. 1977. Maintenance of a 'parasitic' B chromosome in the grasshopper *Melanoplus femur-rubrum*. *Genetics* **87**, 499–512.

Nüsslein-Volhard, C. 1979. Maternal effect mutations that alter the spatial co-ordinates of the embryo of *Drosophila melanogaster*. In *Determinants of spatial organisation*, S. Subtelny & I. R. Konisberg (eds), 185–211. 37th Symposium of the Society for Developmental Biology. New York: Academic Press.

Nüsslein-Volhard, C. & E. Wieschaus 1980. Mutations affecting segment number and polarity in *Drosophila*. *Nature* **287**, 795–801.

Nüsslein-Volhard, C., E. Wieschaus, H. Kluding 1984. Mutations affecting the pattern of the larval cuticle in *Drosophila melanogaster* I. Zygotic loci on the second chromosome. *Roux's Archives of Developmental Biology* **193**, 267–82.

Nüsslein-Volhard, C., H. Kluding & C. Jürgens 1985. Genes affecting the segmental subdivision of the *Drosophila* embryo. *Cold Spring Harbor Symposia on Quantitative Biology* **50**, 145–54.

O'Brien, S. J. 1973. On estimating functional gene number in eukaryotes. *Nature* **242**, 52–4.

O'Brien, S. J. 1984. *Genetic maps*, Vol. 3. Cold Spring Harbor Laboratory.

O'Brien, S. J. & Y. Shimada 1974. The α-glycerophosphate cycle in *Drosophila melanogaster*. *Journal of Cell Biology* **63**, 864–82.

O'Brien, S. J., H. N. Seuanez & J. E. Womack 1985. On the evolution of genome organization in mammals. In *Molecular evolutionary genetics*, R. J. MacIntyre (ed.), 519–89. New York: Plenum Press.

O'Brochta, D. A. & P. J. Bryant 1985. A zone of non proliferating cells at a lineage restriction boundary in *Drosophila*. *Nature* **313**, 138–41.

O'Farrell, P. H., C. Desplan, S. Di Nardo, J. A. Kassis, J. M. Kuner, E. Sher, J. Theis & D. Wright 1985. Embryonic pattern in *Drosophila*: the spatial distribution and sequence specific DNA binding of engrailed protein. *Cold Spring Harbor Symposia on Quantitative Biology* **50**, 235–42.

O'Hare, K. 1985. The mechanism and control of P element transposition in *Drosophila melanogaster*. *Trends in Genetics* **1**, 250–4.

O'Hare, K. 1986. Genes within genes. *Trends in Genetics* **2**, 23.

O'Hare, K. & G. M. Rubin 1983. Structures of P transposable elements and their sites of insertion and excision in the *Drosophila melanogaster* genome. *Cell* **34**, 25–35.

Ohno, S. 1970a. So much 'junk' DNA in our genome. In *Evolution of genetic systems*, H. H. Smith (ed.), 366–70. Brookhaven Symposia in Biology **23**. New York: Gordon & Breach.

Ohno, S. 1970b. *Evolution by gene duplication*. Berlin: Springer.

Ohno, S. 1974. *Animal Cytogenetics 4 Chordata 1: Protochordata, Cyclostomata and Pisces*. Stuttgart: Gebrüder Borntraeger.

Ohno, S. 1982. The common ancestry of genes and spacers in the euchromatic region: *ominis ordinis hereditarium a ordinis priscum minutum*. *Cytogenetics and Cell Genetics* **34**, 102–11.

Ohno, S. 1985. Dispensable genes. *Trends in Genetics* **1**, 160–4.

Ohta, T. 1983. On the evolution of multigene families. *Theoretical Population Biology* **23**, 216–40.

Ohta, T. & G. A. Dover 1984. The cohesive population genetics of molecular drive. *Genetics* **108**, 501–21.

Olmo, E. 1981. Evolution of genome size and DNA base composition in reptiles. *Genetic* **57**, 39–50.

Olmo, E. 1983. Nucleotype and cell size in vertebrates: a review. *Basic and Applied Histochemistry* **27**, 227–56.

Olmo, E. & A. Morescalchi 1975. Evolution of the genome and cell sizes in salamanders. *Experientia* **31**, 804–6.

Olmo, E. & A. Morescalchi 1978. Genome and cell size in frogs: a comparison with salamanders. *Experientia* **34**, 44–66.

Olmo, E. & G. Odierna 1982. Relationship between DNA content and cell morphometric parameters in reptiles. *Basic and Applied Histochemistry* **26**, 27–34.

Olmo, E., V. Stingo, G. Odierna & O. Cobror 1981. Variations in repetitive DNA and evolution in reptiles. *Comparative Biochemistry and Physiology* **69B**, 687–91.

Orgel, L. E. & F. H. C. Crick 1980. Selfish DNA: the ultimate parasite. *Nature* **284**, 604–7.

Orias, E. 1981. Probable somatic DNA rearrangements in mating type determination in *Tetrahymena thermophila*: a review and a model. *Developmental Genetics* **2**, 185–202.

Orr, W., K. Komitopoulou & F. C. Kafatos 1984. Mutants suppressing in *trans* chromosome gene amplification in *Drosophila*. *Proceedings of the National Academy of Sciences* **81**, 3773–7.

Oster, G. & P. Alberch 1982. Evolution and bifurcation of developmental programs. *Evolution* **36**, 444–59.

Östergren, G. 1945. Parasitic nature of extra fragment chromosomes. *Botaniker Notiser* **2**, 157–67.

O'Tousa, J., W. Baehr, R. L. Martin, J. Hirsh, W. L. Pak & M. L. Applebury 1985. The *Drosophila* nina E gene encodes an opsin. *Cell* **40**, 839–50.

Ouwenweel, W. J. 1976. Developmental genetics of homeosis. *Advances in Genetics* **18**, 179–248.

Ozawa, H., E. Kushiya & Y. Takahashi 1980. Complexity of RNA from the neuronal and glial nuclei. *Neuroscience Letters* **18**, 191–6.

Palka, J. & A. Ghysen 1982. Segments, compartments and axon paths in *Drosophila*. *Trends in Neurosciences* **5**, 382–6.

Palmer, J. D. 1985. Comparative organisation of chloroplast genomes. *Annual Review of Genetics* **19**, 325–54.

Palmer, J. D. & L. A. Herbon 1987. Unicellular structure of the *Brassica hirta* mitochondrial genome. *Current Genetics* **11**, 565–70.

Parker, C. S. & J. Topol 1984a. A *Drosophila* RNA polymerase II transcription factor contains a promoter-region-specific DNA binding activity. *Cell* **36**, 357–69.

Parker, C. S. & J. Topol 1984b. A *Drosophila* RNA polymerase II transcription factor binds to the regulatory site of an hsp 70 gene. *Cell* **37**, 273–83.

Parnes, J. R. & R. R. Robinson 1983. Multiple mRNA species with distinct 3' termini are transcribed from the β_2-microglobulin gene. *Nature* **302**, 449–52.

Pasternak, J. & M. Haight 1975. DNA accumulation during oogenesis in the free living nematode *Pangrellus silusiae*. *Chromosoma* **49**, 279–98.

Patel, N. H., P. M. Snow & S. C. Goodman 1986. Cellular and molecular characterization of a cell surface glycoprotein expressed on a subset of axons in the *Drosophila* embryo. In *Molecular neurobiology of* Drosophila, p. 7 (Abst). New York: Cold Spring Harbor Laboratory.

Patterson, J. T. & W. W. Stone 1952. *Evolution in the genus* Drosophila. New York: Macmillan.

Peacock, W. J., D. Brutlag, E. Goldring, R. Appels, C. W. Hinton & D. L. Lindsley 1974. The organisation of highly repeated DNA sequences in *Drosophila melanogaster* chromosomes. *Cold Spring Harbor Symposia on Quantitative Biology* **38**, 405–16.

Peacock, W. J., A. R. Lohe, W. L. Gerlach, P. Dunsmuir, E. S. Dennis & R. Appels 1977. Fine structure and evolution of DNA in heterochromatin. *Cold Spring Harbor Symposia on Quantitative Biology* **42**, 1121–34.

Pearston, D. H.,.M. Gordon & N. Hardman 1985. Transposon-like properties of the major long repetitive sequence family of *Physarum polycephalum*. *The EMBO Journal* **4**, 3557–62.

Perrimon, N. & A. P. Mahowald 1986. The maternal effects of zygotic lethals. Crete Drosophila Meeting, Kolymbari (Abst).

Perrimon, N., L. Engstrom & A. P. Mahowald 1984. The effects of zygotic lethal mutations on female germ-line functions in *Drosophila*. *Developmental Biology* **105**, 404–14.

Petes, T. D. 1980. Unequal meiotic recombination with tandem arrays of yeast ribosomal DNA genes. *Cell* **19**, 765–74.

Piechaczyk, M., J. M. Blanchard, S. R-E. Sabouty, C. Dani, L. Marty & P. Jeanteur 1984. Unusual abundance of vertebrate 3-phosphate dehydrogenase pseudogenes. *Nature* **312**, 469–71.

Pilleri, G. & M. Gihr 1971. Brain-body weight ratio in *Pontoporia blainvillei*. In *Investigations on Cetacea*, vol. 3, G. Pilleri (ed.), 69–73. Berne, Switzerland.

Pilleri, G. & A. Wanderler 1970. Ontogeny and functional morphology of the eye of the fin whale, *Balaenoptera physalis*. In *Investigations on Cetacea*, vol. 2, G. Pilleri (ed.), 179–229. Berne, Switzerland.

Pimpinelli, S., S. Bonaccorsi, M. Gatti & L. Sandler 1986. The peculiar genetic organisation of *Drosophila* heterochromatin. *Trends in Genetics* **2**, 17–20.

Pirrotta, V. 1984. Chromosome microdissection and microcloning. *Trends in Biochemical Science* **9**, 220–1.

Policansky, D. 1982. The asymmetry of Flounders. *Scientific American* **246**, 96–102.

Poole, S. J., L. M. Kauvar, B. Drees & T. Kornberg 1985. The engrailed locus of *Drosophila melanogaster*: structural analysis of an embryonic transcript. *Cell* **40**, 37–43.

Preiss, A., U. B. Rosenberg, A. Kienlin, E. Seifert & H. Jäckle 1985. Molecular genetics of Krüppel, a gene required for segmentation of the *Drosophila* embryo. *Nature* **313**, 27–32.

Price, H. J., K. L. Chambers, K. Bachmann & J. Riggs 1983. Inheritance of nuclear 2C DNA content variation in intraspecific and interspecific hybrids of *Microseris* (Asteraceae). *American Journal of Botany* **70**, 1133–8.

Procunier, J. D. & K. D. Tartof 1975. Genetic analysis of the 5S RNA genes in *Drosophila melanogaster*. *Genetics* **81**, 515–23.

Procunier, J. D. & K. D. Tartof 1978. A genetic locus having trans and contiguous cis functions that control the disproportionate replication of ribosomal RNA genes in *Drosophila melanogaster*. *Genetics* **88**, 67–79.

Proudfoot, N. J. 1980. Pseudogenes. *Nature* **286**, 840–1.

Quinn, W. G., P. P. Sziber & R. Booker 1979. The *Drosophila* memory mutant amnesiac. *Nature* **277**, 212–14.

Raff, R. A. & T. C. Kaufman 1983. Embryos, genes and evolution. New York: Macmillan Publishing.

Raff, R. A., J. A. Anstrom, C. J. Huffman, D. S. Leaf, J-H. Loo, R. M. Showman & D. E. Wells 1984. Origin of a gene regulatory mechanism in the evolution of echinoderms. *Nature* **310**, 312–14.

Raganath, H. A. & K. Hägele 1982. The chromosomes of two *Drosophila* races: *Drosophila nasuta nastua* and *Drosophila nasuta albomicana* I. Distribution and differentiation of heterochromatin. *Chromosoma* **85**, 83–92.

Raganarth, H. A. & N. B. Krishnamurthy 1981. Population genetics of *Drosophila nasuta nasuta*, *Drosophila nasuta albomicana* and their hybrids. *Journal of Heredity* **72**, 19–21.

Raikov, I. B. 1976. Evolution of macronuclear organisation. *Annual Review of Genetics* **10**, 413–40.

Rasch, E. M. 1974. The DNA content of sperm and hemocyte nuclei of the silkworm *Bombyx mori* L. *Chromosoma* **45**, 1–26.

Rasch, E. M., H. J. Barr & W. Rasch 1971. The DNA content of sperm of *Drosophila melanogaster*. *Chromosoma* **33**, 1–18.

Raup, D. M. 1981. Extinction: bad genes or bad luck? *Acta Geologica Hispanica* **16**, 25–33.

Raup, D. M. & J. J. Sepkoski 1984. Periodicity of extinctions in the geologic past. *Proceedings of the National Academy of Sciences* **81**, 801–5.

Razin, A. & A. D. Riggs 1980. DNA methylation and gene function. *Science* **210**, 604–10.

Reddy, P., A. C. Jacquier, N. Abovich, G. Petersen & M. Rosbash 1986. The period clock locus of *Drosophila melanogaster* codes for proteoglycan. *Cell* **46**, 53–61.

Reddy, P., W. A. Zehring, D. A. Wheeler, V. Pirrotta, C. Hadfield, J. C. Hall & M. Rosbash 1984. Molecular analysis of the period locus in *Drosophila melanogaster* and identification of a transcript involved in biological rhythms. *Cell* **38**, 701–10.

Reeder, R. H. 1984. Enhancers and ribosomal gene spacers. *Cell* **38**, 349–51.

Reeder, R. H. & J. G. Roan 1984. The mechanisms of nucleolar dominance in *Xenopus* hybrids. *Cell* **38**, 39–44.

Reeder, R. H., J. G. Roan & M. Dunaway 1983. Spacer regulation of *Xenopus* ribosomal gene transcription: competition in oocytes. *Cell* **35**, 449–56.

Rees, H. & R. K. J. Narayan 1977. Evolutionary DNA variations in *Lathyrus*. *Chromosomes Today* **6**, 131–9.

Rees, H., D. D. Shaw & P. Wilkinson 1978. Nuclear DNA variation among acridid grasshoppers. *Proceedings of the Royal Society, London* **B202**, 517–25.

Rees, H., G. Jenkins, A. G. Seal & J. Hutchinson 1982. Assays of the phenotypic effects of changes in DNA amount. In *Genome evolution*, G. A. Dover & R. B. Flavell (eds), 287–97. London: Academic Press.

Renkawitz, R. 1978. Two highly repetitive DNA satellites of *Drosophila hydei* localised within the α–heterochromatin of specific chromosomes. *Chromosoma* **66**, 237–48.

Rensch, B. 1959. *Evolution above the species level*. New York: John Wiley.

Rheinsmith, E. L., R. Hinegardner & K. Bachmann 1974. Nuclear DNA amounts in Crustacea. *Comparative Biochemistry and Physiology* **48B**, 343–8.

Richter, J. D., L. D. Smith, D. M. Anderson & E. H. Davidson 1984. Interspersed poly (A) RNAs of amphibian oocytes are not translatable. *Journal of Molecular Biology* **173**, 227–41.

Rick, C. M. 1971. Some cytogenetic features of the genome in diploid plant species. *Stadler Symposia 1 and 2*, 153–74.

Rimpau, J. & R. B. Flavell 1975. Characterisation of rye B chromosome DNA by DNA/DNA hybridisation. *Chromosoma* **52**, 207–17.

Rine, J., R. Jensen, D. Hagen, L. Blair & I. Herskowitz 1981. Pattern of switching and fate of the replaced casette in yeast mating type interconversion. *Cold Spring Harbor Symposia on Quantitative Biology* **45**, 951–60.

Rio, D. C., F. A. Laski & G. M. Rubin 1986. Identification and immunochemical analysis of biologically active *Drosophila* P-element transposase. *Cell* **44**, 21–32.

Ripoll, P. & A. Garcia-Bellido 1979. Viability of homozygous deficiencies in somatic cells of *Drosophila melanogaster*. *Genetics* **91**, 443–53.

Roach, A., K. Boylan, S. Horvath, S. B. Prusiner & L. E. Hood 1983. Characterisation of cloned cDNA representing rat myelin basic protein: absence of expression in brain of shiverer mutant mice. *Cell* **34**, 799–806.

Roberts, D. B. & S. Evan-Roberts 1979. The X-linked α-chain gene of *Drosophila* LSP-1 does not show dosage compensation. *Nature* **280**, 691–2.

Robertson, M. 1981. Genes of lymphocytes: diverse means to antibody diversity. *Nature* **290**, 625–7.

Robinson, R. R. & N. Davidson 1981. Analysis of a *Drosophila* tRNA gene cluster: two tRNA[Leu] genes contain intervening sequences. *Cell* **23**, 251–9.

Roeder, G. S. & G. F. Fink 1982. Movement of yeast transposable elements by gene conversion. *Proceedings of the National Academy of Sciences* **79**, 5621–5.

Roehrdanz, R. L., J. M. Kitchens & J. C. Lucchesi 1977. Lack of dosage compensation for an autosomal gene relocated to the X-chromosome in *Drosophila melanogaster*. *Genetics* **85**, 489–96.

Roiha, H., C. A. Read, M. J. Browne & D. M. Glover 1983. Widely differing degrees of sequence conservation of the two types of rDNA insertion within the *melanogaster* sub group of *Drosophila*. *The EMBO Journal* **2**, 721–6.

Rosbash, M. & P. J. Ford 1974. Polyadenylic acid–containing RNA in *Xenopus laevis* oocytes. *Journal of Molecular Biology* **85**, 87–101.

Rosbash, M., P. J. Ford & J. O. Bishop 1974. Analysis of the C-value paradox by molecular hybridisation. *Proceedings of the National Academy of Sciences* **71**, 3746–50.

Rose, M. R. & W. F. Doolittle 1983. Molecular biological mechanisms of speciation. *Science* **220**, 157–62.

Rosenberg, U. B. 1984. Molecular cloning of Krüppel. Crete Drosophila Meeting, Kolymbari (Abst).

Rosenberg, U. B., A. Preiss, E. Seifert, J. Jäckle & D. C. Knipple 1985. Production of phenocopies by Krüppel antisense RNA injection into *Drosophila* embryos. *Nature* **313**, 703–6.

Rosenberg, U. B., C. Schröder, A. Preiss, A. Kienlin, S. Cote, I. Reide & H. Jäckle 1986. Structural homology of the product of the *Drosophila* Krüppel gene with *Xenopus* transcription factor IIIA. *Nature* **319**, 336–9.

Rosenfeld, M. G., S. G. Amara & M. Evans 1984. Alternative RNA processing: determining neuronal phenotype. *Science* **225**, 1315–20.

Roth, G. E. 1979. Satellite DNA properties of the germ line limited DNA and the organization of the somatic genomes in the nematodes *Ascaris suum* and *Parascaris equorum*. *Chromosoma* **74**, 355–71.

Rotman, G., A. Itin & E. Kesher 1984. 'Solo' large terminal repeats (LTR) of an endogenous retrovirus-like gene family (VL30) in the mouse genome. *Nucleic Acids Research* **12**, 2273–82.

Rougon, G., H. Deagostini-Bazin, M. Hun & C. Goridis 1982. Tissue and developmental stage specific forms of a neural cell surface antigen linked to differences in glycosylation of a common peptide. *The EMBO Journal* **1**, 1239–44.

Royden, C. S. & L-Y. Jan 1986. Molecular analysis of the *tko* locus in *Drosophila melanogaster*. In *Molecular neurobiology of* Drosophila, p. 77 (Abst). New York: Cold Spring Harbor Laboratory.

Rozek, C. E. & N. Davidson 1983. *Drosophila* has one myosin heavy chain gene with three developmentally regulated transcripts. *Cell* **32**, 23–34.

Rubin, G. M. 1983. Dispersed repetitive DNAs in *Drosophila*. In *Mobile genetic elements*, J. A. Shapiro (ed.), 329–61. New York: Academic Press.

Rubin, G. M. 1985a. Summary. *Cold Spring Harbor Symposia on Quantitative Biology* **50**, 905–8.

Rubin, G. M. 1985b. P transposable elements and their use as genetic tools in *Drosophila*. *Trends in Neuroscience* **8**, 231–3.

Rubin, G. M. & A. C. Spradling 1982. Genetic transformation of *Drosophila* with transposable element vectors. *Science* **218**, 348–53.

Rubin, G. M., M. G. Kidwell & P. M. Bingham 1982. The molecular basis of P-M hybrid dysgenesis: the nature of induced mutations. *Cell* **29**, 987–94.

Rubin, G. M., T. Hazelrigg, R. E. Karess, F. A. Laski, T. Laverty, R. Levis, D. C. Rio, F. A. Spencer & C. S. Zuker 1985. Germ line specificity of P-element transposition and some novel patterns of expression of transduced copies of the white gene. *Cold Spring Harbor Symposia on Quantitative Biology* **50**, 329–35.

Ruddle, F. H., C. P. Hart, A. Awgulewitsch, A. Fainsod, M. Utset, D. Dalton, N. Kerk, M. Rabin, A. Ferguson-Smith, A. Fienberg & W. McGinnis 1985. Mammalian homeo box genes. *Cold Spring Harbor Symposia on Quantitative Biology* **50**, 277–84.

Ruppert, S., G. Scherer & G. Schütz 1984. Recent gene conversion involving bovine vasopressin and oxytocin precursor genes suggested by nucleotide sequence. *Nature* **308**, 554–7.

Rutishauser, U. & C. Goridis 1986. N-CAM: the molecule and its genetics. *Trends in Genetics* **2**, 72–6.

Sachs, R. I. & U. Clever 1972. Unique and repetitive DNA sequences in the genome of *Chironomus tentans*. *Experimental Cell Research* **74**, 587–91.

Saigo, K., W. Kugimiya, Y. Matsuo, S. Inouye, K. Yoshioka & S. Yuki 1984. Identification of the coding sequence for a reverse transcriptase-like enzyme in a transposable genetic element in *Drosophila melanogaster*. *Nature* **312**, 659–61.

Sakoyama, Y., Y. Yaoita & T. Honjo 1982. Immunoglobulin switch region-like sequences in *Drosophila melanogaster*. *Nucleic Acids Research* **10**, 4203–14.

Salkoff, L. B. & M. A. Tanouye 1986. Genetics of ion channels *Physiological Reviews* **66**, 301–29.

Salkoff, L. & R. Wyman 1983. Ion channels in *Drosophila* muscle. *Trends in Neurosciences* **6**, 128–33.

Salkoff, L., A. Butler, M. Hiken, A. Wei, K. Griffen, C. Ifune, R. Goodman & G. Mandel 1986. A *Drosophila* gene with homology to the vertebrate Na^+ channel. In *Molecular neurobiology of* Drosophila, p. 26 (Abst). New York: Cold Spring Harbor Laboratory.

Salvaterra, P. M., N. Itoh, V. Maines, R. Slemmon & N. Mori 1986. *Drosophila* choline acetyltransferase: molecular biology and evolution. In *Molecular neurobiology of* Drosophila, p. 31 (Abst). New York: Cold Spring Harbor Laboratory.

Samois, D. & H. Swift 1979. Genome organisation in the fleshfly *Sarcophaga bullata*. *Chromosoma* **75**, 129–43.

Sanchez-Herrero, E., J. Casanova, S. Kerridge and G. Morata 1985. Anatomy and function of the Bithorax Complex of *Drosophila*. *Cold Spring Harbor Symposia on Quantitative Biology* **50**, 165–72.

Sanford, J. P., V. M. Chapman & J. Rossant 1985. DNA methylation in extraembryonic lineages of mammals. *Trends in Genetics* **1**, 89–93.

Savakis, C. & M. Ashburner 1985. A simple gene with a complex pattern of transcription: the alcohol dehydrogenase gene of *Drosophila melanogaster*. *Cold Spring Harbor Symposia on Quantitative Biology* **50**, 505–14.

Scavarda, N. J., J. O'Tousa & W. L. Pak 1983. *Drosophila* locus with gene-dosage effects on rhodopsin. *Proceedings of the National Academy of Sciences* **80**, 4441–5.

Schalet, A. & G. Lefevre 1976. The proximal region of the X chromosome. In *The genetics and biology of* Drosophila, vol. 1b, M. Ashburner & E. Novitski (eds), 848–902. New York: Academic Press.

Scheller, R. H., J. F. Jackson, L. B. McAllister, B. S. Rothman, E. Mayeri & R. Axel 1983. A single gene encodes multiple neuropeptides mediating a stereotyped behaviour. *Cell* **32** 7–22.

Scherer, G., C. Tschudi, J. Perera, H. Delius & V. Pirrotta 1982. B104, a new dispersed repeat gene family in *Drosophila melanogaster* and its analogy with retroviruses. *Journal of Molecular Biology* **157**, 435–51.

Schmid, H. P., O. Akhalfat, De Sa C. Martin, F. Puvion, K. Koehler & K. Scherrer 1984. The prosome: an ubiquitous morphologically distinct RNP particle associated with repressed mRNPs and containing a specific ScRNA and a characteristic set of proteins. *The EMBO Journal* **3**, 29–34.

Schmidtke, J. & W. Engel 1975. Gene action in fish of tetraploid origin I. Cellular and biochemical parameters in cyprinid fish. *Biochemical Genetics* **13**, 45–51.

Schmidtke, J. & J. T. Epplen 1980. Sequence organisation of animal nuclear DNA. *Human Genetics* **55**, 1–18.

Schmidtke, J., E. Schmitt, M. Leipoldt & W. Engel 1979. Amount of repeated and non-repeated DNA in the genomes of closely related fish species with varying genome sizes. *Comparative Biochemistry and Physiology* **64B**, 117–20.

Schmidtke, J., B. Schulte, P. Kuhl & W. Engel 1976. Gene action in fish of tetraploid origin V. Cellular RNA and protein content and enzyme activities in cyprinid, clupeoid and salamonoid fishes. *Biochemical Genetics* **14**, 975–80.

Schmidtke, J., M. Brennecke, H. Smid, H. Neitzel & K. Sperling 1981. Evolution of muntjac DNA. *Chromosoma* **84**, 187–93.

Schneidermann, H. A. 1976. New ways to probe pattern formation and determination in insects. In *Insect development*, Symposium no. 8. Royal Entomological Society London, P. A. Lawrence (ed.), 3–34. New York: John Wiley and Sons.

Schneuwly, S. & W. J. Gehring 1986. Molecular and functional analysis of the homeotic gene Antennapedia in *Drosophila melanogaster*. In *Molecular neurobiology of* Drosophila, p. 52 (Abst). New York: Cold Spring Harbor Laboratory.

Scholnick, S. B., B. A. Morgan & J. Hirsch 1983. The cloned dopa decarboxylase gene is developmentally regulated when reintegrated into the *Drosophila* genome. *Cell* **34**, 37–45.

Schubert, I. & U. Wobus 1985. *In situ* hybridization confirms jumping nucleulus organizing regions in *Allium*. *Chromosoma* **92**, 143–8.

Schultz, R. J. 1980. The role of polyploidy in the evolution of fishes. In *Polyploidy*, W. H. Lewis (ed.), 313–40. New York: Plenum Press.

Schüpbach, T. 1982. Autosomal mutations that interfere with sex determination in somatic cells of *Drosophila* have no direct effect on the germ line. *Developmental Biology* **89**, 117–27.

Schwartz, M. B., R. B. Imberski & T. J. Kelly 1984. Analysis of metamorphosis in *Drosophila melanogaster*: characterisation of giant, an ecdysteroid-deficient mutant. *Developmental Biology* **103**, 85–95.

Schweber, M. S 1974. The satellite bands of the DNA of *Drosophila virilis*. *Chromosoma* **44**, 371–82.

Scott, M. P. & A. J. Weiner 1984. Structural relationships among genes that control development: sequence homology between the Antennapedia, Ultrabithorax and fushi tarazu loci of *Drosophila*. *Proceedings of the National Academy of Sciences* **81**, 4115–19.

Scott, M. P., A. J. Weiner, T. I. Hazelrigg, B. A. Polisky, V. Pirrotta, F. Scalenghe & T. C. Kaufman 1983. The molecular organisation of the Antennapedia locus of *Drosophila*. *Cell* **35**, 763–76.

Segraves, W. A., C. Louis, P. Schedl & B. P. Jarry 1983. Isolation of the rudimentary locus of *Drosophila melanogaster*. *Molecular and General Genetics* **189**, 34–40.

Shafit-Zagardo, B., J. J. Maio & F. L. Brown 1982. Kpn-1 families of long, interspersed repetitive DNAs in human and other primate genomes. *Nucleic Acids Research* **10**, 3175–93.

Shepherd, J. C. W., W. McGinnis, A. E. Carrasco, E. M. DeRobertis & W. J. Gehring 1984. Fly and frog homeo domains show homologies with yeast mating type regulatory proteins. *Nature* **310**, 70–1.

Sherwood, S. W. & J. L. Patton 1982. Genome evolution in pocket gophers (genus *Thomomys*) II. Variation in cellular DNA content. *Chromosoma* **85**, 163–79.

Shiba, T. & K. Saigo 1983. Retrovirus-like particles containing RNA homologous to the transposable element *copia* in *Drosophila melanogaster*. *Nature* **302**, 119–24.

Shilo, B-Z. & R. A. Weinberg 1981. DNA sequences homologous to vertebrate oncogenes are conserved in *Drosophila melanogaster*. *Proceedings of the National Academy of Sciences* **78**, 6789–92.

Short, R. V. 1976. The origin of species. In *Reproduction in mammals: Book 6, The evolution of reproduction*, C. R. Austin & R. V. Short (eds), 110–48. Cambridge: Cambridge University Press.

Short, R. V., A. C. Chandley, R. C. Jónes & W. R. Allen 1974. Meiosis in interspecific equine hybrids II. The Przewalski horse/domestic horse hybrid (*Equus przewalski* × *Equus caballus*). *Cytogenetics and Cell Genetics* **13**, 465–78.

Sierra, F., A. Lichtler, F. Marashi, R. Rickles T. Van Dyke, S. Clark, J. Wells & G. Stein 1982. Organisation of human histone genes. *Proceedings of the National Academy of Sciences* **79**, 1795–9.

Silverman, M. & M. Simon 1983. Phase variation and related systems. In *Mobile genetic elements*, A. J. Shapiro (ed.), 537–57. New York: Academic Press.

Simon, M., J. Zieg, M. Silverman, G. Mandel & R. Doolittle 1980. Phase variation: evolution of a controlling element. *Science* **209**, 1370–4.

Singer, M. F. 1982a. Highly repeated sequences in mammalian genomes. *International Review of Cytology* **76**, 67–112.

Singer, M. F. 1982b. SINEs and LINEs: highly repeated short and long interspersed sequences in mammalian genomes. *Cell* **28**, 433–4.

Singh, L., C. Phillips & K. W. Jones 1984. The conserved nucleotide sequences of Bkm, which define *Sxr* in the mouse, are transcribed. *Cell* **36**, 111–20.

Sittman, D. B., I-M. Chiu, C-J. Pan, R. H. Cohn, L. H. Kedes & W. F. Marzluff 1981. Isolation of two clusters of mouse histone genes. *Proceedings of the National Academy of Sciences* **78**, 4078–82.

Siu, G., S. P. Clark, Y. Yoshikai, M. Malissen, Y. Yanagi, E. Strauss, T. W. Mak & L. Hood 1984. The human T cell antigen receptor is encoded by variable, diversity and joining gene segments that rearrange to generate a complete V gene. *Cell* **37**, 393–401.

Slack, J. M. W. 1983. *From egg to embryo*. Cambridge: University Press.

Slightom, J. L., A. E. Blecht & O. Smithies 1980. Human fetal $^G\gamma-$ and $^A\gamma$-globin genes: complete nucleotide sequences suggest that DNA can be exchanged between these duplicated genes. *Cell* **21**, 627–38.

Slijber, E. J. 1976. *Whales and dolphins*. Ann Arbor: University of Michigan Press.

Smith, G. P. 1976. Evolution of repeated DNA sequences by unequal crossover. *Science* **191**, 528–35.

Smith, G. R. 1983. Chi hotspots of generalised recombination. *Cell* **34**, 709–10.

Smith, H. M. 1925. Cell size and metabolic activity in Amphibia. *Biological Bulletin* **48**, 347–78.

Smith-Gill, S. J. & K. A. Berven 1979. Predicting amphibian metamorphosis. *American Naturalist* **113**, 563–83.

Smith-Gill, S. J. & V. Carver 1981. Biochemical characterisation of organ differentiation. In *Metamorphosis, a problem in developmental biology*, 2nd edn, L. I. Gilbert & E. Frieden (eds), Ch. 15, 491–544. New York: Plenum Press.

Smouse, D., N. Perrimon, A. Mahowald & C. Goodman 1986. Topless: a new class of zygotic mutations which cause hypertrophy of the nervous system in the *Drosophila* embryo. In *Molecular neurobiology of* Drosophila, p. 47 (Abst). New York: Cold Spring Harbor Laboratory.

Sohn, U-Ik., K. H. Rothfels, & N. A. Straus 1975. DNA:DNA hybridisation studies in blackflies. *Journal of Molecular Evolution* **5**, 75–85.

Solnick, D. 1985. Trans splicing of mRNA precursors. *Cell* **42**, 157–64.

Soma, H., H. Kada, K. Mtayoshi, Y. Suzuhi, C. Mechvichal, A. Mahannop & B. Vatanaromya 1983. The chromosomes of *Muntiacus feae*. *Cytogenetics and Cell Genetics* **35**, 156–8.

Sonneborn, T. M. 1967. The evolutionary integration of genetic material into genetic systems. In *Heritage from Mendel*, R. A. Brink & E. D. Styles (eds), 375–401. Madison: University Wisconsin Press.

Soreq, H., D. Zevin-Sonkin, A. Anvi, L. C. Hall & P. Spierer 1985. A human acetylcholinesterase gene identified by homology to the *Ace* region of *Drosophila*. *Proceedings of the National Academy of Sciences* **82**, 1827–31.

Sparrow, A. H., H. J. Price & A. G. Underbrink 1972. A survey of DNA content per cell and per chromosome of prokaryotic and eukaryotic organisms: some evolutionary considerations. In *Evolution of genetic systems*, H. H. Smith (ed.), 451–93. New York: Gordon and Breach.

Sparrow, A. H., R. C. Sparrow, K. H. Thompson & L. A. Schauer 1965. The use of nuclear and chromosomal variables in determining and predicting radio-sensitivities. In *The use of induced mutations in plant breeding*, 101–32. London: Pergamon Press.

Spear, B. B. 1974. The genes for ribosomal RNA in diploid and polytene chromosomes of *Drosophila melanogaster*. *Chromosoma* **48**, 159–79.

Spear, B. B. & J. G. Gall 1973. Independent control of ribosomal gene replication in polytene chromosomes of *Drosophila melanogaster*. *Proceedings of the National Academy of Sciences* **70**, 1359–63.

Spencer, F. A., F. M. Hoffmann & W. M. Gelbart 1982. Decapentaplegic: a gene complex affecting morphogenesis in *Drosophila melanogaster*. *Cell* **28**, 451–61.

Spierer, P., A. Spierer, W. Bender & D. S. Hogness 1983. Molecular mapping of genetic and chromomeric units in *Drosophila melanogaster*. *Journal of Molecular Biology* **168**, 35–50.

Spofford, J. B. 1976. Position effect-variegation in *Drosophila*. In *Genetics and biology of* Drosophila, Vol. 1c, M. Ashburner & E. Novitski (eds), 955–1018. London: Academic Press.

Spradling, A. C. 1981. The organisation and amplification of two chromosomal domains containing *Drosophila* chorion genes. *Cell* **27**, 193–201.

Spradling, A. C. & A. P. Mahowald 1981. A chromosome inversion alters the pattern of specific DNA replication in *Drosophila* follicle cells. *Cell* **27**, 203–9.

Spradling, A. C. & G. M. Rubin 1981. *Drosophila* genome organisation: conserved and dynamic aspects. *Annual Review of Genetics* **15**, 219–64.

Spradling, A. C. & G. M. Rubin 1982. Transposition of cloned P elements into *Drosophila* germ line chromosomes. *Science* **218**, 341–7.

Spradling, A. C. & G. M. Rubin 1983. The effect of chromosomal position on the expression of the *Drosophila* xanthine dehydrogenase gene. *Cell* **34**, 47–57.

Stanley, S. M. 1979. *Macroevolution: pattern and process*. San Francisco: W. H. Freeman.

Stebbins, G. L. 1966. Chromosome variation and evolution. *Science* **152**, 1463–9.

Steinbrück, G. 1984. Overamplification of genes in macronuclei of hypotrichous ciliates. *Chromosoma* **88**, 156–63.

Steinbrück G. 1986. Molecular reorganization during nuclear differentiation in ciliates. In *Results and problems in cell differentiation 13, Germ line–soma differentiation*, W. Hennig (ed.), 105–74. Berlin: Springer.

Steinbrück, G., I. Haas, K-H. Hellmer & D, Ammermann 1981. Characterisation of macronuclear DNA in five species of ciliates. *Chromosoma* **83**, 199–208.

Steinmann-Zwicky, M. & R. Nöthiger 1985. A small region on the X-chromosome of *Drosophila* regulates a key gene that controls sex determination and dosage compensation. *Cell* **42**, 877–87.

Steller, H. & V. Pirrotta 1985. A transposable P vector that confers selective G418 resistance to *Drosophila* larvae. *The EMBO Journal* **4**, 167–71.

Steller, H., K. F. Fischbach & G. M. Rubin 1987. Disconnected: a locus required for neuronal pathway formation in the visual system of *Drosophila*. In *Molecular biology of neurogenesis*, 1st European Meeting of *Drosophila* Neurogenetics, Simonswald. *The Journal of Neurogenetics* **4**, 150–1 (Abst).

Stephenson, E. C., H. P. Erba & J. G. Gall 1981. Histone gene clusters of the newt *Notophthalmus* are separated by long tracts of satellite DNA. *Cell* **24**, 639–47.

Steward, R., L. Ambrosio & P. Schedl 1985. Expression of the dorsal gene. *Cold Spring Harbor Symposia on Quantitative Biology* **50**, 223–8.

Steward, R., F. J. McNally & P. Schedl 1984. Isolation of the dorsal locus of *Drosophila*. *Nature* **311**, 262–5.

Stewart, B. & J. Merriam 1980. Dosage compensation. In *Genetics and biology of* Drosophila, Vol. 2d, M. Ashburner (ed.), 107–40. London: Academic Press.

Strachan, T., D. Webb & G. A. Dover 1985. Transition stages of molecular drive in multiple-copy DNA families in *Drosophila*. *The EMBO Journal* **4**, 1701–8.

Strachan, T., E. Coen, D. Webb & G. Dover 1982. Modes and rates of change of complex DNA families of *Drosophila*. *Journal of Molecular Biology* **158**, 37–54.

Straus, N. A. 1971. Comparative DNA renaturation kinetics in amphibians. *Proceedings of the National Academy of Sciences* **68**, 799–802.

Strausfeld, N. J. 1976. *Atlas of an insect brain*. Berlin: Springer.

Strauss, F. & A. Varshavsky 1984. A protein binds to a satellite DNA repeat at three specific sites that would be brought into mutual proximity by DNA folding in the nucleosome. *Cell* **37**, 889–901.

Strobel, E., P. Dunsmuir & G. M. Rubin 1979. Polymorphisms in the chromosomal locations of elements of the 412, copia and 297 dispersed repeated gene families in *Drosophila*. *Cell* **17**, 429–39.

Struhl, G. 1982. Decapentaplegic – hopes held out. *Nature* **298**, 13–14.

Struhl, G. 1984. Splitting the bithorax complex of *Drosophila*. *Nature* **308**, 454–7.

Struhl, G. 1985. Near-reciprocal phenotypes caused by inactivation or indiscriminate expression of the *Drosophila* segmentation gene *ftz*. *Nature* **318**, 677–80.

Stumph, W. E., C. P. Hodgson, M-J. Tsai & B. W. O'Malley 1984. Genomic structure and possible retroviral origin of the chicken CRI repetitive DNA sequence family. *Proceedings of the National Academy of Sciences* **81**, 6667–71.

Stumph, W. E., P. Kristo, M-J Tsai & B. W. O'Malley 1981. A chicken middle-repetitive DNA sequence which shares homology with mammalian ubiquitous repeats. *Nucleic Acids Research* **9**, 5383–97.

Sturtevant, A. H. & E. Novitski 1941. The homologies of the chromosome elements in the genus *Drosophila*. *Genetics* **26**, 517–41.

Sulston, J. E. 1983. Neuronal cell lineages in the nematode *Caenorhabditis elegans*. *Cold Spring Harbor Symposia on Quantitative Biology* **48**, 443–52.

Sulston, J. E. & S. Brenner 1974. The DNA of *Caenorhabditis elegans*. *Genetics* **77**, 95–104.

Sulston, J. E., E. Schierenberg, J. G. White & J. N. Thomson 1983. The embryonic cell lineage of the nematode *Caenorhabditis elegans*. *Developmental Biology* **100**, 64–119.

Sun, L., K. E. Paulson, C. W. Schmid, L. Kadyk & L. Leinwand 1984. Non-Alu family interspersed repeats in human DNA and their transcriptional activity. *Nucleic Acids Research* **12**, 2669–90.

Sutcliffe, J. G., R. J. Milner & F. E. Bloom 1983. Cellular localization and function of the proteins encoded by brain-specific mRNAs. *Cold Spring Harbor Symposia on Quantitative Biology* **48**, 477–84.

Suzuki, Y., L. P. Gage & D. D. Brown 1972. The genes for silk fibroin in *Bombyx mori*. *Journal of Molecular Biology* **70**, 637–49.

Swanton, M. T., J. M. Heumann & D. M. Prescott 1980. Gene-sized DNA molecules of the macronuclei in three species of hypotrichs: size distributions and absence of nicks. DNA of ciliated Protozoa VIII. *Chromosoma* **77**, 217–27.

Swartz, M. N., T. A. Trautner & A. Kornberg 1962. Enzymatic synthesis of deoxyribonucleic acid II. Further studies on nearest neighbour base sequences in deoxyribonucleic acid. *Journal of Biological Chemistry* **237**, 1961–7.

Synder, M. & N. Davidson 1983. Two gene families clustered in a small region of the *Drosophila* genome. *Journal of Molecular Biology* **166**, 101–18.

Szabad, J. & P. Bryant 1982. The mode of action of 'discless' mutations in *Drosophila melanogaster*. *Developmental Biology* **93**, 240–56.

Szostak, J. W. & R. Wu 1980. Unequal crossingover in the ribosomal DNA of *Saccharomyces cerevisiae*. *Nature* **284**, 426–30.

Takeda, Y., D. H. Ohlendorf, W. F. Anderson & B. W. Matthews 1983. DNA-binding proteins. *Science* **221**, 1020–6.

Tamkun, J. W., J. E. Schwarzbauer & R. O. Hynes 1984. A single rat fibronectin gene generates three different mRNAs by alternative splicing of a complex exon. *Proceedings of the National Academy of Sciences* **81**, 5140–4.

Tanouye, M. A. 1986. The Shaker locus. In *Molecular neurobiology of* Drosophila, p. 23. (Abst). New York: Cold Spring Harbor Laboratory.

Tartof, K. D. 1973a. Regulation of ribosomal RNA gene multiplicity in *Drosophila melanogaster*. *Genetics* **73**, 57–71.

Tartof, K. D. 1973b. Unequal mitotic sister chromatid exchange and disproportionate replication as mechanisms regulating ribosomal RNA gene redundancy. *Cold Spring Harbor Symposia on Quantitative Biology* **38**, 491–500.

Tartof, K. D., C. Hobbs & M. Jones 1984. A structural basis for variegating position effects. *Cell* **37**, 869–78.

Tautz, D. & M. Renz 1984. Simple sequences are ubiquitous repetitive components of eukaryotic genomes. *Nucleic Acids Research* **12**, 4127–38.

Tautz, D., M. Trick & G. A. Dover 1986. Widespread cryptic simplicity in DNA: a major source of genetic variation. *Nature* **322**, 652–6.

Tchurikov, N. A., A. K. Naumova, E. S. Zelentsova & G. P. Georgiev 1982. A cloned unique gene of *Drosophila melanogaster* contains a repetitive 3' exon whose sequence is present at the 3' ends of many different mRNAs. *Cell* **28**, 365–73.

Technau, G. M. & J. A. Campos-Ortega 1985. Fate-mapping in wild type *Drosophila melanogaster* II. Injections of horseradish peroxidase in cells of the early gastrula stage. *Roux's Archives of Developmental Biology* **194**, 196–212.

Tempel, B., D. Papazian, T. Schwarz, Y-N. Jan & L-Y. Jan 1986. Molecular analysis of Shaker, a genetic locus that affects A-current. In *Molecular neurobiology of* Drosophila, p. 81. (Abst). New York: Cold Spring Harbor Laboratory.

Tengels, E. & A. Ghysen 1985. Domains of action of bithorax genes in *Drosophila* central nervous system. *Nature* **314**, 558–61.

Thiebaud, C. H. 1979. Quantitative determination of amplified rDNA and its distribution during oogenesis in *Xenopus laevis*. *Chromosoma* **73**, 37–44.

Thiebaud, Ch. H. & M. Fischberg 1977. DNA content in the genus *Xenopus*. *Chromosoma* **59**, 253–7.

Thireos, G., R. Griffin-Shea & F. C. Kafatos 1980. Untranslated mRNA for a chorion protein of *Drosophila melanogaster* accumulates transiently at the onset of specific gene amplification. *Proceedings of the National Academy of Sciences* **77**, 5789–93.

Thoday, J. M. 1975. Non-Darwinian 'evolution' and biological progress. *Nature* **255**, 675–7.

Thomas, J. B. & R. J. Wyman 1984a. Mutations altering synaptic connectivity between identified neurons in *Drosophila*. *Journal of Neurosciences* **4**, 530–8.

Thomas, J. B. & R. J. Wyman 1984b. Duplicated neural structure in bithorax mutant *Drosophila*. *Developmental Biology* **102**, 531–3.

Thomas, J. B., M. J. Bastiani, M. Bate & C. S. Goodman 1984. From grasshopper to *Drosophila*: a common plan for neural development. *Nature* **310**, 203–7.

Thomson, K. S. 1972. An attempt to reconstruct evolutionary changes in the cellular DNA content of lungfish. *Journal of Experimental Zoology* **180**, 363–72.

Thomson, K. S. & K. Muraszko 1978. Estimation of cell size and DNA content in fossil fishes and amphibians. *Journal of Experimental Zoology* **205**, 315–20.

Throckmorton, L. H. 1982. The *virilis* species group. In *The genetics and biology of* Drosophila, Vol. 3b, M. Ashburner, H. L. Carson & J. N. Thompson (eds), 227–96. London: Academic Press.

Timberlake, W. E. 1980. Developmental gene regulation in *Aspergillus nidulans*. *Developmental Biology* **78**, 497–510.

Timberlake, W. E., D. S. Shumard & R. B. Goldberg 1977. Relationship between nuclear and polysomal RNA populations of *Achlya*: a simple eukaryotic system. *Cell* **10**, 623–32.

Tisty, T. D., A. M. Albertini & J. H. Miller 1984. Gene amplification in the *lac* region of *E. coli*. *Cell* **37**, 217–24.

Tobin, S. L., O. Hanson-Painton, C. Courchesne, J. Vigoreaux, L. Sandor & J. Sangstad 1986. Crete Drosophila Meeting, Kolymbari (Abst).

Tonegawa, S. 1985. The molecules of the immune system. *Scientific American* **253**, 104–13.

Tompkins, R. 1978. Genic control of axolotl metamorphosis. *American Zoologist* **18**, 313–19.

Townes, T. M., M. C. Fitzgerald & J. B. Lingrel 1984. Triplication of a four-gene set during evolution of the goat β-globin locus produced three genes now expressed differentially during development. *Proceedings of the National Academy of Sciences* **81**, 6589–93.

Travers, A. 1985. Sigma factors in multitude. *Nature* **313**, 15–16.

Treisman, R. & T. Maniatis 1985. Simian virus 40 enhancer increases number of RNA polymerase II molecules on linked DNA. *Nature* **315**, 72–5.

Trends in Neurosciences, June 1985. Developmental Neurobiology – Special Issue, 300 pp.

Truett, M. A., R. S. Jones & S. S. Potter 1981. Unusual structure of the FB family of transposable elements in *Drosophila*. *Cell* **24**, 753–63.

Truman, J. W. & S. E. Reiss 1976. Dendritric reorganization of an identified motoneuron during metamorphosis of the Tobacco Hornworm moth. *Science* **192**, 477–9.

Tschudi, C., V. Pirrotta & N. Junakovic 1982. Rearrangements of the 5S RNA gene cluster of *Drosophila melanogaster* associated with the insertion of a *B104* element. *The EMBO Journal* **1**, 977–85.

Tymowska, J. & M. Fischberg 1982. A comparison of the karyotype, constitutive heterochromatin, and nucleolar organiser regions of the new tetraploid species *Xenopus epitropicalis* Fischberg and Picard with those of *Xenopus tropicalis* Gray (Anura, Pipidae) *Cytogenetics and Cell Genetics* **34**, 149–57.

Ullu, E. 1982. The human *Alu* family of repeated DNA sequences. *Trends in Biochemical Science* **7**, 216–19.

Ullu, E. & C. Tschudi 1984. *Alu* sequences are processed 7SL RNA genes. *Nature* **312**, 171–2.

Urieli-Shoval, S., Y. Gruenbaum, J. Sedat & A. Razin 1982. The absence of detectable methylated bases in *Drosophila melanogaster* DNA. *FEBS Letters* **146**, 148–52.

Varmus, H. E. 1983. 'Retroviruses'. In *Mobile genetic elements*, A. J. Shapiro (ed.), 411–503. New York: Academic Press.

Vasek, F. C. 1956. Induced aneuploidy in *Clarkia unguiculata* (Onagraceae). *American Journal of Botany* **43**, 366–71.

Vassin, H. & J. A. Campos-Ortega 1987. Genetic characterization and molecular cloning of the Delta locus of *Drosophila melanogaster*. In *Molecular biology of neurogenesis*, 1st European Meeting of *Drosophila* Neurogenetics, Simonswald. *The Journal of Neurogenetics* **4**, 153–4 (Abst).

Velazquez, J. M. & S. Lindquist 1984. hsp 70: nuclear concentration during environmental stress and cytoplasmic storage during recovery. *Cell* **36**, 655–62.

Velissariou, V. & M. Ashburner 1980. The secretory proteins of the larval salivary glands of *Drosophila melanogaster*: cytogenetic correlation of a protein and a puff. *Chromosoma* **77**, 13–27.

Venolia, L., S. M. Gartler, E. R. Wassmann, P. Yen, T. Mohandras & L. J. Shapiro 1982. Transformation with DNA from 5-azacytidine-reactivated X-chromosomes. *Proceedings of the National Academy of Sciences* **79**, 2352–4.

Venter, J. G., B. Eddy, L. M. Hall & C. M. Fraser 1984. Monoclonal antibodies detect the conservation of muscarinic cholinergic receptor structure from *Drosophila* to human brain and detect possible structural homology with α_1-adrenergic receptors. *Proceedings of the National Academy of Sciences* **81**, 272–6.

Verma, R. C. & S. N. Raina 1981. Cytogenetics of *Crotalaria*. Supernumerary nucleoli in *C. agatiflora* (Leguminosae). *Genetica* **56**, 75–80.

Vlad, M. 1977. Quantitative studies of rDNA in Amphibians. *Journal of Cell Science* **24**, 109–18.

von Wettstein, F. 1937. Experimentelle Untersuchungen zum Artbildungsproblem I. Zellgrössenregulation und fertilwerden einer polyploiden *Bryum*-sippe. Zeitschrift für induktive Abstammungs-und Vererbungslehre **74**, 34–53.

Wahlsten, D. 1982. Mode of inheritance of deficient corpus callosum in mice. *Journal of Heredity* **7**, 281–5.

Wald, G. 1981. Metamorphosis: an overview. In *Metamorphosis, a problem in developmental biology*, 2nd edn, L. I. Gilbert & E. Frieden (eds), 1–39. New York: Plenum Press.

Warner, A. H. & J. C. Bagshaw 1984. Absence of detectable 5-methylcytosine in DNA of embryos of the brine shrimp, *Artemia*. *Developmental Biology* **102**, 264–7.

Wasserman, M. 1982. Evolution of the *repleta* group. In *The genetics and biology of Drosophila*, Vol. 3b, M. Ashburner, H. L. Carson & J. N. Thompson (eds), 60–139. London: Academic Press.

Weber, F. & W. Schaffner 1985. Simian virus 40 enhancer increases RNA polymerase density within the linked gene. *Nature* **315**, 75–7.

Weber, K. & M. Osborn 1985. The molecules of the cell matrix. *Scientific American* **253**, 92–102.

Weeden, C., K. Harding & M. Levine 1986. Spatial regulation of Antennapedia and bithorax gene expression by the Polycomb locus in *Drosophila*. *Cell* **44**, 739–48.

Weinberg, R. A. 1985. The molecules of life. *Scientific American* **253**, 34–43.

Weiner, A. J., M. P. Scott & T. C. Kaufmann 1984. A molecular analysis of *fushi tarazu*, a gene in *Drosophila melanogaster* that encodes a product affecting embryonic segment number and cell fate. *Cell* **37**, 843–51.

Weir, M. P. & C. W. Lo 1984. Gap-junctional communication compartments in the *Drosophila* wing imaginal disc. *Developmental Biology* **102**, 130–46.

Weir, M. P. & C. W. Lo 1985. An anterior/posterior communication border in engrailed wing discs: possible implications for *Drosophila* pattern formation. *Developmental Biology* **110**, 84–90.

Weller, P., A. J. Jeffreys, V. Wilson & A. Blanchetot 1984. Organisation of the human myoglobin gene. *The EMBO Journal* **3**, 439–46.

Wessels, N. K. 1982. A catalogue of processes responsible for metazoan morphogenesis. In *Evolution and development*, J. T. Bonner (ed.), 115–54. Dahlem Konferenzen. Berlin: Springer.

Wharton, K. A., K. M. Johansen, T. Xu & S. Artavanis-Tsakonas 1985a. Nucleotide sequences from the neurogenic locus Notch implies a gene product that shares homology with proteins containing EGF-like repeats. *Cell* **43**, 567–81.

Wharton, K. A., B. Yedvobnick, V. G. Finnerty & S. Artavanis-Tsakonas 1985b. *Opa*: a novel family of transcribed repeats shared by the Notch locus and other developmentally regulated loci in *Drosophila melanogaster*. *Cell* **40**, 55–62.

Wheeler, L. L. & L. C. Altenberg 1977. Hoechst 33258 banding of *Drosophila nasutoides* metaphase chromosomes. *Chromosoma* **62**, 351–60.

Wheeler, L. L., F. Arrighi, M. Cordeiro-Stone & C. S. Lee 1978. Localisation of *Drosophila nasutoides* satellite DNAs in metaphase chromosomes. *Chromosoma* **70**, 41–50.

White, B. A. & C. S. Nicoll 1981. Hormonal control of Amphibian metamorphosis. In *Metamorphosis, a problem in developmental biology*, 2nd edn, L. I. Gilbert & E. Frieden (eds), 363–96. New York: Plenum Press.

White, J. G. 1985. Neuronal connectivity in *Caenorhabditis elegans*. *Trends in Neuroscience* **8**, 277–83.

White, K. & D. R. Kankel 1978. Patterns of cell division and cell movement in the formation of the imaginal nervous system in *Drosophila melanogaster*. *Developmental Biology* **65**, 296–321.

White, M. J. D. 1978. *Modes of speciation*. San Francisco: W. H. Freeman.

White, M. M., K. M. Mayne, H. A. Lester & N. Davidson 1985. Mouse-*Torpedo* hybrid acetylcholine receptors: functional-homology does not equal sequence homology. *Proceedings of the National Academy of Sciences* **82**, 4852–6.

White, R. A. H. & M. E. Akam 1985. Contrabithorax mutations cause inappropriate expression of Ultrabithorax products in *Drosophila*. *Nature* **318**, 567–9.

White, R. A. H. & M. Wilcox 1985. Regulation of the distribution of Ultrabithorax proteins in *Drosophila*. *Nature* **318**, 563–7.

Whittaker, J. R. 1979. Ctyoplasmic determinants of tissue differentiation in the ascidian egg. In *Determinants of spatial organisation*, 29–51. *37th Symposium of the Society for Developmental Biology*. New York: Academic Press.

Whittington, H. B. 1975. The enigmatic animal *Opabinia regalis*, Middle Cambrian, Burgess Shale, British Columbia. *Philosophical Transactions of the Royal Society, London* **B270**, 1–43.

Whittington, H. B. & D. E. G. Briggs 1985. The largest Cambrian animal, *Anomalocaris*, Burgess Shale, British Columbia. *Philosophical Transactions of the Royal Society, London* **B309**, 569–609.

Wieben, E. D., S. J. Madore & T. Pedersen 1983. Protein binding sites are conserved in U1 small nuclear RNA from insects and mammals. *Proceedings of the National Academy of Sciences* **80**, 1217–20.

Wiedmann, M., A. Hulth & T. A. Rapoport 1984. *Xenopus* oocytes can secrete bacterial β-lactamase. *Nature* **309**, 637–9.

Wieschaus, E., C. Nüsslein-Volhard & G. Jürgens 1984. Mutations affecting the pattern of the larval cuticle in *Drosophila melanogaster* III. Zygotic loci on the X chromosome and fourth chromosome. *Roux's Archives of Developmental Biology* **193**, 296–307.

Williamson, P. G. 1981. Palaeontological documentation of speciation in Cenozoic molluscs from Turkana basin. *Nature* **293**, 437–43.

Wirth, T., M. Schmidt, T. Baumruker & I. Horak 1984. Evidence for mobility of a new family of mouse middle repetitive DNA elements (LTR-IS). *Nucleic Acids Research* **12**, 3603–10.

Witschi, E. 1956. *Development of vertebrates*. Philadelphia: W. B. Saunders.

Wolf, R. & M. Heisenberg 1986. Visual orientation in motion-blind flies in an operant behaviour. *Nature* **323**, 154–6.

Wood, W. B. 1973. Genetic control of bacteriophage T4 morphogenesis. In *Genetic mechanisms of development*, F. H. Ruddle (ed.), 29–46. 31st Symposium of the Society for Developmental Biology. New York: Academic Press.

Woodland, H. R. 1980. Histone synthesis during the development of *Xenopus*. *FEBS Letters* **121**, 1–7.

Wozney, J., D. Hanahan, V. Tate, H. Boedtker & P. Doty 1981. Structure of the pro α2(I) collagen gene. *Nature* **294**, 129–35.

Wright, T. R. F. 1974. The genetics of embryogenesis in *Drosophila*. *Advances in Genetics* **15**, 261–395.

Wright, T. R. F., E. Y. Wright, B. C. Black, E. S. Pentz, C. P. Bishop, J. Kullman, M. H. Corjay & G. R. Hankins 1984. The genetic and molecular organisation of a gene cluster involved in catecholamine metabolism and sclerotisation. Crete Drosophila Meeting, Kolymbari (Abst).

Wu, C-F., B. Ganetzky, F. N. Haughland & A-X. Liu 1983. Potassium currents in *Drosophila*: different components affected by mutations of two genes. *Science* **220**, 1076–8.

Wurster, D. H. & N. B. Atkin 1972. Muntjac chromosomes: a new karyotype for *Muntiacus muntjak*. *Experientia* **28**, 972–3.

Wurster Hill, D. H. & B. Seidel 1985. The G banded chromosomes of Roosvelt's muntjac, *Muntiacus roosveltorum*. *Cytogenetics and Cell Genetics* **39**, 75–6.

Wyman, R. J., J. B. Thomas, L. Salkoff & D. G. King 1984. The *Drosophila* giant fiber system. In *Neural mechanisms of startle behavior*, R. C. Eaton (ed.), 133–61. New York: Plenum Publishing Corp.

Yamamoto, M., P. Coombe, V. Pirrotta, K. F. Fischbach & G. L. G. Miklos 1987. *In situ* and genetic localization of microclones to the small optic lobes – sluggish area of the *Drosophila melanogaster* X-chromosome. In *Molecular biology of neurogenesis*, 1st European Meeting of *Drosophila* Neurogenetics, Simonswald. *The Journal of Neurogenetics* **4** 158–9 (Abst).

Yedvobnik, B., M. A. J. Muskavitch, K. A. Wharton, M. E. Halpern, E. Paul, B. G. Grimwade & S. Artavanis-Tsakonas 1985. Molecular genetics of *Drosophila* neurogenesis. *Cold Spring Harbor Symposia on Quantitative Biology* **50**, 841–54.

Yelton, D. E. & M. D. Sharff 1980. Monoclonal antibodies. *American Scientist* **68**, 510–16.

Young, B. D., G. Birnie & J. Paul 1976. Complexity and specificity of polysomal poly (A)$^+$ RNA in mouse tissues. *Biochemistry* **15**, 2823–9.

Young, M. W., F. R. Jackson, H. S. Shin & T. A. Bargiello 1985. A biological clock in *Drosophila*. *Cold Spring Harbor Symposia on Quantitative Biology* **50**, 865–75.

Young, R. A., O. Hagenbüchle & U. Schibler 1981. A single mouse α-amylase gene specifies two different tissue-specific mRNAs. *Cell* **23**, 451–8.

Yund, M. A., D. S. King & J. W. Fristrom 1978. Ecdysteroid receptors in imaginal discs of *Drosophila melanogaster*. *Proceedings of the National Academy of Sciences* **75**, 6039–43.

Zachar, Z. & P. M. Bingham 1982. Regulation of white locus expression: the structure of mutant alleles at the white locus of *Drosophila melanogaster*. *Cell* **30**, 529–41.

Zacharias, H. 1979. Underreplication of a polytene chromosome arm in the chironomid *Prodiamesa olivacea*. *Chromosoma* **72**, 23–51.

Zehner, Z. E. & B. M. Paterson 1983. Characterisation of the chicken vimentin gene: a single copy gene producing multiple mRNAs. *Proceedings of the National Academy of Sciences* **80**, 911–15.

Zimmerman, J. L., D. L. Fouts, L. S. Levy & J. E. Manning 1982. Non adenylated mRNA is present as polyadenylated RNA in nuclei of *Drosophila*. *Proceedings of the National Academy of Sciences* **79**, 3148–52.

Zipursky, S. L., T. R. Venkatesh & S. Benzer 1985. From monoclonal antibody to gene for a neuron-specific glycoprotein in *Drosophila*. *Proceedings of the National Academy of Sciences* **82**, 1855–9.

Zuker, C. S., A. F. Cowman & G. M. Rubin 1985. Isolation and structure of a rhodopsin gene from *Drosophila melanogaster*. *Cell* **40**, 851–8.

Zuker, C., A. Cowman, C. Montell & G. Rubin 1986. Studies on photo-transduction: *Drosophila* opsins. Crete Drosophila Meeting, Kolymbari (Abst).

Zusman, S. B. & E. F. Wieschaus 1985. Requirements for zygotic gene activity during gastrulation in *Drosophila melanogaster*. *Developmental Biology* **111**, 359–71.

Index

aardvark 151, 287
accumulation mechanisms 289
Achaete–Scute Complex 70, 131, 223, 317
achiasmatic 274
adaptation 237, 244, 252, 338
adaptations 238, 256, 309
adaptive
    change 239, 245
    radiation 272, 307
algae 312
    Chlamydomonas 195
    Euglena 82, 195
alleles
    amorphic 329
    null 327, 333
allometry 301
alterations
    genome 95
    nucleotide sequence 95
amino acid substitutions 239
amphibians 168, 173, 185, 194, 227, 242, 246,
    248, 279, 281, 282, 287, 306, 311, 320
amplification 23–4, 160, 161, 179–80, 209–11,
    212, 242, 270, 281
    developmentally regulated 208
    differential 22, 105, 107
    efficiency 210
    gene 178
    gradients 209
    levels 209
    multiple 283
    NOR 179
    rDNA 201, 212
    rRNA gene 24, 180
    selective 212, 214
    sequence 22–3
    signals 210
    transient 207
Anabaena 112
aneuploid 288
    gametes 277
    tolerance 227
aneuploidy 28
    segmental 223
angiosperm families
    Anthemidae 253
    Asteraceae 252
    Fabaceae 252
    Liliaceae 252
    Poaceae 252

angiosperm plants 14, 195, 246, 249, 252,
    277, 285, 324
    Allium cepa 182
    Arabidopsis thaliana 320
    Artemisia 253
    Brassica campestris 195
    Brassica hirta 195
    Brassica sp. 194, 318
    Chamaemelum 253
    Clarkia unguiculata 227
    Crotalaria agatiflora 182
    cucurbits 195, 312
    Datura stramonium 226
    Endymnion non-scriptus 253
    Festuca 276
    Fritillaria aurea 253
    Lathyrus 283, 285
    Lathyrus miniatus 285
    Lathyrus visititis 285
    Lolium 255, 276, 319
    Microseris 194
    mung bean 195, 318
    musk melon 195
    Petunia 227
    Pisum 195, 318
    Secale cereale 289
    Trillium erectum 253
    Triticum aestirum 249
    Triticum dicoccum 249
    Triticum monococcum 249
    Tulipa kaufmannia 253
    Vicia faba 182, 312
    Zea mays 215, 227, 289
annelids 307
Antennapedia Complex 46, 53–5, 56, 70, 196
    Antennapedia 43, 53, 56, 60, 61, 163, 165, 313
    proboscipedia 53
    Sex combs reduced 43, 53
antero–postero
    axis 30, 44, 315
    gradient 39–40, 63
antibodies 52, 98, 109, 219
antigen receptors 98
antineurogenic loci 70
    denervated 70
    embryonic lethal abnormal vision 70
    lethal EC4 70
    ventral nervous system defective 70, 71
anuran amphibians (frogs and toads) 28, 151,
    251, 256, 277, 291, 298–9

Ascaphus truei 214
Bolitoglossini 300
Rana catesbiana 228
Rana clamitans 228
arthropods (*see also* insects) 59, 291, 307
Artemia salina 122
crabs 311
Cyclops 103, 311
trilobites 307–8
ascidians 93
atavism 325
avian retroviruses 159

bacteria 5, 14, 22, 116–17, 177, 196, 324
bacterial
DNA binding proteins 53
morphogenesis 233
regulatory proteins 157
repressor protein 13
bacteriophage
λ 3, 9, 13, 21, 297
T4 232
behavioural mutants 77
amnesiac 88
bang senseless 82
bang sensitivity 81
cacophony 77
celibate 77
coitus interruptus 79
dunce 84, 88
easily shocked 82
ether-a-go-go 79
flightless I 88
flightless O 88
Hyperkinetic 77
minibrain 82, 87
no action potential 82, 83
optomotor blind 77
paralytic 83
Passover 88
period 81
reduced optic lobes 87
rutabaga 88
Shaker 79, 82
shaking-B 82, 88
shibhire 77
sluggish-A 88
small optic lobes 87, 88
smell-blind 77
stoned 77, 88
stress sensitive-C 88
stuck 79
technical knockout 82
temperature induced paralysis 83
tetanic 82
turnip 88
uncoordinated 88, 91
uncoordinated-like 88

binary
choices 231
circuits 64
decisions 60, 66, 88–9, 93
binary switch
genes 302
mechanisms 297, 302
binary switches 317
biochemical pathways 30
biological clocks 81, 332
birds 151, 155, 156, 164, 227, 279, 281, 285, 288, 304, 305
chicken 11, 14, 47, 84, 116, 119, 125, 155, 169, 170, 184, 185, 189, 191, 227, 305, 321
duck 13, 328
ostrich 155, 301
Bithorax complex 14, 42, 46, 47–50, 52–3, 56, 71, 196, 200
anterobithorax 42, 50–1, 52, 74
bithorax 32, 47, 51, 52, 74, 163, 165, 294, 313–14
bithoraxoid 42, 47, 51, 52
Contrabithorax 47, 51–2
Haltere mimic 51
infra abdominal 42, 47, 50
post bithorax 47, 50–1, 52
Ultrabithorax 43, 47, 51–2, 53, 61
blastocoel 26
blastocyst 30, 185, 216
blastoderm 27, 34, 40, 42, 44–5, 55, 61, 76
cell-positional identity 315
formation 293
blastomeres 216, 257
blastula 26, 28, 60, 166, 168
blastula stage proteins 184
blood coagulation factor 70
blots
Northern 170
Southern 170, 202
bobbed
allele 204
mutation 206
phenotype 201, 206
ring X-chromosome 207
Bolwig's nerve 86
bone marrow 94
boxes
A 121
B 121
CAT 116, 117
Pribonow 116
TATA 114–16, 117, 130
brain 9, 60, 69, 71, 74, 77, 81, 143, 163, 168–9, 170, 227, 330, 331, 333
DNA 219
function 332
mRNA 170

mutants 82, 86–7
poly (A)⁻ mRNA 170
poly (A)⁺ mRNA 170
poly (A) RNA 9
polysomal RNA 169
RNA populations 79
Bryum caespiticium 248
buffered
    pathways 329
    systems 327, 334
Burgess Shale 307
Burgess Shale fauna
    Anomalocaris 307
    Hallucigenia sparsa 307
    Opabinia regalis 307
    Wiwaxia corrugata 307

cDNA hybridization 199
casette
    mechanism 97
    replacement 98
cell
    adhesion molecules 29
    attachment 125
    clones 27, 37
    contact 336
    cycle 28, 89, 253
    death 321, 337
    division 1, 29, 37, 89, 93, 189, 224, 252, 337
    division rate 246. 256
    fates 90, 91, 301, 317
    interactions 30, 89, 297, 303–4, 309
    lethals 57
    lineage 88, 90–1, 91–3, 95, 309
    lineage restriction 34–5
    membrane 13, 79, 100
    metabolism 248, 256
    mobility 89
    motility 189
    movement 26, 29–30, 34, 89, 224, 231, 337
    migration 94, 125
    necrosis 298
    populations 55, 303, 335
    position 297, 305
    proliferation 34, 337
    rearrangement 337
    recognition 37, 68
    shape 189, 304, 336, 337
    shape changes 337
    size 139, 246–9, 256, 281, 305, 320
    surface 89, 335
    surface antigens 28
    surface glycoproteins 86
    surface proteins 30, 337
cells (*see also* germ cells) 335
    B 9, 98, 101
    blast 90

blastoderm 316
brain 201, 204
    differentiated 62
    ectodermal 69, 85
    embryonal carcinoma founder 37, 90
    embryonal carcinoma stem 60
    embryonic 216
    endodermal 335
    endoploid 36
    epidermal 69, 303
    extra-embryonic 216
    follicle 36, 208, 209, 212
    guide post 72
        hamster V79 218
    Hela 119
    hybrid 9, 219
    hypodermal 91
    imaginal disc 37, 52, 201
    larval brain 204
    mesenchymal 321
    mesodermal 335
    murine 23
    myeloma 9
    neural 91
    neural crest 29
    neuronal precursor 85
    nurse 36, 215
    photoreceptor 9, 80
    plasma 9, 98
    pole 40, 42, 46, 63, 67, 275
    primary pigment 232
    red blood 246, 248
    somatic 68, 90, 121, 136, 178, 201, 216, 226
    stem 93–4
    suspensor 215
    T 98
    thoracic muscle 204
    trichogen 36
centromere 141, 180, 184, 225, 283, 290
cephalization 93
cerebellum 29
chaetae pattern 317
change (*see also* chromosome and evolutionary)
    allometric 272, 301
    concerted 243, 306
    genetic 320
    genome 237
    interspecific 276
    karyotypic 290
    macroevolutionary 292, 302, 309
    microevolutionary 292
    morphological 32, 234–5, 243, 272–3, 281, 290–1, 302, 319, 331
    mutational 273
    phenotypic 235, 321
    phyletic 272
    punctuated 272, 321
    regressive 287

regulatory 296
    structural gene 293
channels 332
    calcium 196, 332
    gap junctions 38
    potassium 79, 82
    protein 74
    sodium 82, 332
chemoreceptors 74
chicken (*see also* birds)
    lysozyme gene 119
    β-tubulin gene 14
    vimentin gene 125
    yeast chimeric gene 14, 233
chloroplast 126
    DNA 318
    genes 82, 195
    genomes 194, 195
cholesterol 328
cholinergic
    synapsis 83
    synaptic transmission 83
chordates 150–1
chorion 208, 322
    gene amplification 24, 208–9, 211, 214
    gene cluster 209, 211
    genes 14, 24, 208–9, 214
    gene transcription 209
    multigene families 243
    protein genes 212
    proteins 14, 208, 211
    protein underproduction 211
    structure 211
choriogenesis 14
chromatid exchange 57
chromatin
    diminution 103, 105
    higher order structure 140–1
    structure 119, 140
    supernumerary 288, 290
    vesicles 106
chromocentre 2, 132, 136, 148, 173, 179
chromosome
    abnormalities 28, 227
    change 234, 272, 322
    deficiencies 85, 90, 223, 231, 296
    deletion 204
    domain 209
    duplication 137, 271
    fusion 222, 260, 263, 268–71, 293
    insertion 148
    inversion 209, 269, 276, 293
    jumping 4
    loss 226
    non-disjunction 266
    pairing 139
    polymorphism 290
    puffing 211

rearrangements 4, 132, 139, 263, 268,
    273–4, 275–6, 292, 294, 296–7, 312, 318,
    322, 324
    recognition phenomena 139
    reorganization 324
    translocation 82, 271, 293
    walk 4, 50, 83, 86, 87, 128, 146, 156, 163
chromosomes (*see also* sex chromosomes)
    double minute 24
    euchromatic 227
    fragmented 105
    homologous 21
    lampbrush 224–5
    micronuclear 157
    mini 137
    mitotic 1, 225
    non-homologous 20
    polytene 2, 4, 32, 45, 105, 129, 140, 204,
        215, 220
    ring 24, 207
    supernumerary 249, 288–9
ciliates 14, 104, 107, 112, 157, 164, 311, 318,
    324
    Colpoda steini 105
    Oxytricha fallax 157
    Oxytricha similis 107, 164
    Paramecium bursaria 105, 107, 164
    Stylonychia lemnae (= mytilus) 105, 107,
        112, 162, 164, 184, 318
    Tetrahymena pyriformis 318
    Tetrahymena thermophila 105, 107, 124, 157
    Trachelonema sulcata 105
circular DNA molecules 207
cis
    acting regulatory elements 14
    inactivation 295
cleavage 28, 90, 93, 103, 166, 185
cloned DNA 45
    DNA fragments 168
    walk 146
clones
    cDNA 9–10, 59, 82, 170
    chicken pro α2(I) 191
    cosmid 5
    cross hybridizing 162
    genomic 7, 9, 191
    λ 159
    orphon 183
cloning 84
coding capacity 140
cohesive
    genomic change 243
    population change 241–2
collagen 29
    fibres 303
comparative genome organization 193
compartment
    border 38

boundary 37–8, 50
  formation 39
compartments 37, 42, 52, 56–7, 61, 316
  anterior 57
  posterior 57
compartmentalization 37, 313
compensatory responder locus 135, 204
complementation groups 183
complex transcription units 125
compound eye 71, 80
conidiospores 165, 226
conjugation 104
consensus
  formula 157
  sequence 116, 117, 124, 129–30
continuous variation 289
contractile protein genes 159, 199
controls
  gene expression 193
  genotypic 245
  post-transcriptional 124
  transcriptional 113–14
conversion 20–1, 161, 242
  gene 21–2
  intergenic 21, 161
coordinate
  gene expression 157
  gene induction 156
coordination
  spatial 89
  temporal 89
copy number 159
corpus callosum 330
courtship behaviour 77
cross genomic raiding 14, 84, 333
crossover hot spot instigator 17, 161
currents
  calcium 79
  glutamate-dependent synaptic 79
  potassium 79
cuticle formation 199
  proteins 191
  synthesis 84
cuticular sensory organs 223
C-value 185, 251, 281, 282
  paradox 153, 169, 173, 195, 290
cycle time
  division 249
  mitotic 249
Cyclops 103, 112
  divulsus 103
  furcifer 103
  strenuus 103
cytidine
  anti-5 methyl 219
  5-Aza 219
cytochrome c 17

cytoplasm
  M 275
  P 275
cytoplasmic
  RNA 13, 179
  RNA populations 163
  ribonucleoprotein particles 13
cytoskeleton 190, 233, 314, 336
cytotype
  M 275
  P 275

deep brain functions 334
defective gene copies 24
deficiencies 130, 231, 296
deletions 19, 23, 112, 117, 313, 315, 328
DNA
  additions 177
  amplification 212, 215
  base changes 177
  binding 45
  binding proteins 52, 61, 114, 119, 139, 140
  binding site 117
  c 9, 13, 16, 42, 53, 59, 83, 164
  centromeric 184
  circular 112, 207
  clones 4
  coding 177, 235
  conformation 162
  content 105, 153, 253, 255, 276
  deletion 51, 177
  differential amplification 208
  duplication 281
  euchromatic 128, 141
  flanking 5, 130, 209–10
  heterochromatic 135–6
  highly repetitive 22, 24, 103, 128, 130, 139, 149, 160, 173, 177, 178, 184, 245, 276
  histone 207
  human 162
  ignorant 243, 245
  insertions 177, 275
  interspersed repetitive 154
  inversion 195
  junk 25, 109, 141, 153, 245, 312
  mediated gene transfer 217
  methylation 122, 219
  misdemeanours 273
  moderately repeated 10, 16, 128, 154, 165, 185, 193, 225, 255, 282
  mouse 162
  nomadic 131, 134, 149
  non-coding 10, 25, 100, 124, 235, 246
  non-coding spacer 185
  non-functional 10
  non-genic 193
  non-repetitive 281
  plasmid 257

probes 7
puffing 212
puffs 211–12
rearrangements 156–7, 255, 281
repeated sequence 18
repetitive 45, 110, 120, 156, 255, 281, 285
replication 135, 208, 214
ribosomal 177, 180–1, 186, 242
rings 103, 106
5S 207
7SL 160
salmon 162
satellite 22, 139–40, 160, 225, 245, 269, 290
selfish 16, 244, 245
sequence analysis 9
sequence conservation 325
sequence organization 290
sequence structure 290
sequencing 59
simple sequence 10, 21, 141, 161, 176
single copy 10, 45, 50, 128, 142, 155, 165,
    166, 168, 170, 255, 260, 281, 285
somatic 109
spacer 120, 241
telomeric 157
unique 129, 156, 166, 172
DNase-I hypersensitive sites 118, 119
determinants 62
determination 27, 62–3, 90
determined state 32, 37, 57, 61, 62
development 26, 230, 310, 311, 314, 318
    differential 280
    embryological 196
    embryonic 167
    epidermal 321
    female 64, 68
    larval 37, 298
    lysogenic 297
    macronuclear 104
    male 64, 68
    postembryonic 142–3
    rate 223
    sexual 63
    somatic 103, 227, 260, 263, 311
    vegetative 104
    vertebrate 324
developmental
    activity 171–2
    blocks 30
    canalization 301
    capability 97
    capacity 171–2
    changes 95, 237, 273
    circuits 30, 32, 46, 55, 168, 201, 231, 234,
        243, 245, 273, 293, 297, 302, 309, 325,
        329, 335
    control 328
    cost 321

defects 32
domains 200
expression 243
gene circuits 245, 260, 329
gene hierarchies 336
hierarchy 335
interactions 292
mechanisms 335
morphogens 323
networks 310, 314, 315, 322, 323, 326, 335
novelty 313
pathways 46, 56, 58, 61, 62, 64, 66, 89, 90,
    193, 199, 326
pattern polytenization 207
potentiality 55
processes 137, 292
programmes 27, 30, 35, 37, 93, 142, 200,
    231, 273, 294, 329
pupal 35
rate 223, 246
regulation 231, 297, 305
reorganization 273, 306
sequence 93
shunts 329
specificity 280
specific regulation 186
switches 326
templates 215
time 137, 246, 249–55, 256
timing 93, 297, 309, 312
timing adjustments 298
variants 293
Dictyostelium discoideum 156, 166, 177
differential
    gene amplification 105, 107
    polytenization 207
    processing 52
    replication 106
    rDNA replication 204
    transcription 221
differentiated states 27, 37
differentiation 37, 61, 70, 93, 95, 98, 139, 298
    cellular 60, 119
    muscle cell 200
    soma-germ line 104
    somatic 109
dihydrofolate reductase 23
    gene 24
diploid segregation units 105
dispersed
    mobile elements 160
    simple sequences 161, 163
dispersed moderately repeated DNA sequences
    Alu 10, 163
    Kpn 10
    LINES 160, 274
    poly (C-A) 10
    SINES 160

dispersion 241
    pattern 159
disproportionate rDNA production 207
divergent karyotypes 260
division cycle time 249, 255
domains
    chromosome 148, 209, 279
    coding 101
    DNA 114
    developmental 200
    extracellular 70
    functional 135, 191, 315, 319
    gene 101, 148
    homeo 98
    internal coding 111
    intracellular 70
    protein 70, 81, 191, 325
    repetitive DNA binding 13
    structural 315
    white locus 131
dominant female sterile technique 33
dorsal group genes 40–2, 49, 313
dorsalizing loci 41
dorso-ventral axes 30
dosage
    compensation 64, 66, 67, 148, 216, 220–2,
        314
    pattern 42
    sensitive loci 223, 230
    sensitivity 228–30
    site 222
Drosophila
    albomicana 270
    americana texana 269
    arizonensis 269
    erecta 180
    ezoana 173
    grimshawi 173
    gymnobasis 173
    hawaiensis 132
    hydei 139, 180, 202, 204
    littoralis 173
    mauritiana 180
    melanogaster 1, 3–5, 9, 13–14, 24, 32, 33–90,
        91, 93, 95, 98, 101, 116–18, 119, 121,
        126–8, 129, 132–49, 151, 154, 158–60,
        162–4, 165, 167, 173, 174, 175, 178, 180,
        182, 184, 185–7, 189–91, 193–4, 196,
        199–203, 207–11, 214, 215, 219–24, 230,
        231–2, 251–2, 271–6, 294–5, 297, 299,
        302, 309, 312–13, 315, 319–20, 323, 325–7,
        331–4, 337
    melanogaster species sub group 24
    mojavensis 269
    nasuta 270
    nasutoides 173
    nebulosa 132
    orena 24

    pseudoobscura 222
    saltans 132
    simulans 132, 134, 180, 194, 312
    silvarentis 173
    virilis 162, 173, 269
    willistoni 132
duplication 17–19, 64, 82, 130, 191, 195–6,
        199, 277–80, 305, 325–6
    chromosome 137
    free 139
    gene 22, 61, 280, 292, 325, 329
    imperfect 120
    localized 17
    products 83
    tandem 61, 199, 277
    target site 161
    X-chromosome 64
duplicative transmission 244
duplicative transposition mechanism 112
dysgenic
    sterility 275
    systems 275

earthworms 277
echinoderm evolution 307
echinoderms 185, 307
ectoblasts 91
ectoderm 27, 28, 30, 35, 37, 42, 53, 69, 70,
        93, 335
    embryonic 69, 216
    primitive 122, 160
egg
    complexity 168
    nucleus 185
    polarity 39, 63
    RNA complexity 172
    spatial coordinates 39
elements
    Alu 159
    cis-acting 210
    FB 131–2
    I 274
    IS 159
    LTR-IS 159
    MR 274
    TE 132
    VL30 160
embryo screens 90
embryogenesis 33, 34, 35, 42, 46–7, 52, 70,
        88, 90, 93, 137, 142, 168, 178, 185, 201,
        217, 224, 228, 251, 279, 309, 316, 335
embryonic
    axes of symmetry 29–30
    death 42, 216
    hierarchies 33
    midline neurons 85
    nervous system 81, 85–6
    protein 140

embryos
  monosomic 266
  polysomic 266
endo
  mitosis 214
  mitotic DNA replication 204
  ploid 36
  ploidy 215, 226
  polyploid 201, 224, 226, 249
  reduplication 214
endoderm 27, 335
  primary 216
  primitive 122
enhancer sequences 296, 314
enhancers 119
enzyme
  activity 148, 248, 327
  activity levels 220
  polymorphisms 245
  systems 325
enzyme loci
  Adh 224
  Ddc 224
  αGpdh 224
  cMdh 224
  Pgk 224
enzymes 201, 325, 327
  acetylcholinesterase 83, 146
  Alu I restriction 159
  acid phosphatase 239, 327
  alkaline phosphatase 327
  cAMP-dependent protein kinase 84
  cAMP phosphodiesterase 84
  calcium calmodulin-dependent cyclic
    nucleotide phosphodiesterase 84
  carbonic anhydrase 293
  choline acetyl transferase 83
  dihydrofolate reductase 23
  glucose-6-phosphate dehydrogenase 327
  hypoxanthine phosphoribosyl transferase
    217
  lysozyme 293, 328
  3-phosphate dehydrogenase 171
  multilocus 277
  6-phosphogluconate dehydrogenase 327
  polymerase 114
  polymerase I 114, 120
  polymerase II 114
  polymerase III 114, 121
  preproelastase 42
  procollagen peptidase 128, 191
  protein kinase 81
  recombinase 112
  reverse transcriptase 158, 163
  RNA polymerase 116–17, 140
  ribozyme 124
  serine protease 42
  tryptophan pyrrolase 222

tyrosine-kinase 135
urea cycle 298
urokinase 70
xanthine dehydrogenase 327
epidermal
  development 321
  growth factor 70, 71, 332
  nerve plexus 331
  placodes 29, 93, 303
epidermal structures
  feathers 304
  hair 304
  scales 304
  skin glands 304
epidermis 42
epidermogenesis 70
epistatic
  hierarchy 66
  network 69
escape response 72
Escherichia coli 13, 17, 22, 82, 107, 110,
    116–17, 141, 149, 161, 200, 294, 325
  lac region 22
  lac repressor 141, 294
euchromatic
  components 270, 285
  DNA 128, 132, 142
  genes 202, 294
  rearrangements 148
  regions 255
  segment 136
euchromatin 2, 134, 137, 143, 204, 215, 270
Euglena gracilis 82
evolution 162, 182, 191, 234–8, 242, 245, 272,
    280, 288, 291, 292, 296, 309–11, 314,
    322, 325, 330, 331, 334–5, 337
  accelerated 272
  behavioural 330
  chordate 277
  concerted 241–2
  dipnoan 282
  genomic 290
  macro 302, 305, 309
  micro 305
  morphological 290
  plant 277
  punctuated 239–41, 272
  punctuational 321
evolutionary
  change 25, 26, 235–7, 241, 243, 245, 273,
    306, 319
  conservation 162
  distance 291
  divergence 277
  diversification 282
  history 149, 231, 306, 311, 320
  implications 21
  novelty 242–3, 308, 318, 330, 338

pattern 321
  significance 276
  strategies 310
  time 281
exchange
  chromatid 57
  mitotic 57, 68
  sites 112
  unequal 161, 207
excision 276
  points 112
exons 52, 53, 84, 119, 124–5, 163, 191, 305, 315
expression site 110, 112
extinctions 306, 310
extracellular
  matrix 29, 30
  proteins 315
extrachromosomal replication 207
extraembryonic
  cell lineages 122
  tissues 60

fasicles
  axon 86
  transverse 42
fate map 34
faulty genic interaction 260
fertility 104, 226, 237, 276, 289
  factors 136, 139
  female 84
  loci 135–6
  male 63
fertilization 22, 26, 33, 43, 69, 167, 185, 216, 226, 244, 275
filaments
  actin 233
  intermediate 233
filopodia 29
fish 150–1, 228, 277, 287, 306
  bony 277, 319
  cyprinid 194, 248
  electric 15
  freshwater 298
  fossil 281
  lobe-finned 282
  marine 298
  puffer 319
  teleost 288, 328
  Torpedo californica 15, 83
flatworms 277
flip-flop
  switch mechanisms 95
  transitions 111
foldback fraction 165
follicles
  meroistic 215
  panoistic 215

fossil
  amphibians 281, 320
  data sets 306–7
  fauna 307
  fish 281
  record 272–3, 302, 306, 309, 320, 322
  sequences 320
functional
  protein domains 315
  redundancy 312, 327
fungi 20, 156, 166, 193, 226
  Achlya bisexualis 156, 165
  Aspergillus nidulans 165–6, 226
  Neurospora 226
  Physarum polycephalum 157

gametes
  aneuploid 277
  hyper-modal 266
  hypo-modal 266
ganglia 29
gap junction(s) 38
  channels 38
  communication 39
gap mutants 45
  giant 314
  hunchback 314
  knirps 314
  Krüppel 13, 45, 314
gastrula 27, 28, 55, 60, 167, 185, 303
  mRNAs 169
gastrulation 26, 28, 30, 34, 40, 47, 63, 93, 166–7, 228
gene
  activation 122
  amplification 178, 208
  buffer systems 327
  circuits 30, 141, 230, 231–2, 236, 245, 273, 288, 296, 329
  clustering 196, 199–200
  clusters 186, 199, 211, 223, 312–13
  complexes 61
  conversion 21, 98, 157–8
  deletion 117
  domains 148
  dosage 200–1, 221, 223, 317, 333
  duplication 21, 61, 200, 239, 280, 292, 325, 329
  expression 85, 98, 107, 112, 122, 139, 155, 157, 168, 193, 210–11, 216, 233, 294, 296, 306, 312, 314
  flow 275
  frequencies 292
  function 318
  fusion 117
  hyperactivation 221
  interaction 233, 294
  isolation 5

landscape 148
location 193
mutation 274, 275
network 294
number 107, 163, 164, 165, 193
pathways 287
regulation 109, 113, 124, 156, 193, 210,
    220, 294, 296, 297, 304, 305, 314, 315,
    338
sequence 17
structure 193
switching mechanisms 312
transcription 119, 141, 231, 245, 318
unemployment 333
variegation 296
gene loci
abnormal oocyte 135, 327
Adh 224
f19Aldox 222
Beadex 223
compensatory responder 135, 204
cut 85
Ddc 224, 317
dominant spotting 94
dorsal 41
dosage sensitive 230
Gart 319
α Gpdh 224, 327
β-globin 279
haplo-insufficient 223, 230, 297
K10 5
lin-12 93, 317, 326
mating type 96
cMdh 224
Notch 162
pelle 42
Pgk 224
rDNA 19, 179, 180
Rp 11215 13
scute 57
Serendipity 13
Sgs-4 118, 222, 295
steel 94
steroid sulphatase 216
Toll 41
triplo-lethal 230
unc-86 91
white 131, 222, 275, 295
Xg 216
gene product
amount 201
interaction 233
genes
acetylcholinesterase 83, 87, 146, 333
actin 116, 190, 199–200
ad 8, 13
adenine phosphoribosyl transferase 116
alcohol dehydrogenase 13, 117, 148, 222

alpha methyl dopa 84
amplified 23
α-amylase 125
ancestral 83
calcitonin 126
calmodulin 14, 319
caudal (= 567) 60
ced-3 90
ced-4 90
chloroplast 195
choline acetyltransferase 83
chorion 208–9
chymotrypsin 119
cloned 10
coding 236
collagen 122, 191
crooked neck 7
cuticle 319
cytochrome c 17, 21, 325
defunct 329
DHFR 24, 116
disconnected 86
Dopa decarboxylase 84, 148, 199, 221, 224,
    317
dunce 84
engrailed 43, 57, 59, 163
engrailed related (= invected) 59
executive 30, 33, 61, 89, 290, 326
extra macrochaetae 223, 317
extra sex combs 49, 56, 61
fascilin 86
folded gastrulation 313
globin 21, 116, 119, 242, 280
glucose-6-phosphate dehydrogenase 116
glue protein 148, 221
glyceraldehyde β-phosphate
    dehydrogenase 11
H_1 95
H_2 95
haemoglobin 278
hairy 223
histocompatibility 242
histone 116, 127, 128, 182, 184–6, 196, 242
homeo box 60, 173
housekeeping 30, 33, 42, 116, 319, 322–3
hsp 70 117
htpR 117
hypersensitive 118–19
hypoxanthine phospho-ribosyl transferase
    116
immunoglobulin 109, 116, 242, 278
incompatibility 242
insulin 119
interrupted 179
kurz 7
larval cuticle protein 196
larval serum protein 200, 222
Leu-2 18–19

lysozyme 119
maternal 39, 43
metabolic 196, 315
microglobulin 125
morphological 293
mutator 274
myoglobin 17, 278–9
myosin heavy chain 223, 315
neomycin 5
nicotinic acetylcholine receptor 83
nina A 81
nina C 81
nina E 13, 80, 81
nitrogen fixation 112
pair rule 44
pecanex 7
period 81
phosphoglycerate kinase 116
6-phosphoguconate dehydrogenase 222
polycomb 49, 52, 56, 61, 112
polymerase I 242
polymerase II 16, 17, 116–19
pyrimidine biosynthesis 14
regulator class 56
regulatory 294, 297
retrovirus 130
ribosomal 107, 120, 135–6, 139, 179–80, 201
rRNA 128, 177–8, 181–2, 195, 200, 201–3, 212, 241, 312
rp 49 5
r-protein 194
5S 128, 178, 194
S8 85–6, 87
selector 46, 56, 61, 313
Sgs 3, 7, 8, 120
single copy 238
Slp 328
snake 41
5S RNA 121, 140, 177–8, 194
structural 18–19, 131, 237, 293–4, 296, 302, 306
switch 58–9, 64, 88–9, 297, 302
tubulin 14, 189, 200
twisted gastrulation 313
variable surface glycoprotein 110–12
variegating 296
vimentin 125
vitellogenin 122
v-onc 134
xanthine dehydrogenase 146, 148, 200, 221
zen 60
zygotic 55
genes within genes 319
genetic
    balance 226–7, 228
    cohesion 242
    cost 273, 321

change 320
map 50
mosaics 57, 74
genome
    alteration 95, 111
    architecture 313
    change 237, 257
    chicken 159
    compartments 334
    complexity 110, 132
    contraction 283
    expansion 17, 283
    flux 15, 276, 280, 313
    function 334
    hurnan 150
    internal readjustments 243
    macronuclear 184
    micronuclear 106
    minimum size 305
    organization 26, 112, 149, 193–4, 296
    plasticity 15
    polyploidization 195
    raiding 84
    rearrangement 312, 318
    reorganization 275, 294, 324
    size 105, 110, 149–55, 159, 165, 166, 169, 172, 176, 183, 194, 225, 232, 236, 238, 246–56, 263, 276–7, 280–2, 283–8, 290, 292, 305, 320, 329
    size change 237–8, 246–5
    size variation 153
    structure 334
    tetraploidization 278
    turnover 244, 305
genomes
    angiosperm 320
    chloroplast 194, 294, 312
    housekeeping 319
    mitochodrial 194, 294, 312
    ontogenetic 319
    polyploid 228
    teleost 320
genomic
    alterations 111
    change 237, 290
    clones 191
    compartments 291
    evolution 290
    flux 15, 276
    landscapes 24, 291, 311
    modification 112
    perturbation 290
    rearrangement 312
    reorganization 234
    segregation 319
    shuffling 322
    topography 318
    turn over processes 245

genotype 245
genotypic control 245
germ
    band extension 45
    cells 5, 63, 67, 94, 103, 201, 217, 227, 275
    layer differentiation 28
    layers 27, 28, 34, 303
    plasm 63
germ line 14, 17, 61, 64, 66, 68, 88, 93, 103,
        132, 147, 173, 180, 207, 217, 227, 236,
        275, 289
    cells 67
    development 66, 260
    genes 103, 107, 236
    genome 16, 103, 107, 236
    precursor 63
    transformation 199
    transposition 132
    variants 237
giant fibre system 72, 74
Giemsa (G) banding 263, 268, 293
Giemsa stained bands 319
glial sling 331
globin
    genes 21, 280
    pseudogenes 280
glycolysis 196
glycoproteins 10, 29, 109, 125
gonadal
    atrophy 275
    dysgenesis 275
    sterility 275
gradients 39, 63
    antero-postero 39, 40, 63
    dorso-ventral 39, 40–1, 63, 305, 313
    morphogen 41
gradualism 239, 241
growth factors
    epidermal 70, 93
    transforming 70
    vaccinia viral 70
gynandromorphs (= sex mosaics) 63, 74,
        76–7, 79
gynogenetic haploid 228

haemoglobin 21, 325
    adult 279
    chains 293
    gene clusters 279, 293
    juvenile 279
halteres 47, 51–2
haplo-diploidy 226
haploid 19, 140, 226
haploid cell types 95
    genome 128
haplo-insufficient 50, 222, 223
haplolethal region 82
hatching time 137

heat shock
    gene cluster 319
    genes 14, 117
    promoter 5, 45, 53, 117
    proteins 14, 45, 117, 127
    response 126
hemopoiesis 94
heredity
    chromosomal 274
    cytoplasmic 274
hermaphrodites 68
heterochromatic
    breakpoints 294–5
    chromocentre 148, 179
    components 285
    DNA sequences 128, 295
    junctions 295
    regions 1, 128, 139, 174, 327
    segments 103, 290, 311
    sequestering 200
heterochromatin 103, 132, 135–9, 174, 207,
        215, 252, 271–2, 274, 285, 295
    additions 269
    content 137, 139, 141, 173–4, 255, 272
    deletions 139, 269
    differences 269
    flanking 295
    pericentromeric 174
    proximal 272
    transcription 173
    sex chromosome 295
    X 141
heterochromatinization 216, 296
heterochronic
    mutations 93, 301
    shifts 95
heterochrony 93, 298, 301
heterocysts 112
heterokaryon 226
heterospecific dimers 239
heterothallic 97
highly repeated 1723 sequence 160
highly repetitive satellite
    DNA families 245
histoblasts 35–7, 57, 66
histone
    classes 183
    gene cluster 182, 296
    gene multiplicity 296
    gene number 185–6
    gene promoter 225, 311
    genes 182–3, 185
    gene transcripts 124, 141
    mRNA 185
    orphon genes 185
    proteins 116, 182, 184, 196, 201, 242, 296
    readthrough 184
    sites 225

system 186
histones 201, 224
homeo box containing genes 59–60
homeo boxes 46–7, 50, 53, 59–60, 61, 66, 70,
    163, 173, 281, 297
homeo box products 297
homeo box sequences 59–60
homeo domains 98
homeosis 55
homeotic genes 39, 45–7, 53, 55–6, 60, 61,
    88–91, 93, 313
  additional sex combs 49
  extra sex combs 49, 56, 61
  Polycomb 49, 61, 315
  Posterior sex combs 49
  probescipedia 53
  Sex combs on mid leg 49
  Sex combs reduced 53
  spineless aristapedia 74
  super sex combs 56
homeotic
  complex 49
  loci 61, 66
  mutations 50, 55, 57, 74, 301, 302
  mutants 55, 62, 91
  regulatory genes 56, 61
  selector genes 56
  transformation 49, 52, 53, 55–6
Homo sapiens 10–11, 14, 21, 47, 59, 84, 93,
    98, 116, 119, 152, 159, 162, 169, 170,
    172, 178, 180, 182, 184, 185, 191, 193–4,
    196, 216, 219, 227–8, 238, 279, 287, 293,
    301–2, 322, 328, 332–3, 336
homogenization 18, 242
homogenizing mechanisms 249
homology comparisons 15
homothallic 97
hormones 30, 119, 196, 299, 315
  ACTH 196
  ecdysteroids 337
  β-endorphin 196
  egg laying 84
  20-hydroxyecdysone 299
  insulin 196
  moulting 299
  neotenin 299
  oestrogen 119
  prolactin 299
  steroid 119, 328
  thyroid 299–300
  throxine 300
hormone treatment 319
HpaII repeats 157
human
  acetylcholinesterase gene 333
  Alu repeats 16, 23, 159
  Alu sequence 159
  brain 332–3

epidermal growth factor 135
eye 238
genome 11, 150
genomic DNA library 184
globin gene cluster 161
globin genes 159, 161
haemoglobin gene
    clusters 279
histone gene cluster 184
leucocyte interferon gene 13
methionine tRNA gene 160
myoglobin gene 17, 161
pro-α1(I) gene 193
pro-α2(I) gene 193
satellite 1-DNA 23
trisomics 227
β-tubulin gene 159
hyaluronic acid 29
hybrid
  dysgenesis 275–6, 374
  instability 275
  sterility 132, 260, 269, 274, 275
  swarms 269
  zones 266
hybridization 267, 277
hybridization probe 191
hybridoma 9, 28
hybrids 266, 271, 275–6, 288
  interspecies 120, 255, 260, 268, 270–1, 276,
    319–20
  interstrain 202, 274–5
  5S-tRNA gene 121
hyperactivation 221
hypermorphosis 301
hypersensitive site 118–19
hypertrophy of the CNS 69, 93
hypervariable regions 161
hypomethylation 122, 124

imaginal disc
  development 337
  perturbation 337
imaginal discs 33, 35–7, 51–2, 55–61, 62, 76,
    88, 90, 143, 167, 201, 232, 298, 337
  antennal 52, 55
  eye 52, 337
  foreleg 66
  genital 36–7, 40, 52, 66
  haltere 36
  leg 36, 52, 337
  prothoracic 36
  wing 36, 37, 51, 62
imaginal rings 36
immunoflourescent staining 52
immunoglobulin
  class switching 98–9, 101
  genes 109
  heavy chain cluster 328

heavy chain genes 101, 119
heavy chains 99
light chain diversity 100
light chains 99–100
protein 98
switching sites 101
system 111
VDJ coding region 119
VDJ joining 101
VJ joining 100, 111
immunoglobulins 9, 21, 98–101
in situ hybridization 2, 4, 9, 24, 52, 71, 128, 173, 183
inactive X-chromosome 216, 219
inbuilt redundancy 329
initiation sites 117, 125
insect families
    Cecidomyidae 103
    Chironomidae 103
    Sciaridae 103, 212
insects 14, 55, 103, 152, 156, 242, 280, 283, 298, 299
    Acheta domestica 215
    Atractomorpha similis 290
    Bombyx mori 212, 215, 243, 322
    Diptera 152, 182, 211, 324
    grasshoppers 72, 86, 152, 330, 332
    Hymenoptera 226
    Lucilia cuprina 215, 251
    Manduca sexta 71
    moths 14
    Musca domestica 152
    Prodiamesa olivacea 152
    Rhyncosciara americana 211
insert families 178
insertion 23, 178, 328
    DNA 51
    elements 159
    lines 221
    mobile elements 51
    mutagenesis 112
    rDNA 178, 207
    sequences 136, 179–80
    X-linked 221
insertions 23, 179, 202, 207, 221
in vitro
    hybridization 80
    transcription 119
interaction
    cell–cell 29–30, 37, 70, 93, 231, 338
    cellular 338
    epidermal–dermal 304
    gene product 233
    hierarchical 313
    inductive 28, 66
    protein 338
    protein–protein 338
    regulatory 313

intergenic
    conversion events 161
    spacers 235
internal
    control region 121
    genomic readjustment 243
    promoter sequence 121
interneuron
    command 72
    peripherally synapsing 72
intersexes 66
interspersed repetitive DNA 154
interspersion pattern 132, 154, 156, 160, 193
    long period 154–5
    short period 154–5
intervening sequences 21, 59, 121, 188–9
intracisternal A type particles 135, 158
intron sizes 191
intron–exon junction 124
introns 10, 17, 119, 124–6, 143, 147, 158–9, 161, 190–1, 195, 315, 319
isoenzymes 277

jump response 77
junctions
    gap 38
    intron–exon 124
    neuromuscular 79
junk
    DNA 25, 109, 141, 245, 312
    RNA 225, 312

karyotype 236, 263, 269, 282, 283, 322
karyotypic
    alteration 290
    changes 290
    disimilarity 258
    variants 249
kinetic complexity 107
kynurenine 232

lac region 22
    repressor 141, 294
lampbrush
    chromosomes 168
    loops 160
landscapes
    coding 24
    DNA 64, 74, 210
    flanking 143, 149, 318
    gene 85, 114, 116, 148
    genomic 83
    regulatory 24
    TATA type 116
    Xdh DNA 148
larvae
    Amphioxus 302
    anuran 298

Drosophila 33
feeding stage 28
pluteus 166–7
tadpole 93, 160, 214, 249, 298
larval
brain 86, 98
development 211, 251
period 251
leader regions 161
leg disc evagination 337
lethals
male specific 67
maternal effect 302
zygotic 33, 302
libraries
bacteriophage 82
cDNA 71, 83, 212
cosmid 82
DNA 1, 5
embryonic CNS 86
genomic 159
invertebrate 84
mini 3
plasmid 132
rat brain cDNA 170
recombinant DNA 4, 9, 159
vertebrate 84
lineage
compartments 38
restriction 37
segregation 90
lineages
neural 91
pluripotent 122
LINES 160, 274
living fossils 272
local DNA duplication 196
localized
cell death 337
highly repetitive sequences 194
tandem arrays 17
Loxoma 320
lungfish 248, 256, 277, 281, 302, 306, 320
evolution 281, 320
lineages 281
macroevolution 302, 305, 309
macronuclear
complexity 107, 164
DNA 107
development 104, 106–7, 109, 157
formation 311
genome 184
macronucleus 104–5, 107, 109, 112, 157, 162, 164, 318
magnification 201
male
courtship behaviour 77
development 64

germ line 226
recombination 274
sexual differentiation 66
specific lethals 67
sterility 267, 271
males
XO 137, 141
XYY 137, 141
XYYY 141
malpighian tubules 36, 40
malsegregation 266, 268
mammalian retroviruses 158
mammals 60, 70, 88, 90, 93–4, 111, 151, 156, 159, 164, 169, 176, 180, 185, 189, 196, 219, 231, 251, 258, 268, 278, 287–8, 304, 327, 330, 331, 332.
aardvark 151, 287
Anotomys leander 263
Aotus trivirgatus 322
armadillos 287
bat-eared foxes 243
bats 287
Bos taurus 22
carnivores 322
cat 322
calf 22–3
cattle 42, 162
Cercopithecus aethiops 159
chimpanzees 293, 302, 336
Chinese hamster 159
Dipodomys 176
dogs 301, 328
Elaphodus cephalophus 260
elephants 251, 301, 320, 325
equuids 260, 301, 324
Galago crassicaudatus 160
gibbons 180, 322
giraffe 236, 301
goat 279
gorilla 336
great apes 159
hamster 13, 116, 194, 218
Irish elk 301
kangaroo rats 173, 176
lynx 288
marsupials 330
mastodons 301
Megaptera 325
monkeys 287, 333
monotremes 330
muntjacs 260, 263, 290, 294, 318, 324
orang utan 336
Pakicetus 306
panda 304
placentals 330
Plantista gangetica 287
Plantista indi 287
pocket gophers 194

primates 178, 180, 274, 288, 322, 330
rabbit 42, 328
Rattus colletti 268
Rattus norvegicus 176
Rattus sordidus 268
Rattus villosissimus 268
rhinoceroses 301
rodents 84, 177, 180, 274, 288, 324
whales 176, 287, 301–2, 306, 320, 326
map
   fate 34
   genetic 50
Marchantia polymorpha 126
maternal effect
   area 82
   genes 33, 39–40, 42, 313
   mutations 40, 42, 63, 327
   region 82
maternal effect dorsalizing loci 41
   dorsal 40–1
   pelle 42
   snake 41–2
   Toll 41
maternal effect type homeotics 56
   extra sex combs 56
   super sex combs 56
maternal gene products 33
maternal inheritance 274
maternal mRNAs 42, 185
maternal mutations 40, 42, 46
   dorsal 40
   oskar 46
   pumilio 46
   staufen 46
   tudor 46
   vasa 46
mating 77
mating type
   genes 98
   locus 96–8
   switching 97–8
   system 112
mating types 97–8, 109
meiosis 17, 18, 21, 22, 96, 104, 164, 190, 214,
   226, 244, 249, 255, 260, 266
meiotic
   block 260
   cycle 207, 249
   cycle times 249
   duration 249, 253
   loss 207
   pairing 276, 288
   pairing sites 135, 276
   recognition processes 137
   system 274
melanoblasts 94
membrane
   channels 79

glycoprotein receptors 135
glycoproteins 86
pumps 79, 86
memory 68, 84, 88, 332
meristic modifications 272
mesoderm 27, 29, 34, 43, 93
mesodermal masses 42
mesothorax 47, 50, 62, 74
message
   coding capability 146
   complexity 163–4, 171, 173
messenger (m)RNA 15, 101, 107, 114, 117,
   124–7, 141–6, 164–6, 170, 172, 189, 220,
   224
   brain specific 170
   cellular 16
   complex 164
   complexity 141, 167
   chimeric 131
   Ddc 84
   fibronectin 126
   globin 164
   glue protein 120
   heat shock protein 128
   histone 185
   Kr^+ 45
   lysozyme 119
   maternal 185
   moderately prevalent 164
   ovalbumin 164
   $poly(A)^-$ 141, 170
   $poly(A)^+$ 60, 142, 162, 170
   polycistronic 196
   polymerase II 114
   populations 30, 169
   precursors 124, 126, 225
   processing 124
   sequence diversity 167
   small nuclear U6 13
   splicing 124
   super-prevalent 164
   synthesis 209
   Ubx^+ 52
metabolic
   genes 196, 315, 327
   diversity 330
   machinery 179
   pathways 196, 231
   rate 246, 248
   shunts 329
   systems 201
metabolism 246, 248, 318
   catecholamine 84
   cAMP 84, 88
   neurotransmitter 74
metameres 43
metameric segmentation 28, 302
metamorphosis 30, 33, 35–7, 61, 71, 118,

167, 249, 251, 298, 299, 302
  amphibian 299, 300, 321
  insect 299
  retrogressive 93
  reverse 300
metathorax 47, 74
methotrexate 23
methyl cytosine 122
methylation 122
micro
  chromosomes 157
  cloning 1, 3, 32, 45, 80, 87–8, 90
  dissection 3, 7, 32, 45, 80, 87, 88, 90
  evolution 305
  injection 132, 147
  nucleus 104–5, 107, 112, 162, 164
  sequencing 7, 9
  tubules 14, 141, 189, 233
  villi 87
micronuclear
  complexity 107
  DNA 106–7
  genome 106, 164
microtubule formation 189
mid-blastula transition 257
middle repetitive (mr) sequences
  Alu 10, 163
  clustered scrambled 10
  dispersed 21
  Kpn 10
  poly(C–A) 10
mini
  chromosomes 137
  nucleolus 200
  satellites 17, 161
  tandem arrays 161
minimum
  genome size 153, 305
  repeat unit 174
minute
  loci 5, 57
  phenotype 5
mitochondrial
  DNAs 318
  genomes 194, 195
mitogenic
  activity 70
  polypeptide 135
mitosis 21, 33, 104, 214, 249
mitotic
  cycle 249
  cycle time 249
  exchange 57, 68
  recombination 231
mobile element families 132, 157
mobile elements 5, 32, 131, 154, 157, 161,
    180, 193, 194, 235, 244, 276, 293, 312,
    324, 334

dispersed 160
  long interspersed
    repeats 193
  short interspersed repeats 193
molecular
  clocks 336
  cloning 70
  drive 22, 237, 239, 241–5, 276, 306, 321–2
  drive mechanisms 244
  turnover processes 276
molluscs 59, 128, 287
  Aplysia californica 84, 215
monoclonal antibodies 7, 9, 42, 49, 52, 79,
    85, 86, 90, 305, 332–3
monosomic 207
morphogen 316
  concentration 317
  developmental 323
  gradient 41
morphogenesis 29, 90, 200, 293, 296, 309,
    314, 336, 337
  abnormal 228
  cell 303
  epithelial 304
  phage 233
  skin 303
  spindle 233
  viral 233
morphogenetic
  movements 29–30, 313
  pathway 64
morphological
  change 147, 234–6, 237, 241, 242–3, 272–3,
    290–1, 293, 305, 319–20, 322
  complexity 177, 189, 276
  conservatism 322
  constancy 272
  differentiation 36, 176
  divergence 292
  diversity 172, 322, 330
  effects 293
  evolution 241, 290
  innovations 329
  novelties 291, 335
  novelty 234, 236, 243, 245, 272, 290, 291,
    292, 297, 306, 319–21, 321–2, 330, 335,
    336
mosaic
  analysis 331
  females 64
  organism 294
mosaics 63, 67
  bilateral 76
  genetic 57, 74, 81
  partial bilateral 76
  sex 63–4, 77
mosses 20
multicircuit systems 329

multigene
  families 177, 194, 238, 241, 242–3, 245,
    280, 305, 309, 322, 324–5, 326–7, 329,
    334
  systems 329
  turnover 242
multilocus
  enzymes 277
  systems 321
multiple
  fragmentation 105
  gene families 20
multisequence
  families 242–3, 309, 321, 322, 326
  systems 325, 329
Mus castaneus 217–18
Mus musculus 9, 11, 13, 14, 15, 29, 30, 32, 42,
    47, 50, 60, 93, 94, 116, 122, 125, 135,
    158, 159–60, 162, 170, 172, 176, 180,
    182, 184, 185, 194, 217–19, 227, 251,
    263, 321, 328, 331, 333
  α-amylase gene 125
  embryo 185
  B1 repeat 159, 162
  B2 sequence family 160, 162
  extra embryonic cell lineages 122
  immunoglobulin genes 162
  intracisternal A-type particles 135, 158
  β-2-microglobulin gene 125
  MIF-1 repeat 159
  primitive ectoderm 122
  primitive endoderm 122
  R sequences 162
  Tailess complex 30
  trophectoderm 122
  visceral yolk sac 122
  VL30 elements 160
muscle cell differentiation 200
muscles
  dorsal longitudinal flight 72, 79
  indirect flight 199, 223
  tergotrochanteral 72
mutagenesis
  insertion 112
  in vitro 80
  in vivo 333
  P element mediated 333
  saturation 87, 313
mutagenic assault 326
  screens 80
mutants
  acallosal 333
  amnesiac 332
  bicaudal 40, 46
  brain 82, 86–7
  connectivity 334
  developmentally significant 32
  discless 37, 143

  ecdysteroid deficient 299
  engrailed 38, 313, 314
  grandchildless 63
  homeotic 91
  lin-22 91
  neurological 88
  nina A–E 81
  null activity 327
  oscelliless 210
  pair rule 44
  pyr B 13
  segmentation 43
  segment polarity 44
  small optic lobe 87
  X-linked 196
mutation 243, 274
  intragenic 296
  lethal 5, 74, 132
  neutral 336
mutational
  change 273
  perturbations 329
  switch control 55
mutations
  abnormal oocyte 327
  acallosal 331
  bang sensitive 82
  behavioural 74, 77
  bendless 72
  bicaudal 39
  BX-C 49–50
  decapentaplegic 337
  dominant flightless 223
  dominant gain of function 53
  female sterile 211
  gfA 72
  giant 299
  grandchildless 63
  hairy 223
  haltere mimic 51
  hedgehog 314
  heterochronic 93, 301
  homeotic 50, 57, 301
  insertion 158
  larval serum protein 222
  lethal 132
  lin-14 93, 301
  loss of function 53, 64, 69, 317, 328
  male specific lethal 220
  maternal 42
  Minute 223
  oscelliless 209
  Passover 72
  paired 71
  patched 314
  selectively neutral 336
  staggerer 29
  subfunction 52

substitution 158
tra-1 68
tudor 63
white 232
white crimson 131
white-Dominant Zeste like 131
white eosin 222
X-linked recessive female sterile 211
myoglobin minisatellite 161
myosin
heavy chain genes 223, 315
heavy chains 81

nematodes (roundworms) 59
Ascaris 195
ascarids 103
Caenorhabditis elegans 32, 55, 68, 90, 93,
191, 302, 315, 317, 321, 323, 326, 332
Parascaris equorum 103
Parascaris univalens 103
neotenic 251, 299
neoteny 298, 300
nervous system 29, 68, 49, 74, 86, 90, 321,
330, 331–2
central 36, 42, 52–3, 68–9, 71–2, 74, 77–84,
321, 330, 332–3
construction 89
embryonic 61, 81, 85
evolution 196
invertebrate 196
peripheral 85, 333
ventral 71
neural
activity 196
cell adhesion molecules 29, 332
cell lineages 90
crests 28–9, 93–4
development 74
folds 28
lineage 91
pathways 321
plate 28
tube 28
neuroactive peptides 128
neuroblasts 35–6, 68–70
neurogenesis 68–70, 71, 90, 162, 314, 331
neurogenic
epidermal placodes 93
mutations 69
proteins 332
region 68–9
neurogenic loci 69–70
almondex 314
amnesiac 88
big brain 70, 314
canoe 69, 91
Delta 70, 314
dunce 84, 88

Enhancer of split 70, 314
kayak 69, 91
mastermind 70, 314
neuralized 314
Notch 70, 93, 162, 314
punt 69, 93
rutabaga 88
topless 69, 93
turnip 88
ventral nervous system 71
defective 71
neurohormones 332
neurological circuits 334
neurological mutations
no action potential 83
temperature induced paralysis 83
neurological pathways 335
neuronal
cell types 330
connections 79
novelties 291
pathfinding 84, 86
precursor cells 85
surface recognition molecules 86
neuron myelination 170
neuron specific glycoproteins 9
neurons 9, 28, 63, 68, 71, 74, 79, 84, 86, 246,
298, 331, 334
chordotonal 85
dopaminergic 91
embyronic midline 85
giant 215
local circuit 330
male lobula giant 79
motor 71, 72, 93
peripheral 71–2
pioneer 72, 335
projection 330
sensory 71, 74, 85
XO 77
neuropeptides 215, 332
neuroregulators 332
neurotransmitters 10, 332
neurulation 28
neutral
characters 238
drift 237, 238–41, 306, 321–2
mutations 336
neutralist-selection debate 241
neutrality 239, 241, 242
nicotinic acetylcholine receptor 14
nomadic DNA 131, 134, 149
nomadic elements 16, 98, 128, 130, 134, 155,
157–8, 160, 161, 182, 236, 273, 274, 283,
310, 312, 322
297 98
412 98, 130
Alu 10, 163

B104 130, 231, 293, 312
copia 98, 129–30, 132, 135, 158, 274, 313
delta 157–9
dispersed 163
foldback 130–2
gypsy 130, 131
interspersed 319
sigma 158
TE 132
Ty 21, 130, 157–8
non-coding DNA sequences 235
novel gene hypothesis 305
nuclear
  division 185
  dimorphism 105
  mRNA 124
  pre-tRNAs 124
  pre-RNA 124
  proteins 140
  RNA 13, 141
  RNA complexity 168, 169, 172
  RNA polymerases 114
  RNA populations 163
  size 246, 256
  transcripts 169, 172
  transplantation 28
  volume 253
nuclei
  macro 104–7, 157, 162, 164, 311, 318
  micro 104–5, 107, 157, 162, 164
  oocyte 168
  posterior silk gland 215
  salivary gland 212, 219
  somatic 214
nucleolar
  dominance 202–4
  expression 206
  volume 257
nucleolus 114, 180, 212
  extrachromosomal 212
  mammalian 319
  organizing region 177, 179, 180, 181, 182,
    201–7, 212, 283
nucleosome
  configuration 122
  positioning 140
nucleotide sequence alteration 95
nucleotype 245, 252, 257, 281
nucleotype hypothesis 246, 252
nucleotypic
  alteration 290
  control 245
  effects 245, 249, 256, 320
null
  activity mutant 327
  alleles 327

oligonucleotide synthesis 7

ommochrome pathway 199
oncogenes 14, 134
  V-erb B 135
  V-onc 134
  src family 135
oocyte 24, 60, 68, 121, 140, 160, 166, 178,
  208–9, 212–15, 217, 225, 248
  abnormal 327
  development 224
  membrane 15
  nuclei 13, 168
  type 5S RNH 178
oogenesis 33, 39, 61, 63, 126, 160, 185–6,
  208–9, 211, 212–15
oogonia 212, 217
operator 13
operon 196
opsin genes
  Rh3 80
  Rh4 80
organization
  axial 28, 30
  dorso-ventral 43
  spatial 39
organogenesis 27, 29, 40, 63, 227
orphon rDNA sequences 182
orphons 183
ovary
  meroistic 215
  panoistic 215

pair rule loci 55, 313, 315, 316, 317
  engrailed 38, 43, 45, 57, 59, 61, 163, 313,
    314
  engrailed-related 59
  even-skipped 313, 314, 316, 317
  fushi tarazu 43–5, 53, 55, 71, 313
  hairy 44, 314, 317
  odd-paired 314
  odd-skipped 314, 316
  paired 71, 314, 317
  runt 314, 316
  sloppy-paired 314
pairing sites 139
parasegmental boundaries 43
parasegments 42–3, 316
parthenogenetic
  eggs 226
  populations 277
particles
  copia 158
  intracisternal A-type 135, 158
  virus-like 135
pattern
  deletions 315
  dorso-ventral 41–2
  formation 37, 90, 317

P element mediated germ line transformation
    5, 14, 33, 53, 55, 74, 80–2, 90, 117, 132,
    148, 193, 200, 209, 221, 222
P element
    system 333
    vector 117, 318
P elements 5, 53, 82, 132, 147, 210, 222,
    274–6
P family 132
pentose shunt pathway 327
phase
    variation 95
    variation mechanism 111
    variation system 101
phenocopies 45, 302
phenotypes 26, 137, 220, 236, 294, 318
    behavioural 5
    discless 337
    lethal 146
    modified 32
    mutant 334
    pair rule 45
    paralytic 81, 83
    permissible 322
    sterility 275
    visible 146
phenotypic
    change 235
    perturbation 290
    rescue 5, 41–2, 45–6
    structural change 292
pheromones 77
Pholidogaster 320
photon capture 196
photopigment conversion 298
photopigments
    ommochromes 231–2
    opsin 13
    porphyropsin 298, 300
    pterins 231–2
    rhodopsin 13, 80, 223, 298, 300
    xanthommatin 232
photoreceptor
    cells 9, 80, 81, 86
    function 80
photoreceptors 13, 80, 223
phylogeny 269, 292
placodes 29
plasmid
    integration 18
    libraries 132
plasmids 18, 257
plasminogen activator 70
pleiotropic effects 94
polar granules 63
polarity 41, 90, 95
pole
    cells 40, 63, 275

plasm 63, 275
poly (A)$^+$
    cytoplasmic RNA 170
    fraction of polysomal RNA 170
    RNA complexity 165
poly (A)$^-$ fraction of polysomal RNA 170
poly A segment 141
polyclones 37, 42, 58, 61, 90, 316
polygenic systems 317
polygenomic 105
polymerase
    readthrough 173
    recognition sites 296
    RNA I complex 242
    I systems 114, 120, 242
    II systems 16–17, 114, 116
    III systems 114, 121
    II transcription 119
polyploidization cycles 105
polyploids 107, 228, 248–9, 277
    sexual 277
    tetraploids 228
    triploids 226–8
polyploidy 277
polyproteins 128, 215
polysomal
    poly (A)$^+$ fraction 170
    poly (A)$^-$ fraction 170
    RNA 142, 167
    RNA complexity 168–9
    RNA population 166
polysomes 126, 141, 166, 167, 211
polytene 201
    analysis 148
    arm 3
    band number 146
    bands 2, 146–7, 183, 186, 199, 211, 332
    banding 270
    banding sequence 146
    cells 201, 204
    chromosomes 2, 4, 32, 45, 105, 129, 140,
        204, 215, 220
    map 117
    nuclei 35, 200
    section 36, 53
    system 119, 179
    tissues 179–80, 202–3, 219, 224
polytene–like chromosomes 105
polytenization 1, 35, 179, 202, 206, 212, 215,
    295
polyteny 215
position effect 296
    variegation 294, 295
positional
    coordinates 313
    cues 317
    information 39, 42, 63, 337

post
  embryonic development 142
  transcriptional control 13, 124
  transcriptional regulation 125
  translational modifications 128, 191
pre
  blastoderm egg 5
  blastoderm embryo 132
  blastoderm stage 231
  somatic elimination 103
preferential insertion 130
primary
  organogenesis 34
  sex determination 68
primitive ectoderm 122
primordial
  germ cells 35, 63, 94
  germ line 36
  gonad 63
  repeats 22
probes
  cloned 163, 244
  cDNA 83
  oligonucleotide 32, 80
  polynucleotide 162
  ribosomal RNA 283
  synthetic DNA 7
  uncloned 139
  unique cloned 283
processed
  gene insertion 17
  transcription 147
  transcripts 163
programmed
  cell death 90–1, 321
  cell movement 335
prokaryote genomes 25
prokaryotes 11, 45, 95, 116–17, 134, 196,
    232, 297
prokaryotic
  DNA binding proteins 52, 61
  promoters 116
  systems 294
  transposable elements 111
promoter
  addition 110
  elements 95, 114
  regions 117, 121, 242
  sequences 121
  sites 114, 242
  systems 122
promoters 13. 16, 95, 114, 117, 119–21, 313
  ADH-1 13
  Bam–islands 120
  B factor regions 121
  heat shock 5, 45; 53
  histone gene 225, 311
  hsp 70 117

internal 121
  polymerase I 306
  polymerase III 140
  rDNA 242
  ribosomal gene 242
  spacer 120
prosome 13
proteolytic cascades 42
protein
  archetypal 325
  binding 162
  DNA binding 140
  DNA binding studies 139
  domains 70, 81, 191
  evolution 325, 330
  expression 10
  synthesis 5
protein–protein interaction 338
proteins 81, 325–6, 338
  actin 190, 196, 223
  activator 116
  albumin 293
  brain specific 9
  C 70
  cell surface 337
  cell surface glycoprotein 86
  chloroplast ribosomal S12 82, 126
  chondroitin sulphate proteoglycan 81
  chorion 14, 126, 211
  chymotrypsin 42
  collagen 29, 128, 191
  cytochrome c 17, 293
  D1 140
  DNA binding 52, 61, 114, 119, 139, 140,
    140–1
  fibrinogen 328
  fibrinopeptides 293, 328
  fibronectin 29, 125
  flagellin 95
  globins 116, 191, 242, 278
  glue secretion 118, 119–20, 212
  glycoproteins 10, 29, 109, 125
  haemoglobin 21, 325
  histone 116, 182, 184, 194–6, 201, 242, 296
  immunoglobulin 21, 98, 116, 191, 242, 278
  larval serum 200
  LDL receptor 70
  membrane 79, 86
  mitogenic 135
  myelin 7
  myoglobin 293
  myosin 223
  myosin heavy chain 81, 126
  nina C 81
  Notch 70, 163
  nuclear 140
  oxytocin 21
  plasminogen activator 70

procollagen 191
regulatory 61, 118, 140–1, 297
ribosomal 82, 195
ribosomal 49 5
ribosomal S12 126
self assembling 233
serum albumin 328
Sgs 4 118
sigma initiation factor 116
silk fibroin 215
toponimic 305
transferrin 293
transmembrane 70
tropomyosin 200, 223
troponin 42, 223
trypsinogen 42
tubulin 14, 189–90, 196, 305
vasopressin 21
Ubx$^+$ 52
protochordates 150
proviruses 130, 134–5
pseudogenes 10, 17, 171, 189, 194, 235, 239, 279, 305, 312, 319
  dispersed 17
  globin 280
  processed 16
  tRNA 160
pteridophytes 277
pumps 332
punctualism 239–41, 306
punctualist-gradualist controversy 241, 292
pupariation 66
puparium formation 37
pupation 211
pyrimidine biosynthesis 14

radiation
  adaptive 272, 307
  vertebrate 277
rat 7, 9, 11, 14, 17, 42, 61, 83, 126, 163, 172, 268, 328
  brain 170, 332
  indentifier family 160
  myelin protein 7
reactivation of X-linked enzymes 219
readthrough products 168
rearrangements
  complex 195
  euchromatic 148
  white locus 295
reassociation
  data 146, 163, 193
  kinetics 132
  methods 163
  reactions 142
  techniques 141, 283
receptors
  acetylcholine 15, 83

antigen 98
human epidermal growth factor 135
muscarinic cholinergic 332–3
nicotinic acetylcholine gene 83
T-cell antigen 101
recombination 17, 24, 111, 235, 255, 276
  distances 50
  male 274
  mitotic 231
  sites 17
  site specific 297
recombinant type events 161
redifferentiation 71
regulation 294, 297, 316
  gene 314, 315
  gene activity 314
  RNA polymerase II 159
  selective translational 126
  transcriptional 231, 319
  translational 127
regulatory
  changes 296
  circuits 141, 193, 211, 235
  elements 14, 47, 156, 296
  function 155
  gene involvement 294
  genes 13, 294, 297
  hierarchies 313
  homeotics 61
  interactions 314
  junctions 38
  loci 66
  networks 296, 314
  proteins 61, 118, 140–1, 157, 297
  regions 52
  sequences 280
  subunit 84
renaturation kinetics 107, 132
repeated centromeric DNA sequences 184
repeated sequence families 193, 241, 245
  clustered scrambled 157
  dispersed 16, 160, 161, 163
repeats
  cloned 182
  direct 17, 129, 159, 161, 274
  dispersed 16
  flanking direct 111, 159
  interspersed 193
  interspersed short 193
  inverted 95, 130–2, 159–60, 274
  long basic 23
  long interspersed 159, 193
  long order periodicity 22
  long tandem 24
  long terminal 129, 130, 159–60
  M 70
  nomadic 149
  Opa 70, 162

poly (TG) myoglobin 163
primordial 22-3
short 159
short direct 312
short flanking direct 111
short inverted terminal 111
short terminal 159
simple sequence 161-2
small inverted 129
tandem 18, 162, 185, 196
telomere type 157
terminal 130
terminal direct 17, 111, 135
terminal inverted 111, 274
repetitive elements
1723 159-60, 225
Alu 159, 163
B1 159
CR1 159
Hpa II 157
MIF-1 159
replication 24, 202, 214, 235, 237, 297, 319
bidirectional 209
chromosome 1
differential 106, 202, 209, 215, 237
extrachromosomal 207
origin 209
partial 105
quirks 17, 161
slippage 17, 161, 242, 312
under 202, 295
replicative machinery 208, 311-12
repressors 95, 223
reproductive
capacity 272
isolating barriers 267
isolation 267, 272, 273, 275-7, 294
reptiles 151, 246, 304, 306
dinosaurs 301
pterodactyls 306, 320
snakes 162
turtles 306, 320
restriction
digest patterns 22, 180, 202
enzyme sites 275
fragments 13
site differences 201
restriction enzymes 3, 122
Alu I 159
EcoRI 202
Kpn-1 159
retina 9, 223, 305
retinal binding site 13
retinoic acid 60
retrovirus genes
env 130
gag 130
pol 130

rhabdomere structure 81
rhodopsin
bovine 13
Drosophila melanogaster 80, 223
genes 13, 80
ribonucleic acid (RNA) 196
antisense 45, 315
chimeric 131
coding sequences 146
complexity 142, 166-8, 170, 172, 330
cytoplasmic 163
cytoplasmic poly (A)$^+$ 170
cytoplasmic populations 163
egg complexity 172
genes 18
glue protein 120
junk 225
locus 18
nuclear 165-6, 168
nuclear populations 163
poly (A)$^+$ 9, 41, 45, 130, 141, 162, 165-6, 212, 225
poly (A)$^-$ 52, 130, 141-2, 160, 164, 165
polymerase 13, 16-17, 114-16, 140
polymerase II 13
polymerase I complex 242
populations 30, 143, 164, 166
processed 53
reassociation data 147
retroviral 158
ribosomal 196, 242
ribosomal precursor 124
5S 140, 178
5S genes 45, 121, 140
5S gene cluster 312-13
7SL 160
5S sequences 177
small nuclear 13
synthesis 114
splicing 124
transfer (t) 21, 114, 186, 196, 201, 224, 312, 324
transcription 113-14, 245
transcripts 72, 146, 168, 224
Ubx$^+$ 52
ribosomal
gene compensation 206
genes 107, 135, 143, 177, 180, 182, 195, 201
proteins 5, 82, 126, 195
spacer sequence 324
ribosomal (r) DNA 136, 177, 180, 186, 201-3, 206-7, 242
amplification 180, 201, 212
assembly 5
circles 212
cluster 19, 196
compensation 201, 204, 206-8
content 178, 201, 203, 206-7, 214

extrachromosomal 212–14
fragment 18
function 200
increase 202, 204
insertions 207
level 203
locus 18, 179
magnification 201, 206–8
over-replication 204
partial deficiency 204
polyadenylated 211
promoter regions 242
redundancy 194
under-replication 204
ribosomal (r) RNA 120, 179, 194, 196
gene number 182, 312
genes 107, 114, 135, 177–8, 180–2, 195, 200, 201–3, 207, 312
locus 179, 242
precursor 124
5S 114
sequences 242
sites 181
synthesis 212
ribosome production 194
ribosomes 201, 224
ring X-chromosome 63, 76

Saccharomyces cerevisiae 13, 14, 18–21, 53, 95–8, 101, 112, 130, 157–8, 162, 177, 186, 190–1, 193–4, 207, 248, 312
ADH1 promoter 13
cassette mechanism 97–8
chicken chimeric β-tubulin gene 14, 233
cytochrome c genes 21
delta elements 159
haploid strains 19
Leu-2 gene 18–19
mating type locus 53
mating type switching 98
mating type system 112
mating types 96–7
promoter 13
RNA polymerase I 242
RNA small nuclear families 242
rDNA clusters 19
rDNA locus 19
rRNA genes 19
sporulation 96, 248
tRNA genes 159
tRNA ^{met} 158
tRNA ^{ser} 21
tRNAs 312
β-tubulin gene 14
Ty element 21, 158
Salmonella typhimurium 95, 101, 111
satellite DNA 22, 140, 160, 225, 245, 269, 290
binding proteins 140

families 24, 245
sequences 106, 112, 136, 137, 139–40, 242–3, 311, 329
saturation mutagenesis 87, 313
schizosaccharomyces pombe 21
scrambled clusters 157
sea urchins 14, 30, 101, 162, 169, 184–5
Lytechinus pictus 239
Strongylocentrotus franciscanus 244, 291
Strongylocentrotus purpuratus 166–9, 185, 225, 239, 244, 291
Tripneuster gratilla 59
secondary diploidization 248
segment
boundaries 37
identity 302
polarity mutations 314
segmental
aneuploidy 223
development 53
diversity 50
primordia 56
transformation 47, 50–1
segmentation 35, 41, 43, 55, 74, 313, 315
adult 36
body 42, 46, 59, 280, 314
larval 41
metameric 28, 302
pattern 315
segmentation genes 39, 43, 45, 57, 196, 313
selection 20, 23, 239, 244
directional 321
intragenomic 237
natural 236, 237–8, 241–2, 243, 245, 256, 306, 321, 338
punctuational 321
stabilizing 321
selectionist–neutralist debate 241, 292, 334
selective
agent 24
axon elimination 321
cell adhesion 29, 37
gene amplification 321
translational regulation 126
selector gene expression 61
self assembly 233
mechanisms 233
processes 314
systems 233, 338
sensory transduction 74
sequence
amplification 22, 276
analysis 136
arrangement 193
complexity 313
conservation 21, 182, 245
deletion 313
duplication 15, 160

flux 242
generalized recombination 17
homogeneity 19, 21
homology 14, 21
internal promoter 121
inversion 95, 112
nucleotide 157
specific binding protein 121
termination 114
transfer 21
variant 306
sequences
   autonomous replicating 13
   AT-rich 140
   Alu 159
   cloned DNA 143
   clustered 21
   coding 117, 235
   c-onc 116
   consensus 114, 116, 117, 124
   constant 100–1
   control 116
   DNA non-coding 235
   DNase 1-hypersensitive 118
   dispersed 10, 21, 157
   dispersed moderately repeated DNA 10,
      160
   dispersed simple 163
   diversity 101
   enhancer 119, 296
   FB 130
   flanking 57, 117, 148, 158
   homeo box 59
   highly repeated 1, 10, 18, 22, 24, 106, 109,
      128–9, 135–9, 174, 311
   identifier 163
   ignorant 195
   initiator 295
   insertion 136, 179
   interspersed 112
   interspersed middle repetitive 156, 159
   intervening 21, 59, 101
   joining 100–1
   leader 100, 116, 143, 147
   localized highly repetitive 194
   long highly repetitive 157
   long interspersed 159
   long terminal repeat 160
   message 226
   mRNA 163
   middle repetitive 16, 109, 129, 134, 159,
      165, 225, 245, 290, 312
   mobile 10, 273–4
   moderately repeated 10, 128, 136, 154, 157,
      166, 255
   multigene 245
   nomadic 10, 128, 134, 136, 140, 178
   non genic 245

orphon rDNA 182
poly (A) 124, 161
recognition 122
regulatory 280
repeated 18, 193
rDNA 243
rDNA spacer 306
rRNA 242
satellite I 140, 255
scrambled clustered 10, 157
selfish 195
short direct 312
short dispersed 160–1, 162
short interspersed middle repetitive 155
signal 11, 157
silencer 314
simple 10, 112, 136
simple dispersed 160–1, 163
simple repeated 161, 162
single copy 10, 128, 156, 226
suffix 163
tandemly arranged 20, 311
tandemly repeated 161
target 130
terminator 296
trailer 143, 147
transposable 16
type I rDNA insertion 136, 178–80
type II rDNA insertion 178–80
unique 2, 17, 128–9, 131
upstream 117
variable 100, 101
sex
   chromatin body 216
   determination 63, 68, 88, 162, 230, 314
   mosaics 77
sex chromosomes 216, 263, 295
   attached X 202–3
   W 68, 162
   X 2, 35, 63–4, 68, 70, 76, 87, 120, 128, 135,
      139, 141, 147–8, 178–80, 196, 201–2,
      204, 206, 209, 216–18, 219–22, 227, 230,
      295, 327
   Y 23, 63–4, 128, 135–6, 139–40, 148, 162,
      178, 180, 201, 202–3, 204, 206–7, 270,
      295, 327
   Z 68
sex determining genes
   daughterless 64, 67
   doublesex 64, 66
   intersex 64, 67
   sex lethal 64, 67
   sisterless 64
   transformer 64–8
sexual
   development 63
   differentiation 64, 66–7
   dimorphism 64, 66

phenotype 68
reproduction 164, 226, 242
Shaker complex 82
sigma initiation factor 116
signal recognition particle 160
silencers 314
SINES 160
single copy complexity 165
site
  acetylcholine binding 84
  cap 116
  dosage compensation 221
  exchange 112
  expression 110, 112
  DNA binding 117
  rDNA 180
  DNase 1-hypersensitive 118–19
  histone 225
  internal cleavage 128
  recognition 101
  splice 124
  switching 101
  termination 125, 209
  upstream 119
sites
  histone 225
  NOR 283
  polymerase recognition 296
  promoter 242
  5S DNA 178
  splicing 315
site specific
  DNA binding proteins 119
  DNA sequences 112
slime molds 14, 59
small nuclear RNA families 242
small optic lobes system 331
solitary LTRs 159
soma 27, 64, 88, 194, 209, 227
soma–germ line differentiation 164
somatic
  cells 64, 68, 90, 103, 121, 136, 216, 218, 219
  DNA 109
  development 95, 103, 141, 227, 258, 260, 263, 300, 311
  differentiation 109
  dimorphism 66
  endoploidy 35, 214–15
  functions 140, 312
  genomes 103, 107
  heredity 216
  joining 101
  junk 153, 313
  message complexity 173
  nuclei 103, 214
  phenotype 194
  5S RNA 178

Southern blot
  comparisons 202
  hybridization 190
spacer
  DNA 120, 185
  length variation 120,
  promoters 120
  regions 121, 182
  sequences 178, 184
spacers
  intergenic 235
  non transcribed 182
  transcribed 182
specialization 286–7
speciation 234, 236, 241, 257–8, 263, 268–9, 272–3, 275–6, 286–7, 288, 292, 294, 312, 319, 324
  divergent 272
  events 139, 324
  mechanisms 324
  multiple 272
  phenomena 334
  processes 324
  rapid 272
  sibling 272
splicing 101, 124–6, 158
  alternative 126
  differential 125
  trans 126
sporogenesis 253
  sporulation 96, 248
spreading effect 295
sterility 227, 274
  dysgenic 275
  female 5, 33
  gonadless 275
  hybrid 132, 269, 274
stored histone 185
strains
  inducer 274
  maternal 274–5
  paternal 274–5
  reactive 274
structural
  change 28, 272
  chromosome changes 269
  chromosome mutation 244
  chromosome rearrangement 2–4, 132, 263, 274, 293, 318
  gene changes 293
  genes 95, 294, 296
suffix sequences 163
supernumerary
  chromatin 288, 290
  chromosomes 249, 288–9
  chromosome segments 290
suppressed developmental programmes 326

surface
    coat antigen 109
    glycoprotein 109
switch
    genes 58, 88–9, 297, 302
    loci 98
    mechanisms 88, 95, 297
    points 112
    sites 111
switching 98, 101, 112
    antigenic 109
    flip-flop 95, 111, 317
    multiple 112
    points 95, 101
    site 101
synapses 72, 333
synaptic
    chemistry 332
    connections 68, 79, 321, 331–2
    connectivity 314
    modelling 333
    reorganization 71
    transmission 83
synaptogenesis 170
syntenic homologies 322
systems
    dysgenic 275
    IR 274
    multicircuit 329
    multigene 329
    multilocus 321
    multisequence 321, 325, 329
    PM 274

tandem
    insertion 221
    repeats 107, 185
    repetition 200
tandem arrays 10, 17, 22, 120, 128, 136, 161, 177, 182, 239
    localized 17, 160
    under-representation 160–1
tandemly repeated sequences 161
target
    sampling 321
    sequences 130
    site duplication 161
telomeres 107, 110, 157, 178, 188, 283
telomeric DNA 157
termination signals 225
tetraploid 96, 228, 248, 277
tetraploid state 204
thoracic musculature 204
threshold effect 326
timing
    adjustments 298
    alterations 302
tissue specific expression 163

T locus complex 30
trailer regions 161
trans
    acting regulators 119
    acting regulatory genes 13, 14
    determination 62
    formation 18, 218
    regulatory loci 47
    regulatory system 49
    splicing 126
transcript
    Ddc 84
    capping 124
    complexity 149
    giant 50, 53
    length 143, 146
    Notch 70
    number 143, 147, 149
    producing units 183
    primary 50, 101, 124–6, 182
    processing 124
    size 146
transcripts
    histone gene 124, 141
    primary 61
    reverse 16
    processed 143
    truncated 131
transcription 16, 113–14, 120, 130, 141, 168, 196, 200, 211, 216, 219, 223, 225, 231, 235, 297
    ADH2 13
    complexes 122
    default 60–1, 95, 161, 184, 224–5
    differential 221
    factors 13, 45, 116, 121
    genomic 164
    hormone stimulated 28
    inefficient 179
    initiation 16, 121
    initiation sites 114, 116, 119, 125
    level 121
    polymerase II 114, 119
    preblastoderm 231
    primary 125
    processing 128
    products 141, 179
    readthrough 168, 184
    reverse 16, 134, 158, 160, 193
    RNA polymerase II 159
    5S RNA 140
    symmetrical 225
    termination 114, 121
    termination sites 125
    units 41, 120, 125, 143, 146, 148, 163, 178–80, 183, 186, 201, 225, 296
    X-chromosome 221
    X-linked gene 219–20

transcriptional
    activation 13
    block 295
    controls 113–14, 122
    defects 131
    machinery 116
    modulation 337
    regulation 231, 319
    unit 120
transcriptive
    activity 168, 214
    capacity 169, 215
transcripts
    brain specific 61, 163
    giant 168
    histone gene 124, 141
    homeo box 60
    polyadenylated 163
    primary 182
    processed 143, 149, 163
    RNA 72, 113, 186, 224
    Ubx⁺ 52
transdetermination 62
transfer (t) RNA
    genes 21, 160, 186
    pseudogenes 160
transfer (t) RNAs 21, 196, 201, 224, 324
transformation 5, 18, 62, 218, 222
    analysis 221
    homeotic 49, 52, 53, 55–6
    intersegmental 55
    intrasegmental 55
    segmental 47, 50
    sensory neuron 85
translation 185
translocation
    attached-X 202–3
    chromosomes 202
    X-3 222
    X-autosome 217
    X.Y 203
    Y.3 295
    Y-autosome 203
transmitter release 79
transposable
    elements 16–17, 98, 111, 157–8, 160, 185,
        225, 274
    mating type 96
transposition 22, 129, 158, 242, 274, 276, 334
    activation 111
    conservative 15
    direct 157
    duplicative 15, 110, 112
    replicative 15
    Ty 158
transposon mutagenesis 5
transposons 132, 134, 148, 209, 221, 275
triplication 279

triplo lethal locus 223
triploids 96, 215, 226–8, 251, 277
trisomics 226–7
trisomy 227, 230
trophectoderm 122, 216
trypanosomes 109–12
tryptophan pyrrolase activity 222
tubulins 14, 189–90, 196, 305
turnover mechanisms 242, 276

under
    methylation 122, 124
    replication 295
unequal
    crossing over 242
    exchange 17–20, 161, 207
units
    complementation 146
    transcription 16, 114–15, 120, 130, 168,
        211, 216, 219, 223, 225, 231, 235, 297
upstream
    controlling regions 180
    signals 231
uridine autoradiography 220
urodele amphibians (newts and salamanders)
        151, 212, 251, 256, 277, 287, 299
    Ambystoma jeffersonianum 251
    Ambystoma mexicanum 185, 299
    Ambystoma platineum 251
    Ambystoma figrinum 299
    Amphiuma means 251
    Chiropterotriton 281
    Necturus maculosus 251, 299
    Notophthalamus viridescens 177, 180, 183–4,
        185, 225, 300
    Oedipina 281
    Plethodon 178, 281, 282, 285, 312
    Plethodon cinereus 180, 281, 282, 312
    Plethodon vehiculum 282
    plethodontid salamanders 153, 225, 251,
        257, 281, 283, 287, 291
    Pleurodeles waltii 227
    Proteus anguinus 299
    Pseudotriton 281
    Scaphiopus holbrooki 256
    Thorius 281
    Triturus cristatus 168, 225
    Triturus cristatus carnifex 182, 251
    Triturus vulgaris meridionalis 181, 312

vaccinia virus growth factor 70
variegating genes 296
variegation 295–6
variegation reversal 295
vector
    expression 86
    P element 117, 318
    pUChsneo 5

Xdh 148
vertebrate
    body plan 93
    collagens 191
    development 324
    embryo 28
    evolution 29, 93, 330, 331
    genome 32, 89, 171, 334
    head 331
    neuromuscular transmission 15
    oncogenes 14
    opsins 13
    3-phosphate dehydrogenase 171
    pro-viruses 130, 134–5
    radiation 277
    retroviruses 14, 111, 130, 135
    sodium channel 82–3
vertebrates 13–14, 59, 61, 82, 88, 89, 99, 122,
        160, 168, 180–2, 191, 193, 246, 251, 277,
        283, 310, 319
viability 237, 255, 289, 328
viral morphogenesis 233
virus-like particles 135, 158
viruses 314
    avian leukaemia sarcoma 130
    simian 40, 119
    spleen necrosis 158–9
    vertebrate RNA tumor 130, 134
visual mutations
    ocelliless 209
    transient receptor potential 80
visual orientation behaviour 87
vital loci 147

wing scalloping 224
worms
    earth 277
    flat 277

X-
    compensation 219–20
    dosage 227
    euchromatin 295
    heterochromatin 139, 141, 204
    inactivation 216–19, 227
X/A ratio 64, 67
X-chromosome
    dosage 68
    transcription 221, 227
X-linked 83, 211
    enzyme activity levels 220
    enzymes 219
    genes 148, 196, 219
    insertion 221
X-ray mutagenesis 295
Xenopus 257
    borealis 120
    laevis 13–14, 15, 32, 45, 47, 60, 120–1, 140,
        154, 156, 160, 168, 178, 180, 182, 184,
        185, 188, 194, 212, 225, 257, 278, 300
    tropicalis 278
XXY females 203

yeast
    genome 21
    mitochondrial genomes 194
yeasts 59, 194
Y-
    heterochromatin 148, 295
    NORs 179, 202–4
yolk sac 94

zygote 27, 33
zygotic
    genes 55
    lethality 33
    lethal loci 33
zymogens 42